Predictive Control

Predictive Control

Fundamentals and Developments

Yugeng Xi
Shanghai Jiao Tong University
Shanghai, China

Dewei Li
Shanghai Jiao Tong University
Shanghai, China

WILEY

国防工業出版社
National Defense Industry Press

This edition first published 2019 by John Wiley & Sons Singapore Pte. Ltd under exclusive licence granted by National Defence Industry Press for all media and languages (excluding simplified and traditional Chinese) throughout the world (excluding Mainland China), and with a non-exclusive license for electronic versions in Mainland China.

Registered Offices
John Wiley & Sons, Inc., 111 River Street, Hoboken, NJ 07030, USA
John Wiley & Sons Singapore Pte. Ltd, 1 Fusionopolis Walk, #07-01 Solaris South Tower, Singapore 138628

Editorial Office
The Atrium, Southern Gate, Chichester, West Sussex, PO19 8SQ, UK

For details of our global editorial offices, customer services, and more information about Wiley products visit us at www.wiley.com.

Wiley also publishes its books in a variety of electronic formats and by print-on-demand. Some content that appears in standard print versions of this book may not be available in other formats.

Library of Congress Cataloging-in-Publication Data

Names: Xi, Yugeng, 1946- author. | Li, Dewei, 1971- author.
Title: Predictive control : fundamentals and developments / Yugeng Xi, Shanghai Jiao Tong University, Shanghai, China, Dewei Li, Shanghai Jiao Tong University, Shanghai, China.
Description: First edition. | Hoboken, NJ : John Wiley & Sons, Inc., [2019] | Includes bibliographical references and index. |
Identifiers: LCCN 2019003713 (print) | LCCN 2019005389 (ebook) | ISBN 9781119119586 (Adobe PDF) | ISBN 9781119119579 (ePub) | ISBN 9781119119548 (hardback)
Subjects: LCSH: Predictive control.
Classification: LCC TJ217.6 (ebook) | LCC TJ217.6 .X53 2019 (print) | DDC 629.8–dc23
LC record available at https://lccn.loc.gov/2019003713

Cover Design: Wiley
Cover Images: © Jesica Leach/EyeEm/Getty Images, © science photo/Shutterstock

Set in 10/12pt Warnock by SPi Global, Pondicherry, India
Printed and bound in Singapore by Markono Print Media Pte Ltd

10 9 8 7 6 5 4 3 2 1

Contents

Preface

Predictive control was developed in the middle of the 1970s. It originally referred to a kind of advanced computer control algorithm that appeared in complex industrial processes. Because of the ability of real-time solving the optimal control under constraints, it has received great attention from the industrial community and been successfully applied to chemical, oil refining, power, and other industries. Since the 1980s, the algorithms and applications of predictive control have been rapidly expanded. The commercial software of predictive control has undergone several generations of version updating and function expansion. Not only has it been applied in thousands of industrial processes around the world and achieved remarkable economic benefits but also the application fields have expanded rapidly from industrial processes to manufacturing, aerospace, transportation, environment, energy, and so on. Compared with applications, the development of predictive control theory lagged behind. However, since the 1990s, it has rapidly become a hotspot in the control field through adopting novel ideas and powerful tools from a new perspective. Especially the systematic progress in the synthesis theory of stable and robust predictive controllers has deepened the understanding of the essential mechanism of predictive control and constructed a rich theoretical system on qualitative synthesis of predictive control. Nowadays, predictive control is not only favored by industrial communities and regarded as the most representative advanced process control algorithm but it has also become a systematic theory of synthesizing stable and robust controllers characteristic of rolling optimization for uncertain systems.

After entering the twenty-first century, with the scientific and technological development and the social progress, the requirement for control is becoming higher and higher. Rather than traditionally satisfying the stabilizing design, optimization is much more incorporated in control design in order to achieve a better control performance. In the meantime, optimization is restricted by more and more factors. In addition to traditional physical constraints such as actuator saturation, various constraints brought by technology, safety, economics, and sociality indices should be incorporated. The contradiction between the higher requirements and the more complicated constraints has become a new challenge in many application fields. Because of remarkable achievements in the field of industrial process control, predictive control naturally becomes the first choice to solve this kind of problem. In recent years, many application reports on adopting predictive control for solving various constrained optimization control problems appeared in the fields of advanced manufacturing, energy, environment, aerospace, and medicine, etc., which reflects the expectations of people for this advanced control technology and also motivates more people to become familiar with or master predictive control.

In order to meet the different needs for research and application of predictive control in a wide range of fields, this book tries to give a brief overview on research and application in the field of predictive control through introducing the basic problems and solutions of predictive control, and unscrambling its representative research progress, so as to provide readers with different needs with basic knowledge and background. Reviewing the development of predictive control during the past 40 years, it can be concluded that predictive control has become a diversified disciplinary branch, including various development tracks from theory and methods to applications, with different purposes and characteristics. Each contains rich contents with its own concerns and the fundamental knowledge that needs to become familiar. For example, the research on predictive control applications focuses on how to apply the predictive control principles and algorithms to implementing optimization-based control for a specific plant. There is a need to become acquainted with predictive control algorithm and its implementation technology, particularly on how to select or formulate a predictive control algorithm for a specific problem, how to solve the optimization problem, and how to tune the parameters, etc. The research on predictive control methods focuses on how to develop effective implementation modes in terms of specific system properties, structural characteristics, or implementation environments, so that predictive control can be practically applicable or applied with a better performance and higher efficiency. It is necessary to understand the existing methods and strategies commonly used in predictive control, as well as the specified structures and research branches deriving from them. It is important that the ideas behind these methods need to be caught from the perspective of information and control, and taken as a reference for extension. The research of predictive control theory focuses on solving difficult or new problems brought on by deep research and background expansion. As a prerequisite, it is necessary to get acquainted with the basic ideas, solving methods, and required tools of mature theory on stable and robust predictive control synthesis. The implied novel ideas for successfully solving various difficulties in the existing literature need to be carefully explored and taken for reference, and the existing difficulties and problems need to be clarified.

In view of the above requirements and characteristics, this book attempts to make a comprehensive introduction to predictive control on the aspects of basic principles and algorithms, system analysis and design, algorithm development and applications according to its historical development process. The purpose is to help readers understand the basic principles and algorithms of predictive control, as well as to get acquainted with the most fundamental contents of predictive control theories, methods, and application techniques, in order to provide the basis and reference for researchers and engineers in the field of predictive control to go deep into theoretical research, to develop high-level industrial applications, and to extend predictive control to more application fields.

The book consists of five parts. The first part (Chapters 1 and 2 and Sections 5.1 and 5.2) gives a brief overview of the development trajectory of predictive control and introduces its methodological principles and basic algorithms. The second part (Chapters 3 and 4) introduces system analysis and design of classical predictive control algorithms. On the one hand, for the dynamic matrix control (DMC) algorithm, which is based on a nonparametric model and is commonly used in industry, its control mechanism and system performance are analyzed and its parameter tuning is discussed. On the other hand, the relationship between DMC and the generalized predictive control (GPC) algorithm is clarified, and then the quantitative relationships between design parameters and the closed-loop system performance are uniformly derived. The third part (Chapters 6 and 7) introduces the qualitative synthesis theory of predictive control, including stable predictive controller synthesis and robust predictive controller synthesis. Emphasis is put on unscrambling basic problems as well as solutions, and introducing some representative works. The fourth part (Chapters 8, 9, and Section 5.3) introduces the development of methods and strategies oriented to the characteristics and requirements of predictive control applications. Control structures, optimization concepts and strategies useful in industrial applications, as well as various decomposition algorithms commonly used in networked large-scale systems, are presented. According to the characteristics of nonlinear systems, some practical and effective algorithms and strategies are also introduced. The fifth part (Chapters 10 and 11) concerns implementation technology and application examples of predictive control, with a detailed introduction of industrial predictive control technology and an overview on predictive control applications in other fields. The universality of predictive control principles is illustrated and potentially extended to general control problems in a dynamic uncertain environment. All of the above parts make a full view of predictive control from general concepts and basic algorithms, quantitative analysis, qualitative synthesis, method and strategy developments to application techniques. They are interdependent and combined organically. The book not only introduces the relevant knowledge on theory and applications but also runs through the methodological principles of predictive control, which will help readers to get rid of the limitations of specific algorithms and problems, deepen their understanding of the essential characteristics of predictive control, and broaden their thinking in predictive control research and application from a higher stand.

This book is based on the book *Prediction Control* (second edition) (Chinese version), published by the National Defense Industry Press in 2013. The major contents come from the research results of our group in the direction of predictive control during the past 30 years. In order to comprehensively reflect the field of predictive control, some pioneering and representative research work in this field has also been included and unscrambled. At the time of publication of this book, the authors would like to thank the National Natural Science Foundation of China for its long-standing support for our predictive control research. We would also like to thank our colleagues in academia and industry. It is the helpful discussions and cooperation with them that have enabled us to deepen our understanding and receive new inspiration. Over the past 30 years, colleagues and students in our group have worked together with us in the field of predictive control, and we do appreciate their contributions. Special thanks is given to those whose work directly contributed to this book, including Professor Xiaoming Xu, Professor Shaoyuan Li, doctoral students Hao Sun, Jian Yang, Jian Fang, Jun Zhang, Xiaoning Du, Ning Li, Chungang Zhang, Baocang Ding, Bing Wang, Shu Lin, and Zhao Zhou, and master students Junyi Li, Hanyu Gu, Hui Qin, Shu Jiang, Nan Yang, Yang Lu, Yunwen Xu, and Yan Luo. It is their efforts and contributions that enrich the content of this book. The Chinese Defense Science and Technology Publishing Fund has provided financial support for the first and second editions of the Chinese version, and the authors would also like to express their deep gratitude.

Yugeng Xi and Dewei Li
Shanghai Jiao Tong University
November, 2018, Shanghai

1

Brief History and Basic Principles of Predictive Control

1.1 Generation and Development of Predictive Control

Predictive control, later also called Model Predictive Control (MPC), is a kind of control algorithm originally rising from industrial processes in the 1970s. Unlike many other control methods driven by theoretical research, the generation of predictive control was mainly driven by the requirements of industrial practice. For quite a long time the industrial process control mainly focused on regulation using the feedback control principle. The well-known PID (Proportional–Integral–Differential) controller can be used for linear or nonlinear processes, even without model information, and has few tuning parameters and is easy to use. These features are particularly suitable for the control environment in industrial processes and make it a "universal" controller that is widely used. However, the advantage of the PID controller is mainly embodied in the loop control. When the control turns from a loop to the whole system, it is difficult to achieve good control performance by using such a single-loop controller without considering the couplings between the loops. Furthermore, a PID controller can handle input constraints but is incapable of handling various real constraints on outputs and intermediate variables. Particularly, when the control goal is promoted from regulation to optimization, this kind of feedback-based controller seems powerless because it lacks understanding of the process dynamics. With the development of industrial production from a single machine or a single loop to mass production, the optimization control for constrained multivariable complex industrial processes became a new challenging problem.

During this period, modern control theory was rapidly developed and went to mature with brilliant achievements in the aerospace, aviation, and many other fields. At the same time the progress of computer technology provided a powerful tool for real-time computing. Both were undoubtedly attractive for the industrial process control engineers when pursuing higher control quality and economic benefits. They began to explore the applications of the mature modern control theory, such as optimal control, pole placement. etc., in optimization control of the complex industrial processes. However, through practice they found that a big gap existed between the perfect theory and the real industrial processes, mainly manifested in the following:

1) Modern control theory is based on an accurate mathematical model of the plant, while for a high-dimensional multivariable complex industrial process, it is hard to get its accurate mathematical model. Even if such a model could be established, it

Predictive Control: Fundamentals and Developments, First Edition. Yugeng Xi and Dewei Li.

should be simplified according to practical applicability and a strictly accurate mathematical model is not available.

2) The structure, parameters, and environment of an industrial plant are often uncertain. Due to the existence of uncertainty, the optimal control designed based on an ideal model would never remain optimal in real applications, and would even result in serious degeneration of the control performance. In an industrial environment, the control system is more focused on robustness, i.e. the ability of keeping better performance under uncertainty, rather than pursing ideal optimality.
3) The industrial control must take the economics of the control tools into account. The control algorithm should satisfy the real-time requirement. But most of the algorithms in modern control theory seem complex and are difficult to implement by economic computers.

Due to the above issues coming from practice, it is hard to directly use modern control theory for complex industrial processes. To overcome the gap between theory and practice, in addition to investigating system identification, robust control, adaptive control, etc., people began to break the constraints of traditional control methods and tried to seek new optimization control algorithms in accordance with the characteristics of the industrial processes, i.e. with a low requirement for the process model, capable of dealing with multivariables and constraints, and an acceptable computational burden for real-time control. The appearance of predictive control just fitted these needs.

The earliest predictive control algorithms arising from industrial processes include Model Predictive Heuristic Control (MPHC) [1], or similarly Model Algorithmic Control (MAC) [2], and Dynamic Matrix Control (DMC) [3]. They use the process impulse or step response as a nonparametric model, which is easy to obtain from the process data, and the controller can be designed directly based on these responses without further identification. Such predictive control algorithms absorb the idea of optimization in modern control theory, but instead of solving an infinite horizon optimization problem off-line, they introduce a rolling horizon mechanism with solving a finite horizon optimization problem on-line, meanwhile at each step making feedback according to real-time information. These features make them avoid the difficulty of identifying a minimal parametric model, reduce the real-time optimization burden, and strengthen the robustness of the control systems against uncertainty, which are particularly suitable to practical requirements of industrial process control. Therefore, after their appearance, they were successfully applied to process control systems in oil refining, chemical industry, electric power, and other industries in the USA and Europe, and attracted wide interest from the industrial control community. Thereafter, various predictive control algorithms based on impulse or step response were proposed in succession, and a number of predictive control software packages for process control of different installations were soon launched and quickly generalized and applied to process control of various industries. For more than 30 years, predictive control has been successfully run in thousands of process control systems globally, and achieved great economic benefits. It has been recognized as the most efficient advanced process control (APC) method with great application potential [4].

Besides the predictive control algorithms based on nonparametric models directly from industrial process control, another kind of predictive control algorithm also appeared in the adaptive control field, motivated by improving the robustness of adaptive control systems. In the 1980s, in adaptive control research it was found that in order to overcome the disadvantages of minimum variance control, it is necessary to absorb the multistep prediction and optimization strategy of industrial predictive control algorithms so as to improve the robustness of control systems against time delay and model parameter uncertainty. Then a number of predictive control algorithms based on identified minimum parametric models and self-adaption also appeared, such as Extended Prediction Self-Adaptive Control (EPSAC) [5], Generalized Predictive Control (GPC) [6], etc. Among those, the GPC algorithm has particularly attracted wide attentions and got applications. The identified minimum parametric model adopted in these predictive control algorithms provides a more rigorous theoretical basis for control system analysis and design, which greatly promoted the theoretical research of predictive control.

With the appearance of the above predictive control algorithms, various methods and strategies were also developed for predictive control applications to suit specific system dynamics and different application scenarios. These studies could be roughly divided into several categories:

- appropriate strategies and algorithms for specific kinds of systems, such as the Hammerstein model [7], hybrid systems [8], etc.;
- reasonable control structures for efficiently utilizing process information or reducing control complexity, such as cascade [9], hierarchical [10], decentralized [11], distributed control schemes [12, 13], etc.;
- heuristic strategies for improving the usability of predictive control algorithms, such as the blocking technique [14], multirate control [15], etc.;
- effective predictive control algorithms and strategies particularly for nonlinear systems, such as fuzzy [16], neural network [17], multimodel [18], etc.

These studies not only strongly supported the applications of predictive control but also facilitated some new sub-branches of predictive control, such as hybrid predictive control [19], hierarchical and distributed predictive control [20, 21], economic predictive control [22], and so on.

While predictive control has found successful applications in industry, its theoretical research naturally became the hotspot of control academia. To meet the application requirements, the first theoretical issue to be considered is how the design parameters in predictive control algorithms are related to control system performance, which can guide the parameter tuning to guarantee stability and tracking performance of predictive control systems and is of great significance to practical applications. This problem was explored in the Z domain and the time domain, respectively, using different tools. In the Z domain, Garcia and Morari [23] proposed Internal Model Control (IMC) in 1982 as a general framework for analyzing predictive control systems. Through transforming the predictive control algorithms into the IMC framework, some quantitative relationships between the main design parameters and the closed-loop performances were established for nonparametric model-based predictive control algorithms such as DMC [24]. Later on more theoretical results uniformly applicable to both DMC and GPC algorithms were obtained by using coefficient mapping of the characteristic polynomials of open-loop and closed-loop systems in IMC structures [25]. In the time domain, Clarke and Mohtadi [26] transformed the GPC algorithm into an optimal control problem described in state space and analyzed the system performance as well as the deadbeat property with the help of existing results in Linear Quadratic (LQ) optimal control. Some later studies on GPC borrowed more ideas from optimal control theory and adopted the monotonically decreasing property of the receding horizon cost and the endpoint equality constraints on the tracking error to guarantee the stability of GPC [27]. All of the above theoretical researches achieved some progress, but the results were quite limited due to the essential difficulties of quantitative analysis for predictive control systems. Firstly, quantitative analysis should be based on analytical expressions of system and control law, but in most cases, predictive control should handle various constraints and the control action can only be obtained through solving a constrained optimization problem. Without analytical expression of the control law it is impossible to make a quantitative analysis. Secondly, even if analytical forms for a predictive controller and the closed-loop system could be derived in an unconstrained case, there are no direct relationships between the design parameters and the closed-loop performances, which prevents more accurate and useful quantitative results to be obtained. Therefore, most obtained analytical results in this period are only available for single-input single-output (SISO) unconstrained predictive control systems, which is far from the need of real applications.

In view of this situation, the theoretical research for predictive control turned from quantitative analysis to qualitative synthesis in the 1990s. This study was no longer limited to existing algorithms but focused on synthesizing new algorithms to guarantee system performance. A key idea of this study is to regard predictive control as traditional optimal control with the only difference being in implementation. After formulating the system model and the rolling optimization procedure in state space, a predictive controller with guaranteed stability can be investigated with the help of mature ideas and tools in optimal control theory, particularly the ideas and methods used in Receding Horizon Control (RHC) [28] proposed in the 1970s. During this period, a great number of new predictive control algorithms with guaranteed stability were proposed by introducing novel techniques, such as terminal equality constraints [29], terminal cost function [30], terminal constraint set [31], terminal cost and constraint set [32], etc. A highlighted feature of such an investigation is to artificially impose some constraints to theoretically guarantee stability of the predictive control system. An excellent survey was comprehensively given by Mayne et al. [33] on the stability and optimality of constrained model predictive control. Promoted by stable predictive control theory and robust control theory, robust predictive control for uncertain systems was also investigated. In fact, it appeared very early [34], but has become a hotspot only since the middle of the 1990s, mainly due to introducing novel ideas and new mathematical tools. A representative work by Kothare et al. [35] in 1996 studied the robust stability of model predictive control for a large class of plant uncertainties using linear matrix inequalities (LMIs), which stimulated a lot of follow-up researches in later years.

Different from the early quantitative analysis theory, these new works concerned quite general systems and problems, including nonlinear systems and systems with constraints and various uncertainties. Taking optimal control theory as a reference, Lyapunov stability analysis as a basic method to guarantee system performance, invariant set, linear matrix inequality (LMI) as fundamental tools, and the performance analysis for rolling horizon optimization as the core of the study, they constituted rich contents and achieved fruitful results, showing academic profundity and methodological innovations. In the last two decades, hundreds of papers have appeared in main control journals and formed the mainstream of current theoretical research on predictive control [33, 36]. However, the conclusions achieved from these studies often lack clear physical meanings and the developed algorithms still have a big gap with the requirements of industrial applications.

Throughout historical development of predictive control, roughly three climax stages appeared: the first stage is characterized by the industrial predictive control algorithms based on step or impulse response models developed in the 1970s. These algorithms are very suitable for the requirements of industrial applications, both in model selection and in control ideas, and therefore became the main algorithms used in industrial predictive control software packages. However, without theoretical guidance, application of these algorithms greatly depends on the specific knowledge and user experiences. The second stage is characterized by the adaptive predictive control algorithms developed from the adaptive control field in the 1980s. The models and control ideas of these algorithms are more familiar to the control community and thus more suitable for investigating the predictive control theory. Indeed, some quantitative analytical results for predictive control systems have been achieved. However, the essential difficulty of quantitative analysis always exists because of the lack of the analytical expression of the optimal solution for constrained optimization. The third stage is characterized by the predictive control qualitative synthesis theory developed since the 1990s. Because of the change in research ideas, this has given the predictive control theory a great leap forward and become the mainstream of current predictive control theoretical research. However, these results still have a big gap with practical applications of predictive control. Although the studies of the above three stages have had their own problems, after development of these stages, predictive control has become a diversified control branch, containing many development paths with different purposes and different characteristics. It is not only regarded as the most representative APC algorithm and favored by the majority of the industrial community, but also forms a theoretical system of stable and robust design for uncertain systems with rolling optimization characteristics. Figure 1.1 roughly gives

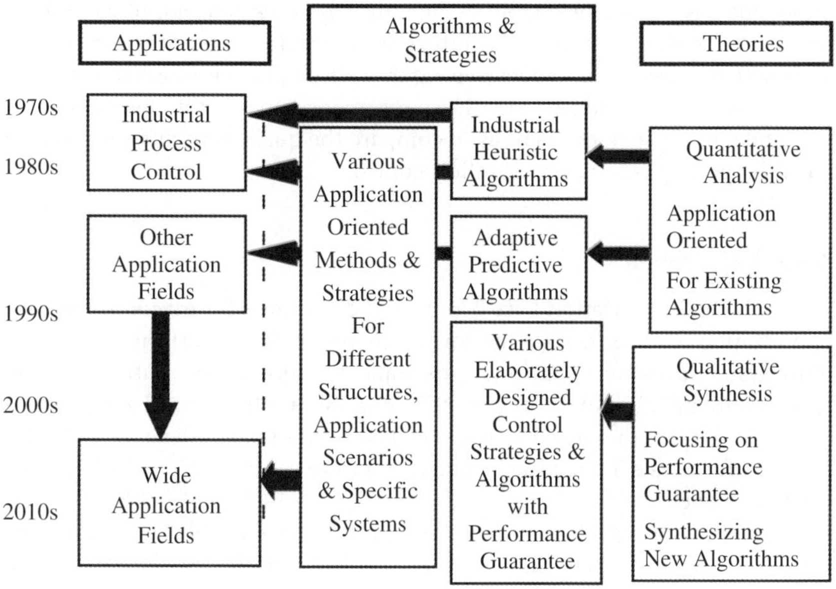

Figure 1.1 The development trajectory of predictive control.

an overall historical trajectory of predictive control where different aspects and interests are shown in parallel.

1.2 Basic Methodological Principles of Predictive Control

Although there exists a wide variety of predictive control algorithms with different forms, they have common characteristics in methodological principles, i.e. to predict future system dynamic behavior according to given control actions by using the system model, to control the plant in rolling style through online solving an optimization problem with required performance subject to given constraints but only implementing current control, and to correct the prediction of future system behavior by using real-time measured information at each rolling step, which can be summarized as three principles, i.e. prediction model, rolling optimization, and feedback correction [37].

1.2.1 Prediction Model

Predictive control is a model-based control, where the model is called a prediction model. A prediction model serves as the basis of optimization control. Its main function is to predict the future dynamic variations of system states or outputs according to historical information and assumed future system inputs. The model is particularly emphasized by its function rather than its structural form. Therefore, traditional models such as a transfer function, state space equation, etc., can be used as a prediction model. For stable linear systems, even nonparametric models such as step or impulse responses can be directly used as a prediction model without further identification. Furthermore, models for nonlinear systems, distributed parameter systems, etc., can also be used as prediction models as long as they have the above function.

A prediction model is capable of exhibiting the future dynamic behavior of the system. Given future control policy arbitrarily, the future system states or outputs can be predicted according to the prediction model (see Figure 1.2) and, furthermore, the satisfaction of constraints can be judged and the performance index can be calculated. A prediction model thus provides the basis to compare the qualities of different control policies and is the prerequisite of optimization control.

1.2.2 Rolling Optimization

Predictive control is an optimization-based control. It determines future control actions through optimization with a certain performance index. This performance index involves future system dynamic behavior, for example, to minimize the output tracking errors at future sampling times or, more generally, to minimize the control energy while keeping the system output in a given range, etc. The future system dynamic behavior involved in the performance index is determined by future control actions and is based on the prediction model.

It should be pointed out that the optimization in predictive control is quite different from that in traditional discrete-time optimal control. In the predictive control

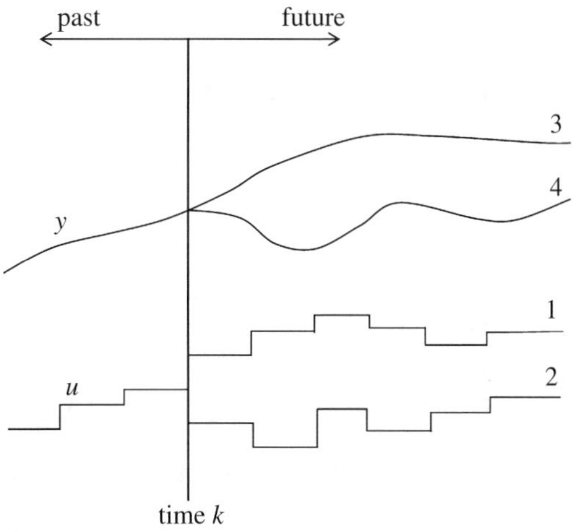

Figure 1.2 Prediction based on model. 1 – control policy I; 2 – control policy II; 3 – output w.r.t. I; 4 – output w.r.t. II.

algorithms used in industrial processes, the optimization is commonly over a finite horizon and implemented in a rolling style. At each sampling time, an open-loop optimization problem with a performance index over a future finite time horizon is solved where a certain number of future control actions are taken as optimization variables. However, the solved optimal control actions do not need to be implemented one by one; instead only the current control action is implemented. At the next sampling time, the optimization horizon is moved one step forward (see Figure 1.3). Therefore, in predictive control, the optimization performance index is time dependent. The relative formulation of the performance index at each time is the same, but the concrete time interval is moved forward and is different. This indicates that the predictive control uses on-line repeatedly performed finite horizon optimization to replace the infinite horizon off-line optimization once solved in traditional optimal control, which characterizes the meaning of rolling optimization and is the specific feature of the optimization in predictive control.

1.2.3 Feedback Correction

Predictive control is also a feedback-based control. The above-mentioned rolling optimization might be an open-loop optimization if it is only based on the prediction model. In practice, there unavoidably exist uncertainties such as model mismatch, unknown disturbance, etc., and the real system behavior may deviate from the ideal optimal one. For compensating the influence of various uncertainties to system behavior to a certain extent, a closed-loop mechanism is introduced into predictive control. At each sampling time, the measured real-time information for system states or outputs should be utilized to update or correct the future system behavior before solving the optimization problem. In this way, prediction and optimization at that time could be put on a basis that is closer to the real status. We call this procedure feedback correction.

The ways of feedback correction are diverse. In industrial predictive control algorithms, each time after the current control action has been calculated and implemented,

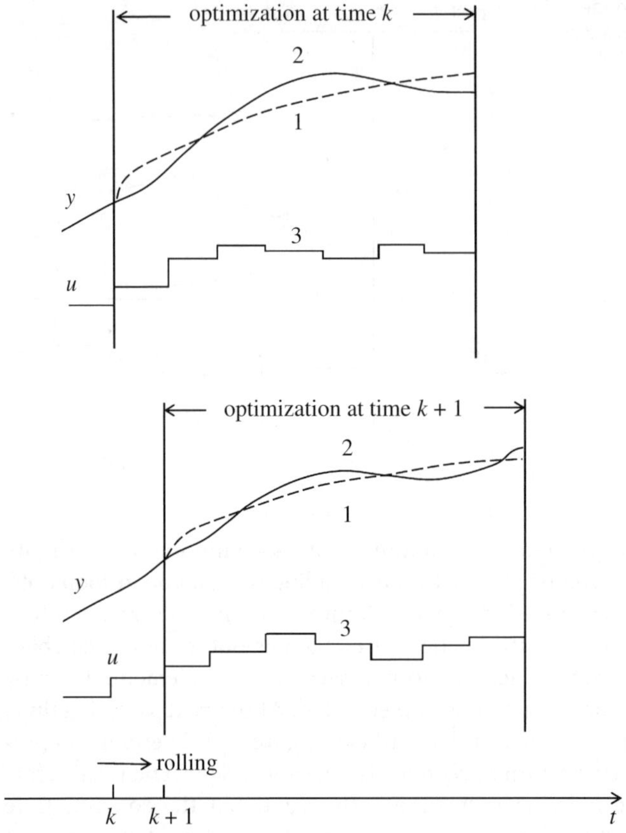

Figure 1.3 Rolling optimization. 1 – reference trajectory; 2 – predicted optimal output; 3 – optimal control action.

future outputs under this control action can be predicted by using the prediction model. Next time, the prediction error constructed by measured real output and predicted output can then be utilized to heuristically correct the prediction of future outputs (see Figure 1.4). By using predictive control algorithms developed from adaptive control, feedback correction can be implied in the adaption mechanism of adaptive control, i.e. by identifying the model using real-time input/output information and online updating the prediction model and the control law. For predictive control algorithms based on the state space model, feedback correction may be taken as different forms according to the available information. If the system states are real-time measurable, they can be directly used as a new basis for prediction and optimization each time. However, if the states are unmeasurable, it is necessary to correct the predicted states through constructing an observer according to the real-time measured system outputs. No matter which form of feedback correction is adopted, each time predictive control always tries to put the optimization on a basis that is closer to the real status by feedback of measured real-time information. Therefore, although each time predictive control performs an

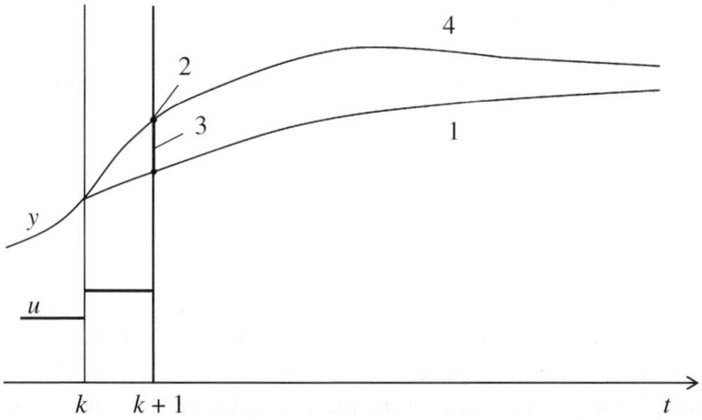

Figure 1.4 Error correction. 1 – predicted output trajectory at k; 2 – real output at $k+1$; 3 – predictive error; 4 – predicted output trajectory after correction at $k+1$.

open-loop optimization, the rolling implementation combined with feedback correction makes the whole process become a closed-loop optimization.

According to this brief introduction of the general principles of predictive control, it is not difficult to understand why predictive control is so favored in complex industrial environments. Firstly, for a complex industrial plant, the big cost of identifying its minimum parametric model always brings difficulty for control algorithms based on transfer function or state space models. However, the model used in predictive control is more convenient to obtain because it emphasizes its prediction function rather than its structural form. In many cases, the prediction model can be directly obtained by measuring the system step or impulse response without further identifying its transfer function or state space model, which is undoubtedly attractive for industrial users. More important is the fact that predictive control absorbs the idea of optimization control, but performs rolling horizon optimization combined with feedback correction instead of solving global optimization. In this way, not only is the big computational burden for solving global optimization avoided but also the uncertainty caused by the inevitable model mismatch or disturbance in industrial environment can be constantly taken into account and their influence can be timely corrected. Thus predictive control can achieve stronger robustness as compared with the traditional optimal control only based on the model. In this sense, predictive control may be regarded as a type of new control algorithm that can amend the shortcomings of traditional optimal control applying to industrial processes.

The basic principles of predictive control actually reflect the general thinking of humans when handling problems where optimization is sought but with uncertainty. For example, when a person goes across a road, there is no need to see whether there are vehicles far away; only the vehicle status a few tens of meters away and the estimation of their velocities (model) are important. However, it is wise to look into the distance while walking in order to avoid accidents caused by new vehicles entering (disturbance) or unpredicted acceleration of vehicles (model mismatch). This repeated decision process indeed includes optimization based on the model and feedback based on real-time information. Actually, this kind of rolling optimization method in an uncertain environment has already appeared in economics and management fields. Both predictive economics and rolling planning in business management adopt the same thinking. In predictive control the methodological principle within was absorbed and combined with control algorithms, so it can be effectively applied to controlling complex industrial processes.

1.3 Contents of this Book

For over 30 years, predictive control has been developed into a diversified research field, including algorithms, strategies, theories, and applications. The contents of predictive control are extremely rich and relevant results and literatures are numerous. This book will comprehensively introduce the fundamentals and new developments of predictive control from a broader perspective, including basic principles and algorithms, system analysis and design, algorithm development, and practical applications, aiming to reflect the development trajectory and core contents of predictive control. The detailed contents of each chapter in the book are as follows.

Chapter 1 gives a brief introduction of the development trajectory of predictive control and its methodological principles.

In Chapter 2, three typical predictive control algorithms, based on the step response model, stochastic linear difference equation model, and state space model, respectively, are introduced for unconstrained SISO systems, focusing on illustrating how the basic principles are embodied in the algorithms when different models are adopted.

Chapter 3 adopts the IMC structure to deduce the analytical expression of the unconstrained SISO DMC algorithm in the Z domain. After analyzing the controller and the filter in the IMC structure, some trending relationships of stability/robustness with respect to design parameters are provided, based on which the parameter tuning method is suggested with illustration examples.

Chapter 4 presents the main results of quantitative analysis theory of predictive control. The quantitative relationships between the design parameters and the closed-loop performances are achieved with the help of the Kleinman controller in the time domain and by establishing the coefficient mapping of the open-loop and the closed-loop characteristic polynomials in the Z domain, respectively.

In Chapter 5, multivariable constrained predictive control algorithms are presented with DMC as an illustration example, focusing on the description of online optimization and the solving algorithms. The decomposition concept to reduce the computational complexity is introduced and implemented by hierarchical, decentralized, and distributed predictive control algorithms.

Chapter 6 illustrates the basic ideas of the qualitative synthesis theory of predictive control. Fundamental approaches of synthesizing stable predictive control systems are presented. General conditions for a stable predictive controller are given and suboptimality is analyzed.

Fundamental materials of synthesizing robust model predictive control (RMPC) are given in Chapter 7, including basic philosophy and main developments of RMPC for polytopic uncertain systems, and typical RMPC algorithms for systems with disturbances. The main difficulties of RMPC synthesis are pointed out, for which some efficient strategies and improved algorithms are introduced.

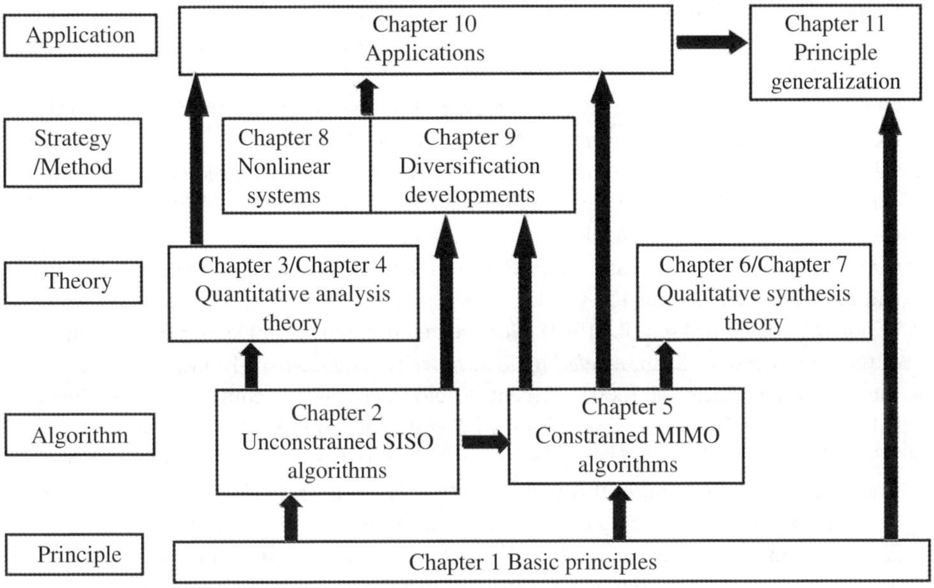

Figure 1.5 The relationship of the chapters in this book.

Chapter 8 focuses on predictive control for nonlinear systems. The general description of predictive control for nonlinear systems is given. Some commonly used strategies or algorithms are introduced, including multilayer, multimodel, neural network methods, and a specific strategy for Hammerstein models.

Chapter 9 briefly introduces the diversified development of algorithms and strategies for predictive control applications, including some effective control structures, different optimization concepts and goals, high efficiency control strategies, etc.

Chapter 10 describes the general status of industrial applications of predictive control with illustration examples. The applications of predictive control for large-scale networked systems, embedded systems, and its extension to other fields are also introduced with examples.

In Chapter 11, the basic principles of predictive control are interpreted from the viewpoints of cybernetics and information theory. These principles are then generalized to other optimization-based dynamic decision problems with illustration examples of production scheduling and robot path planning.

The relationships between the above chapters with the predictive control principles, algorithms, strategies, theories, and applications are roughly shown in Figure 1.5.

References

1 Richalet, J., Rault, A., Testud, J.L. et al. (1978). Model predictive heuristic control: applications to industrial processes. *Automatica* 14 (5): 413–428.
2 Rouhani, R. and Mehra, R.K. (1982). Model algorithmic control (MAC); basic theoretical properties. *Automatica* 18 (4): 401–414.

3 Cutler, C.R., and Ramaker, B.L. Dynamic matrix control – a computer control algorithm. *Proceedings of the 1980 Joint Automatic Control Conference*, San Francisco, USA (13–15 August 1980), WP5-B.

4 Qin, S.J. and Badgwell, T.A. (2003). A survey of industrial model predictive control technology. *Control Engineering Practice* 11 (7): 733–764.

5 De Keyser, R.M.C. and Van Cauwenberghe, A.R. (1985). Extended prediction self-adaptive control. In: *Identification and System Parameter Estimation*, vol. 2 (ed. H.A. Barker and P. C. Young), 1255–1260. Oxford: Pergamon Press.

6 Clarke, D.W., Mohtadi, C., and Tuffs, P.S. (1987). Generalized predictive control – Part 1 and 2. *Automatica* 23 (2): 137–162.

7 Wang, W. and Henriksen, R. (1994). Generalized predictive control of nonlinear systems of the Hammerstein form. *Modeling, Identification and Control* 15 (4): 253–262.

8 Morari, M. and Baric, M. (2006). Recent developments in the control of constrained hybrid systems. *Computers and Chemical Engineering* 30 (10–12): 1619–1631.

9 Maciejowski, J.M., French, I., Fletcher, I., and Urquhart, T. Cascade control of a process plant using predictive and multivariable control. *Proceedings of the 30th IEEE Conference on Decision and Control*, Brighton, UK (11–13 December 1991), 1: 583–584.

10 Zhang, Z., Xi, Y., and Xu, X. (1988). Hierarchical predictive control for large scale industrial systems. In: *Automatic Control Tenth Triennial World Congress of IFAC*, vol. 7 (ed. R. Isermann), 91–96. Oxford: Pergamon Press.

11 Xu, X., Xi, Y., and Zhang, Z. (1988). Decentralized predictive control (DPC) of large scale systems. *Information and Decision Technologies* 14: 307–322.

12 Jia, D., and Krogh, B.H. Distributed model predictive control. *Proceedings of the 2001 American Control Conference*, Arlington, VA, USA (25–27 June 2001), 4: 2767–2772.

13 Du, X.N., Xi, Y.G., and Li, S.Y. Distributed model predictive control for large scale systems. *Proceedings of the 2001 American Control Conference*, Arlington VA, USA (25–27 June 2001), 4: 3142–3143.

14 Ricker, N.L. (1985). Use of quadratic programming for constrained internal model control. *I&EC Process Design and Development* 24 (4): 925–936.

15 Ohshima, M., Hashimoto, I., Takeda, M. et al. Multi-rate multivariable model predictive control and its application to a semi-commercial polymerization reactor. *Proceedings of the 1992 American Control Conference*, Chicago, IL, USA (22–24 June 1992), 2: 1576–1581.

16 Sousa, J.M., Babuška, R., and Verbruggen, H.B. (1997). Fuzzy predictive control applied to an air-conditioning system. *Control Engineering Practice* 5 (10): 1395–1406.

17 Li, J., Xu, X., and Xi, Y. Artificial neural network-based predictive control. *Proceedings of IECON'91*, Kobe, Japan (28 October–1 November 1991), 2: 1405–1410.

18 Nakamori, Y., Suzuki, K., and Yamanaka, T. Model predictive control of nonlinear processes by multi-model approach. *Proceedings of IECON'91*, Kobe, Japan (28 October–1 November 1991), 3: 1902–1907.

19 Camacho, E.F., Ramirez, D.R., Limon, D. et al. (2010). Model predictive control techniques for hybrid systems. *Annual Reviews in Control* 34: 21–31.

20 Scattolini, R. (2009). Architectures for distributed and hierarchical model predictive control – a review. *Journal of Process Control* 19: 723–731.

21 Christofides, P.D., Scattolini, R., de la Pena, D.M., and Liu, J. (2013). Distributed model predictive control: a tutorial review and future research directions. *Computers and Chemical Engineering* 51: 21–41.

22 Ellis, M., Duranda, H., and Christofides, P.D. (2014). A tutorial review of economic model predictive control methods. *Journal of Process Control* 24: 1156–1178.
23 Garcia, C.E. and Morari, M. (1982). Internal model control, 1. A unifying review and some new results. *I&EC Process Design and Development* 21 (2): 308–323.
24 Schmidt, G. and Xi, Y. (1985). A new design method for digital controllers based on nonparametric plant models. In: *Applied Digital Control* (ed. S.G. Tzafestas), 93–109. Amsterdam: North-Holland.
25 Xi, Y. and Zhang, J. (1997). Study on the closed-loop properties of GPC. *Science in China, Series E* 40 (1): 54–63.
26 Clarke, D.W. and Mohtadi, C. (1989). Properties of generalized predictive control. *Automatica* 25 (6): 859–873.
27 Scokaert, P.O.M. and Clarke, D.W. (1994). Stabilising properties of constrained predictive control. *IEE Proceedings D – Control Theory and Applications* 141 (5): 295–304.
28 Kwon, W.H. and Pearson, A.E. (1978). On feedback stabilization of time-varying discrete linear systems. *IEEE Transactions on Automatic Control* 23 (3): 479–481.
29 Keerthi, S.S. and Gilbert, E.G. (1988). Optimal infinite-horizon feedback laws for a general class of constrained discrete-time systems: stability and moving-horizon approximations. *Journal of Optimization Theory and Applications* 57 (2): 265–293.
30 Bitmead, R.R., Gevers, M., and Wertz, V. (1990). *Adaptive Optimal Control – The Thinking Man's GPC*. Englewood Cliffs, New Jersey: Prentice Hall Inc.
31 Michalska, H. and Mayne, D.Q. (1993). Robust receding horizon control of constrained nonlinear systems. *IEEE Transactions on Automatic Control* 38 (11): 1623–1633.
32 Chen, H. and Allgöwer, F. (1998). A quasi-infinite horizon nonlinear model predictive control scheme with guaranteed stability. *Automatica* 14 (10): 1205–1217.
33 Mayne, D.Q., Rawling, J.B., Rao, C.V. et al. (2000). Constrained model predictive control: stability and optimality. *Automatica* 36 (6): 789–814.
34 Campo, P.J., and Morari, M. Robust model predictive control. *Proceedings of the 1987 American Control Conference*, Minneapoils, USA (10–12 June 1987), 2: 1021–1026.
35 Kothare, M.V., Balakrishnan, V., and Morari, M. (1996). Robust constrained model predictive control using linear matrix inequalities. *Automatica* 32 (10): 1361–1379.
36 Xi, Y.G., Li, D.W., and Lin, S. (2013). Model predictive control – status and challenges. *Acta Automatica Sinica* 39 (3): 222–236.
37 Xi, Y.G., Xu, X.M., and Zhang, Z.J. (1989). The status of predictive control and the multilayer intelligent predictive control. *Control Theory and Applications* 6 (2): 1–7. (in Chinese).

2

Some Basic Predictive Control Algorithms

Based on the three methodological principles of predictive control, a variety of predictive control algorithms can be formed by adopting different model types and optimization strategies as well as feedback correction measures. In this chapter, we introduce some typical predictive control algorithms based on these basic principles, with different model types, aiming at illustrating how the predictive control algorithm can be developed by concretizing these principles. The predictive control algorithms are introduced here for unconstrained SISO (Single-Input Single-Output) systems. Predictive control algorithms for MIMO (Multi-Input Multi-Output) systems and for constrained cases will be discussed in later chapters.

2.1 Dynamic Matrix Control (DMC) Based on the Step Response Model

Dynamic Matrix Control (DMC) [1] is one of the most widely used predictive control algorithms in industrial processes. Early in the 1970s, DMC was successfully applied to process control in the oil refinery industry [2]. DMC is based on the step response of the plant, and thus is suitable for asymptotically stable linear systems. For a weakly nonlinear system, step response can be tested at its operating point and then DMC can be adopted. For an unstable system, the DMC algorithm can be used after stabilizing the system by a traditional PID controller.

2.1.1 DMC Algorithm and Implementation

DMC algorithm consists of three parts: prediction model, rolling optimization, and feedback correction.

(1) Prediction model

DMC starts from the sampling values of the unit step response of the plant $a_i = a(iT)$, $i = 1, 2, \ldots$, which should be firstly measured or calculated according to the process input/output data, where T is the sampling period. For an asymptotically stable plant, its step response will tend to a constant after a certain time $t_N = NT$, such that the error between $a_i (i > N)$ and a_N is decreased to the same order as the quantization error or measurement error. Thus a_N can be regarded as approximately equal to the steady value a_∞ of the step

Predictive Control: Fundamentals and Developments, First Edition. Yugeng Xi and Dewei Li.

response when $t \to \infty$. Therefore, the information about the plant dynamics can be approximately described by a limited number of step response coefficients $\{a_1, a_2, \ldots, a_N\}$. They constitute the model parameters of DMC and $\boldsymbol{a} = [a_1 \ \cdots \ a_N]^{\mathrm{T}}$ is called the model vector and N the model length.

Although the step response is a kind of nonparametric model, according to the proportion and superposition properties of linear systems, this set of model parameters $\{a_i\}$ is enough for predicting the future plant outputs. At sampling time k, assume that the initial prediction for plant output at future N sampling instants is $\tilde{y}_0(k+i\,|\,k)$, $i = 1, \ldots, N$ if the control action remains unchanged. For example, when starting from steady state they can be initialized as $\tilde{y}_0(k+i\,|\,k) = y(k)$, where $k+i\,|\,k$ represents the quantity of time $k+i$ predicted at time k. Then, if the control input has an increment $\Delta u(k)$ at time k, the future outputs under control can be calculated according to the proportion and superposition properties

$$\tilde{y}_1(k+i\,|\,k) = \tilde{y}_0(k+i\,|\,k) + a_i \Delta u(k), \qquad i = 1, \ldots, N \tag{2.1}$$

Similarly, if there are M successive control increments starting from time k, denoted as $\Delta u(k), \ldots, \Delta u(k+M-1)$, then the future outputs under these control actions can be predicted by

$$\tilde{y}_M(k+i\,|\,k) = \tilde{y}_0(k+i\,|\,k) + \sum_{j=1}^{\min(M,i)} a_{i-j+1} \Delta u(k+j-1), \qquad i = 1, \ldots, N \tag{2.2}$$

where the subscript of y represents the number of control input variations. It is obvious that at any time k, if the initial prediction for plant output $\tilde{y}_0(k+i\,|\,k)$ is known, the future plant output can be calculated by the prediction model (2.2) according to future control increments, where (2.1) is only a special case of the prediction model (2.2) for $M = 1$.

(2) Rolling optimization

DMC is an algorithm that determines the control input through optimization. At each time k, M current and future control increments $\Delta u(k), \ldots, \Delta u(k+M-1)$ would be determined such that under their control the predicted plant outputs at future P sampling instants $\tilde{y}_M(k+i\,|\,k)$ can be as close as possible to the desired values $w(k+i)$, $i = 1, \ldots, P$ (see Figure 2.1), where M, P are called control horizon and optimization horizon, respectively, with their meanings explained in Figure 2.1. To make the problem meaningful, it is always assumed that $M \le P$.

In addition to the requirement for system output tracking the desired value, during the control process it is also preferred that the control increment Δu does not change excessively. This issue can be addressed by adding soft penalties in the performance index. Then the DMC optimization problem at time k can be formulated as

$$\min_{\Delta u} J(k) = \sum_{i=1}^{P} q_i \left[w(k+i) - \tilde{y}_M(k+i\,|\,k) \right]^2 + \sum_{j=1}^{M} r_j \Delta u^2(k+j-1) \tag{2.3}$$

where q_i, r_j are weighting coefficients, representing the extents of suppressing tracking errors and control variations, respectively.

In the case when constraints on control input and system output exist, the optimization problem (2.3) should be solved combined with constraint conditions and is often a

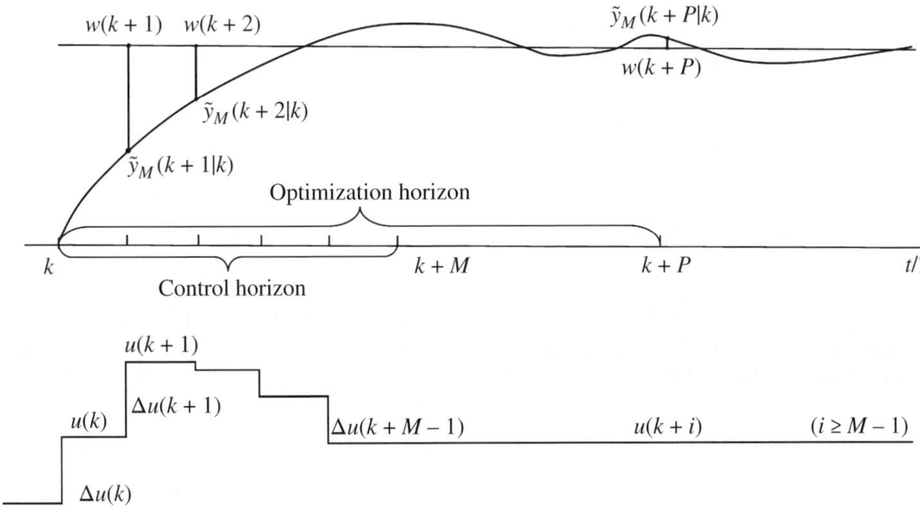

Figure 2.1 Optimization strategy of DMC at time *k*.

typical QP (Quadratic Programming) problem if the constraints are linear, which will be discussed in detail in Chapter 5. For unconstrained cases, the above optimization problem can be formulated as minimizing the performance index (2.3) subject to equality constraints given by the prediction model (2.2), with $\Delta \boldsymbol{u}_M(k) = [\Delta u(k) \ \cdots \ \Delta u(k+M-1)]^{\mathrm{T}}$ as optimization variables. To solve this problem, the relationship of $\tilde{y}_M$ in the performance index and the optimization variable Δu should be at first derived by using the prediction model (2.2). It can be given in the following vector form:

$$\tilde{\boldsymbol{y}}_{PM}(k) = \tilde{\boldsymbol{y}}_{P0}(k) + \boldsymbol{A}\Delta \boldsymbol{u}_M(k) \tag{2.4}$$

where

$$\tilde{\boldsymbol{y}}_{PM}(k) = \begin{bmatrix} \tilde{y}_M(k+1 \mid k) \\ \vdots \\ \tilde{y}_M(k+P \mid k) \end{bmatrix}, \quad \tilde{\boldsymbol{y}}_{P0}(k) = \begin{bmatrix} \tilde{y}_0(k+1 \mid k) \\ \vdots \\ \tilde{y}_0(k+P \mid k) \end{bmatrix}$$

$$\boldsymbol{A} = \begin{bmatrix} a_1 & \cdots & 0 \\ \vdots & \ddots & \vdots \\ a_M & \cdots & a_1 \\ \vdots & \ddots & \vdots \\ a_P & \cdots & a_{P-M+1} \end{bmatrix}$$

and $\boldsymbol{A}$ is a $P \times M$ matrix composed of the unit step response coefficients a_i, called the dynamic matrix. In the above formulae, the former subscript of vector $\tilde{\boldsymbol{y}}$ represents the number of future time instances over which the outputs are predicted while the latter one represents the number of control variations.

Similarly, the performance index (2.3) can be also written in vector form as

$$\min_{\Delta \boldsymbol{u}_M(k)} J(k) = \|\boldsymbol{w}_P(k) - \tilde{\boldsymbol{y}}_{PM}(k)\|_{\boldsymbol{Q}}^2 + \|\Delta \boldsymbol{u}_M(k)\|_{\boldsymbol{R}}^2 \tag{2.5}$$

where

$$\boldsymbol{w}_P(k) = [w(k+1) \ \cdots \ w(k+P)]^{\mathrm{T}}$$
$$\boldsymbol{Q} = \mathrm{diag}(q_1, \ldots, q_P)$$
$$\boldsymbol{R} = \mathrm{diag}(r_1, \ldots, r_M)$$

The diagonal matrices $\boldsymbol{Q}, \boldsymbol{R}$ composed of weighting coefficients q_i, r_j are called the error weighting matrix and the control weighting matrix respectively.

Put (2.4) into (2.5), it follows

$$\min_{\Delta \boldsymbol{u}_M(k)} J(k) = \|\boldsymbol{w}_P(k) - \tilde{\boldsymbol{y}}_{P0}(k) - \boldsymbol{A}\Delta \boldsymbol{u}_M(k)\|_{\boldsymbol{Q}}^2 + \|\Delta \boldsymbol{u}_M(k)\|_{\boldsymbol{R}}^2$$

At time k, both $\boldsymbol{w}_P(k)$ and $\tilde{\boldsymbol{y}}_{P0}(k)$ are known and the optimal $\Delta \boldsymbol{u}_M(k)$ can be solved by using the necessary condition for extreme value $\mathrm{d}J(k)/\mathrm{d}\Delta \boldsymbol{u}_M(k) = 0$. Thus

$$\Delta \boldsymbol{u}_M(k) = (\boldsymbol{A}^{\mathrm{T}}\boldsymbol{Q}\boldsymbol{A} + \boldsymbol{R})^{-1}\boldsymbol{A}^{\mathrm{T}}\boldsymbol{Q}[\boldsymbol{w}_P(k) - \tilde{\boldsymbol{y}}_{P0}(k)] \tag{2.6}$$

which gives all the optimal solutions $\Delta u(k), \ldots, \Delta u(k+M-1)$ for the optimization problem at time k. However, in DMC, instead of implementing all of them, only the current control increment $\Delta u(k)$ is adopted to form the control action that is actually implemented to the plant. $\Delta u(k)$ is the first element of vector $\Delta \boldsymbol{u}_M(k)$ and can be given by

$$\Delta u(k) = \boldsymbol{c}^{\mathrm{T}}\Delta \boldsymbol{u}_M(k) = \boldsymbol{d}^{\mathrm{T}}[\boldsymbol{w}_P(k) - \tilde{\boldsymbol{y}}_{P0}(k)] \tag{2.7}$$

where the M-dimensional row vector $\boldsymbol{c}^{\mathrm{T}} = [1 \ 0 \ \cdots \ 0]$ represents the operation of selecting the first row for the followed matrix, and the P dimensional row vector

$$\boldsymbol{d}^{\mathrm{T}} = \boldsymbol{c}^{\mathrm{T}}(\boldsymbol{A}^{\mathrm{T}}\boldsymbol{Q}\boldsymbol{A} + \boldsymbol{R})^{-1}\boldsymbol{A}^{\mathrm{T}}\boldsymbol{Q} \triangleq [d_1 \ \cdots \ d_P] \tag{2.8}$$

is called the control vector. Once the control strategy has been determined, i.e. $P, M, \boldsymbol{Q}, \boldsymbol{R}$ are all fixed, $\boldsymbol{d}^{\mathrm{T}}$ can be solved off-line according to (2.8).

Equation (2.7) is the analytical form of the DMC control law in the unconstrained case, where online solving of the optimization problem is reduced to explicitly computing the control law (2.7) and thus is very simple. It should be emphasized that if constraints on system input and/or output are taken into account in the optimization problem, the solution will be no longer of the analytical form (2.6) and no analytical control law (2.7) exists.

After obtaining the control increment $\Delta u(k)$, the real control action is calculated by

$$u(k) = u(k-1) + \Delta u(k) \tag{2.9}$$

Implement $u(k)$ on the plant and, at the next sampling time, the same optimization problem with $k+1$ instead of k is formulated and solved for $\Delta u(k+1)$, and $u(k+1)$ is obtained and implemented. The whole procedure continues in this rolling way, which is what "rolling optimization" means.

(3) Feedback correction

At time k, implementing control $u(k)$ means adding a step with amplitude $\Delta u(k)$ on the plant input. Using the prediction model (2.1), the predicted future output caused by it can be given by

$$\tilde{\boldsymbol{y}}_{N1}(k) = \tilde{\boldsymbol{y}}_{N0}(k) + \boldsymbol{a}\Delta u(k) \tag{2.10}$$

It is indeed the vector form of (2.1), where the subscripts of N-dimensional vectors $\tilde{\boldsymbol{y}}_{N1}(k)$ and $\tilde{\boldsymbol{y}}_{N0}(k)$ have the same meanings as defined above. Since the elements of $\tilde{\boldsymbol{y}}_{N1}(k)$ are the predicted future outputs only with $\Delta u(k)$ acted, after shifting they can be used as the initial prediction values for new optimization at time $k+1$. However, due to actual existing unknown factors such as model mismatch or disturbance, the predicted values given by (2.10) might deviate from actual ones. If these values could not be corrected in time by using real-time information, model prediction and optimization at the next time would be established on an inaccurate basis, and the predicted output may become more and more deviated from actual output as the procedure continues. In order to prevent the error caused by open-loop optimization only based on the model, before solving the optimal control actions at time $k+1$, the actual plant output $y(k+1)$ should first be measured and compared with the predicted output $\tilde{y}_1(k+1 \mid k)$ calculated by the model (2.10) to construct the output error

$$e(k+1) = y(k+1) - \tilde{y}_1(k+1 \mid k) \tag{2.11}$$

This error information reflects the effects of uncertain factors not included in the model on the plant output. It can be used to predict future output errors as a supplement of the model-based prediction. Due to the lack of causal information of the error, error prediction is often implemented in a heuristic way. For example, one can use weighted $e(k+1)$ to correct the prediction of future outputs

$$\tilde{\boldsymbol{y}}_{\text{cor}}(k+1) = \tilde{\boldsymbol{y}}_{N1}(k) + \boldsymbol{h}e(k+1) \tag{2.12}$$

where

$$\tilde{\boldsymbol{y}}_{\text{cor}}(k+1) = \begin{bmatrix} \tilde{y}_{\text{cor}}(k+1 \mid k+1) \\ \vdots \\ \tilde{y}_{\text{cor}}(k+N \mid k+1) \end{bmatrix}$$

is the predicted output vector after correction. The N-dimensional vector $\boldsymbol{h} = [h_1 \ \cdots \ h_N]^{\mathrm{T}}$ composed of weighting coefficients h_i is called the correction vector.

At time $k+1$, since the time basis has moved ahead, the future sampling instants to be predicted will be moved to $k+2, \ldots, k+1+N$. Thus the elements of $\tilde{\boldsymbol{y}}_{\text{cor}}(k+1)$ should be shifted to construct the initial predicted outputs at time $k+1$:

$$\tilde{y}_0(k+1+i \mid k+1) = \tilde{y}_{\text{cor}}(k+1+i \mid k+1), \quad i = 1, \ldots, N-1 \tag{2.13}$$

Due to truncation of the model, $\tilde{y}_0(k+1+N \mid k+1)$ does not appear in the output prediction at time k, but can be approximated by $\tilde{y}_{\text{cor}}(k+N \mid k+1)$. Setting the initial prediction at time $k+1$ in a vector form gives

$$\tilde{\boldsymbol{y}}_{N0}(k+1) = \boldsymbol{S}\tilde{\boldsymbol{y}}_{\text{cor}}(k+1) \tag{2.14}$$

where the shift matrix $\boldsymbol{S}$ is defined by

$$\boldsymbol{S} = \begin{bmatrix} 0 & 1 & & 0 \\ \vdots & \ddots & \ddots & \\ \vdots & & 0 & 1 \\ 0 & \cdots & 0 & 1 \end{bmatrix}$$

With $\tilde{\boldsymbol{y}}_{N0}(k+1)$, the optimization at time $k+1$ can be performed as above and $\Delta u(k+1)$ can be solved. The whole control procedure goes repeatedly in the form of online rolling optimization combined with error correction. The algorithmic structure of DMC is shown in Figure 2.2.

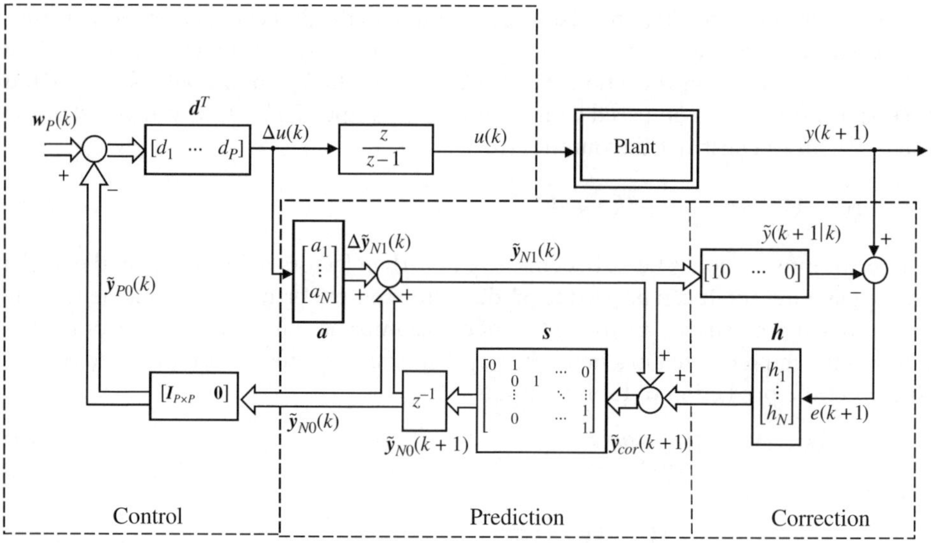

Figure 2.2 The algorithmic structure of DMC.

From Figure 2.2 it is shown that the DMC algorithm is composed of three parts: prediction, control, and correction, where the rough arrow represents vector flow while the thin arrow represents scalar flow. At each sampling time, the error vector constructed by desired output $\boldsymbol{w}_P(k)$ and initial predicted output $\tilde{\boldsymbol{y}}_{P0}(k)$ for future P sampling instants is dot multiplied with the dynamic control vector $\boldsymbol{d}^{\mathrm{T}}$ to obtain a current control increment $\Delta u(k)$; see (2.7). This control increment is on the one hand used to calculate the control action $u(k)$ through digital integration (summation), see (2.9), which is then implemented on the plant, and on the other hand, multiplied with the model vector $\boldsymbol{a}$, giving the predicted output when $\Delta u(k)$ has been acted; see (2.10). At the next sampling time, real output $y(k+1)$ is firstly measured and compared with the predicted output $\tilde{y}_1(k+1 \mid k)$, i.e. the first element in vector $\tilde{\boldsymbol{y}}_{N1}(k)$, to construct the output error $e(k+1)$ according to (2.11). This error is multiplied with correction vector $\boldsymbol{h}$ as the error prediction, added into the model prediction to obtain the predicted output $\tilde{\boldsymbol{y}}_{\mathrm{cor}}(k+1)$ after correction according to (2.12), and then shifted according to (2.14) to get the new initial output prediction $\tilde{\boldsymbol{y}}_{N0}(k+1)$. In the figure, z^{-1} represents the time basis one step back, which means that the new sampling instant is defined again by k. The whole procedure is repeatedly online.

For implementation of the unconstrained DMC algorithm, three groups of parameters should be prepared off-line, i.e. the parameters of the model vector, control vector, and correction vector. They can be obtained as follows.

1) Model parameters $\{a_i\}$ can be obtained by measuring the unit step response and smoothing the parameters. It should be emphasized that the noise and disturbance included in the measured data should be filtered as much as possible such that the obtained model has a smooth dynamic response; otherwise the control quality may be degenerated and even unstable.
2) Control coefficients $\{d_i\}$ can be calculated according to (2.8), where the model parameters $\{a_i\}$ are adopted, and the parameters P, M, $\boldsymbol{Q}$, $\boldsymbol{R}$ constituting the optimization policy should be determined by tuning through simulation.
3) Correction parameters $\{h_i\}$ are independent from the above two groups of parameters and can be selected freely.

After these three groups of parameters are determined, put them in fixed memories for real-time use.

The online computation of DMC is composed of an initialization module and real-time control module. The initialization module is called at the first step of running. It measures the actual plant output $y(k)$ and sets it as the initial values of output prediction $\tilde{y}_0(k+i \mid k)$, $i = 1, \ldots, N$. Then it turns to the real-time control module. The flowchart of online computation at each sampling time is given in Figure 2.3, where only one N-dimensional array $y(i)$ needs to be set for output prediction. The formulae in the chart successively correspond to (2.11), (2.12), (2.14), (2.7), (2.9), and (2.10). It should be noted that in this chart the desired output w is fixed and stored in memory in advance. However, for a tracking problem, an additional tracking trajectory module is needed. During online computation, w in the chart should be replaced by the desired output at future sampling instants in the optimization horizon $w(i)$, $i = 1, \ldots, P$.

Remark 2.1 It should be pointed out that for the unconstrained DMC, due to the existence of the analytical solution for optimization, online optimization is indeed implied in the optimal solution (2.7). During online optimization only three groups of parameters $\{a_i\}$, $\{d_i\}$, $\{h_i\}$ are needed and there is no explicit optimization procedure. However, for a constrained DMC algorithm, the control increment should be solved by the QP algorithm and the analytical solution (2.7) no longer exists. In this case, a QP module should be used to replace the analytical form (2.7) in the chart. Since control vector $\boldsymbol{d}^{\mathrm{T}}$ does not appear in this case, only model parameters $\{a_i\}$ and correction parameters $\{h_i\}$ are needed to be off-line calculated.

2.1.2 Description of DMC in the State Space Framework

Just after DMC as a heuristic control algorithm appeared in industry, researchers tried to explore its control mechanism with respect to existing control schemes. In [3] a state

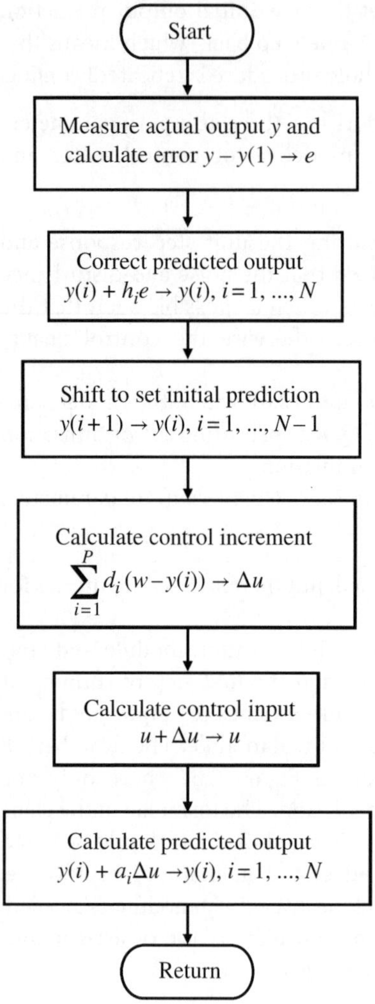
Start
Measure actual output y and calculate error y – y(1) → e
Correct predicted output y(i) + h_ie → y(i), i = 1, ..., N
Shift to set initial prediction y(i + 1) → y(i), i = 1, ..., N – 1
Calculate control increment Σ_{i=1}^{P} d_i (w – y(i)) → Δu
Calculate control input u + Δu → u
Calculate predicted output y(i) + a_iΔu → y(i), i = 1, ..., N
Return

Figure 2.3 Online computation chart of unconstrained DMC.

space design approach starting from the system step response is proposed, resulting in the same algorithm as DMC. The state space description gives a better understanding on the control mechanism of DMC and particularly helps to design the feedback correction of DMC more reasonably. In the following, we briefly introduce the main result of this approach.

As in DMC, for a stable linear time invariant SISO plant, the state space design also starts from the sampling values of its unit step response $\{a_1, a_2, \dots, a_N\}$. According to the linear system theory, a linear system with input—output description/data can be represented in state space form if certain realization conditions are satisfied that characterize the relationship of its input–output response and the parameters in state space form [4]. For a linear system with step response coefficients $\{a_i\}$, if a state space model with control increment as input

$$x(k+1) = Ax(k) + b\Delta u(k)$$
$$y(k) = c^{T}x(k)$$

is adopted to describe it, the corresponding realization condition is given by

$$c^{T}A^{i-1}b = a_i, \quad i = 1,2,\ldots$$

where the triple (c, A, b) is not unique and the dimension of the state space is not necessarily minimum. Thus the following state space representation can be taken as an approximate state space realization of the plant with step response $\{a_1, a_2, \cdots, a_N\}$:

$$\begin{aligned} x(k+1) &= Sx(k) + a\Delta u(k) \\ y(k) &= c^{T}x(k) \end{aligned} \tag{2.15}$$

because it satisfies

$$c^{T}A^{i-1}b = \begin{cases} a_i, & i = 1,\ldots,N-1 \\ a_N \cong a_i, & i \geq N \end{cases}$$

where the matrix S and the vectors a, c^{T} are the same as those introduced in the last subsection. Note that if the plant input $u(k+i)$ is fixed (i.e. $\Delta u(k+i) = 0$) for $i \geq 0$, it follows that

$$y(k+i-1) = c^{T}S^{i-1}x(k) = x_i(k), \quad i = 1,\ldots,N$$

Therefore the N-dimensional state $x(k)$ is called the predictive state because its elements are the future plant outputs if the plant input remains unchanged.

According to the state space model (2.15), at time k, the future outputs can be predicted according to the current state $x(k)$ and the assumed future control inputs

$$\tilde{y}_{PM}(k) = GSx(k) + A\Delta u_M(k) \tag{2.16}$$

where the matrices S, A and vectors $\tilde{y}_{PM}(k)$, $\Delta u_M(k)$ are the same as those in the last subsection, and $G = [I_{P\times P} \vdots 0_{P\times(N-P)}]$ represents the operation of selecting the first P rows of the followed matrix.

At time k, consider the same optimization problem as (2.5):

$$\min_{\Delta u_M(k)} J(k) = \|w_P(k) - \tilde{y}_{PM}(k)\|_Q^2 + \|\Delta u_M(k)\|_R^2$$

where the matrices Q, R and the vector $w_P(k)$ are the same as those in the last subsection. For unconstrained cases, using the prediction model (2.16), the optimal solution can be given in analytical form similar to (2.6):

$$\Delta u_M(k) = \left(A^{T}QA + R\right)^{-1}A^{T}Q[w_P(k) - GSx(k)]$$

and the current control increment is given by

$$\Delta u(k) = d^{T}[w_P(k) - GSx(k)]$$

which is similar to (2.7) with the same d^{T} given by (2.8). However, the elements of the predictive state $x(k)$ are unmeasurable at time k except for $x_1(k) = y(k)$, so the above control law cannot be implemented. However, note that as the system (2.5) is fully observable because its observability matrix is an identity matrix, a predictive state observer can be introduced to reconstruct the predictive state:

$$\begin{aligned}\hat{\boldsymbol{x}}(k+1) &= \boldsymbol{S}\hat{\boldsymbol{x}}(k) + \boldsymbol{a}\Delta u(k) + \boldsymbol{h}(y(k+1) - \hat{y}(k+1)) \\ \hat{y}(k+1) &= \boldsymbol{c}^{\mathrm{T}}\boldsymbol{S}\hat{\boldsymbol{x}}(k) + \boldsymbol{c}^{\mathrm{T}}\boldsymbol{a}\Delta u(k)\end{aligned} \tag{2.17}$$

where $\hat{\boldsymbol{x}}(k)$ and $\hat{y}(k)$ are the reconstructed predictive state and the observer output, respectively, while $\boldsymbol{h} = [h_1 \ \cdots \ h_N]^{\mathrm{T}}$ is the observer feedback vector. With the reconstructed $\hat{\boldsymbol{x}}(k)$, the control law in the above can be rewritten in the feasible form

$$\Delta u(k) = \boldsymbol{d}^{\mathrm{T}}[\boldsymbol{w}_P(k) - \boldsymbol{G}\boldsymbol{S}\hat{\boldsymbol{x}}(k)] \tag{2.18}$$

with which the control action can be calculated by (2.9) and implemented.

The above design starts from the same data set $\{a_1, a_2, \cdots, a_N\}$ as in DMC, but introduces a state space model (2.15) to describe the system dynamics. According to (2.15), the prediction model (2.16) can be derived, which indicates how future system outputs are influenced by the current predictive state $\boldsymbol{x}(k)$ and the assumed future control inputs. Based on this, the state feedback control law with respect to certain optimization criterion can be designed by using the usual method. Since the predictive state $\boldsymbol{x}(k)$ in the control law is not available, a state observer (2.17) is designed to reconstruct it with which the feasible optimal control law (2.18) can be derived. All of the above can be regarded as a standard state space method of designing an optimization-based state feedback control law with a state observer.

Although the above control law is designed in the state space framework, the resultant algorithm is just the same as DMC, which could be shown by the following variable correspondence:

$$\begin{aligned}&\boldsymbol{S}\hat{\boldsymbol{x}}(k) \Leftrightarrow \tilde{\boldsymbol{y}}_{N0}(k), \ \ \boldsymbol{G}\boldsymbol{S}\hat{\boldsymbol{x}}(k) \Leftrightarrow \tilde{\boldsymbol{y}}_{P0}(k), \ \ \boldsymbol{S}\hat{\boldsymbol{x}}(k) + \boldsymbol{a}\Delta u(k) \Leftrightarrow \tilde{\boldsymbol{y}}_{N1}(k) \\ &\hat{y}(k+1) \Leftrightarrow y(k+1 \mid k), \ \ \hat{\boldsymbol{x}}(k+1) \Leftrightarrow \tilde{\boldsymbol{y}}_{\mathrm{cor}}(k+1)\end{aligned}$$

as well as formulae correspondence:

$$(2.16) \Leftrightarrow (2.4), \ \ (2.17) \Leftrightarrow (2.10), (2.11), (2.12), \ \ (2.18) \Leftrightarrow (2.7)$$

Therefore, the design method presented here can be as an interpretation of a realization of the DMC algorithm in the state space framework. It results in the same algorithm as DMC, but uses the standard design method in traditional control theory.

Remark 2.2 The state space representation introduced above brought insight into the running mechanism of the DMC algorithm. The error correction in the original DMC algorithm is heuristic, often using the weighted current output error to compensate the model-based output prediction, while with the state space design, the correction vector $\boldsymbol{h}$ is naturally introduced as the feedback vector of the state observer. On the one hand, it is more clearly recognized that the feedback correction is an indispensable part of the DMC algorithm because it updates the predictive state online, i.e. the initial output prediction, to make the optimization on a more realistic basis. On the other hand, the introduction of the free correction vector $\boldsymbol{h}$ increases the degree of freedom for feedback correction and extends the original simple correction scheme, while the mature theoretical results on observer design can be used to help the design of $\boldsymbol{h}$. According to the separation principle, the design of $\boldsymbol{h}$ is independent of the optimization strategy if the model parameters

$\{a_i\}$ are accurate. As long as the selection of $\boldsymbol{h}$ makes the observer converge, no matter whether the initial state of the observer $\hat{\boldsymbol{x}}(0)$ is correct or not, the observed state $\hat{\boldsymbol{x}}(k)$ can asymptotically tend to the real state $\boldsymbol{x}(k)$. This ensures the correctness of using (2.18) as the control law, and can also make the algorithm work effectively in the case of an unsteady start or with an inaccurate initial prediction.

2.2 Generalized Predictive Control (GPC) Based on the Linear Difference Equation Model

Generalized Predictive Control (GPC) [5] is another kind of predictive control algorithms rising from the research area of adaptive control. It is well known that adaptive control, as an efficient control scheme against model uncertainties (such as unknown parameter variations, etc.), has been successfully used in aerospace and aviation fields, and also in industrial processes. However, such kinds of control algorithms strongly depend on the mathematical model of the plant. Some of them (such as a minimum variance self-tuning regulator) need to know the accurate time delay of the plant, while some others (such as the pole assignment self-tuning regulator) are very sensitive to the order of the model. Therefore, when applying self-tuning control to complex industrial processes where accurate modeling is almost impossible, further improvement for the original algorithm is necessary. Based on the minimum variance self-tuning control algorithm, Clarke et al. proposed the GPC algorithm [5], where multistep prediction and optimization are adopted instead of one-step prediction and optimization in the original algorithm.

As an algorithm developed from adaptive control, GPC is proposed for stochastic linear discrete time systems. In the following, referring to [5], we introduce the GPC algorithm in its original formulation.

(1) Prediction model

In the GPC algorithm, the plant with random disturbance is described by the following CARIMA (Controlled Auto-Regressive Integrated Moving Average) model, which is a stochastic linear difference equation and is popular in minimum variance control

$$A(q^{-1})y(t) = B(q^{-1})u(t-1) + \frac{C(q^{-1})\xi(t)}{\Delta} \tag{2.19}$$

where

$$A(q^{-1}) = 1 + a_1 q^{-1} + \cdots + a_n q^{-n}$$
$$B(q^{-1}) = b_0 + b_1 q^{-1} + \cdots + b_{n_b} q^{-n_b}$$
$$C(q^{-1}) = c_0 + c_1 q^{-1} + \cdots + c_{n_c} q^{-n_c}$$

where t represents the discrete time in sampling control and q^{-1} is a backward shift operator, representing the quantity at the time one sampling period backward; $\Delta = 1 - q^{-1}$ is a differencing operator; $\xi(t)$ is an uncorrelated random sequence, representing the influence of a kind of stochastic noises; A, B, C are all polynomials of q^{-1}. To fix the order of the polynomials it is assumed that $a_n \neq 0$, $b_{n_b} \neq 0$. The first several coefficients in polynomial $B(q^{-1})$, i.e. b_0, b_1, …, may be zero, which indicates the time delay of the system. In order to focus on the methodological principles of the algorithm, assume that $C(q^{-1}) = 1$.

Note that if the stochastic disturbance term in (2.19) is removed, the model (2.19) becomes a deterministic input–output difference model with regressive terms

$$y(t) + a_1 y(t-1) + \cdots + a_n y(t-n) = b_0 u(t-1) + \cdots + b_{n_b} u(t-n_b-1) \tag{2.20a}$$

which is identical to the discrete time transfer function model in the Z domain

$$G(z) = \frac{y(z)}{u(z)} = \frac{z^{-1}B(z^{-1})}{A(z^{-1})} \tag{2.20b}$$

Now consider how to use (2.19) to derive a prediction model. From (2.20a) it is known that the outputs at different future times are correlated and the prediction for $y(t+j)$ involves outputs $y(t+j-1)$, $y(t+j-2)$, etc., which are unknown at t. To make the optimization problem at time t easy to solve, it is always desirable to make the predicted future output $y(t+j)$ expressed only by current available information $\{y(\tau),\ \tau \le t\}$, $\{u(\tau),\ \tau < t\}$ and assumed future control input $\{u(\tau),\ \tau \ge t\}$. According to (2.20a) it is straightforward to give such expression for $y(t+1)$, but for $y(t+2)$, such desired expression can only be obtained if $y(t+1)$ appearing in it is further substituted by the derived expression above. Similarly, since $y(t+1)$, $y(t+2)$ appear in $y(t+3)$, to obtain a desired expression, they must be substituted by their corresponding expressions derived before, and so on. It is obvious that with the increase in the prediction time $t+j$, more and more variables need to be substituted and the derived expression after compounded substitutions would be very complex. It is thus not an appropriate way to get an output prediction. To overcome this difficulty, it is necessary to remove the currently unknown output information $y(t+j-1)$, …, $y(t+1)$, etc. from the expression of $y(t+j)$. To do that, consider the following Diophantine equation:

$$1 = E_j(q^{-1})A\Delta + q^{-j}F_j(q^{-1}) \tag{2.21}$$

where E_j, F_j are polynomials determined only by $A(q^{-1})$ and the prediction length j:

$$E_j(q^{-1}) = e_{j,0} + e_{j,1}q^{-1} + \cdots + e_{j,j-1}q^{-(j-1)}$$
$$F_j(q^{-1}) = f_{j,0} + f_{j,1}q^{-1} + \cdots + f_{j,n}q^{-n}$$

Multiplying $E_j\Delta q^j$ on both sides of (2.19) it follows that

$$E_jA\Delta y(t+j) = E_jB\Delta u(t+j-1) + E_j\xi(t+j)$$

Using (2.21), the output at time $t+j$ can be written by

$$y(t+j) = E_jB\Delta u(t+j-1) + F_jy(t) + E_j\xi(t+j)$$

According to the expressions of E_j and F_j given above, $E_jB\Delta u(t+j-1)$ is only related to $\Delta u(t+j-1)$, $\Delta u(t+j-2)$, etc., $F_jy(t)$ is only related to $y(t)$, $y(t-1)$, etc., while $E_j\xi(t+j)$ is only related to $\xi(t+j)$, …, $\xi(t+1)$. Since future noises $\xi(t+i)$, $i = 1, …, j$, are unknown at t, the most suitable prediction for $y(t+j)$ can be given by

$$\hat{y}(t+j) = E_jB\Delta u(t+j-1) + F_jy(t) \tag{2.22}$$

Thus, using the Diophantine equation (2.21), the future output $y(t+j)$ predicted at t can be represented by currently known input/output information and assumed future input, independent from currently unknown future output information $y(t+j-1)$, $y(t+j-2)$, etc. Equation (2.22) is then the prediction model in GPC.

The polynomials E_j, F_j in the prediction model (2.22) should be known when doing an output prediction using this model. They can be obtained by solving the Diophantine equations for different $j = 1, 2, \ldots$, respectively. To simplify the calculation, Clarke et al. gave a recursive algorithm for solving E_j, F_j. Firstly, write the Diophantine equation (2.21) for j and $j+1$:

$$1 = E_j A\Delta + q^{-j}F_j$$

$$1 = E_{j+1}A\Delta + q^{-(j+1)}F_{j+1}$$

Make a subtraction as follows:

$$\tilde{A}\tilde{E} + q^{-j}\left(q^{-1}F_{j+1} - F_j + \tilde{A}e_{j+1,j}\right) = 0$$

where

$$\tilde{A} = A\Delta = 1 + \tilde{a}_1 q^{-1} + \cdots + \tilde{a}_{n+1}q^{-(n+1)}$$
$$= 1 + (a_1 - 1)q^{-1} + \cdots + (a_n - a_{n-1})q^{-n} - a_n q^{-(n+1)}$$

$$\tilde{E} = E_{j+1} - E_j - e_{j+1,j}q^{-j} = \tilde{e}_0 + \tilde{e}_1 q^{-1} + \cdots + \tilde{e}_{j-1}q^{-(j-1)}$$

$$\tilde{e}_i = e_{j+1,i} - e_{j,i}, \quad i = 0, \ldots, j-1$$

Since the coefficients from the constant term to $q^{-(j-1)}$ in the left-hand side of the above equality should be zero, there must be $\tilde{E} = 0$, and therefore

$$F_{j+1} = q\left(F_j - \tilde{A}\,e_{j+1,j}\right)$$

Note that as the coefficient of the first term of $\tilde{A}$ is 1, we can further obtain

$$f_{j,0} = e_{j+1,j}$$

$$f_{j+1,i} = f_{j,i+1} - \tilde{a}_{i+1}e_{j+1,j} = f_{j,i+1} - \tilde{a}_{i+1}f_{j,0}, \qquad i = 0, \ldots, n-1$$

$$f_{j+1,n} = -\tilde{a}_{n+1}e_{j+1,j} = -\tilde{a}_{n+1}f_{j,0}$$

This recursive form on the coefficients of F_j can also be represented in vector form $\boldsymbol{f}_{j+1} = \bar{\boldsymbol{A}}\boldsymbol{f}_j$, where

$$\boldsymbol{f}_{j+1} = \begin{bmatrix} f_{j+1,0} \\ \vdots \\ f_{j+1,n} \end{bmatrix}, \quad \boldsymbol{f}_j = \begin{bmatrix} f_{j,0} \\ \vdots \\ f_{j,n} \end{bmatrix}, \quad \bar{\boldsymbol{A}} = \begin{bmatrix} 1-a_1 & 1 & & 0 \\ a_1 - a_2 & & \ddots & \\ \vdots & & & \ddots \\ a_{n-1} - a_n & 0 & & 1 \\ a_n & 0 & \cdots & 0 \end{bmatrix}$$

Furthermore, E_j can be recursively obtained by

$$E_{j+1} = E_j + e_{j+1,j}q^{-j} = E_j + f_{j,0}q^{-j}$$

For $j = 1$, $E_1 = e_{1,0}$, (2.21) becomes

$$1 = E_1\tilde{A} + q^{-1}F_1$$

which gives $E_1 = e_{1,0} = 1$, $F_1 = q(1-\tilde{A})$, i.e. the coefficient vector $\boldsymbol{f}_1$ of F_1 is just the first column of matrix $\bar{\boldsymbol{A}}$. Take $E_1, \boldsymbol{f}_1$ as initial values; then E_j, F_j, $j = 2, 3, \ldots$, can be recursively calculated as follows:

$$\begin{aligned} &\boldsymbol{f}_{j+1} = \bar{\boldsymbol{A}}\boldsymbol{f}_j, && \boldsymbol{f}_1 = \bar{\boldsymbol{A}}\,[1\ \ 0\ \ \cdots\ \ 0]^{\mathrm{T}} \\ &E_{j+1} = E_j + f_{j,0}q^{-j}, && E_1 = 1 \end{aligned} \tag{2.23}$$

(2) Rolling optimization

In GPC, the performance index of optimization at time t has the following form:

$$\min J(t) = E\left\{\sum_{j=N_1}^{N_2}[w(t+j) - y(t+j)]^2 + \sum_{j=1}^{NU}\lambda(j)\Delta u^2(t+j-1)\right\} \tag{2.24}$$

where E means mathematical expectation, $w(t+j)$ could be a constant w equal to the current set-point, or considered as the values of a smoothed reference trajectory as suggested in [6], which starts from the current output $y(t)$ and tends to the set-point w in the form of a typical first-order curve

$$\begin{aligned} &w(t+j) = \alpha w(t+j-1) + (1-\alpha)w, \quad j = 1, 2, \ldots \\ &w(t) = y(t), \qquad 0 \le \alpha < 1 \end{aligned} \tag{2.25}$$

N_1 and N_2 are the minimum and maximum optimization horizons (in [5] also called costing horizons), respectively, and NU is the control horizon, which means that the control variable will be unchanged after NU steps:

$$u(t+j-1) = u(t+NU-1), \quad j \ge NU$$

$\lambda(j)$ is a control weighting sequence. For simplification it is often assumed that they are equally set as constant λ.

Compared with the minimum variance self-tuning regulator, prediction over a long time horizon is adopted in the GPC performance index (2.24). The variance to be optimized is extended from one sampling time to several sampling times over a horizon, where N_1 should be greater than the time delay of the plant and N_2 should be large enough such that the plant dynamic behavior can be fully exhibited. Since long-range optimization is adopted instead of one-step optimization, reasonable control can be achieved even when the estimation for time delay is inaccurate or time delay changes, which makes GPC robust against model uncertainty. The concept of using long-range prediction and optimization was motivated by predictive control algorithms such as IDCOM (Identification–Command) [6] and DMC [1]. Particularly, when comparing the performance indices of GPC and DMC, it is obvious that except for the difference caused by stochastic formulation, the GPC performance index (2.24) can be viewed as a variation of the DMC performance index (2.5), corresponding to the selection of P, M with N_2, NU, weighting coefficients q_i before N_1 with zero and after N_1 with 1, for all $r_j = \lambda$ and $\alpha = 0$ for the reference trajectory (2.25).

In order to predict the future outputs appearing in the performance index (2.24) using (2.22), let $G_j = E_jB$, which is an $n_b + j - 1$-order polynomial of q^{-1}. Denote

$$G_j = g_{j,0} + g_{j,1}q^{-1} + \cdots + g_{j,n_b+j-1}q^{-(n_b+j-1)}$$

According to (2.21), it follows that

$$G_j = \frac{B(1-q^{-j}F_j)}{A\Delta} = \frac{B}{A\Delta} - q^{-j}\frac{BF_j}{A\Delta} \tag{2.26}$$

This indicates that the coefficients of the first j terms of the polynomial G_j can be fully determined by $B/A\Delta$. Meanwhile, the plant transfer function (2.20) can be represented by

$$G(z) = \frac{z^{-1}B(z^{-1})}{A(z^{-1})} = g_1z^{-1} + (g_2 - g_1)z^{-2} + \cdots$$

where $\{g_i\}$ are the coefficients of the plant unit step response. Then it follows that

$$\frac{B}{A\Delta} = g_1 + g_2z^{-1} + \cdots$$

Thus the coefficients of the first j terms of G_j are just the first j sampling values of the plant unit step response, i.e. $g_{j,\,i} = g_{i+1}\ \ (i < j)$.

According to prediction model (2.22), now the prediction for future outputs $y(t+j)$ in the performance index (2.24) can be given by

$$\begin{aligned}
\hat{y}(t+N_1 \mid t) &= G_{N_1}\Delta u(t+N_1-1) + F_{N_1}y(t) \\
&= g_{N_1,0}\Delta u(t+N_1-1) + \cdots + g_{N_1,N_1-1}\Delta u(t) + f_{N_1}(t) \\
&\ \ \vdots \\
\hat{y}(t+N_2 \mid t) &= G_{N_2}\Delta u(t+N_2-1) + F_{N_2}y(t) \\
&= g_{N_2,0}\Delta u(t+N_2-1) + \cdots + g_{N_2,N_2-1}\Delta u(t) + f_{N_2}(t)
\end{aligned}$$

where

$$\begin{aligned}
f_{N_1}(t) &= q^{N_1-1}\left[G_{N_1}(q^{-1}) - g_{N_1,0} - \cdots - g_{N_1,N_1-1}q^{-(N_1-1)}\right]\Delta u(t) + F_{N_1}y(t) \\
&\ \ \vdots \\
f_{N_2}(t) &= q^{N_2-1}\left[G_{N_2}(q^{-1}) - g_{N_2,0} - \cdots - g_{N_2,N_2-1}q^{-(N_2-1)}\right]\Delta u(t) + F_{N_2}y(t)
\end{aligned} \tag{2.27}$$

can be calculated by known information $\{y(\tau),\ \tau \le t\}$ and $\{u(\tau),\ \tau < t\}$ available at t. Since $\Delta u(t + NU + i) = 0$ for $i \ge 0$ and $g_{N_1,N_1-NU} = \cdots = g_{N_1,-1} = 0$ for $N_1 < NU$, the above output prediction can be uniformly written by

$$\begin{aligned}
\hat{y}(t+N_1 \mid t) &= g_{N_1,N_1-NU}\Delta u(t+NU-1) + \cdots + g_{N_1,N_1-1}\Delta u(t) + f_{N_1}(t) \\
&\ \ \vdots \\
\hat{y}(t+N_2 \mid t) &= g_{N_2,N_2-NU}\Delta u(t+NU-1) + \cdots + g_{N_2,N_2-1}\Delta u(t) + f_{N_2}(t)
\end{aligned}$$

Denote

$$\hat{\boldsymbol{y}} = \begin{bmatrix} \hat{y}(t+N_1 \mid t) \\ \vdots \\ \hat{y}(t+N_2 \mid t) \end{bmatrix}, \quad \tilde{\boldsymbol{u}} = \begin{bmatrix} \Delta u(\mathrm{t}) \\ \vdots \\ \Delta u(\mathrm{t}+NU-1) \end{bmatrix}, \quad \boldsymbol{f} = \begin{bmatrix} f_{N_1}(t) \\ \vdots \\ f_{N_2}(t) \end{bmatrix}$$

Then we have

$$\hat{\boldsymbol{y}} = \boldsymbol{G}\tilde{\boldsymbol{u}} + \boldsymbol{f} \tag{2.28}$$

where $\boldsymbol{G}$ is an $(N_2 - N_1 + 1) \times NU$ matrix

$$\boldsymbol{G} = \begin{bmatrix} g_{N_1,N_1-1} & \cdots & g_{N_1,N_1-NU} \\ \vdots & & \vdots \\ g_{N_2,N_2-1} & \cdots & g_{N_2,N_2-NU} \end{bmatrix}$$

Note that $g_{j,\, i} = g_{i+1}$ $(i < j)$ are coefficients of the unit step response and $g_i = 0$ $(i \le 0)$, so for $N_1 \ge NU$,

$$\boldsymbol{G} = \begin{bmatrix} g_{N_1} & \cdots & g_{N_1-NU+1} \\ \vdots & & \vdots \\ g_{N_2} & \cdots & g_{N_2-NU+1} \end{bmatrix}$$

while for $N_1 < NU$,

$$\boldsymbol{G} = \begin{bmatrix} g_{N_1} & \cdots & g_1 & 0 & 0 \\ \vdots & & & \ddots & 0 \\ g_{NU} & & \cdots & & g_1 \\ \vdots & & & & \vdots \\ g_{N_2} & & \cdots & & g_{N_2-NU+1} \end{bmatrix}$$

Similarly, if the input/output constraints are not taken into account, the optimal solution minimizing the performance index (2.24) can be analytically given by

$$\tilde{\boldsymbol{u}} = \left(\boldsymbol{G}^{\mathrm{T}}\boldsymbol{G} + \lambda \boldsymbol{I}\right)^{-1}\boldsymbol{G}^{\mathrm{T}}(\boldsymbol{w} - \boldsymbol{f}) \tag{2.29}$$

where $\boldsymbol{w} = [w(t+N_1) \ \cdots \ w(t+N_2)]^{\mathrm{T}}$, and the current optimal control increment is

$$\Delta u(t) = \boldsymbol{d}^{\mathrm{T}}(\boldsymbol{w} - \boldsymbol{f}) \tag{2.30}$$

with

$$\boldsymbol{d}^{\mathrm{T}} = (1 \ \ 0 \ \ \cdots \ \ 0)\left(\boldsymbol{G}^{\mathrm{T}}\boldsymbol{G} + \lambda \boldsymbol{I}\right)^{-1}\boldsymbol{G}^{\mathrm{T}} \triangleq [d_{N_1} \ \cdots \ d_{N_2}]$$

and the current control action can be given by

$$u(t) = u(t-1) + \boldsymbol{d}^{\mathrm{T}}(\boldsymbol{w} - \boldsymbol{f})$$

(3) Online identification and correction

As the control algorithm developed from self-tuning control, GPC keeps its basic mechanism, i.e. GPC online estimates the model parameters using real-time input/output information and adjusts the control law accordingly. This is also a kind of feedback correction in a general sense, where only one model is adopted for prediction, but it should be online updated to make the prediction accurate. But in DMC, a fixed prediction model together with an error prediction model make the future output prediction more accurate.

Rewrite the plant model (2.19) as

$$\Delta y(t) = -A_1\left(q^{-1}\right)\Delta y(t) + B\left(q^{-1}\right)\Delta u(t-1) + \xi(t)$$

where $A_1(q^{-1}) = A(q^{-1}) - 1$. Denote the model parameters and available data in vector form as

$$\boldsymbol{\theta} = [a_1 \ \cdots \ a_n \mid b_0 \ \cdots \ b_{n_b}]^{\mathrm{T}}$$
$$\boldsymbol{\varphi}(t) = [-\Delta y(t-1) \ \ \cdots \ \ -\Delta y(t-n) \mid \Delta u(t-1) \ \ \cdots \ \ \Delta u(t-n_b-1)]^{\mathrm{T}}$$

Then it follows that

$$\Delta y(t) = \boldsymbol{\varphi}^{\mathrm{T}}(t)\cdot\boldsymbol{\theta} + \xi(t)$$

The parameter vector can be estimated using the recursive least square method with fading memory

$$\begin{aligned}
\boldsymbol{K}(t) &= \boldsymbol{P}(t-1)\boldsymbol{\varphi}(t)[\boldsymbol{\varphi}^{\mathrm{T}}(t)\boldsymbol{P}(t-1)\boldsymbol{\varphi}(t) + \mu]^{-1}\\
\hat{\boldsymbol{\theta}}(t) &= \hat{\boldsymbol{\theta}}(t-1) + \boldsymbol{K}(t)\left[\Delta y(t) - \boldsymbol{\varphi}^{\mathrm{T}}(t)\hat{\boldsymbol{\theta}}(t-1)\right]\\
\boldsymbol{P}(t) &= \frac{1}{\mu}[\boldsymbol{I} - \boldsymbol{K}(t)\boldsymbol{\varphi}^{\mathrm{T}}(t)]\boldsymbol{P}(t-1)
\end{aligned} \tag{2.31}$$

where $0 < \mu < 1$ is the fading factor, often chosen by $0.95 < \mu < 1$, $\boldsymbol{K}(t)$ is the weighting factor, and $\boldsymbol{P}(t)$ is a positive definite covariance matrix. When control starts, the initial values of parameter vector $\boldsymbol{\theta}$ and covariance matrix $\boldsymbol{P}$ should be set up; usually let $\hat{\boldsymbol{\theta}}(0) = 0$ and $\boldsymbol{P}(0) = \alpha^2\boldsymbol{I}$, where α is a large enough positive number. At each step during control, a data vector is first formed; then $\boldsymbol{K}(t)$, $\hat{\boldsymbol{\theta}}(t)$, and $\boldsymbol{P}(t)$ can be calculated using (2.31).

After the parameters of polynomials $A(q^{-1})$ and $B(q^{-1})$ have been obtained through identification, $\boldsymbol{d}^{\mathrm{T}}$ and $\boldsymbol{f}$ in control law (2.30) can be recalculated and the optimal control action can be obtained. Such a mechanism of online identification combined with control law adaption constitutes the feedback correction of the GPC algorithm with adaption.

However, due to the computational complexity of online identification, in many cases the GPC algorithm does not introduce an adaption mechanism. For such a case, feedback correction is implemented only through introducing the reference trajectory. Note that the reference trajectory (2.25) is given by

$$w(t+j) = \alpha^j y(t) + \left(1-\alpha^j\right)w, \quad j = 1, \ldots, N \tag{2.32}$$

It is obvious that the desired values of the reference trajectory should be reset each time with real-time measured output information $y(t)$. Note that the output error term in the performance index (2.24) is constituted by both $w(t+j)$ and $y(t+j)$; thus correcting $w(t+j)$ is equivalent to correcting the predicted output $y(t+j)$. In this sense, introducing the reference trajectory is also a kind of feedback correction.

Remark 2.3 The GPC algorithm given above is in principle also suitable for predictive control based on discrete time transfer function models. As noted above, if the stochastic characteristic in GPC is removed, (2.19) is identical to the discrete time transfer function model (2.20b), and the prediction model (2.22) can also be established by introducing the Diophantine equation. In the rolling horizon optimization, if the stochastic optimization is replaced by a deterministic one, control law (2.30) can also be obtained for unconstrained cases.

2.3 Predictive Control Based on the State Space Model

The early predictive control algorithms generally adopted an input–output model, either nonparametric such as MPHC or DMC, or parametric such as GPC. However, since the 1990s, the state space model has been widely adopted in theoretical research of model predictive control because such research needs to take the mature theoretical results of modern control theory (particularly the optimal control theory) as reference. Recently, with the expansion of predictive control to more and more fields out of the process industry, the state space model is also in common use in the predictive control of many application fields.

There are various formulations for predictive control based on the state space model. In the following, we focus on the predictive control algorithm for linear systems with measurable states. Consider an SISO linear system described by the following state space model

$$\begin{aligned} \boldsymbol{x}(k+1) &= \boldsymbol{A}\boldsymbol{x}(k) + \boldsymbol{b}u(k) \\ y(k) &= \boldsymbol{c}^{\mathrm{T}}\boldsymbol{x}(k) \end{aligned} \tag{2.33}$$

where the state variable $\boldsymbol{x}(k) \in \mathrm{R}^n$ is real-time measurable and $u(k)$ and $y(k)$ are system input and output, respectively.

(1) Prediction model

In the case where $\boldsymbol{x}(k)$ is measurable, the state space model (2.33) can be directly used as a prediction model. Assume that starting from time k the system input has M step changes, i.e. $u(k), \ldots, u(k+M-1)$, and then remains constant, i.e. $u(k+i) = u(k+M-1)$, $i \geq M$. Consider two prediction cases with these control actions: state prediction and output prediction.

State prediction

The system states at future $P(P \geq M)$ sampling times can be predicted by using (2.33):

$$\begin{aligned}
\boldsymbol{x}(k+1) &= \boldsymbol{A}\boldsymbol{x}(k) + \boldsymbol{b}u(k) \\
\boldsymbol{x}(k+2) &= \boldsymbol{A}^2\boldsymbol{x}(k) + \boldsymbol{A}\boldsymbol{b}u(k) + \boldsymbol{b}u(k+1) \\
&\vdots \\
\boldsymbol{x}(k+M) &= \boldsymbol{A}^M\boldsymbol{x}(k) + \boldsymbol{A}^{M-1}\boldsymbol{b}u(k) + \cdots + \boldsymbol{b}u(k+M-1) \\
\boldsymbol{x}(k+M+1) &= \boldsymbol{A}^{M+1}\boldsymbol{x}(k) + \boldsymbol{A}^M\boldsymbol{b}u(k) + \cdots + (\boldsymbol{A}\boldsymbol{b}+\boldsymbol{b})u(k+M-1) \\
&\vdots \\
\boldsymbol{x}(k+P) &= \boldsymbol{A}^P\boldsymbol{x}(k) + \boldsymbol{A}^{P-1}\boldsymbol{b}u(k) + \cdots + \left(\boldsymbol{A}^{P-M}\boldsymbol{b} + \cdots + \boldsymbol{b}\right)u(k+M-1)
\end{aligned} \tag{2.34}$$

or in vector form

$$\boldsymbol{X}(k) = \boldsymbol{F}_x\boldsymbol{x}(k) + \boldsymbol{G}_x\boldsymbol{U}(k) \tag{2.35}$$

where

$$\boldsymbol{X}(k) = \begin{bmatrix} \boldsymbol{x}(k+1) \\ \vdots \\ \boldsymbol{x}(k+P) \end{bmatrix}_{(nP\times 1)}, \quad \boldsymbol{U}(k) = \begin{bmatrix} u(k) \\ \vdots \\ u(k+M-1) \end{bmatrix}_{(M\times 1)}$$

$$\boldsymbol{F}_x = \begin{bmatrix} \boldsymbol{A} \\ \vdots \\ \boldsymbol{A}^P \end{bmatrix}_{(nP\times n)}, \quad \boldsymbol{G}_x = \begin{bmatrix} \boldsymbol{b} & \boldsymbol{0} & \boldsymbol{0} \\ \vdots & \ddots & \boldsymbol{0} \\ \boldsymbol{A}^{M-1}\boldsymbol{b} & \cdots & \boldsymbol{b} \\ \vdots & & \vdots \\ \boldsymbol{A}^{P-1}\boldsymbol{b} & \cdots & \sum_{i=0}^{P-M}\boldsymbol{A}^i\boldsymbol{b} \end{bmatrix}_{(nP\times M)}$$

Output prediction

If the system outputs at future P sampling times need to be predicted, then combined with the output equation in (2.33), it follows that

$$\boldsymbol{Y}(k) = \boldsymbol{F}_y\boldsymbol{x}(k) + \boldsymbol{G}_y\boldsymbol{U}(k) \tag{2.36}$$

where

$$\boldsymbol{Y}(k) = \begin{bmatrix} y(k+1) \\ \vdots \\ y(k+P) \end{bmatrix}_{(P\times 1)}, \quad \boldsymbol{F}_y = \begin{bmatrix} \boldsymbol{c}^{\mathrm{T}}\boldsymbol{A} \\ \vdots \\ \boldsymbol{c}^{\mathrm{T}}\boldsymbol{A}^P \end{bmatrix}_{(P\times n)}, \quad \boldsymbol{G}_y = \begin{bmatrix} \boldsymbol{c}^{\mathrm{T}}\boldsymbol{b} & 0 & 0 \\ \vdots & \ddots & 0 \\ \boldsymbol{c}^{\mathrm{T}}\boldsymbol{A}^{M-1}\boldsymbol{b} & \cdots & \boldsymbol{c}^{\mathrm{T}}\boldsymbol{b} \\ \vdots & & \vdots \\ \boldsymbol{c}^{\mathrm{T}}\boldsymbol{A}^{P-1}\boldsymbol{b} & \cdots & \sum_{i=1}^{P-M+1}\boldsymbol{c}^{\mathrm{T}}\boldsymbol{A}^{i-1}\boldsymbol{b} \end{bmatrix}_{(P\times M)}$$

Since the sampling values of the system unit impulse response $g_i = \boldsymbol{c}^{\mathrm{T}}\boldsymbol{A}^{i-1}\boldsymbol{b}$, $i \geq 1$, $\boldsymbol{G}_y$ in (2.36) can be also written as

$$G_y = \begin{bmatrix} g_1 & 0 & 0 \\ \vdots & \ddots & 0 \\ g_M & \cdots & g_1 \\ \vdots & & \vdots \\ g_P & \cdots & \sum_{i=1}^{P-M+1} g_i \end{bmatrix}$$

(2) Rolling optimization

The optimization problem to be solved at time k is discussed for two cases: state optimization and output optimization, according to the practical requirements.

State optimization

At time k, the state optimization problem can be formulated as: to determine M control actions $u(k), \ldots, u(k+M-1)$ such that, under their control, the system states at future P times are stabilized, i.e. approaching $\boldsymbol{x} = \mathbf{0}$; meanwhile, excessive control actions should be suppressed. This optimization performance index can be represented in vector form as

$$\min_{\boldsymbol{U}(k)} J(k) = \|\boldsymbol{X}(k)\|^2_{\boldsymbol{Q}_x} + \|\boldsymbol{U}(k)\|^2_{\boldsymbol{R}_x} \tag{2.37}$$

where $\boldsymbol{Q}_x$, $\boldsymbol{R}_x$ are state and control weighting matrices with proper dimensions, respectively. In unconstrained cases, combined with the state prediction model (2.35), the optimal solution can be analytically given by

$$\boldsymbol{U}(k) = -\left(\boldsymbol{G}_x^{\mathrm{T}}\boldsymbol{Q}_x\boldsymbol{G}_x + \boldsymbol{R}_x\right)^{-1}\boldsymbol{G}_x^{\mathrm{T}}\boldsymbol{Q}_x\boldsymbol{F}_x\boldsymbol{x}(k) \tag{2.38}$$

and the current control action can be given by the state feedback control law

$$u(k) = -\boldsymbol{k}_x^{\mathrm{T}}\boldsymbol{x}(k) \tag{2.39}$$

where the feedback gain is

$$\boldsymbol{k}_x^{\mathrm{T}} = (1\ \ 0\ \ \cdots\ \ 0)\left(\boldsymbol{G}_x^{\mathrm{T}}\boldsymbol{Q}_x\boldsymbol{G}_x + \boldsymbol{R}_x\right)^{-1}\boldsymbol{G}_x^{\mathrm{T}}\boldsymbol{Q}_x\boldsymbol{F}_x$$

Output optimization

For the case of output optimization, M control actions $u(k), \ldots, u(k+M-1)$ should be determined such that under their control, the system outputs at future P sampling times are close to the given desired values $w(k+i)$, $i = 1, \ldots, P$, as far as possible; meanwhile, excessive control actions should be suppressed. The performance index can be similarly given in vector form

$$\min_{\boldsymbol{U}(k)} J_y(k) = \|\boldsymbol{W}(k) - \boldsymbol{Y}(k)\|^2_{\boldsymbol{Q}_y} + \|\boldsymbol{U}(k)\|^2_{\boldsymbol{R}_y} \tag{2.40}$$

where $\boldsymbol{W}(k) = [w(k+1)\ \cdots\ w(k+P)]^{\mathrm{T}}$ is the vector form of the desired output, $\boldsymbol{Q}_y$, $\boldsymbol{R}_y$ are error and control weighting matrices with proper dimensions, respectively. Similarly, in the unconstrained case, combined with the output prediction model (2.36), the optimal solution can be analytically given by

$$\boldsymbol{U}(k)=\left(\boldsymbol{G}_y^{\mathrm{T}}\boldsymbol{Q}_y\boldsymbol{G}_y+\boldsymbol{R}_y\right)^{-1}\boldsymbol{G}_y^{\mathrm{T}}\boldsymbol{Q}_y\left(\boldsymbol{W}(k)-\boldsymbol{F}_y\boldsymbol{x}(k)\right) \tag{2.41}$$

and the current control action can be given by

$$u(k)=\boldsymbol{d}_y^{\mathrm{T}}\left(\boldsymbol{W}(k)-\boldsymbol{F}_y\boldsymbol{x}(k)\right) \tag{2.42}$$

where

$$\boldsymbol{d}_y^{\mathrm{T}}=(1\ \ 0\ \ \cdots\ \ 0)\left(\boldsymbol{G}_y^{\mathrm{T}}\boldsymbol{Q}_y\boldsymbol{G}_y+\boldsymbol{R}_y\right)^{-1}\boldsymbol{G}_y^{\mathrm{T}}\boldsymbol{Q}_y$$

(3) Feedback correction

In the case where system states are fully measurable, $\boldsymbol{x}(k)$ measured at each step can be directly used to reinitialize prediction and optimization at that time. This means that both prediction and optimization are based on real-time feedback information and feedback correction is naturally realized. Therefore, additional correction is not necessary in this case.

The above predictive control algorithms based on the state space model are only with respect to unconstrained SISO linear systems with states fully measurable. The case discussed here is relatively simple, just aiming to illustrate how to use the basic principles and the state space model to develop predictive control algorithms. It is straightforward to extend the scheme of these algorithms to MIMO linear systems and nonlinear systems. However, for theoretical study or in real applications, the predictive control algorithms based on the state space model may have various forms of expression. In the following, some discussions are given.

1) **Constraints**. In rolling optimization of the above algorithms, constraints on the system state, input and output are not considered. Then the optimal solution can be given in analytical form as (2.38) or (2.41). However, when these constraints are embedded into the optimization problem, the online optimization should be solved by mathematical programming and in general the optimal solution would be no longer given in closed analytical form.
2) **Control increment form**. The state space model (2.33) can also be represented by the control increment form, such as the state space description of DMC [3]. It is often used in the case when the control law needs to be designed with an integrator. With Δu as input, model (2.33) can be rewritten into the state space equation

$$\begin{aligned}\tilde{\boldsymbol{x}}(k+1)&=\tilde{\boldsymbol{A}}\tilde{\boldsymbol{x}}(k)+\tilde{\boldsymbol{b}}\Delta u(k)\\ y(k)&=\tilde{\boldsymbol{c}}^{\mathrm{T}}\tilde{\boldsymbol{x}}(k)\end{aligned} \tag{2.43}$$

where

$$\tilde{\boldsymbol{x}}(k)=\begin{bmatrix}\boldsymbol{x}(k)\\ u(k-1)\end{bmatrix},\quad \tilde{\boldsymbol{A}}=\begin{bmatrix}\boldsymbol{A} & \boldsymbol{b}\\ \boldsymbol{0} & 1\end{bmatrix},\quad \tilde{\boldsymbol{b}}=\begin{bmatrix}\boldsymbol{b}\\ 1\end{bmatrix},\quad \tilde{\boldsymbol{c}}^{\mathrm{T}}=\begin{bmatrix}\boldsymbol{c}^{\mathrm{T}} & 0\end{bmatrix}$$

Accordingly, the control input u at future sampling times can be expressed by Δu:

$$\boldsymbol{U}(k)=\boldsymbol{l}u(k-1)+\boldsymbol{B}\Delta\boldsymbol{U}(k) \tag{2.44}$$

where

$$\boldsymbol{U}(k)=\begin{bmatrix}u(k)\\ \vdots\\ u(k+M-1)\end{bmatrix},\quad \boldsymbol{l}=\begin{bmatrix}1\\ \vdots\\ 1\end{bmatrix},\quad \boldsymbol{B}=\begin{bmatrix}1&0&0\\ 1&\ddots&0\\ 1&\cdots&1\end{bmatrix},\quad \Delta\boldsymbol{U}(k)=\begin{bmatrix}\Delta u(k)\\ \vdots\\ \Delta u(k+M-1)\end{bmatrix}$$

3) **Optimization variable and model form**. Some remarks are given here on the selection of the optimization variable and the state space model form. Firstly, either control input u or its increment Δu can be chosen as the optimization variable, depending on which one can make the optimization problem more direct and easy to handle. In general, the selection of the optimization variable is in accordance with the control weighting term in the performance index. For the performance index (2.37) or (2.40), the control input u is selected as the optimization variable because its quadratic weighting term appears in the performance index. However, if the weighting term of the control input u in the performance index (2.37) or (2.40) is replaced by the weighting term of the control increment Δu, as in the case of DMC or GPC, it would be reasonable to select the control increment Δu as the optimization variable. Secondly, after the optimization variable has been selected, the state space model as well as the prediction model should be established with the optimization variable as the model input because it should reflect the causal relationship between the optimization variable and the system state or output. In the case when Δu is selected as the optimization variable, the state space model (2.33) should be rewritten in the control increment form (2.43), and based on it, the state prediction model (2.35) or the output prediction model (2.36) should be derived again with the control increment as input. The control input appearing in constraint conditions will also be rewritten in control increment form by using (2.44). Thirdly, in the case when both control input u and control increment Δu appear in the performance index or in the constraints, the first thing to determine is whether u or Δu is selected as the optimization variable; then the corresponding state space model is established, and finally the prediction model and the optimization problem are rewritten in terms of the selected optimization variable.
4) **State observer**. In the above discussion, it is assumed that the system state $\boldsymbol{x}(k)$ is measurable in real time. In the case when only the system output $y(k)$ is measurable, a state observer should be introduced to reconstruct the system state. The reconstructed state $\hat{\boldsymbol{x}}(k)$ is adopted to replace $\boldsymbol{x}(k)$ in model prediction and rolling optimization. The state observer is a typical kind of feedback correction, where the error between the measured output and observer output is fed back to correct the reconstructed state to make it approach the actual state. The state space formulation of predictive control with the observer is closely related to the predictive control algorithms based on the input–output parametric model such as GPC [7]. Even for the nonparametric model, recall the state space design of DMC [3] in Section 2.1.2, it is shown that through approximate realization of the system step response in state space, state feedback control with a state observer can also be designed to obtain the same algorithm as DMC, with which the feedback correction in the original DMC is no longer heuristic and can be designed by referring to the error feedback in the state observer.

Remark 2.4 In this section we introduced predictive control algorithms for SISO unconstrained linear systems with measurable states, but indicated possible variations of the algorithms in the above discussions. Since the state space model can describe both linear and nonlinear systems, predictive control based on it greatly depends on the specific description of the model and the requirements on optimization. It is impossible to give its uniform or representative algorithms similar to DMC or GPC. Particularly in the qualitative synthesis theory of predictive control, most of the studies adopted the state space model. The optimization performance index can be either with a finite horizon or an infinite horizon. In order to achieve stable control, an additional term such as the terminal cost function usually needs to be added to the performance index of the finite horizon optimization, and some artificial constraints such as the terminal invariant set should be imposed. In this case, each new synthesis approach might bring several predictive control algorithms, and each of them is only a special one with respect to the given control strategy.

At the early stage, predictive control was mostly applied to industrial processes where the step/impulse response model was particularly preferred. Although the state space model was not the first choice, it was still used to study predictive control for mechanical systems (such as aircraft [8]) and nonlinear systems. Since the 1990s, there has been an increasing acceptance and use of the state space model in predictive control. On the one hand, the qualitative synthesis theory of predictive control has been developed mainly referring to the traditional optimal control theory in the state space framework. Therefore most of the researches adopted the state space model. On the other hand, with the rapid expansion of the application fields of predictive control, the controlled plants are never limited to industrial processes. Some of them may already have mature state space models (such as aircraft) and some may need to be described in a state space framework (such as traffic network, gas transportation, etc.). Thus the state space model has become more and more commonly used as the basic model for predictive control in these fields, such as in advanced manufacturing, power electronics, energy, environment, traffic, and so on.

2.4 Summary

In this chapter, taking unconstrained predictive control for SISO linear systems as an example, we introduced predictive control algorithms based on different kinds of models, i.e. the step response model, the linear difference equation (or transfer function) model and the state space model, respectively, aiming at illustrating how to use the basic principles to construct a predictive control algorithm according to given model information. It was shown that the prediction model, rolling optimization, and feedback correction as the essential factors of predictive control, are actually embodied in all these algorithms, but have specific forms in model prediction and feedback correction.

The prediction model aims at providing the basis for solving the optimization problem. It should accurately indicate the causal relationship of optimization variables with other unknown variables appearing in the performance index or in the constraints, thus greatly depending on the optimization problem formulation. As shown by the above predictive control algorithms, as well as those discussed in Section 2.3, the form of the prediction model should be in accordance with the requirement of optimization. Its input can be either control input or control increment, depending on which weighting term appears in the performance index, while its output can be either system state or system output, depending on which one needs to be optimized. Furthermore, in order to support the optimization problem solving, the prediction model should have the function of determining the future state or output according to available information and assumed future control input or increment. It is obvious that the original model always has such a function; see the model vector $\boldsymbol{a} = [a_1 \ \cdots \ a_N]^{\mathrm{T}}$ in DMC, the CARIMA model (2.19) in GPC, and the state space model (2.33) in Section 2.3. However, it is more desirable to derive the prediction model as an explicit expression of the causal relationship of the predicted state or output with available information and future control actions, such as (2.2) in DMC, (2.28) in GPC, and (2.35) for state prediction or (2.36) for output prediction in the state space framework. To do that, sometimes the intermediate variables need to be eliminated, for example in GPC the Diophantine equation (2.21) was introduced to remove the currently unknown information from the output prediction.

Compared with the other two principles, rolling optimization focuses on how to formulate and to solve the online optimization problem, which is independent from the basic models although its specific formulation may be model dependent. In principle, the performance index of the online optimization should be formulated according to practical requirements. The quadratic performance index discussed in this chapter is only a usual choice. Furthermore, for simplification the constraints on system input and output are not considered here. Only in this case, the optimal solution can be given in analytical form as shown by (2.7) in DMC, (2.29) in GPC, and (2.38) or (2.41) in state space. In real applications, according to various requirements, the performance index may have quite different forms, including nonlinear or min-max formulation. Different kinds of constraints on the system state, input and output should also be included in the optimization problem. For such optimization problems, an analytical solution is generally impossible and a mathematical programming problem with a limited number of control inputs or control increments as optimization variables needs to be solved online. This will be discussed later in Chapters 5, 8, and 9. It should also be mentioned that in the qualitative synthesis theory of predictive control, the online optimization problem is often reformulated with additional cost function terms and constraints, and sometimes the optimal solution may be assumed as a feedback control law in closed form whose gain should be solved online or offline; see Chapter 6 or 7 in this book.

Feedback correction aims at establishing model prediction and optimization at each step on the basis of being as close as possible to actual system status. To do that, it is necessary to introduce feedback by real-time measured information. If the system state is measurable, feedback can be a simple way of refreshing the system state and no additional correction is needed. Particularly in the research field of qualitative synthesis theory of predictive control, the system state is often assumed to be measurable, so feedback correction is not specially taken into account. However, if only the system output is measurable, the actual output needs to be fed back and combined with the predicted output to construct the error. This error can then be used either to correct the originally predicted future output heuristically, such as (2.12) in DMC, or to reconstruct the observer state to make it approach the actual one. The form of feedback correction is diverse. The reference trajectory introduced in the MPHC or GPC algorithm, see (2.32), makes a correction for the error of the future system output deviate from the desired output through updating the desired output values based on feedback information. In a more general sense, direct identification and correction of the model parameters using real-time input/output information, as sometimes performed in GPC, is also a kind of feedback correction.

References

1 Cutler, C.R. and Ramaker, B.L. Dynamic matrix control – a computer control algorithm. *Proceedings of the 1980 Joint Automatic Control Conference*, San Francisco, CA, USA (13–15 August 1980), WP5-B.

2 Prett, D.M. and Gillete, R.D. Optimization and constrained multivariable control of a catalytic creaking unit. *Proceedings of the 1980 Joint American Control Conference*, San Francisco, CA, USA (13–15 August 1980), WP5-C.

3 Schmidt, G. and Xi, Y. (1985). A new design method for digital controllers based on nonparametric plant models. In: *Applied Digital Control* (ed. S.G.Tzafestas), 93–109. Amsterdam: North-Holland.

4 Chen, C.T. (1984). *Linear System Theory and Design*. New York: Holt, Rinehart and Winston.

5 Clarke, D.W., Mohtadi, C., and Tuffs, P.S. (1987). Generalized predictive control – Parts 1 and 2. *Automatica* 23 (2): 137–162.

6 Richalet, J., Rault, A., Testud, J.L. et al. (1978). Model predictive heuristic control: applications to industrial processes. *Automatica* 14 (5): 413–428.

7 Kwon, W.H., Lee, Y.I., and Noh, S. Partition of GPC into a state observer and a state feedback controller, *Proceedings of the 1992 American Control Conference*, Chicago, IL, USA (24–26 June 1992), 2032–2036.

8 Reid, J.G., Chaffin, D.E., and Silverthorn, J.T. (1981). Output predictive algorithmic control: precision tracking with application to terrain following. *Journal of Guidance and Control* 4 (5): 502–509.

3

Trend Analysis and Tuning of SISO Unconstrained DMC Systems

Predictive control algorithms arising from industrial processes, such as DMC and MPHC or MAC, have been widely accepted and used in industry since their appearance in the 1970s. With increasing applications it was naturally expected that the theoretical research on these algorithms could provide guidance for their applications such that people could reasonably select the design parameters to achieve desired control performance instead of a choice made fully from experience. From the 1970s to the early 1990s, the theoretical research of predictive control was mostly motivated by the need of application. It focused on exploring the relationship between the design parameters (optimization and control horizon, weighting matrices, etc.) and the closed-loop system performances for the existing predictive control algorithms such as DMC, MAC, and later also GPC. Although this study only achieved limited results for unconstrained SISO predictive control systems and could not give satisfactory solution for complex industrial applications, the resultant quantitative or trend results on how the design parameters affect the system performance still provide valuable reference for parameter tuning in predictive control applications. In this and the next chapter, we will introduce some methods and results of the theoretical research of predictive control during this stage, with emphasis on the predictive control system analysis based on Internal Model Control (IMC) structure in the Z domain.

3.1 The Internal Model Control Structure of the DMC Algorithm

IMC is a special kind of control system structure proposed by Garcia and Morari in 1982 [1], which can be efficiently used to analyze predictive control systems. The typical IMC structure is shown in Figure 3.1, where w, u, y, v are, respectively, reference input, control variable, output variable, and disturbance. Z transfer functions $G_P(z)$, $G_M(z)$, $G_C(z)$, $G_F(z)$, and $G_W(z)$ represent the plant, model, controller, filter, and reference input model, respectively, and are all represented by the rational fraction of z^{-1}.

Predictive Control: Fundamentals and Developments, First Edition. Yugeng Xi and Dewei Li.

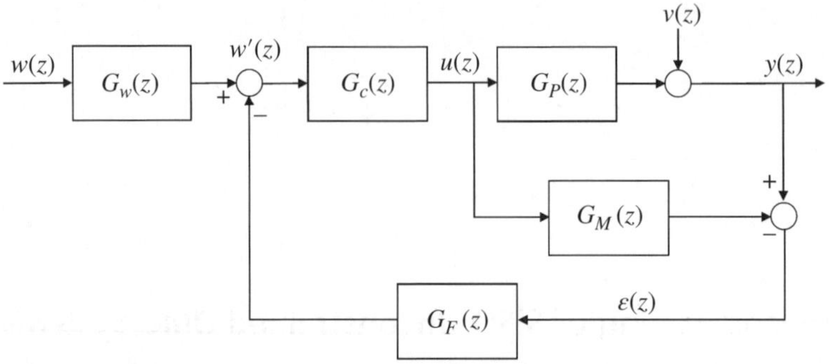

Figure 3.1 Internal Model Control (IMC) structure [1]. *Source:* Reproduced with permission from Garcia and Marari of ACS Publications.

To discuss the closed-loop characteristics, assume that $G_W(z) = 1$. From Figure 3.1 it follows that

$$y(z) = \frac{G_C(z)G_P(z)}{1 + G_C(z)G_F(z)[G_P(z) - G_M(z)]} w(z) + \frac{1 - G_C(z)G_F(z)G_M(z)}{1 + G_C(z)G_F(z)[G_P(z) - G_M(z)]} v(z) \tag{3.1}$$

Consider the transfer relationship from input w to output y. If there is no model–plant mismatch, i.e. $G_M(z) = G_P(z)$, the input–output transfer function of the closed-loop system will only depend on the forward path in the IMC structure

$$F_0(z) = G_C(z)G_P(z) \tag{3.2}$$

Therefore, the stability analysis when a model is exact is actually the open-loop (forward path) stability in IMC, while the closed-loop stability in IMC is indeed the robust stability of the whole system if the model is mismatched or disturbance exists.

Based on (3.1), [1] gave the following important properties of the IMC structure.

Property 1 Dual Stability Criterion

If the model is exact, i.e. $G_M(z) = G_P(z)$, the system is stable if both the plant $G_P(z)$ and the controller $G_C(z)$ are stable. Here a transfer function is said to be stable if all its poles, i.e. all the roots of its characteristic polynomial, are located in the unit circle.

This property is directly shown by (3.2). It indicates that for an open-loop stable plant, the closed-loop stability can be guaranteed by a stable controller and an exact model. Even when the controller is nonlinear, closed-loop stability can be also guaranteed if the controller is input–output stable.

Property 2 Perfect Controller

For a stable plant and an exact model, if the controller is selected as

$$G_C(z) = \frac{1}{G_-(z)} \tag{3.3}$$

then the controlled system has minimum output variance both for stabilization or tracking, where $G_-(z)$ comes from the following model decomposition:

$$G_M(z) = G_+(z)G_-(z) \tag{3.4}$$

with

$$G_+(z) = z^{-(l+1)} \prod_{i=1}^{p} \left(\frac{z - z_i}{z - \hat{z}_i}\right)\left(\frac{1 - \hat{z}_i}{1 - z_i}\right)$$

where l is the step number of the plant time delay, $l + 1$ includes one step additional time delay due to sampling and holding, z_i is the zero out of the unit circle, $\hat{z}_i = 1/z_i$ is its mapping in the unit circle, and p is the number of plant zeros out of the unit circle. The controller (3.3) is called the perfect controller.

The above concept of perfect control is motivated by the expectation for the controller to fully compensate the plant dynamics. From (3.2), an ideal perfect controller would have the form

$$G_C(z) = \frac{1}{G_P(z)}$$

However, after sampling and holding, the plant Z transfer function $G_P(z)$ has at least one step time delay and $G_C(z)$ calculated by above inverse would have a lead term z and is not realizable. Furthermore, for a nonminimal phase plant $G_P(z)$, $G_C(z)$ calculated by the above inverse would have unstable poles because $G_P(z)$ has zeros out of the unit circle. Thus, according to Property 1, the controlled system will be unstable.

The controller (3.3) is just a modification of the above ideal but unrealistic perfect controller. When the model is exact, i.e. $G_P(z) = G_M(z)$, after decomposition (3.4), $G_+(z)$ contains the time delay of $G_P(z)$ and all the zeros out of the unit circle in $G_P(z)$. The former's inverse needs lead prediction and is unrealizable, while the inverse of the latter will cause the controller to be unstable. These are essentially the parts in the model transfer function that cannot be changed by any controller due to physical restrictions, while the other parts in $G_+(z)$ are introduced only to make the inverse of $G_-(z)$ realizable and to keep the steady state value of $G_+(z)$ as a unit. Therefore, the perfect controller (3.3) is indeed a realizable stable controller which fully compensates $G_-(z)$, i.e. the part of the plant that can be changed by control.

When the plant is described by a nonparametric model, it is impossible to get the perfect controller (3.3) by factorization (3.4). However, according to the above illustration, one can separate the time delay from the model and then search for an approximate inverse for the remaining part as the controller $G_C(z)$, which is realizable and stable, is an inverse of the model at least in the steady state, i.e. $G_C(1) = 1/G_M(1)$.

Then consider the influence of the model mismatch and disturbance in the IMC structure.

Property 3 Zero offset

If the closed-loop system is stable and the controller satisfies $G_C(1) = 1/G_M(1)$ and the filter satisfies $G_F(1) = 1$, then the system has a zero output offset for both step input w and constant disturbance v, regardless of whether or not the model is mismatched.

The above property can be directly deduced from (3.1). Assume that the closed-loop system is stable. If $G_C(1) = 1/G_M(1)$ and $G_F(1) = 1$; then for step input $w(z) = w_0/(1 - z^{-1})$ and constant disturbance $v(z) = v_0/(1 - z^{-1})$, according to (3.1) and the final value theorem in the Z domain, it follows that

$$\begin{aligned}\lim_{t\to\infty} y(t) &= \lim_{z\to 1}\left(1-z^{-1}\right)y(z)\\ &= \frac{G_C(1)G_P(1)}{1+G_C(1)G_F(1)[G_P(1)-G_M(1)]}w_0 + \frac{1-G_M(1)G_C(1)G_F(1)}{1+G_C(1)G_F(1)[G_P(1)-G_M(1)]}v_0\\ &= w_0\end{aligned}$$

Thus the system output has zero offset.

In the IMC structure, the error constructed by actual output and model output is fed back through filter $G_F(z)$. From Figure 3.1 it is obvious that the filter only works when the output error caused by model mismatch or disturbance exists. Therefore the filter plays an important role in the robustness and disturbance rejection of the closed-loop system. For a stable plant, according to (3.1), the closed-loop characteristic equation can be written as

$$G_C^{-1}(z) + G_F(z)[G_P(z) - G_M(z)] = 0 \tag{3.5}$$

For a given model mismatch, it is possible to keep the closed-loop system stable by properly selecting $G_F(z)$. Furthermore, the feedback path with the filter may affect disturbance rejection. For example, when the model is exact, the output caused by disturbance can be given by

$$y(z) = [1 - G_C(z)G_F(z)G_M(z)]v(z)$$

For some kinds of disturbance $v(z)$, the influence of $v(z)$ to output $y(z)$ could be partly eliminated by $G_F(z)$ design such that the disturbance can be rejected quickly. Thus one can see that in the IMC structure the selection of filter $G_F(z)$ should consider various requirements, such as zero offset, disturbance rejection, robustness against model mismatch, etc. However, sometimes these requirements are conflicting. For example, selecting $G_F(z) = 0$ is the best for robustness, but causes offset and weakens the ability of disturbance rejection. Therefore, how to synthesize a filter according to specific requirements is a critical issue in an IMC structure. Later in Section 3.3, we will discuss how the closed-loop robustness and disturbance rejection are affected by different filters caused from different feedback correction strategies in the DMC algorithm in more detail.

IMC provides a structural scheme to analyze model-based control algorithms. In order to investigate the performances of predictive control systems by using the basic properties of IMC, in the following, the DMC algorithm is taken as an example and its IMC structure is deduced.

The DMC algorithm introduced in Section 2.1 can be described by the structural block diagram as shown in Figure 3.2a, where the $P \times N$ matrix $\boldsymbol{G} = [\boldsymbol{I}_{P\times P} \quad 0_{P\times(N-P)}]$ represents the operation of selecting the first P elements from the N-dimensional vector

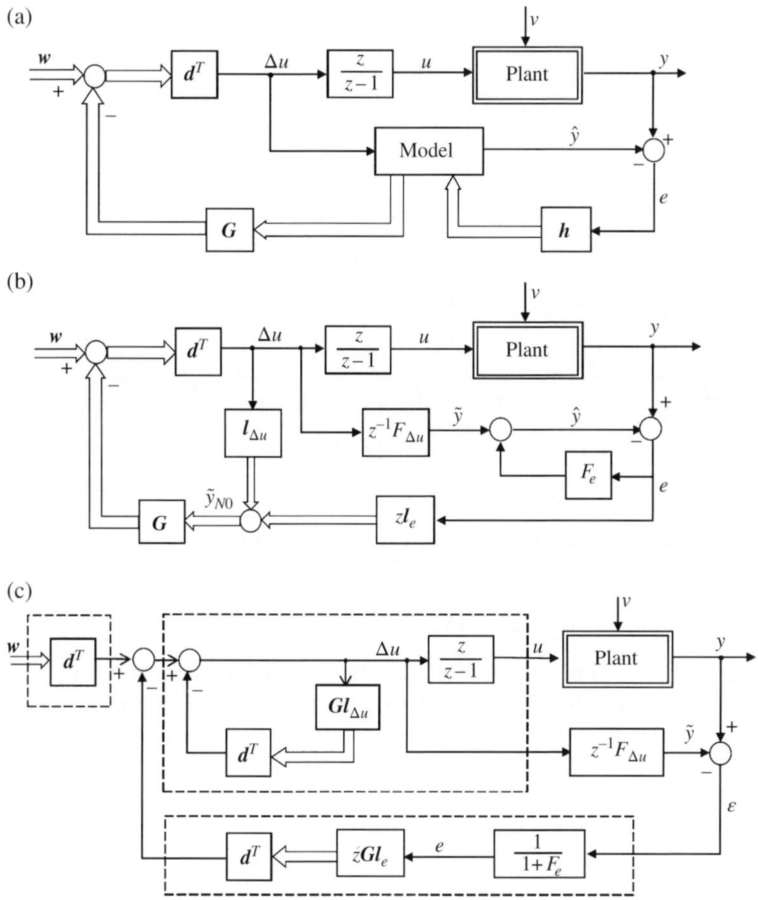

Figure 3.2 Transfer DMC algorithm in the IMC structure.

$\tilde{\boldsymbol{y}}_{N0}(k)$ as $\tilde{\boldsymbol{y}}_{P0}(k)$. In the following, this algorithm is transferred into an IMC structure in the Z domain, which will provide the basis for further analyzing the performance of DMC closed-loop systems.

Firstly, (2.14), (2.12), and (2.10) give

$$\tilde{\boldsymbol{y}}_{N0}(k+1) = \boldsymbol{S}[\tilde{\boldsymbol{y}}_{N0}(k) + \boldsymbol{a}\Delta u(k) + \boldsymbol{h}e(k+1)]$$

Make a Z transformation and it follows that

$$z\tilde{\boldsymbol{y}}_{N0}(z) = \boldsymbol{S}[\tilde{\boldsymbol{y}}_{N0}(z) + \boldsymbol{a}\Delta u(z) + \boldsymbol{h}(ze(z))]$$

After simple deduction we have

$$\tilde{\boldsymbol{y}}_{N0}(z) = \boldsymbol{l}_{\Delta u}\Delta u(z) + \boldsymbol{l}_e(ze(z)) \tag{3.6}$$

where

$$\boldsymbol{l}_{\Delta u} = (z\boldsymbol{I}-\boldsymbol{S})^{-1}\boldsymbol{S}\boldsymbol{a} = \begin{bmatrix} a_2z^{-1}+a_3z^{-2}+\cdots+a_{N-1}z^{-(N-2)}+a_Nz^{-(N-2)}/(z-1) \\ \vdots \\ a_{N-1}z^{-1}+a_Nz^{-1}/(z-1) \\ a_N/(z-1) \\ a_N/(z-1) \end{bmatrix}$$

$$\boldsymbol{l}_e = (z\boldsymbol{I}-\boldsymbol{S})^{-1}\boldsymbol{S}\boldsymbol{h} = \begin{bmatrix} h_2z^{-1}+h_3z^{-2}+\cdots+h_{N-1}z^{-(N-2)}+h_Nz^{-(N-2)}/(z-1) \\ \vdots \\ h_{N-1}z^{-1}+h_Nz^{-1}/(z-1) \\ h_N/(z-1) \\ h_N/(z-1) \end{bmatrix}$$

Next, let $\boldsymbol{c}^{\mathrm{T}} = [1 \quad 0 \quad \cdots \quad 0]$ be the operation of taking the first element from the followed vector; then

$$\tilde{y}_1(k+1\mid k) = \boldsymbol{c}^{\mathrm{T}}\tilde{\boldsymbol{y}}_{N1}(k) = \boldsymbol{c}^{\mathrm{T}}[\tilde{\boldsymbol{y}}_{N0}(k)+\boldsymbol{a}\Delta u(k)]$$

Combining this with (3.6), it can be seen that due to the feedback correction on $\tilde{\boldsymbol{y}}_{N0}(k)$, $\tilde{y}_1(k+1\mid k)$ depends not only on the model but also on the feedback correction at time k. Rewrite it as $\hat{y}(k+1)$, make the Z transformation for (2.11), and then substitute (3.6) into it, giving

$$\begin{aligned} ze(z) &= zy(z)-z\hat{y}(z) = zy(z)-\boldsymbol{c}^{\mathrm{T}}[\tilde{\boldsymbol{y}}_{N0}(z)+\boldsymbol{a}\Delta u(z)] \\ &= zy(z)-F_{\Delta u}\Delta u(z)-F_e(ze(z)) \end{aligned} \tag{3.7}$$

where

$$\begin{aligned} F_{\Delta u} &= \boldsymbol{c}^{\mathrm{T}}\boldsymbol{l}_{\Delta u}+\boldsymbol{c}^{\mathrm{T}}\boldsymbol{a} \\ &= a_1+a_2z^{-1}+\cdots+a_{N-1}z^{-(N-2)}+a_Nz^{-(N-2)}/(z-1) \\ F_e &= \boldsymbol{c}^{\mathrm{T}}\boldsymbol{l}_e = h_2z^{-1}+\cdots+h_{N-1}z^{-(N-2)}+h_Nz^{-(N-2)}/(z-1) \end{aligned}$$

In order to correspond to the model output in the IMC structure, denote the part in $\hat{y}(k+1)$ that is not related to feedback, i.e. one that fully depends on the model, as $\tilde{y}(k+1)$, i.e. $z\tilde{y}(z) = F_{\Delta u}\Delta u(z)$, then (3.7) can be represented as

$$e(z) = y(z)-\hat{y}(z) = y(z)-\tilde{y}(z)-F_e e(z) \tag{3.8}$$

In this way, the DMC algorithm is transferred into the transitive relations in the Z domain, as shown in Figure 3.2b. Denote

$$\varepsilon(z) = y(z)-\tilde{y}(z), \qquad d_s = \sum_{i=1}^{P} d_i$$

From (3.8) it follows that $e(z) = \varepsilon(z)/(1+F_e)$. Meanwhile, make an equivalent transformation for the left part in Figure 3.2b; then Figure 3.2c can be obtained. Furthermore, in this figure, move the model input from Δu to u, add d_s into the control feed-forward path, and divide all $\boldsymbol{d}^{\mathrm{T}}$ by d_s. After rearranging and calculating each block, the IMC structure as shown in Figure 3.1 can be obtained, where the transfer function for each block can be expressed as follows:

1) **Model**

$$G_M(z) = \frac{z-1}{z} z^{-1} F_{\Delta u}$$

Substitute $F_{\Delta u}$ with the expression in (3.7). It then follows that

$$G_M(z) = a(z) = a_1 z^{-1} + (a_2 - a_1) z^{-2} + \cdots + (a_N - a_{N-1}) z^{-N} \tag{3.9}$$

2) **Controller**

$$G_C(z) = \frac{z}{z-1} \cdot \frac{d_s}{1 + \boldsymbol{d}^{\mathrm{T}} \boldsymbol{G} \boldsymbol{l}_{\Delta u}}$$

Note that $\boldsymbol{d}^{\mathrm{T}}\boldsymbol{G} = [d_1 \quad \cdots \quad d_P \quad 0 \quad \cdots \quad 0]$. Substitute $\boldsymbol{l}_{\Delta u}$ with the expression in (3.6) and after a simple deduction, the transfer function of the controller can be given by

$$G_C(z) = \frac{d_s}{b(z)} \tag{3.10}$$

where

$$b(z) = 1 + (b_2 - 1) z^{-1} + (b_3 - b_2) z^{-2} + \cdots + (b_N - b_{N-1}) z^{-(N-1)}$$

$$b_i = \sum\nolimits_{j=1}^{P} d_j a_{i+j-1}, \quad i = 1, \cdots N, \text{ and } \quad a_i = a_N, \quad i \geq N$$

3) **Filter**

$$G_F(z) = \frac{\boldsymbol{d}^{\mathrm{T}} \boldsymbol{G} \boldsymbol{l}_e z}{d_s} \cdot \frac{1}{1 + F_e}$$

where $\boldsymbol{d}^{\mathrm{T}}\boldsymbol{G}\boldsymbol{l}_e$ can be calculated similarly as $\boldsymbol{d}^{\mathrm{T}}\boldsymbol{G}\boldsymbol{l}_{\Delta u}$. F_e is substituted with the expression in (3.7) and after rearrangement it follows that

$$G_F(z) = \frac{c(z)}{d_s h(z)} \tag{3.11}$$

where

$$c(z) = c_2 + (c_3 - c_2) z^{-1} + \cdots + (c_N - c_{N-1}) z^{-(N-2)}$$

$$h(z) = 1 + (h_2 - 1) z^{-1} + (h_3 - h_2) z^{-2} + \cdots + (h_N - h_{N-1}) z^{-(N-1)}$$

$$c_i = \sum\nolimits_{j=1}^{P} d_j h_{i+j-1}, \quad i = 1, \ldots, N, \text{ and } \quad h_i = h_N, \quad i \geq N$$

4) **Reference input model**
After performing the above transformations, the reference input in Figure 3.1 is $w' = (\boldsymbol{d}^{\mathrm{T}}\boldsymbol{w})/d_s$. For set point control, $\boldsymbol{w} = [w \quad \cdots \quad w]^{\mathrm{T}}$, while for tracking control, $\boldsymbol{w} = [w(k+1) \quad \cdots \quad w(k+P)]^{\mathrm{T}}$. Since $w'(z) = G_w(z)w(z)$, the reference input model can be given by

$$G_w(z) = \begin{cases} 1 & \text{(set point control)} \\ (d_1 z + \cdots + d_P z^P)/d_s & \text{(tracking control)} \end{cases} \tag{3.12}$$

3.2 Controller of DMC in the IMC Structure

According to the properties of the IMC structure given in Section 3.1, a stable controller $G_C(z)$ is critical for the closed-loop stability of predictive control systems. In the DMC algorithm, since the plant is stable, in a nominal case where model mismatch and disturbance are not considered, the closed-loop stability will only depend on the stability of the controller $G_C(z)$ in IMC. According to (3.10) the characteristic polynomial of $G_C(z)$ is given by

$$b(z) = 1 + (b_2 - 1)z^{-1} + (b_3 - b_2)z^{-2} + \cdots + (b_N - b_{N-1})z^{-(N-1)} \tag{3.13}$$

where

$$b_i = \sum_{j=1}^{P} d_j a_{i+j-1}, \quad i = 1, \ldots, N, \text{ and } \quad a_i = a_N, \quad i \geq N$$

In this section, through analysis of the characteristic polynomial of the DMC controller in the IMC structure, the stability conditions of the DMC controller are explored and its specific representations for some special cases are discussed.

3.2.1 Stability of the Controller

Since all the above deduced Z transfer functions are expressed in the form of z^{-1}, to avoid confusion, in the following it is assumed that for

$$f(z) = 1 + f_1 z^{-1} + \cdots + f_n z^{-n} \tag{3.14}$$

if all the roots of the corresponding polynomial

$$F(z) = z^n + f_1 z^{n-1} + \cdots + f_n \tag{3.15}$$

are located in the unit circle, $f(z)$ is stable.

Some lemmas are now given.

Lemma 3.1 For $f(z)$ given by (3.14), if $f(z)$ is stable, then

$$f(1) = 1 + f_1 + \cdots + f_n > 0$$

Lemma 3.2 For $f(z)$ given by (3.14), it is stable if one of the following conditions is satisfied:

1) $1 > f_1 > \cdots > f_i > 0, \quad f_{i+1} = \cdots = f_n = 0;$

2) $|f_1| + \cdots + |f_{n-1}| + |f_n| < 1.$

Lemmas 3.1 and 3.2 give the necessary and sufficient condition for stable $f(z)$, respectively; see the literature on discrete control systems such as [2].

Lemma 3.3 Assume $f(z) = 1 + f_1 z^{-1} + \cdots + f_n z^{-n}$ is stable, let $\sigma > 0$, and

$$h(z) = \left(1 - z^{-1}\right) f(z) + \sigma g(z)$$

where

$$g(z) = g_0 z^{-(n-m+1)} + g_1 z^{-(n-m+2)} + \cdots + g_m z^{-(n+1)}, \quad m \le n$$

Then the necessary condition for a stable $h(z)$ is

$$g(1) > 0$$

Furthermore, it is also the necessary and sufficient condition of $h(z)$ stable when $\sigma \to 0$.

Proof: According to the expression of $h(z)$ and $h(1) = \sigma g(1)$, $\sigma > 0$, it is obvious from Lemma 3.1 that $g(1) > 0$ is the necessary condition of $h(z)$ stable. In the following we prove that it is also sufficient for $h(z)$ stable when $\sigma \to 0$.

At first denote

$$F(z) = z^n + f_1 z^{n-1} + \cdots + f_n$$
$$G(z) = g_0 z^m + g_1 z^{m-1} + \cdots + g_m$$
$$H(z) = (z-1)F(z) + \sigma G(z)$$

When $\sigma = 0$, $H(z) = (z-1)F(z)$. It has the roots $\lambda_1 = z_1, \ldots, \lambda_n = z_n, \lambda_{n+1} = 1$, where $z_1, \ldots, z_n$ are the roots of $F(z)$, all located in the unit circle. Since all the roots of $H(z)$ vary continuously with σ, there exists a limited σ_0 such that for $0 < \sigma < \sigma_0$ the first n roots of $H(z)$ are still located in the unit circle, while due to symmetry λ_{n+1} must be located at the positive real axis near to 1, i.e. $\lambda_{n+1} = 1 + \varepsilon$. To determine the symbol of ε, substitute it into $H(z)$:

$$\begin{aligned} H(\lambda_{n+1}) &= \varepsilon\left[(1+\varepsilon)^n + f_1(1+\varepsilon)^{n-1} + \cdots + f_n\right] \\ &\quad + \sigma\left[g_0(1+\varepsilon)^m + g_1(1+\varepsilon)^{m-1} + \cdots + g_m\right] \\ &= \varepsilon[(1 + f_1 + \cdots + f_n) + F_1(\varepsilon)] + \sigma[(g_0 + \cdots + g_m) + G_1(\varepsilon)] = 0 \end{aligned}$$

where both $F_1(\varepsilon)$ and $G_1(\varepsilon)$ are high-order terms of ε. When $\sigma \to 0$, it follows that $\varepsilon \to 0$; after ignoring $F_1(\varepsilon)$ and $G_1(\varepsilon)$, it follows that

$$\varepsilon = -\sigma \frac{g_0 + \cdots + g_m}{1 + f_1 + \cdots + f_n} = -\sigma \frac{g(1)}{f(1)}$$

Since $\sigma > 0$ and $f(1) > 0$, because $f(z)$ is stable, it is obvious that if $g(1) > 0$ then $\varepsilon < 0$, i.e. $\lambda_{n+1} = 1 + \varepsilon$ in the unit circle, or $h(z)$ is stable. Thus the condition $g(1) > 0$ is also sufficient for $h(z)$ stable when $\sigma \to 0$.

Based on the above three lemmas, the following theorems concerning the stability of the DMC controller $G_C(z)$ can be obtained.

Theorem 3.1 If the unit step response sequence $\{a_i\}$ after the k_{th} step increases monotonically with convex form, i.e.

$$a_k > a_{k+1} - a_k > a_{k+2} - a_{k+1} > \cdots > 0$$

then the DMC controller $G_C(z)$ is stable when the following optimization strategy is adopted:

$$\boldsymbol{Q} = \text{block diag}\,[0_{(k-1)\times(k-1)},\ \boldsymbol{I}_{(P-k+1)\times(P-k+1)}],\ \boldsymbol{R} = 0$$
$$M = 1,\ P \geq k$$

Proof: Using the above optimization strategy, according to (2.8), it is easy to obtain

$$[d_1 \ \cdots \ d_P] = s\,[\underbrace{0 \ \cdots \ 0}_{k-1} \ \ a_k \ \cdots \ a_P], \qquad s = 1/\sum_{i=k}^{P} a_i^2$$

With arbitrary P satisfying $P \geq k$,

$$b_1 = d_k a_k + \cdots + d_P a_P = 1$$

$$b_i = d_k a_{i+k-1} + \cdots + d_P a_{i+P-1} = s\sum_{j=k}^{P} a_j a_{i+j-1}, \qquad i = 1, \ldots, N$$

It follows that

$$b_2 - 1 = s[a_k(a_{k+1} - a_k) + \cdots + a_P(a_{P+1} - a_P)] < s[a_k a_k + \cdots + a_P(a_P - a_{P-1})] < s\sum_{i=k}^{P} a_i^2 = 1$$

and

$$b_2 - 1 = s[a_k(a_{k+1} - a_k) + \cdots + a_P(a_{P+1} - a_P)]$$
$$b_3 - b_2 = s[a_k(a_{k+2} - a_{k+1}) + \cdots + a_P(a_{P+2} - a_{P+1})]$$
$$\vdots$$
$$b_{N-k} - b_{N-k-1} = s[a_k(a_{N-1} - a_{N-2}) + a_{k+1}(a_N - a_{N-1})]$$
$$b_{N-k+1} - b_{N-k} = s[a_k(a_N - a_{N-1})]$$

Note that $a_k > a_{k+1} - a_k > \cdots > a_N - a_{N-1} > 0$, together with $b_2 - 1 < 1$, easily show that

$$1 > b_2 - 1 > b_3 - b_2 > \cdots > b_{N-k+1} - b_{N-k} > 0$$

Furthermore,

$$b_{N-k+1+i} - b_{N-k+i} = 0, \qquad i = 1, \ldots, k-1$$

According to condition (1) in Lemma 3.2, $b(z)$ in (3.13) is stable, i.e. $G_C(z)$ is stable.

Theorem 3.1 gives the parameter setting condition of the stable DMC controller for a fairly extensive class of systems whose step response after several steps increases monotonically with convex form, such as the first-order inertia system with time delay, which frequently appears in the industrial processes.

Theorem 3.2 No matter what shape the step response $\{a_i\}$ has, a stable DMC controller $G_C(z)$ can always be achieved by selecting P large enough with the optimization strategy $\boldsymbol{Q} = \boldsymbol{I}$, $M = 1$.

Proof: If the condition P is large enough in the theorem this means that P can be possibly over the model length N such that quite a part of a_i already reaches the steady value a_N, i.e. $a_i = a_N$, $i \geq N$. For the given optimization strategy, according to (2.8) it follows that

$$d_i = a_i \Big/ \left(r + \sum_{j=1}^{P} a_j^2 \right), \quad i = 1, \ldots, P$$

$$\begin{aligned} b_i &= \sum_{j=1}^{P} d_j a_{i+j-1} \\ &= \left(\sum_{j=1}^{P} a_j a_{i+j-1} \right) \Big/ \left(r + \sum_{j=1}^{P} a_j^2 \right) \\ &= \frac{\sum_{j=1}^{N-1} a_j a_{i+j-1} + (P-N+1)\, a_N^2}{r + \sum_{j=1}^{N-1} a_j^2 + (P-N+1)\, a_N^2} \to 1, \quad P \to +\infty, \quad i = 2, \ldots, N \end{aligned}$$

It is obvious that all the coefficients of $b(z)$ tend to zero when $P \to +\infty$, which satisfies the condition (2) in Lemma 3.2. Thus there must exist a sufficiently large P_0 such that $G_C(z)$ is stable when $P > P_0$.

Theorem 3.2 reveals the influence of the optimization horizon to the stability of the DMC controller, where the condition of P sufficiently large means physically that the optimization focuses on the steady-state part of the plant response. It may lead to a stable control, but is in general smooth and sluggish. At the extreme case of $P \to +\infty$, $b(z) \to 1$,

$$d_s = \sum_{i=1}^{P} d_i = \frac{\sum_{i=1}^{P} a_i}{r + \sum_{i=1}^{P} a_i^2} = \frac{\sum_{i=1}^{N-1} a_i + (P-N+1)a_N}{r + \sum_{i=1}^{N-1} a_i^2 + (P-N+1)\, a_N^2} \to \frac{1}{a_N}$$

It follows from (3.10) that

$$G_C(z) \to \frac{1}{a_N}$$

The controller will be degenerated to a proportional component. It only functions as a steady-state controller and can hardly change the system dynamics. In the literature this kind of control with $M = 1$ and large P is called "mean level control" [3].

Remark 3.1 It should be noted that both Theorem 3.1 and Theorem 3.2 only give the sufficient conditions for the stability of the DMC controller. The significance of them lies in reflecting the trend of how the design parameters affect the stability of the controller, rather than the extreme selections as described in the theorems. The conservativeness of these conditions can also be seen in the proof of the theorems. In many cases, a stable controller could also be obtained even when the control horizon M in both theorems is selected small enough but not as 1, or the $\{a_i\}$ in Theorem 3.1 does not increase monotonically with a convex form in a strict sense, or the optimization horizon P in Theorem 3.2 is selected large enough but not necessarily larger than the model length N.

In the above two theorems, the weighting matrices $\boldsymbol{Q}$ and $\boldsymbol{R}$ are set with special values. The following theorem shows how they affect the stability of the controller.

Theorem 3.3 Set the control weighting matrix $\boldsymbol{R} = r\boldsymbol{I}$ where $r \geq 0$; if

$$\left(\sum_{i=1}^{P} a_i q_i\right) a_N > 0 \tag{3.16}$$

a stable DMC controller can be obtained through increasing r. Otherwise the controller must be unstable.

Proof: For $\boldsymbol{R} = r\boldsymbol{I}$, when $r \to +\infty$, the control vector can be given by

$$[d_1 \ \cdots \ d_P] = \frac{1}{r}[1 \ 0 \ \cdots \ 0]\left(\frac{1}{r}\boldsymbol{A}^{\mathrm{T}}\boldsymbol{Q}\boldsymbol{A} + \boldsymbol{I}\right)^{-1}\boldsymbol{A}^{\mathrm{T}}\boldsymbol{Q} \to \begin{bmatrix}\frac{a_1 q_1}{r} & \cdots & \frac{a_P q_P}{r}\end{bmatrix}$$

It follows that

$$b_i = \sum_{j=1}^{P} d_j a_{i+j-1} \to \frac{1}{r}\sum_{j=1}^{P} a_j a_{i+j-1} q_j, \quad i = 2, \ldots, N$$

$$d_s = \sum_{i=1}^{P} d_i \to \frac{1}{r}\sum_{i=1}^{P} a_i q_i$$

Denote

$$b(z) = (1 - z^{-1}) f(z) + \sigma g(z)$$

where

$$f(z) = 1, \qquad \sigma = 1/r$$
$$g(z) = r b_2 z^{-1} + r(b_3 - b_2) z^{-2} + \cdots + r(b_N - b_{N-1}) z^{-(N-1)}$$

We can get

$$g(1) = r b_N = r d_s a_N = \left(\sum_{i=1}^{P} a_i q_i\right) a_N$$

Thus the theorem can be directly concluded by Lemma 3.3.

Theorem 3.3 provides the condition with which the DMC controller can be stabilized by increasing r. Intuitively, it seems that stable control could be achieved by sufficiently suppressing the control increment, i.e. increasing r. However, the theorem points out that this is not always the case. Only when the condition (3.16) is satisfied is it possible to get stable control by increasing r. The summation term in (3.16) can be interpreted as the weighted center of the step response over the optimized horizon, and the condition means that this weighted center must be of the same symbol as the steady-state value of the step response. Therefore, for a time-delay system or nonminimal phase system, the weighting coefficients q_i corresponding to the delay or the reverse part at the beginning of the system response should be selected as zero, which is equivalently to select the initial value of the optimization horizon N_1 in GPC.

Remark 3.2 The condition (3.16) in Theorem 3.3 is both necessary and sufficient for a stable controller when r is sufficiently large, but is only necessary and not sufficient for a stable controller in the global region $r \geq 0$. This means that even when (3.16) is satisfied, stability of the controller may be changed with r. For example, it can be proved that for a second-order system, under certain parameter settings, the closed-loop system is stable when $r = 0$ and $r \to \infty$, but is unstable for r in a certain region $[r_1, \ r_2]$.

The above theorems discuss the relationship between the stability of the controller $G_C(z)$ and the DMC design parameters, where the characteristic polynomial $b(z)$ of the controller is of a nonminimal form. Due to the lack of an analytical relationship between the stability and the nonminimal controller, the conclusions of these theorems have to be achieved through a specific, even an extreme, setting of the design parameters so that the coefficients of $b(z)$ can be analytically expressed and are tractable. As indicated in Remark 3.1, the importance of these theorems is not the conditions for extreme cases (also called limiting cases in [4]), but the trend of changing the corresponding parameter to achieve a stable controller, which is helpful for tuning the DMC parameters and will be discussed in Section 3.4.

3.2.2 Controller with the One-Step Optimization Strategy

In unconstrained DMC algorithms, if the plant has no time delay, we call the optimization strategy with $\boldsymbol{Q} = \boldsymbol{I}$, $\boldsymbol{R} = \boldsymbol{0}$, $P = M$ a One-Step Optimization Strategy (OSOS).

Theorem 3.4 For the DMC algorithm with OSOS, the DMC controller (3.10) in the IMC structure is exactly the model inverse plus a one-step time delay.

Proof: For an OSOS with $\boldsymbol{Q} = \boldsymbol{I}$, $\boldsymbol{R} = \boldsymbol{0}$, $P = M$, it follows from (2.8) that

$$\boldsymbol{d}^{\mathrm{T}}\boldsymbol{A} = [1 \ \ 0 \ \ \cdots \ \ 0]$$

i.e.

$$\begin{aligned} d_1 a_1 + d_2 a_2 + \cdots + d_P a_P &= 1 \\ d_2 a_1 + \cdots + d_P a_{P-1} &= 0 \\ &\vdots \\ d_P a_1 &= 0 \end{aligned}$$

It follows that

$$d_1 = 1/a_1, \ \ d_2 = \cdots = d_P = 0, \ \ d_s = d_1 = 1/a_1$$

and then

$$b_i = d_1 a_i = a_i / a_1, \quad i = 2, \ldots, N$$

Substitute them into (3.10) and note the model transfer function (3.9); it follows that

$$G_C(z) = \frac{d_s}{b(z)} = z^{-1} G_M^{-1}(z) \tag{3.17}$$

Therefore, with OSOS the controller is a one-step delay of the model inverse. Moreover, without model mismatch, the Z transfer function of the controlled system (3.2) is

$$F_0(z) = G_C(z) G_P(z) = z^{-1}$$

It only contains the inherent one-step delay in the Z transfer function of the plant that is caused by sampling and holding. After control starts, the system output will exactly reach the set point immediately at the next sampling time.

Remark 3.3 Note that while defining OSOS, the condition $P = M$ is not restricted as $P = M = 1$, but this strategy is still called one-step optimization. The reason is that no matter how large $P = M$ is selected, the resultant controller is the same as that for $P = M = 1$, as shown by the above proof procedure. Furthermore, since the model inverse appears in the resultant controller (3.17), the OSOS is only suitable for a minimal phase plant without time delay. In this case, the resultant controller (3.17) is a perfect controller (3.3).

3.2.3 Controller for Systems with Time Delay

Consider the plant with l steps time delay

$$\bar{G}_M(z) \triangleq z^{-l} G_M(z) \tag{3.18}$$

where $G_M(z)$ is expressed by (3.9) with $a_1 \neq 0$ and represents the remaining part of $\bar{G}_M(z)$ after removing pure time delays. In the following, we discuss the relationship of DMC controllers for the time delay system (3.18) and for the system (3.9) without time delay.

Theorem 3.5 In the DMC algorithm for the time delay system (3.18), the optimization strategy

$$\bar{P} = P + l, \quad \bar{M} = M$$
$$\bar{\boldsymbol{Q}} = \text{block diag}[\boldsymbol{0}_{l \times l}, \ \boldsymbol{Q}], \quad \bar{\boldsymbol{R}} = \boldsymbol{R}$$

results in the same controller in the IMC structure as that for the delay free system (3.9) with the optimization strategy (2.3).

Proof: Due to the l steps time delay, there exists a relationship between the coefficients of the step responses for $\bar{G}_M(z)$ and $G_M(z)$ as follows:

$$\bar{a}_i = \begin{cases} 0, & i \leq l \\ a_{i-l}, & i > l \end{cases}$$

With the given optimization strategy and using (2.8), it follows that

$$\bar{\boldsymbol{d}}^{\mathrm{T}} = \boldsymbol{c}^{\mathrm{T}}\left(\bar{\boldsymbol{A}}^{\mathrm{T}}\bar{\boldsymbol{Q}}\bar{\boldsymbol{A}} + \bar{\boldsymbol{R}}\right)^{-1}\bar{\boldsymbol{A}}^{\mathrm{T}}\bar{\boldsymbol{Q}}$$

According to the relationship of $\bar{a}_i$ and a_i, we have

$$\bar{\boldsymbol{A}} = \begin{bmatrix} 0_{l\times M} \\ \boldsymbol{A}_{P\times M} \end{bmatrix}$$

and then

$$\bar{\boldsymbol{d}}^{\mathrm{T}} = \boldsymbol{c}^{\mathrm{T}}\left(\boldsymbol{A}^{\mathrm{T}}\boldsymbol{Q}\boldsymbol{A} + \boldsymbol{R}\right)^{-1}\left[\boldsymbol{0}_{M\times l} \quad \boldsymbol{A}^{\mathrm{T}}\boldsymbol{Q}\right] = \left[\boldsymbol{0}_{1\times l} \quad \boldsymbol{d}^{\mathrm{T}}\right]$$

where $\boldsymbol{d}^{\mathrm{T}}$ is the DMC control vector for the delay free system (3.9) with optimization performance (2.3); see (2.8). Thus it follows that

$$\bar{d}_i = \begin{cases} 0, & i \le l \\ d_{i-l}, & i = l+1, \ \ldots, \ \bar{P} \end{cases}$$

Then the coefficients of the characteristic polynomial (3.13) of the DMC controller for the time delay system (3.18) can be calculated as

$$\bar{b}_i = \sum_{j=1}^{\bar{P}} \bar{d}_j \bar{a}_{i+j-1} = \sum_{j=l+1}^{P+l} d_{j-l} a_{i+j-l-1} = \sum_{j=1}^{P} d_j a_{i+j-1} = b_i$$

Moreover, it is obvious that $\bar{d}_s = d_s$, so the controllers for both systems are identical, i.e.

$$\bar{G}_C(z) = G_C(z)$$

When no model mismatch exists, the closed-loop transfer function will have the form

$$\bar{F}_0(z) = \bar{G}_C(z)\bar{G}_P(z) = z^{-l}G_C(z)G_P(z) = z^{-l}F_0(z)$$

This has a fully identical dynamic behavior to that of the DMC control for the delay-free system, except for the additional l steps time delay. Therefore, its output will be postponed for l steps as compared with the delay-free case.

Substituting $\bar{d}_i$ obtained from the above deduction into the control law (2.7), it is known that for a system with l steps time delay, the control law is calculated for future P sampling times after crossing the l steps time delay, which is equivalent to pushing back the whole optimization horizon for l steps.

The above analysis shows that the predictive control algorithm has unique advantages when dealing with time delay systems. It can naturally take the delay into account without the need for an additional control structure, while the control effect is equivalent to that of controlling a delay-free system plus an output delay. Reasonably tuning the design parameters can make the DMC control law for a time delay system fully identical with that for a delay-free system, which could simplify the analysis and design of predictive control for time delay systems to some extent.

3.3 Filter of DMC in the IMC Structure

It was pointed out in Section 3.1 that the filter in the IMC structure works only when the output error exists due to model mismatch or disturbance, so it has a crucial influence on the robustness and disturbance rejection of the closed-loop system. In this section, combined with the selection of the feedback correction strategies in the DMC algorithm, the influence of the filter on the robustness and disturbance rejection of the closed-loop system is discussed in detail.

3.3.1 Three Feedback Correction Strategies and Corresponding Filters

In Section 3.1, the filter of the DMC algorithm in the IMC structure is derived with the form

$$G_F(z) = \frac{c(z)}{d_s h(z)} \tag{3.19}$$

where

$$c(z) = c_2 + (c_3 - c_2)z^{-1} + \cdots + (c_N - c_{N-1})z^{-(N-2)}$$
$$h(z) = 1 + (h_2 - 1)z^{-1} + (h_3 - h_2)z^{-2} + \cdots + (h_N - h_{N-1})z^{-(N-1)}$$
$$c_i = \sum_{j=1}^{P} d_j h_{i+j-1}, \quad i = 1, \ldots, N, \text{ and } \quad h_i = h_N, \quad i \geq N$$

Note that $h(z)$, the denominator polynomial of the filter, only depends on the feedback correction coefficients h_i, which could be freely selected, and the stability of the filter $G_F(z)$ can be easily guaranteed. However, using a large number of independent h_i is not convenient for system analysis and it is often reasonable to parameterize the h_i as simple functions of a single parameter. In the following, some typical selections for h_i and the expressions of the corresponding filters are given.

(1) Equal-value correction
Select

$$h_1 = 1, \quad h_i = \alpha, \qquad 0 < \alpha \leq 1 \tag{3.20}$$

which means equally correcting all the predicted future outputs using the current error with a scale coefficient α. It follows that

$$h(z) = 1 + (\alpha - 1)z^{-1}$$
$$c(z) = \alpha d_s$$

and then

$$G_F(z) = \frac{\alpha}{1 - (1 - \alpha)z^{-1}} \tag{3.21}$$

This filter is indeed a first-order filter. If $0 < \alpha \leq 1$, $G_F(z)$ is stable and without oscillation. The gain in the steady state is $G_F(1) = 1$. Particularly when $\alpha = 1$, $G_F(z) = 1$ degenerates to a constant value filter.

(2) Attenuation correction
Select

$$h_1 = 1, \quad h_i = \alpha^{i-1}, \quad i = 2, \ldots, N, \quad 0 < \alpha < 1 \tag{3.22}$$

The error correction for the predicted future outputs is gradually attenuated with the current error multiplied by α, α^2, ..., α^{N-1}. Since $0 < \alpha < 1$, in the case of N being sufficiently large, it approximately follows that

$$\begin{aligned} h(z) &= 1 + (\alpha - 1)z^{-1} + \cdots + \alpha^{N-2}(\alpha - 1)z^{-(N-1)} \\ &\approx 1 + \frac{(\alpha - 1)z^{-1}}{1 - \alpha z^{-1}} = \frac{1 - z^{-1}}{1 - \alpha z^{-1}} \end{aligned}$$

Denote

$$s \triangleq c_1 = d_1 h_1 + \cdots + d_P h_P = d_1 + \cdots + d_P \alpha^{P-1}$$

and in the case of N sufficiently large, assume $\alpha^i \approx \alpha^N \approx 0, \quad i > N$. Then it follows that

$$\begin{aligned} c_i &\approx \alpha^{i-1} s, \quad i = 2, \ldots, N \\ c(z) &\approx \alpha s + \alpha(\alpha - 1)sz^{-1} + \cdots + \alpha^{N-2}(\alpha - 1)sz^{-(N-2)} \\ &\approx \alpha s \left(1 + \frac{(\alpha - 1)z^{-1}}{1 - \alpha z^{-1}}\right) = \frac{\alpha s(1 - z^{-1})}{1 - \alpha z^{-1}} \end{aligned}$$

and the filter has the form

$$G_F(z) = \frac{\alpha s}{d_s} = \frac{d_1 \alpha + \cdots + d_P \alpha^P}{d_1 + \cdots + d_P} \tag{3.23}$$

In the case of N sufficiently large, this filter can be regarded as a constant value filter, and $G_F(z) \to 0$ when $\alpha \to 0$. Note that in general the gain of this filter is not one.

(3) Increasing correction
Select

$$h_1 = 1, \quad h_{i+1} = h_i + \alpha^i, \quad i = 1, \ldots, N - 1, \quad 0 < \alpha < 1 \tag{3.24}$$

which means increasingly correcting the predicted future outputs with the current error multiplied by $1 + \alpha$, $1 + \alpha + \alpha^2$, ... It follows that

$$h_{i+1} - h_i = \alpha^i, \quad i = 1, \ldots, N - 1$$

In the case of N sufficiently large, we have

$$h(z) = 1 + \alpha z^{-1} + \cdots + \alpha^{N-1} z^{-(N-1)} \approx \frac{1}{1 - \alpha z^{-1}}$$

Denote

$$s \triangleq c_2 - c_1 = d_1 \alpha + \cdots + d_P \alpha^P$$

and in the case of N sufficiently large, assume that $\alpha^i \approx \alpha^N \approx 0,\quad i > N$. Then it follows that

$$c_{i+1} - c_i \approx d_1\alpha^i + \cdots + d_P\alpha^{i+P-1} = \alpha^{i-1}s,\quad i = 1,\ \ldots,\ N-1$$

$$c(z) \approx c_2 + \alpha s z^{-1} + \cdots + \alpha^{N-2} s z^{-(N-2)} \approx \frac{c_2 - \alpha c_1 z^{-1}}{1 - \alpha z^{-1}}$$

The filter (3.19) can be approximately given by

$$G_F(z) \approx \frac{1}{d_s}\left(c_2 - \alpha c_1 z^{-1}\right) \tag{3.25}$$

This filter is always stable. It is equivalent to introducing a zero, which can compensate the pole of the disturbance. Note that

$$h_{i+1} - \alpha h_i = 1,\qquad c_2 - \alpha c_1 = d_s$$

so (3.25) can be expressed by

$$G_F(z) \approx \frac{d_s + \alpha c_1\left(1 - z^{-1}\right)}{d_s}$$

The steady-state gain of this filter is $G_F(z) = 1$.

In the next example, we take the OSOS given in Section 3.2.2 as an example and quantitatively discuss the influence of the above three correction strategies on the robustness and disturbance rejection of the DMC closed-loop system.

Example 3.1 For OSOS, i.e. $\boldsymbol{Q} = \boldsymbol{I}$, $\boldsymbol{R} = \boldsymbol{0}$, $P = M$, the controller was given by (3.17):

$$G_C(z) = z^{-1}G_M^{-1}(z)$$

Since $d_2 = \cdots = d_P = 0$, $d_s = d_1$, the three filters discussed above could be expressed by

Filter (3.21): $G_F(z) = \dfrac{\alpha}{1-(1-\alpha)z^{-1}},\qquad 0 < \alpha \le 1$

Filter (3.23): $G_F(z) = \alpha,\qquad 0 < \alpha < 1$

Filter (3.25): $G_F(z) \approx (1+\alpha) - \alpha z^{-1},\qquad 0 < \alpha < 1$

In the following, two cases are discussed where the influences of different filters on system robustness and disturbance rejection are quantitatively studied, respectively.

Case 1. Robustness against model gain mismatch with different filters

Consider the gain mismatch between the plant and the model:

$$G_P(z) = \mu G_M(z),\qquad \mu > 0$$

The closed-loop stability depends on

$$1 + G_C(z)G_F(z)\left[G_P(z) - G_M(z)\right] = 1 + (\mu - 1)z^{-1}G_F(z)$$

1) With filter (3.21), the closed-loop characteristic polynomial is given by

$$f(z) = 1 - (1-\alpha)z^{-1} + \alpha(\mu - 1)z^{-1} = 1 - (1 - \alpha\mu)z^{-1}$$

The condition for closed-loop stability is

$$0 < \mu < 2/\alpha$$

Since $0 < \alpha \le 1$, it is obvious that if the actual gain is smaller than the model gain, i.e. $0 < \mu < 1$, the closed loop is always stable, while if the actual gain is larger than the model gain, the admissible range of the model mismatch μ would increase as α decreases.

2) With filter (3.23), the closed-loop characteristic polynomial is given by

$$f(z) = 1 + \alpha(\mu - 1)z^{-1}$$

The condition for the closed-loop stability is

$$0 < \mu < 1 + 1/\alpha$$

It is obvious that the admissible range of the model mismatch μ would also increase as α decreases. However, for the same α, the upper bound of the admissible mismatch range for μ is smaller than that of the case with filter (3.21).

3) With filter (3.25), the closed-loop characteristic polynomial is now given by

$$f(z) = 1 + (\mu - 1)(1 + \alpha)z^{-1} - \alpha(\mu - 1)z^{-2}$$

The condition for the closed-loop stability is

$$0 < \mu < 1 + 1/(2\alpha + 1)$$

With this filter, the upper bound of the admissible range for model mismatch μ is smaller than that of the above two cases and is at most 2.

Case 2. Disturbance rejection with different filters

Now the ability of disturbance rejection is discussed when using different filters where the model is assumed to be accurate. For the given OSOS, the disturbance response of the closed-loop system can be written as follows according to (3.1):

$$y(z) = [1 - G_C(z)G_F(z)G_M(z)]v(z) = \left[1 - z^{-1}G_F(z)\right]v(z)$$

Consider a disturbance acting on the system output (see Figure 3.1), which has the form of a unit step passing through a first-order inertial unit, i.e.

$$v(z) = \frac{(1-\sigma)z^{-1}}{(1-\sigma z^{-1})(1-z^{-1})}, \qquad 0 < \sigma < 1$$

Without feedback correction, i.e. $G_F(z) = 0$, the system output y will gradually increase with $1 - \sigma, 1 - \sigma^2, \ldots$, and the disturbance will be completely reflected at the output without attenuation. However, taking the feedback correction strategy, the disturbance could be rejected efficiently.

1) With filter (3.21), the disturbance response of the closed-loop system is given by

$$y(z) = \frac{(1-\sigma)z^{-1}}{(1-\sigma z^{-1})(1-(1-\alpha)z^{-1})}$$

The values of y at the sampling times can be given by

$$y(k) = (1-\sigma)\left(\sigma^{k-1} + \sigma^{k-2}(1-\alpha) + \cdots + (1-\alpha)^{k-1}\right)$$

It is obvious that $y(k) \to 0$ when $k \to \infty$. Furthermore, if $\alpha \to 0$, y will change at a value of $1-\sigma$, $1-\sigma^2$, ..., close to the case without feedback correction, while if $\alpha = 1$, y will change at a value of $1-\sigma$, $\sigma(1-\sigma)$, $\sigma^2(1-\sigma)$, ... and converge to zero.

2) With filter (3.23), the disturbance response of the system is given by

$$y(z) = \frac{(1-\sigma)z^{-1}(1-\alpha z^{-1})}{(1-\sigma z^{-1})(1-z^{-1})}$$

The disturbed output y has a pole at $z = 1$ and will not converge.

3) With filter (3.25), the disturbance response is approximately given by

$$y(z) \approx \frac{(1-\sigma)z^{-1}(1-\alpha z^{-1})}{1-\sigma z^{-1}}$$

where y will gradually attenuate to zero with

$$y(1) = 1-\sigma, \quad y(k) = (1-\sigma)(\sigma-\alpha)\sigma^{k-2}, \quad k = 2, 3, \ldots$$

Particularly when $\alpha = \sigma$,

$$y(z) = (1-\sigma)z^{-1}$$

The disturbed output y seems to be a one-step deadbeat response, i.e. the disturbance affects the system output only for one step and with amplitude $1-\sigma$, and then will be completely suppressed.

From the above example it is shown that with the equal-value and attenuation correction strategies, the robustness to model mismatch is strong. However, the ability of disturbance rejection of the attenuation correction strategy is quite poor, with the disturbance almost unable to be effectively suppressed. While the increasing correction strategy may have a better ability to reject disturbance, its robustness to the model mismatch is much lower than the other two strategies. Therefore, the equal-value correction strategy (3.20) and the corresponding first-order filter (3.21) might be a suitable choice as a trade-off for both robustness and disturbance rejection. Since the filter is only related with the feedback correction parameters and does not depend on the control parameters, it can be freely designed independent of the control strategy.

3.3.2 Influence of the Filter to Robust Stability of the System

The DMC algorithm is designed based on the model $G_M(z)$ rather than the actual plant $G_P(z)$. However, in real processes, model mismatch inevitably exists due to inaccurate estimation of the step response, parameter variation, or nonlinear character of the plant, etc. In this case, the actual step response of the plant $\{\tilde{a}_i\}$ may be inconsistent with the model parameters $\{a_i\}$.

In Section 3.1, Property 3 of the IMC structure involves a zero offset condition when disturbance or model mismatch exists, with the precondition that the closed-loop system is stable. If $G_M(z) \neq G_P(z)$, it is known from the IMC structure and (3.1) that the stability of the closed-loop system depends on the zero polynomial of

$$1 + G_C(z)G_F(z)[G_P(z) - G_M(z)]$$

if the four units $G_M(z)$, $G_P(z)$, $G_C(z)$, $G_F(z)$ are all stable. It is obvious that the coefficients of this characteristic polynomial are generally related to both the optimization and the correction strategy, but if the model is accurate, the system dynamics will only be affected by the controller $G_C(z)$ and is independent of the filter $G_F(z)$. Therefore, in DMC design, $G_C(z)$ is often determined by proper selection of the optimization strategy to meet the requirements on stability and dynamic characteristics in the case without model mismatch (i.e. the nominal case), and the robustness against the model mismatch can be improved through the correction strategy, i.e. the selection of the filter $G_F(z)$. In the following, how the filter $G_F(z)$ affects the robust stability of the closed-loop system is discussed.

According to (3.9) to (3.11), the closed-loop characteristic polynomial of DMC in the IMC structure can be given by

$$f(z) = b(z)h(z) + c(z)[\tilde{a}(z) - a(z)] \tag{3.26}$$

where

$$b(z) = 1 + (b_2 - 1)z^{-1} + (b_3 - b_2)z^{-2} + \cdots + (b_N - b_{N-1})z^{-(N-1)}$$
$$h(z) = 1 + (h_2 - 1)z^{-1} + (h_3 - h_2)z^{-2} + \cdots + (h_N - h_{N-1})z^{-(N-1)}$$
$$c(z) = c_2 + (c_3 - c_2)z^{-1} + \cdots + (c_N - c_{N-1})z^{-(N-2)}$$

$$\tilde{a}(z) = \tilde{a}_1 z^{-1} + (\tilde{a}_2 - \tilde{a}_1)z^{-2} + \cdots + (\tilde{a}_N - \tilde{a}_{N-1})z^{-N}$$
$$a(z) = a_1 z^{-1} + (a_2 - a_1)z^{-2} + \cdots + (a_N - a_{N-1})z^{-N}$$

Without a quantitative description of the model mismatch, it is difficult to make a general robustness analysis for the DMC system according to (3.26). However, the following theorem gives the trend of the influence of the equal-value correction strategy, i.e. the filter (3.21), on robust stability.

Theorem 3.6 Set the filter (3.21) with $0 < \alpha \leq 1$. If $G_M(z)$, $G_P(z)$, $G_C(z)$ are stable, as long as $a_N \tilde{a}_N > 0$, there is always a $\alpha_0 > 0$, such that for $0 < \alpha < \alpha_0$ the closed-loop system is both stable and offset free for an arbitrary model mismatch and constant disturbance.

Proof: For filter (3.21)

$$h(z) = 1 + (\alpha - 1)z^{-1}, \quad c(z) = d_s \alpha$$

the characteristic polynomial (3.26) has the form

$$\begin{aligned} f(z) &= b(z)(1 + (\alpha - 1)z^{-1}) + d_s\alpha(\tilde{a}(z) - a(z)) \\ &= b(z)(1 - z^{-1}) + \alpha(b(z)z^{-1} + d_s(\tilde{a}(z) - a(z))) \end{aligned}$$

Note that $b(z)$ is a stable polynomial, $\alpha > 0$. According to Lemma 3.3, if α is sufficiently small, the necessary and sufficient condition for $f(z)$ stable is

$$b(1) + d_s(\tilde{a}(1) - a(1)) = b(1) + d_s(\tilde{a}_N - a_N) > 0$$

Since $b(1) = b_N = d_s a_N$, the above condition can be written as $d_s \tilde{a}_N > 0$, while $b(z)$ stable implies $b(1) = d_s a_N > 0$. Multiplying both inequalities, the achieved condition is equivalent to

$$a_N \tilde{a}_N > 0$$

If this condition is satisfied, there is always a sufficiently small α_0, and as long as $0 < \alpha < \alpha_0$ is selected, the closed-loop system is stable.

Under the precondition that the closed-loop system is stable, note that the DMC controller and filter satisfy

$$G_C(1) = \frac{d_s}{b(1)} = \frac{1}{a_N} = \frac{1}{G_M(1)}, \qquad G_F(1) = 1$$

It is known from Property 3 of the IMC structure that the closed-loop system is offset-free.

Remark 3.4 Theorem 3.6 reveals the role of the filter $G_F(z)$, or equivalently the feedback correction, in predictive control systems. On the one hand, the existence of it makes the DMC system in closed-loop form such that the steady-state error can be eliminated when model mismatch or disturbance exists. On the other hand, the strength of the feedback correction should be properly selected to ensure robust stability. During the proof of the theorem it also shows that the condition $a_N \tilde{a}_N > 0$ is also necessary for robust stability when using small α. Intuitively, for an arbitrary model mismatch, as long as the feedback α is sufficiently small such that the controlled system is almost open loop, a stable response could always be achieved. However, from the proof of the theorem this conclusion is only correct when the steady-state values of the plant and the model have the same symbols. If both symbols are opposite, when α is sufficiently small, the closed-loop system still has a pole on the real axis close to $z = 1$, but outside the unit circle, whose dynamic response will be diverging slowly.

3.4 DMC Parameter Tuning Based on Trend Analysis

DMC starts from the sampling values of the unit step response $\{a_i\}$, which are related to system dynamics and the sampling period. The selection of the sampling period T should follow the general rules in sampling control. In addition to meeting Shannon's sampling theorem, the dynamic characteristics and physical character of the process should be considered. There are many literatures for reference in the field of process control. After the sampling period T is determined, the sampling values of the unit-step response $\{a_i\}$ are then determined.

In the above sections, the controller and the filter of DMC in the IMC structure are derived and how the design parameters affect the stability and robustness of the closed-loop system are discussed through specific controller and filter setting. Although the conclusions were mostly derived for extreme setting of specific parameters, they reflect the trend on how to change these parameters to make the closed-loop system stable or robustly stable, which would be helpful for tuning the DMC design parameters for general cases where only the step response is available and the minimal system model is unknown. In the following, the selection or tuning rules of various DMC design parameters are discussed with the help of the conclusions obtained in Sections 3.2 and 3.3.

(1) Model length N

Since DMC uses a nonparametric model, in order to make the model more accurately characterize the plant dynamics, the model length N should be selected large enough such that a_N can be approximately equal to the steady-state value of the step response a_∞. In this way, the step response as an infinite dimensional model can be truncated at NT with acceptable truncation errors. After the sampling period T is determined, if the plant dynamic response converges slowly, the dimension of N may be very high, which not only increases the computational burden, but is unreasonable due to redundancy of information. In this case, different strategies according to the plant dynamics can be adopted that truncate the step response at an early time and thus reduce the model length N.

If the plant has a dominant eigenvalue with a large time constant, its step response would possibly exhibit complex dynamic variations at the beginning but after that rise slowly in exponential form. For this case, the model length N does not need to extend to the time where the step response value is almost equal to its steady-state value. Instead, the step response can be truncated early if it has crossed the complex dynamics part. However, this early truncation may lose the information after model length N and results in inaccuracy if the last element of the predicted future output is still obtained by the shifting formula given by (2.14):

$$\tilde{y}_0(k+1+N \mid k+1) = \tilde{y}_{\text{cor}}(k+N \mid k+1)$$

It is therefore suggested that this formula be revised into an exponential recursive form:

$$\tilde{y}_0(k+1+N \mid k+1) = (1+\sigma)\tilde{y}_{\text{cor}}(k+N \mid k+1) - \sigma\tilde{y}_{\text{cor}}(k+N-1 \mid k+1)$$

This is equivalent to revising $\boldsymbol{S}$ in (2.14) as

$$\boldsymbol{S} = \begin{bmatrix} 0 & 1 & & 0 \\ \vdots & \ddots & \ddots & \\ & & 0 & 1 \\ 0 & \cdots & -\sigma & 1+\sigma \end{bmatrix}$$

where $\sigma = \exp(-T/T_0)$ and T and T_0 are the sampling period and the time constant of the plant dominant dynamics, respectively. This revision means that the information of exponentially raising a part after truncation has been concentrated on the parameter σ and the prediction model used here is essentially a combination of directly taking the first $N-1$ data of the step response plus a parametric model for the remaining part. In the case where T_0 cannot be estimated exactly, σ can be approximately taken as $0.9 \sim 1$.

If the plant has a dominant pair of conjugate complex poles with a large time constant, the step response will oscillate around its steady-state value for a long time. In this case, the step response can be truncated early as well, and the model length N should be selected such that a_N is approximately equal to the steady-state value of the step response.

Following the above strategies, the model length can be kept at 20 ~ 50. Although early truncation of the model may result in errors, the resultant prediction error can be effectively compensated because the general trend of the step response after truncation has been taken into account.

(2) Optimization horizon *P* and error weighting matrix *Q*

The optimization horizon P and the error weighting matrix $\boldsymbol{Q}$ correspond to the following items in the optimization performance index (2.3):

$$\sum_{i=1}^{P} q_i[w(k+i) - \tilde{y}_M(k+i\mid k)]^2$$

and have clear physical meanings. P indicates how many future steps of the errors between the predicted output and desired output should be minimized, while q_i as weighting coefficients reflect the degree of importance of errors at different future times in optimization.

To make the dynamic optimization meaningful, the optimization horizon P should be selected large enough to include the main characters of the plant dynamics. Since the time delay and nonminimum phase characteristics are not changeable, the plant output is unable to track the desired value during the time of delay or with reverse dynamics, so the weighting coefficients q_i corresponding to the part of time delay or reverse response could be set as zero. This is in accordance with the analysis given in Theorem 3.3, i.e. the selection of P and q_i should meet the necessary condition (3.16) for a stable controller

$$\left(\sum_{i=1}^{P} a_i q_i\right) a_N > 0$$

This indicates that the weighting center of the step response over the optimization horizon should have the same symbol as its steady-state value, which is easy to satisfy if we set all $q_i = 0$ for $a_i \leq 0$. In this way the control is at least correct in direction and thus possible to get a stable control.

The optimization horizon P has a larger influence on stability and dynamic response of the controlled system. Two extreme cases are considered here. The first one is with sufficiently small P, for example $P = 1$. If the control weighting is not considered, the optimization problem will degenerate into the control problem of selecting $\Delta u(k)$ to make $y(k+1) = w(k+1)$, which is the OSOS discussed in Section 3.2.2. With this strategy the system output could strictly track the desired values at the sampling instants and the inverse control with a one-step time delay could be achieved. However, this strategy also have some drawbacks: ripple may exist between sampling instants; robustness against model mismatch and disturbance is very poor; and it cannot be applied to the time-delay system and will lead to unstable control for the nonminimal phase system. Another extreme case is to select P sufficiently large while keeping the control horizon M limited, as described in Theorem 3.2. In this case, a stable controller is easy to obtain but the dynamic control indeed degenerates to static control. The closed-loop dynamics is very close to the original plant response and the rapidity of the response could not be essentially improved. Therefore, during selection of P, a trade-off should be made between stability (robustness) and dynamic rapidity. The conclusions of the above two extreme cases provide the trend of tuning P. Although stability and dynamic rapidity do not generally vary monotonically with P, for the first-order inertia plant with a time delay, which is popular in industrial processes, it will be shown in the next chapter that the stability and the dynamic rapidity vary with P monotonically and it is not difficult to find a suitable P to balance these two kinds of requirements.

According to the above analysis, in general P can be selected such that the optimization horizon covers the main dynamics of the step response. Then let

$$q_i = \begin{cases} 0, & a_i \le 0 \\ 1, & a_i > 0 \end{cases}$$

where the weights corresponding to the part of time delay and reverse dynamics are taken to be zero. Use the initially selected result for simulation, decrease P if the rapidity is not enough and enlarge P if the stability is poor.

The weighting coefficients q_i in the optimization performance index (2.3) of the DMC algorithm can be freely chosen. Different selections for them can lead to different strategies of predictive control algorithms, such as the single value predictive control in [5], etc.

(3) Control horizon *M*

The control horizon M in the optimization performance index (2.3) represents the number of future control increments to be determined. Since the optimization is with respect to the output errors at future P time instants, which are affected by at most P control increments, it is reasonable to assume that $M \le P$.

M is the number of optimization variables, reflecting the degree of freedom of control. Generally, using M free variables to optimize output errors at P future time instants implies distribution of P tasks to M variables. If P is fixed, the smaller the M, the more difficult it is to guarantee that the output closely tracks the desired value at each sampling instant and the worse the control performance will be. For example, $M = 1$ means using only one control increment $\Delta u(k)$ to make the system output tracking the desired value at the time instant $k + 1, \dots, k + P$, which is almost impossible and the resultant control performance may be poor. However, a smaller M may be helpful for closed-loop stability; see Theorems 3.1 and 3.2 where both stability conditions were deduced under the condition $M = 1$. In order to improve the tracking performance, the number of control variables M needs to be increased to enhance the control ability to some extent, but should be balanced with the stability requirement.

As the number of optimization variables, M, is critical to the computational complexity of solving optimization problems. For predictive control algorithms with constraints or nonlinearity, a nonlinear programming problem should be solved online at each step and the size of M would be directly related to the online computational burden. While for predictive control algorithms of linear unconstrained systems with which an analytical control law is available, as mentioned above, M corresponds to the dimension of the matrix $(\boldsymbol{A}^{\mathrm{T}}\boldsymbol{Q}\boldsymbol{A} + \boldsymbol{R})$ to be inverted off-line for calculating the control vector $\boldsymbol{d}^{\mathrm{T}}$, and is thus related to the complexity of off-line computation. Whether it is for online or off-line computation, it is always desirable to decrease M, or even to set $M = 1$, to reduce the computational burden. Therefore, during tuning, M should be selected by considering both dynamic rapidity and stability, as well as the computational burden, and a compromise between these three factors is often needed.

It has been shown by the above analysis and simulation experiences, that increasing (reducing) P has a similar effect to reducing (increasing) M. To simplify the tuning process, one can at first select M according to the complexity of the plant dynamic characters, and then only P needs to be tuned.

(4) Control weighting matrix $\boldsymbol{R}$

In the control weighting matrix $\boldsymbol{R} = \text{diag}(r_1, \ldots, r_M)$, r_j are often taken as the same value, denoted by r. From the performance index (2.3) it is known that the value of r is relative to that of q_i. If q_i has been selected as 0 or 1 as above, r becomes a tunable parameter.

The control weighting matrix $\boldsymbol{R}$ is used to suppress excessive changes of control increments and is added to the performance index (2.3) as a soft constraint. However, this does not mean that increasing r can always improve the stability of the control system. For example, it will be shown in the next chapter that for a second-order system, under certain parameter settings, the controlled system will be stable when r is sufficiently large or small, but unstable for r with middle values in a certain region. This indicates that the effect of r on stability is not monotonic. The stability of the controlled system cannot be improved only simply by increasing r.

From Theorem 3.3 it is known that for any system, as long as (3.16) is satisfied, a stable control can always be achieved by increasing r. However, from the proof of this theorem it is also known that when r is sufficiently large, the closed-loop system, although stable, has a pole very close to $z = 1$, with which the closed-loop response will be sluggish. Certainly this extreme case cannot provide satisfactory control results.

According to the above analysis, during tuning the parameter r, focus should not be put on stability of the controlled system, which can be satisfied through adjusting P and M. The main purpose of introducing r is to prevent the control variables having excessive changes. Therefore, during tuning it is suggested to firstly set $r = 0$. If the system is stable but the control variables change greatly, r could be slightly increased. In practice, using a small r is often enough to make the variation of the control variable go to gentle.

(5) Correction parameter h_i

The elements h_i in the error correction vector $\boldsymbol{h}$ can be selected independently from other design parameters and thus are the only directly tunable design parameters in the DMC algorithm. From the analysis in Section 3.1 it is known that they only work when the predicted output is not consistent with the actual one, i.e. the plant is disturbed or a model mismatch exists, but do not significantly affect the dynamic response of the controlled system.

It was pointed out in Section 3.3 that it is often reasonable to parameterize h_i as simple functions of a single parameter α to avoid irregularity of the parameter selection. Three typical alternatives for h_i are given, i.e. equal-value correction (3.20), attenuation correction (3.22), and increasing correction (3.24), also in the IMC structure, with a corresponding filter in first-order form (3.21), constant form (3.23), and first-order polynomial form (3.25). These filters have different effects on disturbance rejection and system robustness, which have been analyzed in detail in Section 3.3.1. In general, the equal-value correction strategy (3.20) is suggested because it can better satisfy the requirements on disturbance rejection and robustness against a model mismatch.

However, if the system response to disturbance cannot be quickly suppressed while the model mismatch is not so critical, the increasing correction strategy (3.24) is also possible for selection.

The greatest advantage of h_i that is different from other design parameters lies in the fact that they, as directly tunable parameters, can be set online and adjusted during the control process. If an auxiliary unit is introduced into the DMC algorithm, with which the prediction error caused by model mismatch or disturbance can be analyzed, then the disturbance rejection and the robustness can be improved through online switching between different error correction strategies.

The above analysis shows the trends on how the various design parameters affect the dynamic performance, stability, robustness, and disturbance rejection of the predictive control systems. Since the physical meanings of these design parameters are clear and intuitive, through the tuning process combined with simulation, the corresponding parameters could be adjusted according to the simulation results. It should be noted that selection of DMC design parameters has great redundancy, so for general plants it is not difficult to tune them according to the above rules such that the desired requirements could be achieved. For parameter tuning of predictive control for general plants with a known step response, the following steps are suggested:

1) Firstly, the sampling values of the unit step response should be smoothed to reduce the effect of the measurement noise and disturbance. Experience shows that non-smooth model parameters $\{a_i\}$ may bring serious stability problem and degenerate control performance. Thus a smooth model should be constructed to approximately match the measured data to replace the model completely composed of non-smooth measured data. The dimension of the model N is generally taken as $20 \sim 50$. If necessary, appropriately truncate the model according to the strategies discussed above.
2) Select P to make the optimization horizon cover the main dynamics of the step response. This means that the main characters of the plant dynamics, such as increasing monotonically, periodical oscillation, etc., are fully exhibited in the optimization horizon. It is not necessary to set the optimization horizon P so large that within it the dynamic variation of the step response is almost over. The value of P can be selected according to the series 1, 2, 4, 8, ... After P is initially selected, set the weighting coefficients q_i to meet the condition (3.16),

$$q_i = \begin{cases} 0, & a_i \le 0 \\ 1, & a_i > 0 \end{cases}$$

3) Set $r = 0$ initially. For plants with relatively simple dynamics such as an S shape response, select the control horizon M as $1 \sim 2$, and for plants with relatively complex dynamics including oscillation response, increase M properly, taken as $4 \sim 8$.
4) Test the dynamic response of the controlled system through simulation. If it is unstable or with very slow dynamics, adjust P.
5) If a satisfactory result has been achieved after the above tuning procedure but the variation of the control variable is too large, increase r slightly.

6) Determine the type of the correction coefficients h_i according to the control requirements. Select the parameter α through simulation such that both the robustness and disturbance rejection can be properly taken into account.

In the following, two examples of the DMC design using the above steps are given, with which the influence of design parameters to control performance is also established.

Example 3.2 A minimal phase system

$$G(s) = \frac{8611.77}{\left((s+0.55)^2+6^2\right)\left((s+0.25)^2+15.4^2\right)}$$

has the step response

$$a(t) = 1 - 1.1835e^{-0.55t}\sin(6t+1.4973) - 0.18038e^{-0.25t}\sin(15.4t-1.541)$$

(see Figure 3.3). This is a minimal phase plant with weakly damped oscillation. The steady-state value of $a(t)$ is $a_s = 1$, the maximal overshoot $c_{\max} = 0.93$, and the transient time $T_{95\%} = 6.4$ s.

The DMC design starts from the step response shown in Figure 3.3 without knowing the transfer function of the plant $G(s)$. Firstly, after determining the sampling period $T = 0.2$s, the model length can be selected as $N = 40$ because it basically meets the requirement $a_i \approx a_s$ $(i \geq N)$. It is known from Figure 3.3 that the optimization horizon P should be selected at least covering one period of the oscillation of $a(t)$ such that the main dynamics of the plant can be included. Thus $P = 6$ (corresponding to 1.2s) is selected. Furthermore, since the plant is minimal phase without time delay, one can set $\boldsymbol{Q} = \boldsymbol{I}$, $r = 0$. With the above selections, if the control horizon is set as $M = 1$, the closed-loop response is shown in Figure 3.4a.

It is obvious that $M = 1$ cannot lead to a satisfactory response due to the complexity of the plant dynamics. Thus the control horizon is enlarged to $M = 4$ and the dynamic response in this case is shown in Figure 3.4b. The steady-state value of the step response of the controlled system is still $a_s^* = 1$, the maximal overshoot $c_{\max}^* = 0.165$, reduced to

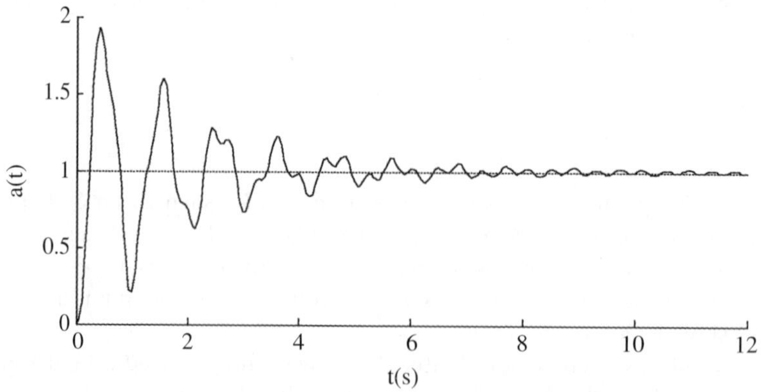

Figure 3.3 The unit step response of the plant in Example 3.2.

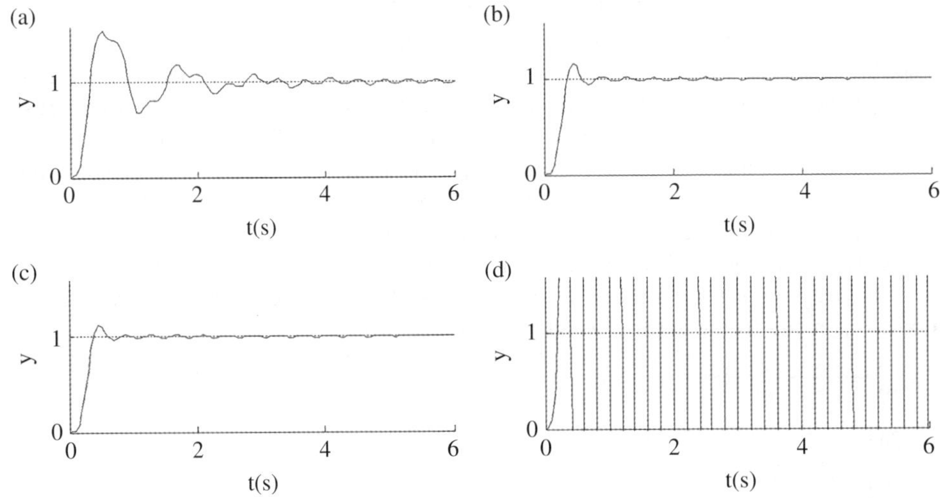

Figure 3.4 Dynamic responses of the control system ($\boldsymbol{Q} = \boldsymbol{I}$, $r = 0$ in all cases). (a) $P = 6$, $M = 1$; (b) $P = 6$, $M = 4$; (c) $P = 20$, $M = 4$; (d) $P = 4$, $M = 4$.

1/5.6 of the original, and the transient time $T^*_{95\%} = 0.72\text{s}$, shortened as 1/9 of the original. The dynamic performance of the system is then significantly improved.

The optimization horizon $P = 6$ covers dynamic variations of one oscillation period in the step response and includes the main dynamic information of the plant. Further enlarging P cannot significantly improve the control result. In Figure 3.4c the response curve for $P = 20$ (corresponding to 4 s) is shown, which is quite close to Figure 3.4b. However, if P is selected to be very large, the closed-loop response will be very similar to the open-loop one shown in Figure 3.3, which is close to the case of static control.

Although the plant is minimal phase without time delay, after sampling and holding its Z transfer function has a zero out of the unit circle $z = -3.946$. Therefore, stable control cannot be achieved with OSOS. The control result for $P = M = 4$ is shown in Figure 3.4d, where $y = 1$ at every sampling instant but the system is unstable. The reason is that the controller, as a one-step delay of the plant inverse, has an unstable pole.

Example 3.3 Consider the wing flutter dynamic model [6]

$$G(s) = \frac{-28705(s-0.3)}{\left((s+0.55)^2+6^2\right)\left((s+0.25)^2+15.4^2\right)}$$

It has an additional zero $s = 0.3$ at the right half plane compared with the transfer function of the plant in Example 3.2, which makes the system exhibit a nonminimal phase character and large range oscillations. Its transient time (see Figure 3.5) is also much larger than that of the plant in Example 3.2.

The unit step response of this nonminimal phase system is given by

$$a(t) = 1 - 23.9\mathrm{e}^{-0.55t}\sin(6t+0.067) + 9.265\mathrm{e}^{-0.25t}\sin(15.4t+0.065)$$

with the steady-state value $a_s = 1$. However, the oscillation amplitude reached $a_{max} = 20$ and $a_{min} = -27$, with the transient time $T_{95\%} = 20$ s.

Starting from the step response shown in Figure 3.5, set the sampling period $T = 0.16$ s. Select the model length $N = 65$, i.e. truncate the step response $a(t)$ at $NT = 10.4$ s because the sampling value of the step response at that time $a_N = a(10.4) \approx 1.0254$ is very close to the steady-state value a_s although the entire dynamic response has not yet reached steady-state. Take $M = 4$ due to the complexity of the plant dynamics and select $P = 8$ (corresponding to 1.28 s) in order to cover the dynamics of at least one oscillation period.

With the above setting, if we let all $q_i = 1$, then the condition (3.16) will not be satisfied. From Figure 3.6 it is shown that the system cannot be stabilized no matter how large an increase is made to the control weighting coefficient r.

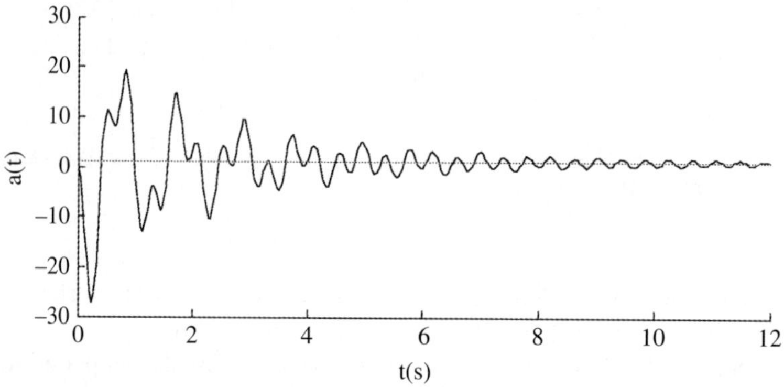

Figure 3.5 The unit step response of the plant in Example 3.3.

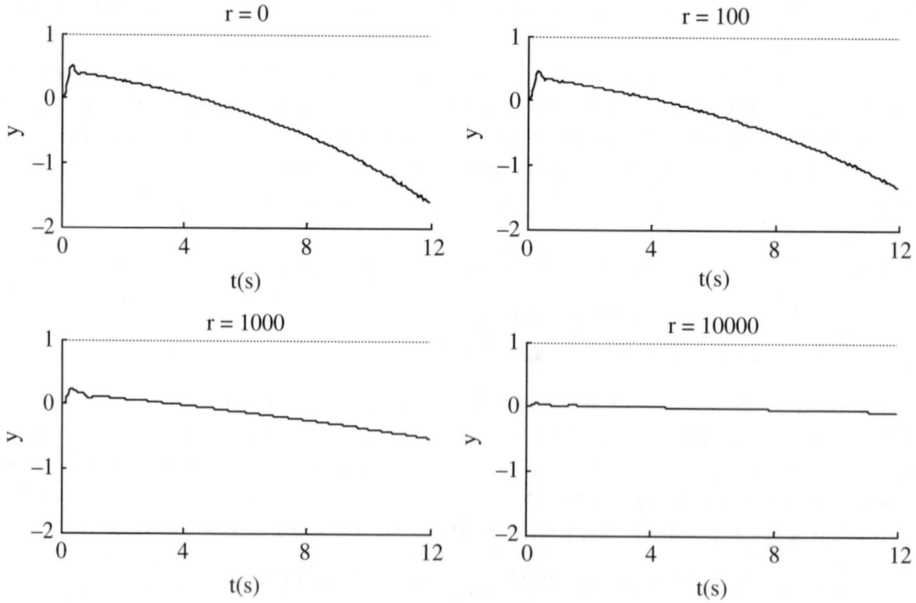

Figure 3.6 The controlled system cannot be stabilized through increasing r.

To make the system stable, as analyzed above, the error weighting coefficients q_i can be selected as follows in order to guarantee that condition (3.16) is satisfied:

$$q_i = \begin{cases} 0, & i = 1 \sim 4 \\ 1, & i = 5 \sim 8 \end{cases}$$

In this way, a stable response can be achieved, even for $r = 0$ (see curve 1 in Figure 3.7). The steady-state value of the step response of the controlled system is still $a_s^* = 1$, but without overshoot in the forward direction. The maximal value in the reverse direction due to the nonminimal phase character is −1 and the transient time is shortened to about 10 s.

As a comparison, an optimal PID controller is well designed for this plant by utilizing the knowledge of the plant transfer function. The optimal PID parameters are obtained as follows. Firstly, assume the transfer function of the PID controller to be

$$G_{\mathrm{PID}}(s) = K\left(1 + T_{\mathrm{D}}s + \frac{1}{T_{\mathrm{I}}s}\right) = KT_{\mathrm{D}}\frac{s^2 + \frac{1}{T_{\mathrm{D}}}s + \frac{1}{T_{\mathrm{I}}T_{\mathrm{D}}}}{s}$$

Let the quadratic polynomial in its numerator cancel $((s + 0.25)^2 + 15.4^2)$ in the plant denominator, which corresponds to plant dynamics with weaker damping and strong oscillations; this leads to $T_{\mathrm{D}} = 2$, $T_{\mathrm{I}} = 0.00210773$. According to these values and the closed-loop transfer function expressed by the parameter K, the optimal $K = 0.000277525$ can be calculated by minimizing the error of the closed-loop response with the unit step input over an infinite horizon. The obtained optimal PID controller can then be implemented as a digit controller through discretization with the sampling period $T = 0.04$s, four times frequently as the DMC controller.

The result of controlling the plant by using this optimal PID controller is also shown in Figure 3.7 (see curve 2). Despite a careful design based on a known transfer function and more frequent control, the performance of PID control is still inferior to DMC control. With the same reverse amplitudes, the DMC curve approaches the steady-state value fast and smoothly, while the PID, as a low-order controller, can only cancel at most one pair of plant poles and the oscillation caused by another pair of the complex poles still remains in its response.

The DMC parameter tuning described above is applicable to general plants. It is mainly based on trend analysis of the influence of the design parameters on system performances, with which the parameters are tuned through experiments or simulations. This

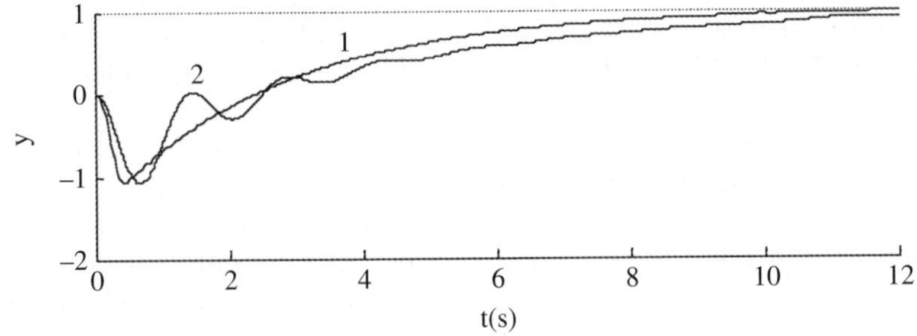

Figure 3.7 Comparison of the control results by DMC (line 1) and PID (line 2).

is essentially a trial-and-error method for system design and in many cases we need to use this method due to a lack of knowledge on the plant minimal model. However, if the minimal transfer function is known or approximately identified by the known step response, it is possible to achieve more precise results in theoretical analysis that could provide analytical guidance for the system parameter setting. This will be discussed in the next chapter.

3.5 Summary

In this chapter, the IMC structure is first introduced and its properties are presented, which are helpful when analyzing the predictive control systems. Taking the DMC algorithm as an example, the concrete expressions of all the components in its IMC structure are derived. In the unconstrained case where the analytic solution can be derived, the coefficients of these transfer components can be expressed by model parameters, control parameters, and correction parameters.

In the IMC structure of the DMC algorithm, the controller has a critical influence on system stability and control performance. Especially in the nominal case without model mismatch and disturbance, the closed-loop stability and dynamic characteristics of the DMC system only depend on the controller. Through analysis of the characteristic polynomial of the controller, some trend stability theorems of the controller are given, which reflect how the step response, the horizons, and the weighting matrices affect the stability of the controller. However, these sufficient conditions are greatly conservative. Based on the transfer function of the controller, it is also proved that the controller with OSOS is the one-step delay of the model inverse. After properly selecting the optimization parameters, the controller for plant with time delay is the same as that for a delay-free system. In this case, if no model mismatch exists, the system output of the controlled system is equivalent to putting the same delay time to the system output obtained by controlling the delay-free system.

For the filter of the DMC algorithm in the IMC structure, we mainly considered its influence on closed-loop stability and disturbance rejection when a model mismatch or disturbance exists. Three typical parameterized feedback correction strategies in the DMC algorithm and the transfer functions of corresponding filters in the IMC structure are given. Different capabilities of these strategies to handle model mismatch and external disturbances are analyzed through a specific case study. Two conclusions were drawn. Firstly, the choice of filter should balance robustness and disturbance rejection. Secondly, it is an appropriate choice to make such a trade-off when equal-value correction strategy in the DMC algorithm and the first-order filter in the corresponding IMC structure are adopted.

According to the discussion on the controller and the filter, the trends of the influence of DMC design parameters on system performances are analyzed and tuning steps for design parameters are suggested. These steps are suitable to general plants and can be adopted as reference to adjust the predictive control design parameters combined with experiments or simulations in the case when a minimal parametric model is unavailable. It is also pointed out that if a minimal parametric model can be obtained or estimated, more theoretical results with less conservativeness could be expected.

References

1 Garcia, C.E. and Morari, M. (1982). Internal model control, 1. A unifying review and some new results. *Industrial and Engineering Chemistry Process Design and Development* 21 (2): 308–323.

2 Ackermann, J. (1985). *Sampled-Data Control Systems: Analysis and Synthesis, Robust System Design*. Berlin: Springer-Verlag.

3 Clarke, D.W., Mohtadi, C., and Tuffs, P.S. (1987). Generalized predictive control. Part 1. The basic algorithm. *Automatica* 23 (2): 137–148.

4 Clarke, D.W. and Mohtadi, C. (1989). Properties of generalized predictive control. *Automatica* 25 (6): 859–873.

5 Yuan, P., Zheng, H.T., and Zuo, X. (1993). State feedback predictive control. *Acta Automatica Sinica* 19 (5): 569–577 (in Chinese). Also in *Chinese Journal of Automation*, 1994, 6 (3): 113–121, Allerton Press Inc.

6 Reid, J.G., Mehra, R.K., and Kirkwood, W. Robustness properties of output predictive deadbeat control: SISO case, *Proceedings of the 1979 IEEE Conference on Decision and Control*, Fort Lauderdale, FL, USA (12–14 December 1979), 307–314.

4

Quantitative Analysis of SISO Unconstrained Predictive Control Systems

In Chapter 3, some trend conclusions on the stability of the DMC controller and on closed-loop robust stability affected by the DMC filter were given by transforming the DMC algorithm into the IMC structure and analyzing the corresponding characteristic polynomials. Here "trend" means that in the theorem conditions some design parameters are set as extreme values, such as $M = 1$, $P \to \infty$, $r \to \infty$, α sufficiently small, etc. (see Theorems 3.1 to 3.3 and 3.6, respectively), which is sometimes also called the "limiting case" or "marginal stability" in the literature and is caused by the lack of an explicit relationship between the design parameters and the characteristic polynomial.

Indeed, since the appearance of predictive control algorithms such as DMC and GPC, in order to get more guidance for applications, many efforts have been put into exploring the relationship between the design parameters and the control performance. Except for the trend conclusions mentioned above, more explicit and analytical relationships between design parameters and system performance were required. During the 1980s and a little later, some interesting quantitative results for analyzing GPC and DMC algorithms were achieved. Since these studies focused on quantitative analysis for existing predictive control algorithms, we call the theoretical study for this purpose quantitative analysis theory of predictive control.

In this chapter, we will introduce some important results of quantitative analysis theory of predictive control. Firstly, theoretical analysis of the GPC algorithm given by Clarke and Mohtadi [1] is briefly introduced, which is based on the Kleinman controller in the time domain. The main part of this chapter is devoted to the Z domain analysis of predictive control systems. With the help of the IMC structure, the minimal form of the DMC controller is derived and proved to be identical to the GPC controller in the IMC structure. Through exploration of the intrinsic relationship between the control parameters and the coefficients of the characteristic polynomial, the coefficient mapping of open-loop and closed-loop characteristic polynomials is established, based on which the quantitative relationship between the design parameters and the closed-loop performance can be analyzed. The derived results hold for both DMC and the GPC algorithm and cover the result derived by theoretical analysis in the time domain.

Quantitative analysis needs analytical expressions of the controlled system, which is impossible for constrained predictive control algorithms because the control law should be obtained by solving an optimization problem and could not be expressed in a closed form in general. Even in the unconstrained case, for multivariable predictive control

Predictive Control: Fundamentals and Developments, First Edition. Yugeng Xi and Dewei Li.

systems, the analytical expression of the control law in state space or the controller in the IMC structure is difficult to derive and would be very complex. Therefore, the quantitative analysis in this chapter is mainly for SISO unconstrained predictive control systems. Furthermore, the study will focus on investigating the quantitative influence of the design parameters on stability and dynamic behavior of the closed-loop system in the nominal case, i.e. without model mismatch and disturbance. Therefore the model and the plant will be uniformly represented by the same state equation or transfer function.

4.1 Time Domain Analysis Based on the Kleinman Controller

Some stability conditions were given for the GPC system in [1] in 1989. In order to utilize the mature theoretical results in optimal control theory, Clarke and Mohtadi firstly transformed the GPC control law into an LQ (Linear Quadratic) control law in state space. Their main contribution is that they deduced the stability condition of the GPC closed-loop system through investigating under which condition the GPC control law can be equivalently transformed into the Kleinman controller, a specific class of controllers with guaranteed stability in the LQ control theory. This method is novel and the result directly describes a quantitative relationship between the design parameters and the closed-loop stability. In this section, referring to [1, 2], we briefly introduce the critical issues of this work with some expansions.

(1) The LQ control law of GPC

As described in Section 2.2, in the GPC algorithm, when the stochastic disturbance $\xi(t)$ is removed, the system model (2.19) is equivalent to the transfer relationship (2.20b) between u and y. The transfer function between Δu and y can accordingly be obtained as follows:

$$G_{\Delta u}(z) = \frac{z^{-1}B(z^{-1})}{\Delta A(z^{-1})} = \frac{b_0 z^{-1} + \cdots + b_{n_b} q^{-(n_b+1)}}{1 + (a_1 - 1)q^{-1} + \cdots - a_n q^{-(n+1)}}$$

where $a_n \neq 0, b_{n_b} \neq 0$. Note that this transfer function is of the order of max $(n+1, n_b+1)$. To simplify the symbols, assume that $n \geq n_b$. Also assume that the numerator polynomial and the denominator polynomial are coprime. Then the above system can be transformed into a realization in the state space,

$$\begin{aligned} \boldsymbol{x}(k+1) &= \boldsymbol{A}\boldsymbol{x}(k) + \boldsymbol{b}\Delta u(k) \\ y(k) &= \boldsymbol{c}^{\mathrm{T}}\boldsymbol{x}(k) \end{aligned} \tag{4.1}$$

where $\boldsymbol{x}(k) \in \mathrm{R}^{n+1}$, $\boldsymbol{A}, \boldsymbol{b}, \boldsymbol{c}^{\mathrm{T}}$ are related to the coefficients of the transfer function but are not unique. Note that $(\boldsymbol{c}^{\mathrm{T}} \;\; \boldsymbol{A} \;\; \boldsymbol{b})$ is both controllable and observable.

With respect to the performance index of GPC (2.24), consider the deterministic case and not lose generality; let $w = 0$. Denote

$$\boldsymbol{Q}_i = \begin{cases} \boldsymbol{c}\boldsymbol{c}^{\mathrm{T}}, & N_1 \leq i \leq N_2 \\ \boldsymbol{0}, & \text{other } i \end{cases}, \qquad \lambda_i = \begin{cases} \lambda, & 0 \leq i \leq NU-1 \\ \infty, & \text{other } i \end{cases} \tag{4.2}$$

Then the GPC performance index (2.24) can be rewritten as

$$J = \|\boldsymbol{x}(k+N_2\,|\,k)\|^2_{\boldsymbol{c}\boldsymbol{c}^{\mathrm{T}}} + \sum_{i=0}^{N_2-1}\left[\|\boldsymbol{x}(k+i\,|\,k)\|^2_{\boldsymbol{Q}_i} + \lambda_i \Delta u(k+i\,|\,k)^2\right] \tag{4.3}$$

This is a standard LQ control problem. According to the standard solution of LQ control, the control law can be obtained as

$$\Delta u(k) = -\left(\lambda + \boldsymbol{b}^{\mathrm{T}}\boldsymbol{P}(k+1)\boldsymbol{b}\right)^{-1}\boldsymbol{b}^{\mathrm{T}}\boldsymbol{P}(k+1)\boldsymbol{A}\boldsymbol{x}(k) \tag{4.4}$$

where $\boldsymbol{P}(k+1)$ can be solved by the following Riccati iterative formula

$$\begin{aligned}
&\boldsymbol{P}(k+N_2) = \boldsymbol{c}\boldsymbol{c}^{\mathrm{T}}\\
&\boldsymbol{P}(k+i) = \boldsymbol{Q}_i + \boldsymbol{A}^{\mathrm{T}}\boldsymbol{P}(k+i+1)\boldsymbol{A}\\
&\qquad\qquad -\boldsymbol{A}^{\mathrm{T}}\boldsymbol{P}(k+i+1)\boldsymbol{b}\left(\lambda_i + \boldsymbol{b}^{\mathrm{T}}\boldsymbol{P}(k+i+1)\boldsymbol{b}\right)^{-1}\boldsymbol{b}^{\mathrm{T}}\boldsymbol{P}(k+i+1)\boldsymbol{A},\\
&\qquad\qquad i = N_2-1,\ldots,1
\end{aligned} \tag{4.5}$$

Here (4.4) is the control law when GPC is described as an LQ problem, called the LQ control law of GPC. This transformation is significant because there already exist quite good results on the control law and stability of the LQ control problem, which can be utilized to investigate the stability of the GPC control law. For the sake of simplification, in the following we use $\boldsymbol{P}_i$ to represent $\boldsymbol{P}(k+i)$.

(2) The Kleinman controller with stability

Early in 1974, Kleinman proposed a specific controller as he investigated the LQ control problem. For an n-dimensional state space model with m inputs

$$\boldsymbol{x}(k+1) = \boldsymbol{\Phi}\boldsymbol{x}(k) + \boldsymbol{B}\boldsymbol{u}(k) \tag{4.6}$$

he gave the following result.

Lemma 4.1 [3] If system (4.6) is completely controllable and $\boldsymbol{\Phi}$ is nonsingular, then for any $N \geq n$, the control law

$$\boldsymbol{u}(k) = -\boldsymbol{R}^{-1}\boldsymbol{B}^{\mathrm{T}}\left(\boldsymbol{\Phi}^{\mathrm{T}}\right)^N\left(\sum_{i=0}^{N}\boldsymbol{\Phi}^i\boldsymbol{B}\boldsymbol{R}^{-1}\boldsymbol{B}^{\mathrm{T}}\left(\boldsymbol{\Phi}^{\mathrm{T}}\right)^i\right)^{-1}\boldsymbol{\Phi}^{N+1}\boldsymbol{x}(k) \tag{4.7}$$

makes the system (4.6) closed-loop stable, where $\boldsymbol{R} > \boldsymbol{0}$.

The controller (4.7) in Lemma 4.1 is called the Kleinman controller. Note that the LQ control law of GPC (4.4) also has the form of state feedback. In order to compare it with the Kleinman controller (4.7), it is needed to firstly give the explicit expression of $\boldsymbol{P}(k+1)$, i.e. $\boldsymbol{P}_1$ in (4.4).

(3) The explicit expression of $\boldsymbol{P}_1$ in the control law (4.4)

From (4.2) it is known that $\boldsymbol{Q}_i$ and λ_i in the Riccati iterative formula are not constant and depend on the optimization horizon and the control horizon. If we assume that $N_1 = NU$, according to the values of $\boldsymbol{Q}_i$ and λ_i in different ranges, the iterative formula (4.5) can be written as

$$P_i = \begin{cases} cc^{\mathrm{T}} + A^{\mathrm{T}} P_{i+1} A, & i = N_2 - 1, \cdots, N_1 \\ A^{\mathrm{T}} P_{i+1} A - A^{\mathrm{T}} P_{i+1} b \left(\lambda_i + b^{\mathrm{T}} P_{i+1} b\right)^{-1} b^{\mathrm{T}} P_{i+1} A, & i = N_1 - 1, \cdots, 1 \end{cases} \tag{4.8}$$

Based on (4.8), start from $P_{N_2} = cc^{\mathrm{T}}$ and iterate from $i = N_2 - 1$ to $i = N_1$; it follows that

$$\begin{aligned} P_{N_2-1} &= cc^{\mathrm{T}} + A^{\mathrm{T}} P_{N_2} A = cc^{\mathrm{T}} + A^{\mathrm{T}} cc^{\mathrm{T}} A \\ P_{N_2-2} &= cc^{\mathrm{T}} + A^{\mathrm{T}} P_{N_2-1} A = cc^{\mathrm{T}} + A^{\mathrm{T}} cc^{\mathrm{T}} A + \left(A^{\mathrm{T}}\right)^2 cc^{\mathrm{T}} A^2 \\ &\vdots \\ P_{N_1} &= cc^{\mathrm{T}} + A^{\mathrm{T}} cc^{\mathrm{T}} A + \left(A^{\mathrm{T}}\right)^2 cc^{\mathrm{T}} A^2 + \cdots + \left(A^{\mathrm{T}}\right)^{N_2-N_1} cc^{\mathrm{T}} A^{N_2-N_1} \end{aligned}$$

P_{N_1} can be also represented by

$$P_{N_1} = \begin{bmatrix} c^{\mathrm{T}} \\ c^{\mathrm{T}} A \\ \vdots \\ c^{\mathrm{T}} A^{N_2-N_1} \end{bmatrix}^{\mathrm{T}} \begin{bmatrix} c^{\mathrm{T}} \\ c^{\mathrm{T}} A \\ \vdots \\ c^{\mathrm{T}} A^{N_2-N_1} \end{bmatrix}$$

Since (c^{T} A) is observable, if $N_2 - N_1 + 1 \geq n + 1$, then the rank of P_{N_1} is $n + 1$, i.e. P_{N_1} is invertible.

Then start from P_{N_1} and continue to iterate from $i = N_1 - 1$ to $i = 1$. According to the iterative formula (4.8), at $i = N_1 - 1$ we have

$$P_{N_1-1} = A^{\mathrm{T}} \left(P_{N_1} - P_{N_1} b \left(\lambda + b^{\mathrm{T}} P_{N_1} b\right)^{-1} b^{\mathrm{T}} P_{N_1} \right) A$$

Since P_{N_1} is invertible, using the matrix inversion formula

$$(W + XYZ)^{-1} = W^{-1} - W^{-1} X \left(Y^{-1} + Z W^{-1} X\right) Z W^{-1}$$

the above formula can be rewritten as

$$P_{N_1-1} = A^{\mathrm{T}} \left(P_{N_1}^{-1} + \frac{bb^{\mathrm{T}}}{\lambda} \right)^{-1} A$$

If A is invertible, from above it is known that P_{N_1-1} is also invertible and

$$A P_{N_1-1}^{-1} A^{\mathrm{T}} = P_{N_1}^{-1} + \frac{bb^{\mathrm{T}}}{\lambda}$$

Then the following can be similarly deduced:

$$A P_i^{-1} A^{\mathrm{T}} = P_{i+1}^{-1} + \frac{bb^{\mathrm{T}}}{\lambda}, \qquad i = N_1 - 2, \ldots, 1$$

and furthermore

$$A^{N_1-1} P_1^{-1} \left(A^{\mathrm{T}}\right)^{N_1-1} = P_{N_1}^{-1} + \sum_{i=0}^{N_1-2} A^i \frac{bb^{\mathrm{T}}}{\lambda} \left(A^{\mathrm{T}}\right)^i$$

When $\lambda \to 0^+$, this can be approximately written by

$$A^{N_1-1}P_1^{-1}(A^{\mathrm{T}})^{N_1-1} = \sum_{i=0}^{N_1-2} A^i \frac{bb^{\mathrm{T}}}{\lambda}(A^{\mathrm{T}})^i \tag{4.9}$$

where (4.9) actually gives the expression of P_1 (i.e. $P(k+1)$).

(4) The control law (4.4) in the form of the Kleinman controller (4.7)

The control law (4.4) can be rewritten as

$$\Delta u(k) = -k^{\mathrm{T}}x(k) \tag{4.10}$$

where

$$k^{\mathrm{T}} = (\lambda + b^{\mathrm{T}}P_1 b)^{-1} b^{\mathrm{T}}P_1 A$$

Using the matrix inversion formula it follows that

$$k^{\mathrm{T}} = (\lambda + b^{\mathrm{T}}P_1 b)^{-1} b^{\mathrm{T}}P_1 A = \left(\frac{1}{\lambda} - \frac{1}{\lambda}b^{\mathrm{T}}\left(P_1^{-1} + \frac{bb^{\mathrm{T}}}{\lambda}\right)^{-1} b\frac{1}{\lambda}\right) b^{\mathrm{T}}P_1 A$$

$$= \frac{1}{\lambda}b^{\mathrm{T}}P_1 A - \frac{1}{\lambda}b^{\mathrm{T}}\left(P_1^{-1} + \frac{bb^{\mathrm{T}}}{\lambda}\right)^{-1}\left(\frac{bb^{\mathrm{T}}}{\lambda} + P_1^{-1} - P_1^{-1}\right)P_1 A$$

$$= \frac{1}{\lambda}b^{\mathrm{T}}\left(P_1^{-1} + \frac{bb^{\mathrm{T}}}{\lambda}\right)^{-1} A$$

$$= \frac{1}{\lambda}b^{\mathrm{T}}(A^{\mathrm{T}})^{N_1-1}\left(A^{N_1-1}\left(P_1^{-1} + \frac{bb^{\mathrm{T}}}{\lambda}\right)(A^{\mathrm{T}})^{N_1-1}\right)^{-1} A^{N_1-1}A$$

Using (4.9), this can be written as

$$k^{\mathrm{T}} = \frac{1}{\lambda}b^{\mathrm{T}}(A^{\mathrm{T}})^{N_1-1}\left(\sum_{i=0}^{N_1-1} A^i \frac{bb^{\mathrm{T}}}{\lambda}(A^{\mathrm{T}})^i\right)^{-1} A^{N_1} \tag{4.11}$$

Obviously it has the same form as the Kleinman controller (4.7).

(5) Deadbeat and stability conditions for GPC

In the last two steps, the assumed condition $N_1 = NU$ leads to the iterative formula (4.8), and the assumed condition $N_2 - N_1 + 1 \geq n + 1$ guarantees that P_{N_1} is invertible and deduces the expression of (4.9) when $\lambda \to 0^+$. With these conditions, the GPC control law (4.4) can be equivalent to the Kleinman controller with feedback gain (4.11). Since the system (4.1) is $n + 1$ dimensional, from Lemma 4.1 it is known that the system (4.1) is closed-loop stable with the GPC control law if $N_1 - 1 \geq n + 1$.

In addition to making the GPC control law (4.4) equivalent to the Kleinman controller, Clarke and Mohtadi also discussed the following special case.

Lemma 4.2 [1] If system (4.1) is controllable and observable, and the GPC design parameters are selected as

$$N_1 = NU = n + 1, \quad N_2 \geq 2n + 1, \quad \lambda = 0 \tag{4.12}$$

then the system (4.1) controlled by GPC is closed-loop stable, and exhibits deadbeat property, i.e. all the states tend to zero in limited steps, and all the eigenvalues of the closed-loop system are located at origin.

The proof of this lemma can be referred to [1]. Note that during above discussion, all the results are derived in terms of the system order $n + 1$ because we assume $n \geq n_b$ for simplification of symbols. It is easy to find that if $n \leq n_b$, the system (4.1) will be of the order $n_b + 1$, and all the results can be similarly derived with n_b instead of n. According to the above discussion including Lemma 4.2, the following theorem can be obtained.

Theorem 4.1 [1] If system (4.1) is controllable and observable and $\boldsymbol{A}$ is invertible, then the GPC closed-loop system is stable if the following parameters are selected:

$$N_1 = NU \geq \max(n, n_b) + 1, \quad N_2 - N_1 \geq \max(n, n_b), \quad \lambda = \varepsilon \to 0^+ \tag{4.13}$$

Deadbeat control in particular can be achieved for the following parameter setting

$$N_1 = NU = \max(n, n_b) + 1, \quad N_2 \geq 2\max(n, n_b) + 1, \quad \lambda = 0 \tag{4.14}$$

Remark 4.1 In [1], the condition "system controllable and observable" for (4.13) in Theorem 4.1 is also relaxed to "system stabilizable and detectable," which means that even when zero-pole cancellation appears due to the common factors in the numerator and the denominator polynomials of the transfer function, the stability conclusion of Theorem 4.1 still holds as long as the cancellation is stable.

Theorem 4.1 quantitatively provides the selection conditions of GPC design parameters with which the closed-loop stability or deadbeat property can be guaranteed. These conditions are only related to the order of the model and are independent from specific model parameters, and are therefore easy to use. Among those, λ in the stability condition (4.13) is very small, but cannot be equal to zero because the solvability of the GPC control law or the invertibility of the matrix $(\boldsymbol{G}^{\mathrm{T}}\boldsymbol{G} + \lambda\boldsymbol{I})$ in (2.29) should be guaranteed. However, $\lambda = 0$ appears in the deadbeat condition (4.14) because the invertibility of the matrix $(\boldsymbol{G}^{\mathrm{T}}\boldsymbol{G})$; i.e. the solvability of the GPC control law, can be guaranteed by other conditions in (4.14).

(6) Extension of the results

Since the condition of Theorem 4.1 is very demanding (such as asking the matrix $\boldsymbol{A}$ to be invertible) and the range of the design parameters involved is quite limited, there were some subsequent works along this line in the 1990s. For various possible cases of denominator and numerator polynomial orders n and n_b in the GPC model (2.19), Ding and Xi [2] extended the Kleinman controller for possible singular $\boldsymbol{A}$ and gave the closed-loop stability or deadbeat conditions for it. They also discussed the equivalence of the LQ control law of GPC and the extended Kleinman controller for sufficiently small λ and obtained the following conclusions.

Theorem 4.2 [2] If the following condition is satisfied, there exists a sufficiently small λ_0 such that GPC is closed-loop stable for $0 < \lambda < \lambda_0$:

$$N_1 \geq n_b + 1, \quad NU \geq n + 1, \quad N_2 - NU \geq n_b, \quad N_2 - N_1 \geq n \tag{4.15}$$

This condition can also be decomposed as

$$n \geq n_b,\ \ N_1 \geq NU \geq n+1,\ \ N_2 - N_1 \geq n;$$
$$n \leq n_b,\ \ NU \geq N_1 \geq n_b+1,\ \ N_2 - NU \geq n_b;$$
$$n \leq n_b,\ \ N_1 \geq NU \geq n+1,\ \ N_1 \geq n_b+1,\ \ N_2 - N_1 \geq n,\ \ N_2 - NU \geq n_b;$$
$$n \geq n_b,\ \ NU \geq N_1 \geq n_b+1,\ \ NU \geq n+1,\ \ N_2 - N_1 \geq n,\ \ N_2 - NU \geq n_b$$

Theorem 4.3[2] The GPC controller is deadbeat if one of the following conditions is satisfied:

$$\begin{aligned} &1)\ \lambda = 0,\ \ NU = n+1,\ \ N_1 \geq n_b+1,\ \ N_2 - N_1 \geq n \\ &2)\ \lambda = 0,\ \ NU \geq n+1,\ \ N_1 = n_b+1,\ \ N_2 - NU \geq n_b \end{aligned} \tag{4.16}$$

It should be pointed out that the stability condition and the deadbeat condition in Theorem 4.1 can be covered by the special case $N_1 = NU$ in (4.15) and in (4.16), respectively. Therefore Theorems 4.2 and 4.3 have a wider range of design parameters.

The quantitative analysis on GPC stability and the deadbeat property described above is based on the Kleinman controller in the time domain and utilizes the mature research results of LQ control theory. Although the system model needs to be transformed into the state space form, the derivation and conclusion does not depend on specific model parameters, only on the order of the system, which makes the conclusion concise and universal. However, this method cannot provide detailed information on dynamic behavior of the controlled system if it is stable. In the following we will quantitatively analyze the performances of the predictive control systems based on the coefficient mapping in the Z domain, which can provide more information on the closed-loop performances than that obtained in the time domain.

4.2 Coefficient Mapping of Predictive Control Systems

In this section, the coefficient mapping of the characteristic polynomials between the plant and the controlled predictive control system will be deduced based on the IMC structure in the Z domain, which is the basis of quantitatively analyzing the predictive control systems in the next section. Under the IMC framework, the expression of the GPC controller is firstly given, the minimal form of the DMC controller is revealed, and the uniformity of both controllers is proved. Then the coefficient mapping of the characteristic polynomials between the plant and the controlled system is established, based on which stability and dynamic performance can be investigated uniformly for both DMC and GPC control systems.

4.2.1 Controller of GPC in the IMC Structure

Similar to deducing the IMC structure of the DMC algorithm in Section 3.2, the expressions of all the components of the GPC algorithm in the IMC structure can also be deduced (see [4]). Since the focus of this chapter is on analyzing the closed-loop performance without model mismatch and disturbance, according to (3.2), only the expression of the GPC controller in the IMC structure needs to be deduced. In order to facilitate a comparison with the DMC algorithm, the model online identification and adaption is not considered here. Assume $n = n_b + 1$ in the model (2.19), and rewrite the coefficients of $B(q^{-1})$ as

$$B(q^{-1}) = b_1 + b_2 q^{-1} + \cdots + b_n q^{-(n-1)}$$

Then the plant transfer function can be written as

$$G(z) = \frac{z^{-1}B(z)}{A(z)} = \frac{b_1 z^{-1} + \cdots + b_n z^{-n}}{1 + a_1 z^{-1} + \cdots + a_n z^{-n}} \tag{4.17}$$

The GPC control law (2.30) is given by

$$\Delta u(k) = \boldsymbol{d}^{\mathrm{T}}(\boldsymbol{w} - \boldsymbol{f}) \tag{4.18}$$

where

$$\boldsymbol{d}^{\mathrm{T}} = (1 \ \ 0 \ \ \cdots \ \ 0)\left(\boldsymbol{G}^{\mathrm{T}}\boldsymbol{G} + \lambda \boldsymbol{I}\right)^{-1}\boldsymbol{G}^{\mathrm{T}} \triangleq [d_{N_1} \ \cdots \ d_{N_2}]$$

$$\boldsymbol{G} = \begin{bmatrix} g_{N_1} & \cdots & g_{N_1 - NU + 1} \\ \vdots & & \vdots \\ g_{N_2} & \cdots & g_{N_2 - NU + 1} \end{bmatrix}$$

The reference trajectory $\boldsymbol{w}$ in (4.18) is given by (2.32) as

$$\boldsymbol{w} = \boldsymbol{L}y(k) + \boldsymbol{M}c \tag{4.19}$$

where c is the reference input

$$\boldsymbol{w} = \begin{bmatrix} w(k+N_1) \\ \vdots \\ w(k+N_2) \end{bmatrix}, \quad \boldsymbol{L} = \begin{bmatrix} \alpha^{N_1} \\ \vdots \\ \alpha^{N_2} \end{bmatrix}, \quad \boldsymbol{M} = \begin{bmatrix} 1-\alpha^{N_1} \\ \vdots \\ 1-\alpha^{N_2} \end{bmatrix}$$

and according to (2.27) the vector $\boldsymbol{f}$ can be written as

$$\boldsymbol{f} = \begin{bmatrix} f_1(k) \\ \vdots \\ f_P(k) \end{bmatrix} = \boldsymbol{H}\Delta u(k) + \boldsymbol{F}y(k) \tag{4.20}$$

where

$$\boldsymbol{H} = \begin{bmatrix} z^{N_1 - 1}\left(G_{N_1} - g_1 - \cdots - g_{N_1} z^{-(N_1 - 1)}\right) \\ \vdots \\ z^{N_2 - 1}\left(G_{N_2} - g_1 - \cdots - g_{N_2} z^{-(N_2 - 1)}\right) \end{bmatrix}, \quad \boldsymbol{F} = \begin{bmatrix} F_{N_1} \\ \vdots \\ F_{N_2} \end{bmatrix}$$

Substitute (4.19) and (4.20) into (4.18), and note that $y(z)/u(z) = z^{-1}B(z)/A(z)$; the Z transfer function of the GPC controller (the transfer function from reference input c to control action u) can then be given by

$$G_C(z) = \frac{\boldsymbol{d}^{\mathrm{T}}\boldsymbol{M}A}{\left(1+\boldsymbol{d}^{\mathrm{T}}\boldsymbol{H}\right)A\Delta + z^{-1}B\boldsymbol{d}^{\mathrm{T}}(\boldsymbol{F}-\boldsymbol{L})} \tag{4.21}$$

It is clear that the vectors $\boldsymbol{F}$ and $\boldsymbol{H}$ appearing in the transfer function of the GPC controller $G_C(z)$ are related to E_j and F_j, and need to be calculated recursively. In order to analyze the system performance, in the following E_j and F_j will be eliminated such that the resultant expression of the controller is only related to the plant parameter A, $\{g_i\}$, and the control parameter d_i. Firstly, the following Lemma is given.

Lemma 4.3 The coefficients of the unit step response of system (4.17) are given by $\{g_i\}$; then the following relationships hold:

$$\begin{aligned}
&g_1 = b_1\\
&(g_2-g_1)+g_1a_1 = b_2\\
&\quad\vdots\\
&(g_n-g_{n-1})+(g_{n-1}-g_{n-2})a_1+\cdots+(g_2-g_1)a_{n-2}+g_1a_{n-1} = b_n\\
&(g_{n+1}-g_n)+(g_n-g_{n-1})a_1+\cdots+(g_2-g_1)a_{n-1}+g_1a_n = 0\\
&(g_i-g_{i-1})+(g_{i-1}-g_{i-2})a_1+\cdots+(g_{i-n}-g_{i-n-1})a_n = 0,\quad i\ge n+2
\end{aligned} \tag{4.22}$$

This lemma can be directly deduced from

$$G(z) = g_1z^{-1}+(g_2-g_1)z^{-2}+(g_3-g_2)z^{-3}+\cdots$$

Next, consider the expression of the GPC controller. Denote the denominator part of $G_C(z)$ as $A_C(z)$; then it follows that

$$\begin{aligned}
A_C(z) &= \left(1+\boldsymbol{d}^{\mathrm{T}}\boldsymbol{H}\right)A\Delta + z^{-1}B\boldsymbol{d}^{\mathrm{T}}(\boldsymbol{F}-\boldsymbol{L})\\
&= A\Delta + \boldsymbol{d}^{\mathrm{T}}\begin{bmatrix} z^{N_1-1}G_{N_1}A\Delta\\ \vdots\\ z^{N_2-1}G_{N_2}A\Delta\end{bmatrix} - A\Delta\boldsymbol{d}^{\mathrm{T}}\begin{bmatrix} g_{N_1}+\cdots+g_1z^{N_1-1}\\ \vdots\\ g_{N_2}+\cdots+g_1z^{N_2-1}\end{bmatrix} + z^{-1}B\boldsymbol{d}^{\mathrm{T}}\begin{bmatrix}F_{N_1}\\ \vdots\\ F_{N_2}\end{bmatrix} - z^{-1}B\boldsymbol{d}^{\mathrm{T}}\boldsymbol{L}
\end{aligned}$$

According to the expression of $\boldsymbol{d}^{\mathrm{T}}$ and $\boldsymbol{L}$, denote $\boldsymbol{d}^{\mathrm{T}}\boldsymbol{L} = d_{N_1}\alpha^{N_1}+\cdots+d_{N_2}\alpha^{N_2} \triangleq -d_0$. At the same time, from (2.21) it follows that $G_iA\Delta = B(1-z^{-i}F_i)$; substituting this into the above gives

$$A_C(z) = A\Delta + B\boldsymbol{d}^{\mathrm{T}}\begin{bmatrix} z^{N_1-1}\\ \vdots\\ z^{N_2-1}\end{bmatrix} - A\Delta\boldsymbol{d}^{\mathrm{T}}\begin{bmatrix} g_{N_1}+\cdots+g_1z^{N_1-1}\\ \vdots\\ g_{N_2}+\cdots+g_1z^{N_2-1}\end{bmatrix} + z^{-1}Bd_0$$

Note that

$$\begin{aligned} z^{-1}B &= A(g_1 z^{-1} + (g_2 - g_1)z^{-2} + (g_3 - g_2)z^{-3} + \cdots) \\ &= A\Delta(g_1 z^{-1} + g_2 z^{-2} + g_3 z^{-3} + \cdots) \end{aligned}$$

It follows that

$$\begin{aligned} A_C(z) &= A\Delta + A\Delta \boldsymbol{d}^{\mathrm{T}} \begin{bmatrix} g_{N_1+1}z^{-1} + g_{N_1+2}z^{-2} + \cdots \\ \vdots \\ g_{N_2+1}z^{-1} + g_{N_2+2}z^{-2} + \cdots \end{bmatrix} + A\Delta d_0 (g_1 z^{-1} + g_2 z^{-2} + \cdots) \\ &= A\Delta(1 + c_2 z^{-1} + c_3 z^{-2} + \cdots) \end{aligned}$$

where

$$c_i = \sum_{j=N_1}^{N_2} d_j g_{i+j-1} + d_0 g_{i-1}, \qquad i = 2, 3, \ldots$$

Furthermore,

$$\begin{aligned} A_C(z) &= \left(1 + (a_1 - 1)z^{-1} + (a_2 - a_1)z^{-2} + \cdots + (a_n - a_{n-1})z^{-n} - a_n z^{-(n+1)}\right) \cdot \\ &\quad \left(1 + c_2 z^{-1} + c_3 z^{-2} + \cdots\right) \\ &= 1 + a_1^* z^{-1} + a_2^* z^{-2} + \cdots \end{aligned}$$

where

$$\begin{aligned} a_1^* &= (a_1 - 1) + c_2 \\ a_2^* &= (a_2 - a_1) + (a_1 - 1)c_2 + c_3 \\ &\vdots \\ a_n^* &= (a_n - a_{n-1}) + (a_{n-1} - a_{n-2})c_2 + \cdots + c_{n+1} \\ a_{n+1}^* &= -a_n + (a_n - a_{n-1})c_2 + \cdots + (a_1 - 1)c_{n+1} + c_{n+2} \\ a_i^* &= -a_n c_{i-n} + (a_n - a_{n-1})c_{i-n+1} + \cdots + (a_1 - 1)c_i + c_{i+1}, \qquad i \geq n+2 \end{aligned}$$

Note that if $i \geq n + 2$, from (4.22) it follows that

$$\begin{aligned} a_i^* &= -a_n c_{i-n} + (a_n - a_{n-1})c_{i-n+1} + \cdots + (a_1 - 1)c_i + c_{i+1} \\ &= (c_{i+1} - c_i) + (c_i - c_{i-1})a_1 + \cdots + (c_{i-n+1} - c_{i-n})a_n \\ &= \sum_{j=N_1}^{N_2} d_j \left[(g_{i+j} - g_{i+j-1}) + (g_{i+j-1} - g_{i+j-2})a_1 + \cdots + (g_{i+j-n} - g_{i+j-n-1})a_n \right] \\ &\quad + d_0 \left[(g_i - g_{i-1}) + (g_{i-1} - g_{i-2})a_1 + \cdots + (g_{i-n} - g_{i-n-1})a_n \right] = 0 \end{aligned}$$

Then

$$A_C(z) = 1 + a_1^* z^{-1} + \cdots + a_{n+1}^* z^{-(n+1)}$$

where

$$\begin{bmatrix} 1 \\ a_1^* \\ \vdots \\ a_{n+1}^* \end{bmatrix} = \begin{bmatrix} 1 & & \mathbf{0} \\ c_2 - 1 & \ddots & \\ \vdots & \ddots & 1 \\ c_{n+2} - c_{n+1} & \cdots & c_2 - 1 \end{bmatrix} \begin{bmatrix} 1 \\ a_1 \\ \vdots \\ a_n \end{bmatrix} \tag{4.23}$$

Furthermore, the numerator part of the GPC controller is

$$\boldsymbol{d}^{\mathrm{T}} \boldsymbol{M} A = (d_s + d_0) A$$

where $d_s \triangleq d_{N_1} + \cdots + d_{N_2}$. Then the expression of the GPC controller can be given by

$$G_C(z) = \frac{(d_s + d_0) A(z)}{A_C(z)} \tag{4.24}$$

Following (3.2), the transfer function of the closed-loop system is

$$F_0(z) = G_C(z) G_P(z) = \frac{(d_s + d_0) z^{-1} B(z)}{A_C(z)}$$

This means that the GPC controller cannot change the zeros of the system, but can change its poles, which is achieved by cancelling the plant characteristic polynomial $A(z)$ and setting a new characteristic polynomial $A_C(z)$ by the controller (4.24), where (4.23) gives the coefficients of the characteristic polynomial of the controlled system according to that of the plant, and is called the coefficient mapping between the open-loop and the closed-loop characteristic polynomials of the GPC system [5].

According to the coefficient mapping given by (4.23), the following corollary is easy to obtain.

Corollary 4.1 The coefficients of the denominator polynomial of the GPC controller satisfy

$$1 + a_1^* + \cdots + a_{n+1}^* = (d_s + d_0)(b_1 + \cdots + b_n) = (d_s + d_0) B(1) \tag{4.25}$$

and the closed-loop steady-state gain $F_0(1) = 1$ when the closed-loop system is stable.

Proof: Multiply the $n+2$-dimensional vector $[1 \ \cdots \ 1]$ on both sides of (4.23). It follows that

$$1 + a_1^* + \cdots + a_{n+1}^* = [c_{n+2} \ \cdots \ c_2] \begin{bmatrix} 1 \\ a_1 \\ \vdots \\ a_n \end{bmatrix}$$

$$= [d_{N_1} \ \cdots \ d_{N_2}] \begin{bmatrix} g_{N_1+n+1} & \cdots & g_{N_1+1} \\ \vdots & & \vdots \\ g_{N_2+n+1} & \cdots & g_{N_2+1} \end{bmatrix} \begin{bmatrix} 1 \\ a_1 \\ \vdots \\ a_n \end{bmatrix} + d_0 [g_{n+1} \ \cdots \ g_1] \begin{bmatrix} 1 \\ a_1 \\ \vdots \\ a_n \end{bmatrix}$$

Add the first $n + i$ $(i \geq 1)$ equations in (4.22) together:

$$g_{n+i} + g_{n+i-1}a_1 + \cdots + g_i a_n = b_1 + \cdots + b_n = B(1), \quad i \geq 1$$

Then (4.25), i.e. $A_C(1) = (d_s + d_0)B(1)$, can be immediately obtained. Substitute it into the above closed-loop transfer function and then $F_0(1) = 1$ holds.

In order to compare this with the DMC controller, in the following, the reference trajectory will not be considered, i.e. $\alpha = 0$ and $d_0 = 0$. The GPC controller (4.24) then has the form

$$G_C(z) = \frac{d_s A(z)}{A_C(z)} \tag{4.26}$$

4.2.2 Minimal Form of the DMC Controller and Uniform Coefficient Mapping

The GPC controller (4.26) in the IMC structure deduced in the last section clearly explains the compensation mechanism of GPC for system dynamics, while the DMC controller (3.10) in the IMC structure shows nontrivial characters because it adopts a nonminimum model with step response coefficients. For example, the order of the model depends on the model length N with great arbitrariness and the control result is insensitive to the model order. The transfer function of such a model has all the zeros at the origin, which makes it difficult to explain the compensation mechanism of DMC for system dynamics. However, although different models are adopted by DMC and GPC, in principle the same control result should be achieved if the rolling optimization strategy of GPC (2.24) is coincident with that of DMC (2.3) and the additional reference trajectory in the GPC algorithm is out of consideration. In this section, the relationship of the DMC controller (3.10) and the GPC controller (4.26) is explored.

Assume that the minimum model of a stable plant with the step response $\{a_i\}$ is given by

$$G(z) = \frac{m(z)}{p(z)} = \frac{m_1 z^{-1} + m_2 z^{-2} + \cdots + m_n z^{-n}}{1 + p_1 z^{-1} + p_2 z^{-2} + \cdots + p_n z^{-n}} \tag{4.27}$$

For simplicity it is assumed that $p_n \neq 0$, $m_n \neq 0$, i.e. the model order is n. For the same plant, it is obvious that the model (3.9) is an approximation of the above model under the assumption that $a_i \approx a_N$, $i \geq N$. Note that there exists the same relationship between the model parameters m_i and p_i and the step response parameters a_i as given by Lemma 4.3; only the symbols a_i, b_i, and g_i in (4.22) should be replaced by p_i, m_i, and a_i here.

Theorem 4.4 Without the reference trajectory in the GPC algorithm, the GPC controller (4.26) is identical to the DMC controller (3.10) in the IMC structure.

Proof: According to (3.2) and the DMC controller (3.10) deduced in Section 3.2, the closed-loop transfer function of DMC without model mismatch is given by

$$F_0(z) = G_C(z) G_M(z) = \frac{d_s m(z)}{p(z) b(z)}$$

Its denominator part can be written as

$$p(z)b(z) = (1 + p_1 z^{-1} + p_2 z^{-2} + \cdots + p_n z^{-n}) \cdot \left(1 + (b_2 - 1)z^{-1} + \cdots + (b_N - b_{N-1})z^{-(N-1)}\right)$$
$$\triangleq 1 + p_1^* z^{-1} + \cdots + p_{N+n-1}^* z^{-(N+n-1)}$$

where

$$\begin{bmatrix} 1 \\ p_1^* \\ \vdots \\ p_{n+1}^* \\ p_{n+2}^* \\ \vdots \\ p_{N-1}^* \\ p_N^* \\ \vdots \\ p_{N+n-1}^* \end{bmatrix} = \begin{bmatrix} 1 & 0 & \cdots & 0 \\ b_2 - 1 & 1 & \cdots & 0 \\ \vdots & \vdots & & \vdots \\ b_{n+2} - b_{n+1} & b_{n+1} - b_n & \cdots & b_2 - 1 \\ b_{n+3} - b_{n+2} & b_{n+2} - b_{n+1} & \cdots & b_3 - b_2 \\ \vdots & \vdots & & \vdots \\ b_N - b_{N-1} & b_{N-1} - b_{N-2} & \cdots & b_{N-n} - b_{N-n-1} \\ 0 & b_N - b_{N-1} & \cdots & b_{N-n+1} - b_{N-n} \\ \vdots & \vdots & & \vdots \\ 0 & 0 & \cdots & b_N - b_{N-1} \end{bmatrix} \begin{bmatrix} 1 \\ p_1 \\ \vdots \\ p_n \end{bmatrix}$$

With the assumption that $a_i \approx a_N$, $i \geq N$, it follows that $b_{N+k} \approx d_s a_N$, $k = 0, 1, \ldots$. Then the zeros in the last n rows in the above matrix can be approximately replaced by the corresponding $b_{N+k+1} - b_{N+k}$. Therefore the rows below the row corresponding to p_{n+1}^* can be approximately represented by

$$p_i^* \approx [b_{i+1} - b_i \ \cdots \ b_{i-n+1} - b_{i-n}]\boldsymbol{p}, \quad i = n+2, \ldots, N+n-1$$

where $\boldsymbol{p} = [1 \ \ p_1 \ \ \cdots \ \ p_n]^{\mathrm{T}}$. According to the expression of b_i in (3.10) and Lemma 4.3, it is easy to get

$$p_i^* \approx \boldsymbol{d}^{\mathrm{T}} \begin{bmatrix} a_{i+1} - a_i & \cdots & a_{i-n+1} - a_{i-n} \\ \vdots & & \vdots \\ a_{i+P} - a_{i+P-1} & \cdots & a_{i+P-n} - a_{i+P-n-1} \end{bmatrix} \begin{bmatrix} 1 \\ p_1 \\ \vdots \\ p_n \end{bmatrix} = 0, \quad i = n+2, \ldots, N+n-1$$

Then it follows that

$$F_O(z) \approx \frac{d_s m(z)}{p^*(z)} = \frac{d_s(m_1 z^{-1} + \cdots + m_n z^{-n})}{1 + p_1^* z^{-1} + \cdots + p_{n+1}^* z^{-(n+1)}} \tag{4.28}$$

Furthermore, the controller can be expressed by

$$G_C(z) = \frac{d_s p(z)}{p^*(z)} \tag{4.29}$$

where

$$\begin{bmatrix} 1 \\ p_1^* \\ \vdots \\ p_{n+1}^* \end{bmatrix} = \begin{bmatrix} 1 & & \mathbf{0} \\ b_2-1 & \ddots & \\ \vdots & \ddots & 1 \\ b_{n+2}-b_{n+1} & \cdots & b_2-1 \end{bmatrix} \begin{bmatrix} 1 \\ p_1 \\ \vdots \\ p_n \end{bmatrix} \tag{4.30}$$

The DMC controller (4.29) is obviously identical to the GPC controller (4.26) when the reference trajectory is not adopted in GPC. Furthermore, the coefficient mapping in DMC is the same as that in GPC (see (4.30) and (4.23), respectively), which means that with the same control strategies, both DMC and GPC algorithms have the same controller in the IMC structure.

Remark 4.2 It should be pointed out that there are two expressions of the DMC controller in the IMC structure, i.e. (3.10) and (4.29). The former is directly expressed by the step response coefficients, while the latter is expressed by the minimum model parameters and does not explicitly use the step response. However, the latter reveals the control mechanism of the DMC algorithm implied in the former controller. Thus (4.29) is called the minimal form of the DMC controller [6]. Compared with the controller (3.10) based on the step response model, the distribution of the zeros and poles of this minimal controller deeply reveals the compensation mechanism of the DMC control. It is also advantageous to use it to analyze the control performance.

Example 4.1 The Z transfer function (4.27) of a plant is given by

$$G(z) = \frac{z^{-1} + 0.835z^{-2}}{1 - 0.478z^{-1} + 0.599z^{-2}}$$

It has a pair of conjugate poles $0.239 \pm 0.736\mathrm{j}$ and a zero -0.835, all in the unit circle, as shown in Figure 4.1, where the poles are represented by × and the zero by ○.

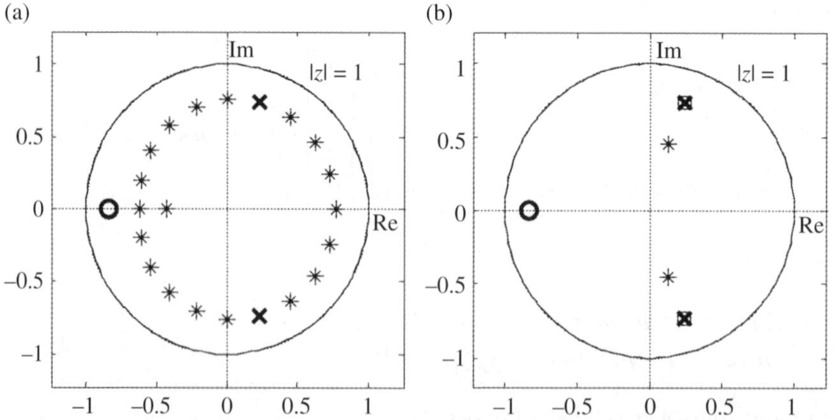

Figure 4.1 The compensation property of the DMC controller: (a) in the form of a nonminimal controller, (b) in the form of a minimal controller [6]. *Source:* Reproduced with permission from Taylor and Francis and Copyright clearance center.

With DMC control, take the model horizon $N = 20$, we obtain the nonminimal model (3.9) represented by the impulse response coefficients

$$G(z) = z^{-1} + 1.313z^{-2} + 0.029z^{-3} + \cdots - 0.008z^{-20}$$

with which the coefficients of the unit step response can be calculated.

Setting the DMC design parameters $P = 2$, $M = 1$, $\boldsymbol{Q} = \boldsymbol{I}$, the DMC control vector can be obtained as $\boldsymbol{d}^{\mathrm{T}} = [0.1575 \quad 0.3643]$. Then the DMC controller (3.10) can be calculated as

$$G_C(z) = \frac{d_s}{b(z)} = \frac{0.5217}{1 + 0.2172z^{-1} - 0.277z^{-2} + \cdots - 0.0013z^{-19}}$$

This controller has 19 poles in the unit circle (represented by $*$). In addition to a single pole located in the real axis, the rest poles together with the plant poles uniformly distribute in a closed curve around the origin of the Z plane (see Figure 4.1a). If the DMC controller is represented by the minimal form (4.29) as

$$G_C(z) = \frac{d_s p(z)}{p^*(z)} = \frac{0.5217(1 - 0.478z^{-1} + 0.599z^{-2})}{1 - 0.2608z^{-1} + 0.2182z^{-2}}$$

then its two zeros (represented by □) cancel the two poles of the plant. Meanwhile it reallocates a pair of conjugate poles $0.1304 \pm 0.4485j$ (represented by $*$) that constructs the closed-loop poles (see Figure 4.1b).

With traditional control theory, it is difficult to understand and judge the dynamic properties of the controlled system by using the pole-zero distribution of the plant and the nonminimal controller, as shown in Figure 4.1a. However, if the DMC controller is represented by its minimal form, it will be clear that the control in Figure 4.1a indeed implies the control mechanism shown in Figure 4.1b. With the minimal form (4.29) instead of its real form (3.10), the zeros of the DMC controller just cancel the plant poles, and its poles then construct the poles of the closed-loop system. The compensation mechanism seems much clearer in Figure 4.1b.

Remark 4.3 Due to the nonminimal representation of the DMC algorithm, sometimes it is unable to use traditional concepts to understand its dynamic behaviors, such as:

1) The model length N directly affects the order of the DMC controller (3.10), and thus affects the number and the distribution of the poles of the controlled system (3.2) and Figure 4.1a. However, the closed-loop dynamics is almost unaffected by N as long as N is selected large enough. In general the DMC controlled system is of order $n + 1$, and the closed-loop dynamics depends only on the plant and the control strategy, almost independent of the model length N.
2) In the DMC algorithm, the poles of the controlled system are composed of the plant poles and the poles of the nonminimal controller. However, the dynamic convergence rate of the controlled system cannot be measured by the maximal distance of the poles in Figure 4.1a to the origin. Particularly when N is large, the poles of the controller (3.10) may be sufficiently close to the boundary of the unit circle but the closed-loop dynamic response may still be quick.

4.3 *Z* Domain Analysis Based on Coefficient Mapping

According to Theorem 4.4, it is now possible to uniformly analyze the system performance of the two main predictive control algorithms, DMC and GPC. As mentioned above, we will focus on the stability analysis without model mismatch and disturbance. In this case, the closed-loop predictive control system can be uniformly described by (4.28), and the closed-loop stability will be uniformly analyzed by using the coefficient mapping (4.30) between the open-loop and the closed-loop characteristic polynomials. In this section, quantitative analysis for the DMC algorithm will be given based on coefficient mapping in the *Z* domain. In the meantime, corresponding conclusions for the GPC algorithm are also given in parallel.

4.3.1 Zero Coefficient Condition and the Deadbeat Property of Predictive Control Systems

The DMC closed-loop system for the nth-order plant (4.27) can be described by (4.28). It is known from the coefficient mapping relationship (4.30) that the closed-loop system is of order $n+1$, and the coefficients of the characteristic polynomial are given by

$$p_i^* = [\overbrace{b_{i+1}-b_i \quad \cdots \quad b_2-1}^{i} \quad \overbrace{1 \quad 0 \quad \cdots \quad 0}^{n+1-i}]\boldsymbol{p}, \qquad i=1,\ldots,n+1 \tag{4.31}$$

where $\boldsymbol{p} = [1 \quad p_1 \quad \cdots \quad p_n]^{\mathrm{T}}$. Firstly, consider how to select the design parameters to make some $p_i^* = 0$. Note that when deducing the minimal form of the DMC controller in Section 4.2.2, Lemma 4.3, i.e. (4.22), is utilized to obtain

$$p_i^* \approx [b_{i+1}-b_i \quad \cdots \quad b_{i-n+1}-b_{i-n}]\boldsymbol{p}$$

$$= \boldsymbol{d}^{\mathrm{T}} \begin{bmatrix} a_{i+1}-a_i & \cdots & a_{i-n+1}-a_{i-n} \\ \vdots & & \vdots \\ a_{i+P}-a_{i+P-1} & \cdots & a_{i+P-n}-a_{i+P-n-1} \end{bmatrix} \begin{bmatrix} 1 \\ p_1 \\ \vdots \\ p_n \end{bmatrix} = 0,$$

for $i = n+2, \ldots, N+n-1$, while each p_i^* in (4.30), $i = 1, \ldots, n+1$, cannot be eliminated to zero in the same way because both 0 and 1 appear in the corresponding expression (4.31). In order to solve the problem, the following lemma is given, showing an important relationship between the parameters in the DMC algorithm.

Lemma 4.4 In the DMC algorithm, if $\boldsymbol{R} = \boldsymbol{0}$, then

$$\begin{aligned} b_1 &\triangleq d_1a_1 + d_2a_2 + \cdots + d_Pa_P = 1 \\ b_0 &\triangleq d_2a_1 + \cdots + d_Pa_{P-1} = 0 \\ &\vdots \\ b_{-(M-2)} &\triangleq d_Ma_1 + \cdots + d_Pa_{P-M+1} = 0 \end{aligned} \tag{4.32}$$

If $\boldsymbol{R} = \boldsymbol{0}$, the above formula can be directly deduced by

$$\boldsymbol{d}^{\mathrm{T}}\boldsymbol{A} = [1 \ 0 \ \cdots \ 0]\left(\boldsymbol{A}^{\mathrm{T}}\boldsymbol{Q}\boldsymbol{A} + \boldsymbol{R}\right)^{-1}\boldsymbol{A}^{\mathrm{T}}\boldsymbol{Q}\boldsymbol{A} = [1 \ 0 \ \cdots \ 0]$$

Note that there are altogether M equations in (4.32), among which $M - 1$ equations are equal to zero.

According to (4.32), in the case $\boldsymbol{R} = \boldsymbol{0}$, if $n - i + 1 \le M - 1$, all the 1s and 0s in the expression (4.31) of p_i^* can be replaced by the corresponding b_j in (4.32). After expansion and utilizing (4.22), it follows that

$$p_i^* = \left[b_{i+1} - b_i \ \cdots \ b_2 - b_1 \ \ b_1 - b_0 \ \cdots \ b_{-(n-i)+1} - b_{-(n-i)}\right]\boldsymbol{p}$$

$$= \boldsymbol{d}^{\mathrm{T}} \begin{bmatrix} a_{i+1} - a_i & \cdots & a_1 & 0 & \cdots & 0 \\ \vdots & & \ddots & \ddots & & \vdots \\ a_n - a_{n-1} & \cdots & & a_1 & & 0 \\ a_{n+1} - a_n & \cdots & & \cdots & & a_1 \\ \vdots & & & & & \vdots \\ a_{i+P} - a_{i+P-1} & \cdots & & & & a_{i+P-n} - a_{i+P-n-1} \end{bmatrix} \begin{bmatrix} 1 \\ p_1 \\ \vdots \\ p_n \end{bmatrix}$$

$$= [d_1 \ \cdots \ d_P] \begin{bmatrix} m_{i+1} \\ \vdots \\ m_n \\ \boldsymbol{0} \end{bmatrix} = d_1 m_{i+1} + \cdots + d_{n-i} m_n, \quad i \ge n - M + 2 \tag{4.33}$$

Let $q_1 = \cdots = q_{n-i} = 0$; then $d_1 = \cdots = d_{n-i} = 0$ and thus $p_i^* = 0$. The following lemma on the zero coefficients in the closed-loop characteristic polynomial can be derived.

Lemma 4.5 In the DMC algorithm, select $\boldsymbol{R} = \boldsymbol{0}$, $q_1 = \cdots = q_{n-i} = 0$, $M \ge n - i + 2$; then $p_i^* = 0$.

Corollary 4.2 In the DMC algorithm, select $\boldsymbol{R} = \boldsymbol{0}$; then

1) $p_{n+1}^* = 0$,
2) $p_n^* = 0$ for $M \ge 2$,
3) all $p_i^* = 0$, $i = 1, \ldots, n + 1$ for $M \ge n + 1$ and $q_1 = \cdots = q_{v-1} = 0$, $v \ge n$.

The above corollary seems to give the conditions of the DMC closed-loop system as deadbeat control (i.e. the denominator polynomial as 1). However, note that when $\boldsymbol{R} = \boldsymbol{0}$ the matrix $\boldsymbol{A}^{\mathrm{T}}\boldsymbol{Q}\boldsymbol{A}$ may be irreversible when utilizing (2.8) to solve the control vector $\boldsymbol{d}^{\mathrm{T}}$. Therefore the solvability problem of the DMC control law should also be discussed.

For the sake of discussion, first of all denote

$$Q = \text{block diag}(Q_1,\ Q_2),\ \ Q_1 = \mathbf{0}_{(v-1)\times(v-1)},\ \ Q_2 = \text{diag}(q_v,\ \ldots,\ q_P) > \mathbf{0}$$

$$A = \begin{bmatrix} A_1 \\ \\ A_2 \end{bmatrix},\ \ A_1 : (v-1)\times M,\ \ A_2 : (P-v+1)\times M$$

where

$$A_2 = \begin{bmatrix} a_v & \cdots & a_1 & & \mathbf{0} \\ & & & \ddots & \\ \vdots & & & & a_1 \\ & & & & \vdots \\ a_P & & \cdots & & a_{P-M+1} \end{bmatrix} \tag{4.34}$$

When $R = \mathbf{0}$, according to (2.8), it follows that

$$d^{\mathrm{T}} = c^{\mathrm{T}}\left(A^{\mathrm{T}}QA + R\right)^{-1}A^{\mathrm{T}}Q = c^{\mathrm{T}}\left(A_2^{\mathrm{T}}Q_2A_2\right)^{-1}\left[\mathbf{0}\ \ A_2^{\mathrm{T}}Q_2\right]$$

It is obvious that the solvability condition for d^{T} is the matrix A_2 of full column rank.

Lemma 4.6 The necessary and sufficient conditions for full column rank of the matrix A_2 given by (4.34) are as follows:

1) $P - v + 1 \geq M,\quad M \leq n + 1$ or
2) $P - v + 1 \geq M,\ \ M \geq n + 2,\ \ v \leq n,$

and in this case the DMC control law is solvable.

Proof: Firstly, note that the matrix A_2 is $(P - v + 1) \times M$-dimensional and the necessary condition for its full column rank is $P - v + 1 \geq M$. With this condition, the following cases are discussed.

1) When $M \leq n + 1$, A_2 must be of full column rank according to the linear system theory.
2) When $M \geq n + 2$, if $v \geq n + 1$, according to (4.22), we have

$$\begin{bmatrix} a_v & \cdots & a_1 & & \mathbf{0} \\ \vdots & & & \ddots & \\ a_M & & \cdots & & a_1 \\ \vdots & & & & \vdots \\ a_P & & \cdots & & a_{P-M+1} \end{bmatrix} \begin{bmatrix} 1 \\ p_1 - 1 \\ \vdots \\ -p_n \\ \mathbf{0} \end{bmatrix} = \begin{bmatrix} 0 \\ \vdots \\ \vdots \\ \vdots \\ 0 \end{bmatrix}$$

The column vectors of $\boldsymbol{A}_2$ are linear dependent and the matrix is not of full column rank.

If $v \le n$, make a column transformation by using the relationship (4.22) starting from the first column of $\boldsymbol{A}_2$, i.e. multiply its second column by $(p_1 - 1)$, its third column by $(p_2 - p_1)$, etc., and add all of them to the first column. This procedure is repeated starting from the second column and so on, listed as follows:

$$\boldsymbol{A}_2 = \begin{bmatrix} a_v & \cdots & a_1 & & \mathbf{0} \\ \vdots & & & \ddots & \\ a_M & & & & a_1 \\ \vdots & & & & \vdots \\ a_P & & \cdots & & a_{P-M+1} \end{bmatrix} \to \begin{bmatrix} m_v & a_{v-1} & \cdots & * \\ \vdots & \vdots & & \\ m_n & a_{n-1} & \cdots & \\ 0 & \vdots & & \\ \vdots & a_{M-1} & \cdots & a_1 \\ & \vdots & & \vdots \\ 0 & a_{P-1} & \cdots & a_{P-M+1} \end{bmatrix}$$

$$\to \cdots \to \begin{bmatrix} m_v & \cdots & m_{v-M+n+1} & & & \\ \vdots & & & & * & \\ m_n & & & & & \\ & \ddots & & & & \\ & & m_n & & & \\ & \mathbf{0} & 0 & a_n & \cdots & a_1 \\ & & \vdots & \vdots & & \vdots \\ & & 0 & a_{P-M+n} & \cdots & a_{P-M+1} \end{bmatrix}$$

It can be shown that $\boldsymbol{A}_2$ is of full column rank because its bottom-right matrix composed of a_i is of full column rank and $m_n \neq 0$.

By combining (3) in the Corollary 4.2 and the Lemma 4.6, the following theorem is given.

Theorem 4.5 In the DMC algorithm for the nth-order plant (4.27), select $\boldsymbol{R} = \mathbf{0}$, $q_1 = \cdots = q_{v-1} = 0$, if

1) $M = n+1, \quad v \ge n, \quad P \ge v+n$ or
2) $M \ge n+1, \quad v = n, \quad P \ge M+n-1.$

Then the controller $G_C(z)$ is a deadbeat controller, with the closed-loop transfer function

$$F_O(z) = d_s m(z) = m(z)/m(1)$$

Proof: Combining (3) in Corollary 4.2 with the condition (1) or (2) in Lemma 4.6, the above condition (1) or (2) can be obtained, respectively. Note that in the latter case, although the condition directly obtained is $M \geq n + 2$, it can be still written in the form of (2) above because the case $M = n + 1$ is already included in the condition (1).

Note that since in the deadbeat case all $p_i^* = 0$, $i = 1, \ldots, n + 1$, and from Corollary 4.1, $d_s = 1/m(1)$, then the conclusion of the above theorem follows.

Remark 4.4 For the GPC algorithm, the following statement corresponding to Theorem 4.5 [5] holds.

In the GPC algorithm for the nth-order plant (4.27), select $\lambda = 0$, if

1) $NU = n + 1, \quad N_1 \geq n, \quad N_2 \geq N_1 + n$ or
2) $NU \geq n + 1, \quad N_1 = n, \quad N_2 \geq NU + n - 1.$

Then the controller $G_C(z)$ is a deadbeat controller, with the closed-loop transfer function

$$F_0(z) = d_s m(z) = m(z)/m(1)$$

Note that the plant (4.27) discussed here corresponds to that in Theorem 4.3 with the assumption that $n = n_b + 1$, so the deadbeat condition of the GPC system given above is fully consistent with condition (4.16) given in Theorem 4.3. Moreover, compared with Theorem 4.3, Theorem 4.5 provides not only the parameter setting conditions for predictive control to achieve deadbeat, but also the concrete response of the closed-loop system.

4.3.2 Reduced Order Property and Stability of Predictive Control Systems

Combining Lemma 4.5 and Lemma 4.6, the relationship between the closed-loop order and the design parameters can be further derived.

Theorem 4.6 For the nth-order plant (4.27), select $q_1 = \cdots = q_{v-1} = 0$, $P - v + 1 \geq M$, $\boldsymbol{R} = \boldsymbol{0}$ in the DMC algorithm. Then the closed-loop system has a reduced order as follows:

$$n_c = \begin{cases} n - M + 1, & 1 \leq M \leq n, \quad v \geq M - 1 \\ n - v, & 1 \leq v \leq n - 1, \quad M \geq v + 1 \\ 0, \quad M = n + 1, \quad v \geq n \ \ or \ \ M \geq n + 1, \quad v = n & \end{cases}$$

and the closed-loop transfer function is given by

$$F_0(z) = \frac{d_s m(z)}{p^*(z)} = \frac{d_s(m_1 z^{-1} + \cdots + m_n z^{-n})}{1 + p_1^* z^{-1} + \cdots + p_{n_c}^* z^{-n_c}}$$

where the coefficients of $p^*(z)$, i.e. p_i^*, $i = 1, \ldots, n_c$, can be calculated by (4.30).

Proof: Firstly note that with the given conditions, DMC control law is always solvable and $p_{n+1}^* = 0$ when $\boldsymbol{R} = 0$.

1) If $1 \leq M \leq n$, $v \geq M - 1$, for all $n - M + 2 \leq i \leq n$, then $M \geq n - i + 2$, $v - 1 \geq M - 2 \geq n - i$. From Lemma 4.5 it follows that $p_i^* = 0$, $n - M + 2 \leq i \leq n$, and thus we have $n_c = n - M + 1$.

2) If $1 \le v \le n-1$, $M \ge v+1$, for all $n-v+1 \le i \le n$, then $v-1 \ge n-i$, $M \ge v+1 \ge n-i+2$. From Lemma 4.5 it follows that $p_i^* = 0$, $n-v+1 \le i \le n$, and thus we have $n_c = n-v$.
3) If $M = n+1$, $v \ge n$ or $M \ge n+1$, $v = n$, it is the deadbeat case in Theorem 4.4 and $n_c = 0$

The remaining coefficients $p_i^*, i = 1, \ldots, n_c$, for case (1) or (2) can be calculated by (4.30).

Remark 4.5 In the above theorem "reduced order" means that the highest order for z^{-1} of the closed-loop characteristic polynomial $p^*(z)$ is reduced from $n+1$ to n_c. If the transfer function is represented by polynomials of z, the order of the characteristic polynomial will always be $n+1$ and the so-called "reduced order" indeed means that the closed-loop system has $n+1-n_c$ poles at the origin $z=0$ while other n_c poles are the roots of the characteristic polynomial $p^*(z)$. Note that Theorem 4.6 only gives the closed-loop transfer function of the DMC control system and does not involve stability. In addition to the deadbeat case, i.e. $n_c = 0$, where all the closed-loop poles are located at the origin $z=0$ and stability can be guaranteed, in other "reduced order" cases, closed-loop stability will depend on whether all the roots of the polynomial $p^*(z)$ are located in the unit circle.

Remark 4.6 For the GPC algorithm, the following statement corresponding to Theorem 4.6 [5] holds.

In the GPC algorithm for the nth-order plant (4.27), select $\lambda = 0$, $N_2 - N_1 + 1 \ge NU$. Then the closed-loop system has a reduced order as follows:

$$n_c = \begin{cases} n - NU + 1, & 1 \le NU \le n, \ \ N_1 \ge NU - 1 \\ n - N_1, & 1 \le N_1 \le n-1, \ \ NU \ge N_1 + 1 \\ 0, & NU = n+1, \ \ N_1 \ge n \ \text{ or } \ NU \ge n+1, \ \ N_1 = n. \end{cases}$$

and the closed-loop transfer function is given by

$$F_0(z) = \frac{d_s m(z)}{p^*(z)} = \frac{d_s(m_1 z^{-1} + \cdots + m_n z^{-n})}{1 + p_1^* z^{-1} + \cdots + p_{n_c}^* z^{-n_c}}$$

where the coefficients of $p^*(z)$, i.e. p_i^*, $i = 1, \ldots, n_c$, can be calculated by (4.30).

Example 4.2 Consider the fourth-order plant as presented in Example 3.2:

$$G(s) = \frac{8611.77}{\left((s+0.55)^2 + 6^2\right)\left((s+0.25)^2 + 15.4^2\right)}$$

Take the sampling period $T_0 = 0.2$ s; after sampling and holding we get the discrete time model

$$G(z) = \frac{0.3728z^{-1} + 1.9069z^{-2} + 1.7991z^{-3} + 0.3080z^{-4}}{1 + 1.2496z^{-1} + 0.4746z^{-2} + 0.9364z^{-3} + 0.7261z^{-4}}$$

Use the GPC algorithm to control this plant. Since $n = 4$, select $\lambda = 0$, $N_2 = 9$. According to Remark 4.6, for different N_1 and NU, the closed-loop characteristic polynomials $p^*(z)$ with corresponding orders can be obtained.

$$N_1 = 2,\ NU = 4,\quad p^*(z) = 1 + 1.0879z^{-1} + 0.1245z^{-2}$$
$$N_1 = 3,\ NU = 4,\quad p^*(z) = 1 + 0.9655z^{-1}$$
$$N_1 = 3,\ NU = 6,\quad p^*(z) = 1 + 1.085z^{-1}$$
$$N_1 = 4,\ NU = 5,\quad p^*(z) = 1$$

Simulation results for these four cases are shown in Figure 4.2. Particularly in the case $N_1 = 4$, $NU = 5$, according to Remark 4.3, $G_C(z)$ is a deadbeat controller and the closed-loop system has the transfer function

$$F_O(z) = d_s m(z) = m(z)/m(1) = 0.0850z^{-1} + 0.4347z^{-2} + 0.4101z^{-3} + 0.0702z^{-4}$$

Therefore, the system output at the first three sampling instants should have the value 0.085, 0.5197, 0.9298, and reaches the steady state value 1 at the fourth sampling instant $t = 0.8$ s and then remains unchanged. The simulation result is fully in accord with the theoretical analysis. Furthermore, in the case $N_1 = 3$, $NU = 6$, the closed-loop characteristic polynomial is of the first order with a root $z = -1.085$ out of the unit circle; thus the closed-loop response in the figure seems unstable, as indicated in Remark 4.5.

Theorem 4.6 and Remark 4.6 give the order of the closed-loop system for all possible selections of $v(N_1)$ and $M(NU)$ when $\boldsymbol{R} = \boldsymbol{0}$, which can be visually shown in Figure 4.3.

In the deadbeat case given by Theorem 4.5 all the roots of the closed-loop characteristic polynomial are located at the origin $z = 0$ when $r = 0$. If $\boldsymbol{R} = r\boldsymbol{I}$, since these characteristic roots vary continuously with r, the following stability theorem can be obtained.

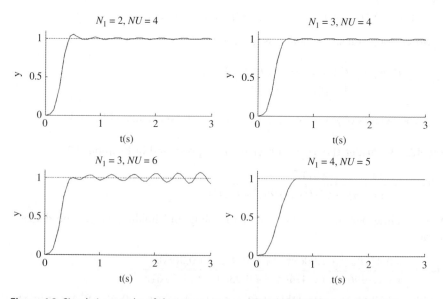

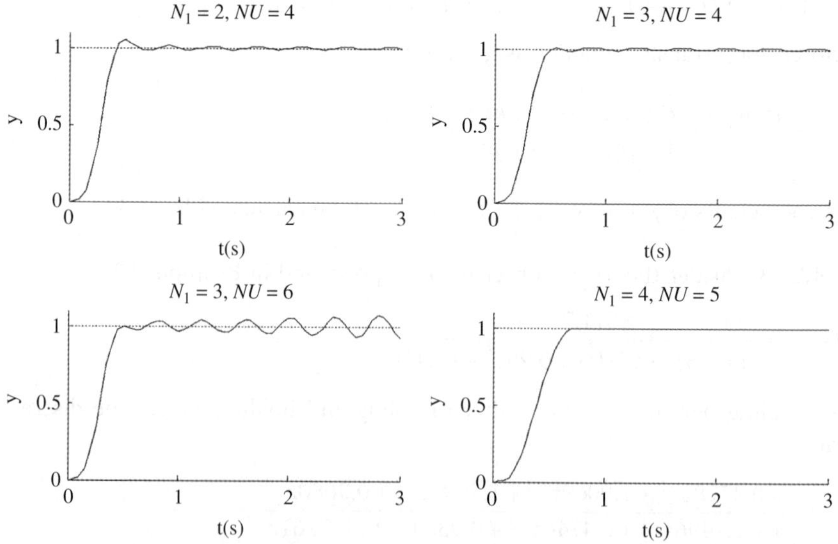

Figure 4.2 Simulation results of the system in Example 4.2 with different parameters.

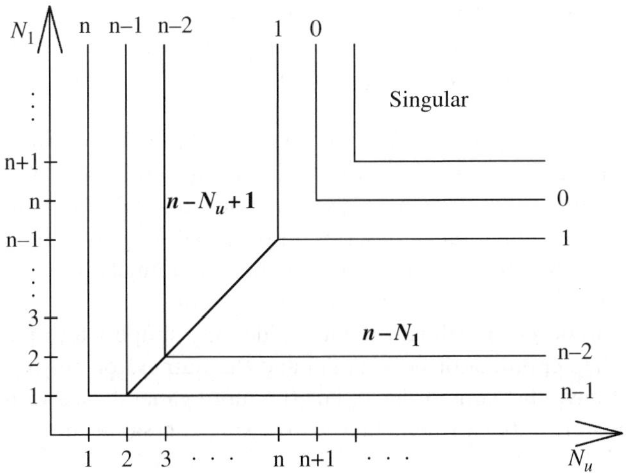

Figure 4.3 The relationship of the GPC closed-loop order and N_1, NU [7].

Theorem 4.7 In the DMC algorithm for the nth-order plant (4.27), select $\boldsymbol{R} = r\boldsymbol{I}$, $q_1 = \cdots = q_{v-1} = 0$, $P - v + 1 \geq M$. If

1) $M = n + 1, \quad v \geq n$ or
2) $M \geq n + 1, \quad v = n,$

there must exist a constant $r_0 > 0$ such that the closed-loop system is stable for $0 \leq r \leq r_0$.

Remark 4.7 For the GPC algorithm, the following statement corresponding to Theorem 4.7 holds.

In the GPC algorithm for the nth-order plant (4.27), select $N_2 - N_1 + 1 \geq NU$. If

1) $NU = n + 1, \quad N_1 \geq n$ or
2) $NU \geq n + 1, \quad N_1 = n$

there must exist a constant $\lambda_0 > 0$ such that the closed-loop system is stable for $0 \leq \lambda \leq \lambda_0$.

Remark 4.8 Note that the plant (4.27) discussed here corresponds to that in Theorem 4.2 with the assumption that $n = n_b + 1$. Compare the stability condition for the GPC closed-loop system given above with the condition (4.15) in Theorem 4.2, the range for parameter selection is somewhat reduced. The reason is that the condition here is also suitable for the case $\lambda = 0$ and thus the solvability of the GPC control law should also be taken into account, while Theorem 4.2 is only suitable for the case $0 < \lambda < \lambda_0$, where λ_0 is a sufficiently small positive number. Due to the existence of λ, the GPC control law is always solvable. It is easy to show that if the solvability of the control law is not considered, then the GPC parameter condition corresponding to the DMC condition (3) in Corollary 4.2 is exactly $N_1 \geq n_b + 1$, $NU \geq n + 1$, as in (4.15).

The above theorems characterize the performances of the DMC closed-loop system more deeply than that given in Section 3.2.1. They are suitable for general plants and the parameter selection is not limited to extreme cases, giving not only the stability result of the closed-loop system but also a quantitative description of closed-loop dynamics. This indicates that the coefficient mapping of the characteristic polynomials plays an important role in the closed-loop analysis of predictive control systems.

It should be pointed out that in order to highlight the main idea and simplify formula deduction, it is assumed that the denominator polynomial and the numerator polynomial in the plant model (4.27) have the same order n. For the more general case with possible different orders, similar results can be achieved by using the same method and details can be referred to [7].

4.4 Quantitative Analysis of Predictive Control for Some Typical Systems

The coefficient mapping relationship (4.30) is also useful to the predictive control design for some typical systems, such as the first-order inertia process and the second-order oscillation system. In this section, based on coefficient mapping, predictive control for these systems is quantitatively analyzed in more detail, which provides the basis for the analytical design of predictive control systems.

4.4.1 Quantitative Analysis for First-Order Systems

A large number of industrial processes can be approximately described by the following typical plant – a first-order inertia process with time delay

$$G_0(s) = \frac{Ke^{-\tau s}}{1 + Ts} \tag{4.35}$$

where K is the plant gain, τ is the time delay, and T is the time constant of the first-order inertia process. All these parameters are easy to identify by directly using its step response. To simplify the discussion it is assumed that $K = 1$. After sampling with period T_0 and holding, the Z transfer function of this typical plant can be given by

$$G_0(z) = z^{-l} \frac{(1-\sigma)z^{-1}}{1-\sigma z^{-1}} \tag{4.36}$$

where $\sigma = \exp(-T_0/T)$, $l = \tau/T_0$ is a nonnegative integer. For the case that τ is not an integer multiple of T_0, the method presented here can also be used to deduce corresponding results and thus it is not discussed here.

The typical plant (4.36) has a delay time of l steps. According to Section 3.2.3, if model mismatch is not considered, select the optimization horizon as $P + l$ and the error weighting matrix as block diag$[\mathbf{0}_{l\times l}, \boldsymbol{Q}_{P\times P}]$; then its DMC design can be attributed to the design for the following delay-free first-order plant with the performance index (2.3):

$$G(z) = \frac{(1-\sigma)z^{-1}}{1-\sigma z^{-1}} \tag{4.37}$$

The closed-loop response of plant (4.36) under DMC control coincides with that of plant (4.37), except that an additional l steps delay time will appear in the output. Therefore, in the following, it is only required to discuss the DMC parameter selection for the first-order inertia plant (4.37).

The coefficients of the unit step response of the first-order inertia plant (4.37) are given by

$$a_i = 1-\sigma^i, \quad i = 0, 1, 2, \ldots \tag{4.38}$$

which satisfy

$$a_{i+1} - \sigma a_i = 1-\sigma$$
$$a_{i+1} - a_i = \sigma(a_i - a_{i-1}) = (1-\sigma)\sigma^i$$

In order to investigate the quantitative relationship between the design parameters and the system performances, firstly consider the case of the control horizon $M \geq 2$. Using Theorem 4.5 to the first-order inertia plant (4.37), the following theorem can be obtained.

Theorem 4.8 In the DMC algorithm for the first-order inertia plant (4.37), select $\boldsymbol{R} = \mathbf{0}$, $M \geq 2$, $P \geq M$. Then the controlled system has a deadbeat property and the closed-loop transfer function is given by

$$F_0(z) = z^{-1}$$

This result is identical to that of using the OSOS given in Section 3.2.2. As discussed there, this kind of control is sensitive to the parameters and may result in large control actions. Although it can be used for the first-order plant (4.37), it is however not a good choice. In the following, for the DMC algorithm, the parameters in the performance index (2.3) are set as follows:

Control horizon $M = 1$,
Error weighting matrix $\boldsymbol{Q} = \boldsymbol{I}$,
Control weighting matrix $\boldsymbol{R} = r$.

Then the free design parameters are only P and r.

With the above parameter setting, the DMC control coefficients can be calculated by

$$d_i = a_i \Big/ \left(r + \sum_{j=1}^{P} a_j^2 \right)$$
$$d_s = \left(\sum_{j=1}^{P} a_j \right) \Big/ \left(r + \sum_{j=1}^{P} a_j^2 \right)$$

To facilitate the analysis, introduce the notations

$$s_1 \triangleq \sum_{j=1}^{P} a_j, \qquad s_2 \triangleq \sum_{j=1}^{P} a_j^2$$

Both s_1 and s_2 are related to σ and the optimization horizon P. It is obvious that $s_1 > 0$, $s_2 > 0$, $s_1 > s_2$, and s_1/s_2 decreases with increasing P because

$$\frac{s_1(P+1)}{s_2(P+1)} = \frac{s_1(P) + a_{P+1}}{s_2(P) + a_{P+1}^2} < \frac{s_1(P)}{s_2(P)}$$

due to

$$a_{P+1}s_2(P) - a_{P+1}^2 s_1(P) = \sum_{j=1}^{P} a_j a_{P+1}\left(a_j - a_{P+1}\right) < 0$$

Furthermore, introduce

$$\beta \triangleq \frac{\sum_{j=1}^{P} a_j^2}{r + \sum_{j=1}^{P} a_j^2} = \frac{s_2}{r + s_2}$$

$$\lambda \triangleq \frac{\sum_{j=1}^{P} a_j\left(a_{j+1} - a_j\right)}{\sum_{j=1}^{P} a_j^2} = \frac{\sum_{j=1}^{P} a_j(1-\sigma)\left(1-a_j\right)}{\sum_{j=1}^{P} a_j^2} = \frac{(1-\sigma)(s_1 - s_2)}{s_2}$$

It then follows that

$$d_i = \beta a_i / s_2$$
$$d_s = \beta s_1 / s_2 = \beta(1 - \sigma + \lambda)/(1 - \sigma)$$

The physical meanings and characters of above critical symbols are illustrated in Table 4.1.

Now a detailed analysis of the properties of the closed-loop system is given as follows.

Table 4.1 The correlations and characters of σ, λ, and β.

	Correlation with				
	Modeling	Control strategy			
Symbols	T_0/T	P	r	Ranges and characters	
σ	√	×	×	decreases with increasing T_0/T,	$0_{T_0/T\to+\infty} < \sigma < 1_{T_0/T\to 0}$
λ	√	√	×	decreases with increasing P,	$0_{P\to+\infty} < \lambda \le \sigma_{P=1} < 1$
β	√	√	√	decreases with increasing r,	$0_{r\to+\infty} < \beta \le 1_{r=0}$

(1) Closed-loop characteristic polynomial and stability
According to (4.29), (4.30), and (4.37), the DMC controller can be written as

$$G_C(z) = \frac{d_s(1-\sigma z^{-1})}{1+p_1^* z^{-1} + p_2^* z^{-2}} \tag{4.39}$$

where

$$\begin{bmatrix} 1 \\ p_1^* \\ p_2^* \end{bmatrix} = \begin{bmatrix} 1 & 0 \\ b_2-1 & 1 \\ b_3-b_2 & b_2-1 \end{bmatrix} \begin{bmatrix} 1 \\ -\sigma \end{bmatrix}$$

With the above parameter setting, it follows that

$$b_2 = \sum_{j=1}^{P} d_j a_{j+1} = \beta \sum_{j=1}^{P} a_j a_{j+1}/s_2 = \beta(1+\lambda)$$

$$b_3 - b_2 = \sum_{j=1}^{P} d_j\left(a_{j+2} - a_{j+1}\right) = \sigma\beta \sum_{j=1}^{P} a_j\left(a_{j+1} - a_j\right)/s_2 = \sigma\lambda\beta$$

and

$$p_1^* = (b_2-1) - \sigma = \beta(1+\lambda) - 1 - \sigma$$
$$p_2^* = (b_3-b_2) - \sigma(b_2-1) = \sigma(1-\beta)$$

Then we have the following theorem.

Theorem 4.9 Using the DMC algorithm to the first-order inertia plant (4.37), select $M = 1$, $\boldsymbol{Q} = \boldsymbol{I}$. The closed-loop system is always stable.
Proof: Since plant (4.37) is stable and there is no model mismatch, closed-loop stability depends only on the stability of the controller (4.39) with the following necessary and sufficient conditions:

$$1 + p_1^* + p_2^* = \beta(1+\lambda-\sigma) > 0$$
$$1 - p_1^* + p_2^* = 2(1+\sigma) - \beta(1+\lambda+\sigma) \geq 1 + \sigma - \lambda > 0$$
$$-1 < p_2^* = \sigma(1-\beta) < 1$$

which are obviously satisfied. Therefore, the closed-loop system is always stable.

(2) Dynamic performance of the closed-loop system
In the case without a model mismatch, it is known from (3.2) that the Z transfer function of the closed-loop system is given by

$$F_O(z) = G_C(z)G_P(z) = \frac{d_s(1-\sigma)z^{-1}}{1+p_1^* z^{-1} + p_2^* z^{-2}}$$

The following two cases are discussed:

1) The case where $r = 0$. In this case $\beta = 1$, $p_2^* = 0$, and $p_1^* = \lambda - \sigma$. Note that $d_s = \beta(1 - \sigma + \lambda)/(1 - \sigma)$, the Z transfer function of the closed-loop system is given by

$$F_0(z) = \frac{d_s(1-\sigma)z^{-1}}{1-(\sigma-\lambda)z^{-1}} = \frac{(1-\sigma+\lambda)z^{-1}}{1-(\sigma-\lambda)z^{-1}} \tag{4.40}$$

This indicates that when $r = 0$ the closed-loop system is still a first-order inertia component, with the steady-state gain 1 and the time constant

$$T^* = -T_0/\ln(\sigma-\lambda) = T\ln\sigma/\ln(\sigma-\lambda) \tag{4.41}$$

The controlled system has a faster response than the original one because $T^* < T$. Furthermore, T^* increases with P because $\sigma - \lambda$ increases with P. Particularly for $P = 1$, $\lambda = \sigma$, $T^* \to 0^+$, the system output reaches the steady-state value only after a one-step delay. For sufficiently large P, $\lambda \to 0$, $T^* \to T$, close to the original response of the plant.

Note that $\sigma - \lambda$ in (4.41) can also be explicitly expressed by P and σ:

$$\sigma-\lambda = \sigma\frac{(1-\sigma^2)P-(1-\sigma^P)(1+\sigma+\sigma^2-\sigma^{P+1})}{(1-\sigma^2)P-\sigma(1-\sigma^P)(2+\sigma-\sigma^{P+1})} \tag{4.42}$$

where σ is only related to the sampling period T_0 and the plant time constant T. Scaled with T, the relationship between T^* with the sampling period T_0 and the optimization horizon P is given in Figure 4.4. It is shown that T^* increases monotonically both with P and T_0. Particularly for small T_0 or small P, T^* changes almost linearly with T_0.

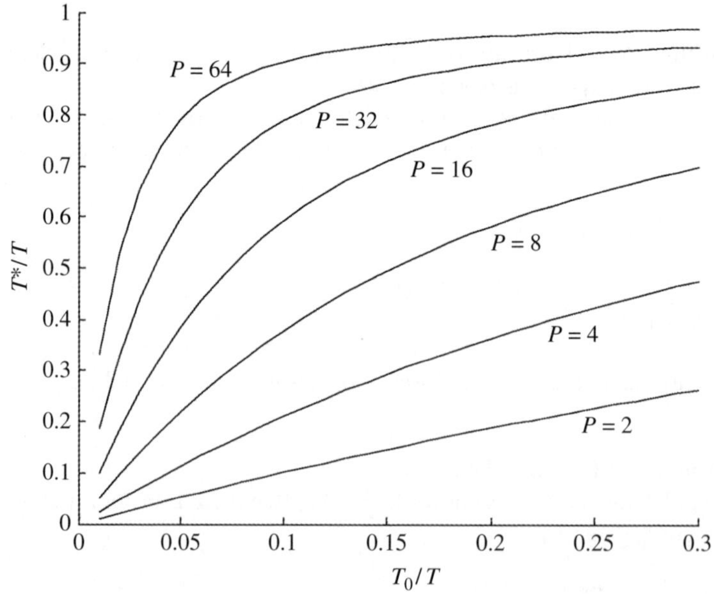

Figure 4.4 The relationship between the closed-loop time constant T^* and P, T_0.

2) The case where $r > 0$. In this case the closed-loop Z transfer function is given by

$$F_0(z) = \frac{\beta(1-\sigma+\lambda)z^{-1}}{1+(\beta(1+\lambda)-(1+\sigma))z^{-1}+\sigma(1-\beta)z^{-2}} \tag{4.43}$$

This is a second-order component with steady-state gain of 1. Denote

$$\begin{aligned}\Delta &= (\beta(1+\lambda)-(1+\sigma))^2-4\sigma(1-\beta)\\ &= (1+\lambda)^2\beta^2-2((1+\lambda)(1+\sigma)-2\sigma)\beta+(1-\sigma)^2\end{aligned}$$

The roots of $\Delta = 0$ are given by

$$\beta_1 = \left(\frac{\sqrt{1+\lambda-\sigma}-\sqrt{\lambda\sigma}}{1+\lambda}\right)^2, \quad \beta_2 = \left(\frac{\sqrt{1+\lambda-\sigma}+\sqrt{\lambda\sigma}}{1+\lambda}\right)^2 \tag{4.44}$$

It is easy to show that $\beta_2 > \beta_1 > 0$ and

$$\begin{aligned}&\Delta \geq 0, \quad \beta \leq \beta_1 \text{ or } \beta \geq \beta_2\\ &\Delta < 0, \quad \beta_1 < \beta < \beta_2\end{aligned}$$

Due to the correspondence between β and r, denote $r_i = (1-\beta_i)s_2/\beta_i$, $i = 1, 2$; then

$$\begin{aligned}&\Delta \geq 0, \quad r \geq r_1 \text{ or } r \leq r_2\\ &\Delta < 0, \quad r_1 > r > r_2\end{aligned}$$

Given σ and P, β_1 and β_2 can be calculated according to (4.44), and then r_1 and r_2 are easy to obtain. In the range $r_1 > r > r_2$, $\Delta < 0$, the closed-loop system (4.43) has two conjugate complex poles and is stable but with oscillation. If $r \geq r_1$ or $r \leq r_2$, $\Delta \geq 0$, the closed-loop system (4.43) has two real poles, corresponding to two first-order inertia dynamics, and the time constant will depend on that of the dominant one

$$T^* = -T_0/\ln\frac{1}{2}\left((1+\sigma)-\beta(1+\lambda)+\sqrt{\Delta}\right) \tag{4.45}$$

The above quantitative analysis for the first-order system is summarized in Table 4.2.

Table 4.2 Performance analysis of DMC for typical first-order systems.

Model parameters T, T_0 given **Design parameters setting: $M = 1$, $Q = I$.**		**Free design parameters: P, r** **Critical variables: σ, $\lambda(P)$, $\beta(P, r)$**
	$r = 0$	**$r > 0$**
Stability	Controller: (4.39) Guaranteed stability: Theorem 4.9	
Closed-loop dynamics	First-order inertia component: (4.40) Time constant T^*: (4.41), $T^* < T$ Increasing with P, T_0: Figure 4.4	Second-order component: (4.43) Two roots of $\Delta = 0$ β_1, β_2: (4.44) $\Rightarrow r_1, r_2$ $r_1 > r > r_2$: with oscillation $r \geq r_1$ or $r \leq r_2$: monotonically rising dominant T^*: (4.45)

4.4.2 Quantitative Analysis for Second-Order Systems

In this section, we briefly give some results on the quantitative analysis of predictive control for typical oscillation systems. Consider the typical second-order oscillation plant with natural frequency ω_0 and damping coefficient $\xi(<1)$ described by the following transfer function:

$$G(s) = \frac{\omega_0^2}{s^2 + 2\xi\omega_0 s + \omega_0^2} \tag{4.46}$$

This system has a damped natural frequency $\omega_n = \omega_0\sqrt{1-\xi^2}$, oscillatory period $T = 2\pi/\omega_n$, and its unit step response is given by

$$a(t) = 1 - \frac{e^{-\eta t}}{f_1}\sin\left(\omega_n t + \arctan\frac{1}{f_2}\right) \tag{4.47}$$

where

$$f_1 = \sqrt{1-\xi^2}, \qquad f_2 = \xi/\sqrt{1-\xi^2}, \qquad \eta = \omega_0\xi = \omega_n f_2$$

Denote the sampling period as $T_0(T_0 < \pi/2\omega_n)$, $\Omega = \omega_n T_0$, $\sigma = \exp(-\eta T_0) = \exp(-\Omega f_2)$. It is obvious that $\cos\Omega > 0$, $0 < \sigma < 1$, and the sampled values of the system step response can be given by

$$a_i = 1 - \sigma^i(f_2 \sin i\Omega + \cos i\Omega), \quad i = 1, 2, \ldots \tag{4.48}$$

It is easy to prove the following lemma.

Lemma 4.7 For a_i in (4.48), it follows that

$$\begin{aligned} &1)\ a_i > 0,\ i = 1, 2, \ldots ;\\ &2)\ (a_{i+1} - a_i) - 2\sigma\cos\Omega\,(a_i - a_{i-1}) + \sigma^2(a_{i-1} - a_{i-2}) = 0,\ i \geq 2 ;\\ &3)\ a_i - 2\sigma(\cos\Omega)\,a_{i-1} + \sigma^2 a_{i-2} = 1 - 2\sigma\cos\Omega + \sigma^2,\ i \geq 2 . \end{aligned} \tag{4.49}$$

After sampling with period T_0 and holding, the plant (4.46) has the Z transfer function

$$G(z) = \frac{m_1 z^{-1} + m_2 z^{-2}}{1 + p_1 z^{-1} + p_2 z^{-2}} \tag{4.50}$$

where

$$\begin{aligned} p_1 &= -2\sigma\cos\Omega, \quad p_2 = \sigma^2\\ m_1 &= a_1 = 1 - \sigma(f_2\sin\Omega + \cos\Omega)\\ m_2 &= (a_2 - a_1) - 2a_1\sigma\cos\Omega = \sigma^2 + \sigma(f_2\sin\Omega - \cos\Omega) \end{aligned}$$

It is easy to prove that $m_1 > |m_2|$.

To simplify the quantitative analysis, the design parameters here are also set the same as those for the first-order system in the last section, i.e. $M = 1$, $\boldsymbol{Q} = \boldsymbol{I}$, $\boldsymbol{R} = r$. It is then focused on how the closed-loop performances are affected by the optimization horizon P and the control weighting coefficient r.

Using the above parameter setting, define the same symbols as in the last section, i.e.

$$d_i = \beta a_i / s_2, \quad d_s = \beta s_1 / s_2$$

where

$$s_1 = \sum_{j=1}^{P} a_j, \quad s_2 = \sum_{j=1}^{P} a_j^2, \qquad \beta = s_2/(r+s_2)$$

According to (4.28), the closed-loop transfer function is now given by

$$F_0(z) = \frac{d_s m(z)}{p^*(z)} = \frac{d_s(m_1 z^{-1} + m_2 z^{-2})}{1 + p_1^* z^{-1} + p_2^* z^{-2} + p_3^* z^{-3}} \tag{4.51}$$

where

$$\begin{bmatrix} p_1^* \\ p_2^* \\ p_3^* \end{bmatrix} = \begin{bmatrix} b_2 - 1 & 1 & 0 \\ b_3 - b_2 & b_2 - 1 & 1 \\ b_4 - b_3 & b_3 - b_2 & b_2 - 1 \end{bmatrix} \begin{bmatrix} 1 \\ -2\sigma\cos\Omega \\ \sigma^2 \end{bmatrix}$$

with

$$b_i = \sum_{j=1}^{P} d_j a_{i+j-1} = \beta \sum_{j=1}^{P} a_j a_{i+j-1}/s_2, \qquad i = 2, 3, 4 \tag{4.52}$$

Define

$$b_1 \triangleq \sum_{j=1}^{P} d_j a_j = \beta \sum_{j=1}^{P} a_j a_j / s_2 = \beta, \; b_0 \triangleq \sum_{j=1}^{P} d_j a_{j-1} = \beta \sum_{j=1}^{P} a_j a_{j-1}/s_2$$

We then have the following lemma.

Lemma 4.8 For b_i in (4.52), the following equations hold:

$$\begin{aligned} &1)\; (b_{i+1} - b_i) - 2\sigma\cos\Omega(b_i - b_{i-1}) + \sigma^2(b_{i-1} - b_{i-2}) = 0, \qquad i \geq 2; \\ &2)\; b_i - 2\sigma(\cos\Omega)b_{i-1} + \sigma^2 b_{i-2} = d_s(1 - 2\sigma\cos\Omega + \sigma^2), \qquad i \geq 2. \end{aligned} \tag{4.53}$$

This lemma is easy to prove by Lemma 4.7 with the expressions of b_i. Now we can get

$$\begin{aligned} p_1^* &= \beta \sum_{j=1}^{P} a_j a_{j+1}/s_2 - 1 - 2\sigma\cos\Omega \\ p_2^* &= \beta\sigma^2 \sum_{j=1}^{P} a_j a_{j-1}/s_2 + \left(\sigma^2 + 2\sigma\cos\Omega\right)(1-\beta) \\ p_3^* &= \sigma^2(\beta - 1) \end{aligned} \tag{4.54}$$

In the following, stability and dynamic performance of the closed-loop system are discussed in two cases where the control action in the performance index (2.3) is not weighted ($r = 0$) and is weighted ($r > 0$), respectively.

(1) Closed-loop performance analysis without control weighting ($r = 0$)

Without control weighting in the performance index (2.3), i.e. $r = 0$, then $\beta = 1$, which leads to $p_3^* = 0$ and the closed-loop characteristic polynomial

$$p^*(z) = 1 + p_1^* z^{-1} + p_2^* z^{-2} \tag{4.55}$$

1) **Stability.** For the stability of the controlled system when $r = 0$, the following theorem is given.

Theorem 4.10 Using the DMC algorithm to the second-order oscillation plant (4.50), select $M = 1$, $\boldsymbol{Q} = \boldsymbol{I}$, $r = 0$; then the closed-loop system is always stable.

Proof: When $P = 1$, this is the OSOS case analyzed in Section 3.2.2. The controller is the model inverse with a one-step time delay, which makes the zero-pole cancellation with the plant. The closed-loop response is $F_0(z) = z^{-1}$. Since $m_1 > |m_2|$, this cancellation is stable.

When $P \geq 2$, the necessary and sufficient condition for (4.55) stable is given by

$$1)\ C_1 \triangleq 1 + p_1^* + p_2^* = 1 + \sum_{j=1}^{P} a_j a_{j+1}/s_2 - 1 - 2\sigma\cos\Omega + \sigma^2 \sum_{j=1}^{P} a_j a_{j-1}/s_2 > 0;$$

$$2)\ C_2 \triangleq 1 - p_1^* + p_2^* = 1 - \sum_{j=1}^{P} a_j a_{j+1}/s_2 + 1 + 2\sigma\cos\Omega + \sigma^2 \sum_{j=1}^{P} a_j a_{j-1}/s_2 > 0;$$

$$3)\ -1 < C_3 \triangleq p_2^* = \sigma^2 \sum_{j=1}^{P} a_j a_{j-1}/s_2 < 1.$$

Using Lemma 4.7 and through some elementary mathematical operations, these conditions can be proved true, the closed-loop system is thus always stable.

2) **Dynamic performance of the closed-loop system.** Denote $\Delta = p_1^{*^2} - 4p_2^*$. Then (4.55) has a pair of conjugate complex roots if $\Delta < 0$ and the closed-loop system exhibits a second-order oscillation response. In this case, the closed-loop system can be fitted by a standard second-order oscillation component whose corresponding parameters are marked by an asterisk, i.e.

$$p^*(z) = 1 - 2\sigma^* \cos\Omega^* z^{-1} + \sigma^{*2} z^{-2}$$

Compared with (4.55), it follows that

$$\sigma^* = \sqrt{p_2^*}$$
$$\Omega^* = \arccos\left(-p_1^*/2\sigma^*\right) = \arccos\left(-p_1^*/2\sqrt{p_2^*}\right)$$

Then the damped natural frequency and the damping coefficient can be given by

$$\begin{aligned} &\omega_n^* = \Omega^*/T_0 = \arccos\left(-p_1^*/2\sqrt{p_2^*}\right)/T_0 \\ &\xi^* = \eta^*/\sqrt{\eta^{*2} + \omega_n^{*2}}, \qquad \eta^* = -\ln\sigma^*/T_0 \end{aligned} \tag{4.56}$$

The closed-loop response can be fitted using the step response

$$a^*(t) = 1 - \frac{e^{-\eta^* t}}{\sqrt{1-\xi^{*2}}} \sin\left(\omega_n^* t + \arctan\frac{\omega_n^*}{\eta^*}\right) \tag{4.57}$$

and the maximal overshoot can be obtained by

$$c_{\max}^* = \exp\left(-\eta^* \pi / \omega_n^*\right) \tag{4.58}$$

For a given second-order oscillation plant (4.46) with known ω_0 and ξ, the closed-loop predictive control system may still be a second-order oscillation one. Its characteristic parameters of the oscillation response (damped natural frequency, damping coefficient, maximal overshoot) can be directly estimated when the sampling period T_0 and the design parameter P are given.

Example 4.3 The transfer function of an oscillation plant is given by

$$G(s) = \frac{41.1234}{s^2 + 2.5651s + 41.1234}$$

With $\omega_0 = 6.4127$ and $\xi = 0.2$, we can get the damped natural frequency, $\omega_n = 6.2832 = 2\pi$, and the oscillatory period $T = 1$. The quantitative relationship between the damping coefficient and maximal overshoot of the controlled system with the optimization horizon P and the sampling period T_0 is shown in Figure 4.5. Simulations for various sampling periods and $P = 4$ in Figure 4.6 show that the fitted second-order response curve obtained using the analytic calculation is almost identical with the actual dynamic response curve.

(2) Closed-loop performance analysis with control weighting ($r > 0$)

With control weighting in the performance index (2.3), i.e. $r > 0$, the closed-loop characteristic polynomial is given by

$$p^*(z) = 1 + p_1^* z^{-1} + p_2^* z^{-2} + p_3^* z^{-3} \tag{4.59}$$

1) **Stability**

Combined with (4.54), the necessary and sufficient conditions for the stable closed-loop system are given by

$$\begin{aligned}
&1)\ \ 1 + p_1^* + p_2^* + p_3^* > 0 && \Rightarrow\ \ \beta C_1 > 0, \\
&2)\ \ 1 - p_1^* + p_2^* - p_3^* > 0 && \Rightarrow\ \ 2(1-\beta)(1 + 2\sigma\cos\Omega + \sigma^2) + \beta C_2 > 0, \\
&3)\ \ |p_3^*| < 1 && \Rightarrow\ \ |\sigma^2(\beta - 1)| < 1, \\
&4)\ \ p_3^{*2} - 1 < p_1^* p_3^* - p_2^* && \Rightarrow\ \ P(\beta) = A\beta^2 + B\beta + C > 0
\end{aligned} \tag{4.60}$$

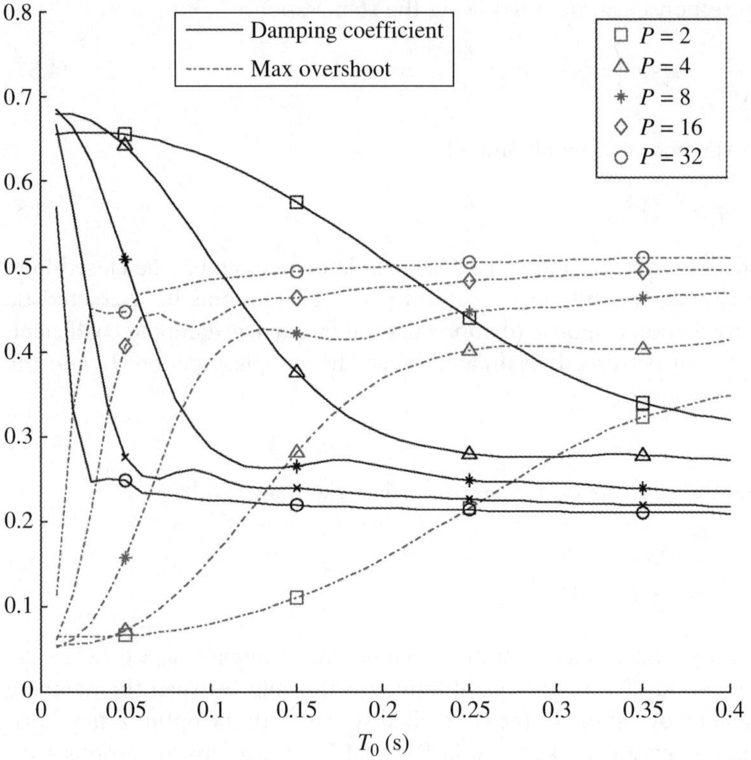

Figure 4.5 Relationship between ξ^* (solid line), c^*_{max} (dotted line), and P, T_0.

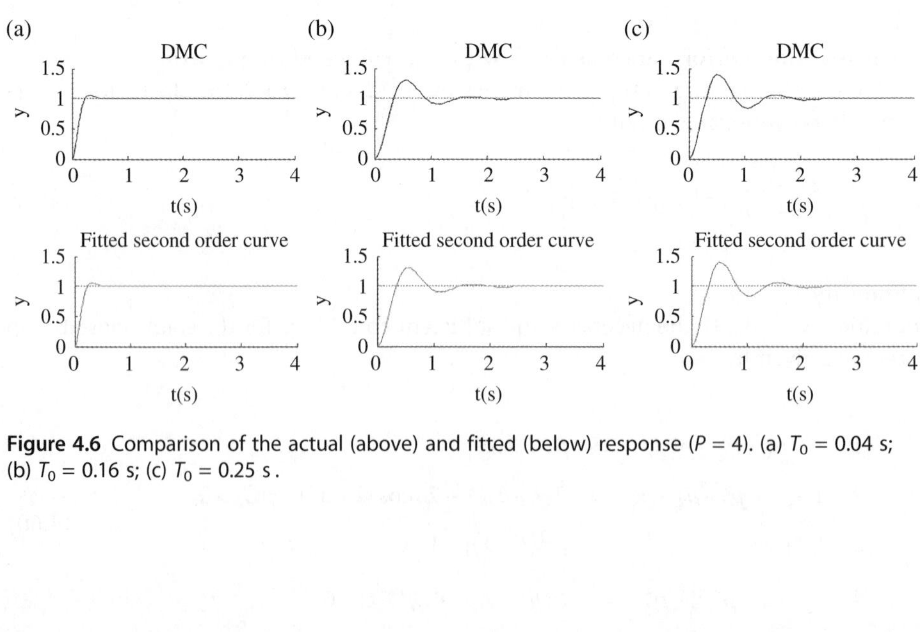

Figure 4.6 Comparison of the actual (above) and fitted (below) response (P = 4). (a) T_0 = 0.04 s; (b) T_0 = 0.16 s; (c) T_0 = 0.25 s .

where C_1, C_2 are referred to the proof of Theorem 4.10 and the parameters in (4) are given by

$$\begin{aligned} A &= \sigma^2\left(\sum_{j=1}^{P} a_j a_{j+1}/s_2 - \sigma^2\right) \\ B &= 2\sigma^4 + 2\sigma\cos\Omega(1-\sigma^2) - \sigma^2\left(\sum_{j=1}^{P} a_j a_{j+1} + \sum_{j=1}^{P} a_j a_{j-1}\right)\Big/ s_2 \\ C &= (1-\sigma^2)(1-2\sigma\cos\Omega+\sigma^2) \end{aligned} \tag{4.61}$$

which are only related to the plant parameters, sampling period, and optimization horizon, and are independent of β or the control weighting r.

In Theorem 4.10 it has been proved that $C_1 > 0$, $C_2 > 0$; thus the conditions (1) to (3) obviously hold. The necessary and sufficient condition for the stability of the closed-loop system can thus be summed up as to whether the condition (4) is established or not. Note that

$$P(0) = C = (1-\sigma^2)(1-2\sigma\cos\Omega+\sigma^2) > 0$$

$$P(1) = A + B + C = 1 - \sigma^2\sum_{j=1}^{P} a_j a_{j-1}/s_2 = 1 - C_3 > 0$$

where C_3 are referred to the proof of Theorem 4.10. Since $0 < \beta \le 1$, if $A \le 0$, then $P(\beta) > 0$ holds for all $0 < \beta \le 1$, i.e. the closed-loop system is stable for all r. If $A > 0$, there are two possible cases:

Case 1: $B^2 - 4AC < 0$. In this case $P(\beta) > 0$ holds for all $0 < \beta \le 1$, i.e. the closed-loop system is stable for all r.

Case 2: $B^2 - 4AC \ge 0$. In this case $P(\beta)$ has two real roots, denoted by β_1, β_2, and $\beta_1 \le \beta_2$. According to $0 < \beta \le 1$ and $P(\beta) > 0$ for $\beta = 0$ or $\beta = 1$, we have the following possibilities for the positions of β_1 and β_2.

When $1 < \beta_1 \le \beta_2$ or $\beta_1 \le \beta_2 < 0$, $P(\beta) > 0$ holds for all $0 < \beta \le 1$ and the closed-loop system is stable for all r.

When $0 < \beta_1 \le \beta_2 < 1$, $P(\beta) \le 0$ holds for $\beta_1 \le \beta \le \beta_2$, and the closed-loop system is unstable at the corresponding range of the design parameter $r_2 \le r \le r_1$.

Based on the above analysis, the following theorem can be obtained.

Theorem 4.11 Using the DMC algorithm for the second-order oscillation plant (4.50), select $M = 1$, $\boldsymbol{Q} = \boldsymbol{I}$, $r > 0$. Then the closed-loop system is stable except when P, r are selected such that the parameters in (4.61) $A > 0$ and $B^2 - 4AC \ge 0$, the solution of $P(\beta) = 0$ satisfies $0 < \beta_1 \le \beta_2 < 1$ and the corresponding $r_2 \le r \le r_1$.

Remark 4.9 The above theorem indicates that for the typical second-order oscillation plant (4.50), when using the DMC control strategy and $r \neq 0$, increasing r is not always helpful for the stability of the system. With r increasing from 0, the controlled system starts from a stable status, then possibly becomes unstable, exhibits oscillation and diverges, and then gradually restores to a stable status as r continues to increase. Since β_1, β_2 depend on the coefficients A, B, C, and only depend on the optimization horizon P if the plant and the sampling period are given, the range for r to guarantee the closed-loop system stable with respect to different P values can be easily calculated using the above equations.

Example 4.4 For the plant in Example 4.3, the relationship between the stable region of r and T_0, P, is given in Figure 4.7. It is shown that if $T_0 = 0.04$, select $P = 2$; then the system is unstable for $0.03 \le r \le 1$. The simulation results in Figure 4.8 exactly verify the correctness of the above theoretical analysis.

2) **Dynamic performance of the closed-loop system.** When $r > 0$, the closed-loop system exhibits a third-order dynamic response. According to the expression of $p^*(z)$ in (4.59), it is known that for sufficiently small r (i.e. $\beta \to 1$), the dynamic response is close to the second-order process. With the increase of r, another characteristic motion is markedly enhanced. When $r \to +\infty$ (i.e. $\beta \to 0$), the (1) in stability condition (4.60) tends to zero, indicating that the system has a real root close to $z = 1$, corresponding to the main characteristic motion of the system. From the simulation results for r varying in Figure 4.8, the correctness of the above analysis can be verified.

The main results of the above quantitative analysis for second-order oscillation systems are summarized in the following Table 4.3.

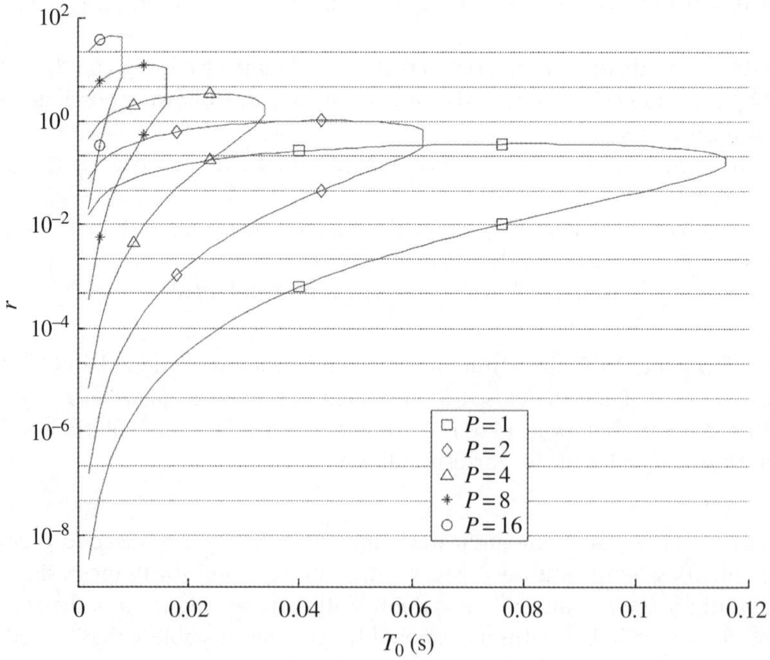

Figure 4.7 The stable region of r in relationship with T_0, P for Example 4.4.

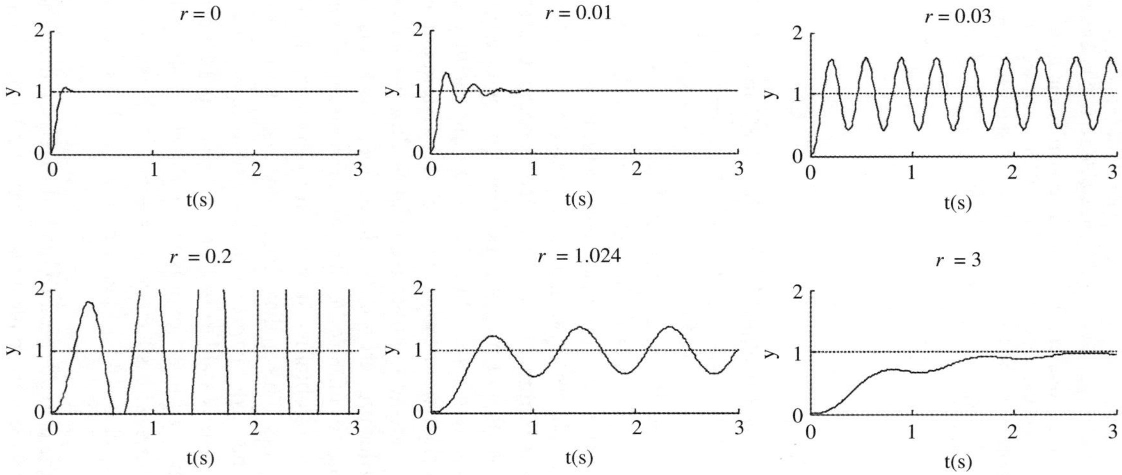

Figure 4.8 Closed-loop response with r for Example 4.4 ($T_0 = 0.04$, $P = 2$).

Table 4.3 Performance analysis of DMC for typical second-order oscillation systems.

Model parameters $\xi, \omega_0 \Rightarrow \omega_n, T$ Sampling period T_0 Design parameters setting: $M = 1$, $Q = I$.		Free design parameters: P, r. Critical variables: σ, Ω, f_2, η(depend on modeling) $s_2(P)$, $\beta(P, r)$(depend on control)
	$r = 0$	$r > 0$
Stability	Char. polynomial: (4.55) with (4.54) Guaranteed stability: Theorem 4.10	Char. polynomial: (4.59) with (4.54) Stability condition: Theorem 4.11 $\Rightarrow$ unstable only $A > 0$ and $B^2 - 4AC \geq 0$ and $0 < \beta_1 \leq \beta_2 < 1$ with A, B, C (4.61), $\beta_1, \beta_2 \Leftarrow P(\beta) = 0$ (4.60) Instable region: $r_2 \leq r \leq r_1$
Closed-loop dynamics	Second-order component $\Delta < 0 \Rightarrow$ oscillation response Fitted step response a_i^*: (4.57) with $\omega_n^*, \xi^*, \eta^*, c_{\max}^*$: (4.56), (4.58)	Third-order component $r \rightarrow 0^+$: near to second-order process $r \rightarrow +\infty$: sluggish response with a pole near to 1

4.5 Summary

In this chapter, for predictive control algorithms such as GPC and DMC, the quantitative relationships between the design parameters and the closed-loop performances are explored both in the time domain and in the Z domain. In the time domain, the GPC control law is firstly transformed into the LQ control law in the state space. By studying under which conditions the GPC control law is equivalent to the stable Kleinman controller, stability/deadbeat conditions of the closed-loop GPC system are derived in relation to the design parameters. In the Z domain, using the IMC structure, the expression of the GPC controller and the minimal form of the DMC controller are derived and both are proved identical. The coefficient mapping between the open-loop and closed-loop characteristic polynomials is then established. Based on the coefficient mapping, deadbeat/stability conditions of the closed-loop DMC or GPC system are derived in relation to the design parameters, and the quantitative relationship between the design parameters and the reduced-order property of the closed-loop system is also given.

By analyzing the performances of the GPC system based on the Kleinman controller in the state space, conclusions are obtained in relation to the system order, but are independent of specific parameters. The analysis based on coefficient mapping of the characteristic polynomials in the Z domain can not only give qualitative conclusions on whether the controlled system is stable/deadbeat but also the reduced-order property of the closed-loop system. The concrete expression of the closed-loop dynamic response can be given when combining with specific parameters of the system, which provides more information for the closed-loop performance analysis of predictive control systems.

The coefficient mapping is also adopted to quantitatively analyze the control performances of predictive control for some typical plants: the first-order inertial plant and the second-order oscillation plant. When setting partial design parameters, more concrete and explicit results on the stability and closed-loop dynamic response can be derived that are only dependent on a few remaining design parameters. This may be helpful to predictive control design. Particularly, a great number of systems can be approximated by these two typical plants and only some key parameters of the plant should be known beforehand, which are very easy to identify using the step response or according to the process data.

References

1 Clarke, D.W. and Mohtadi, C. (1989). Properties of generalized predictive control. *Automatica* 25 (6): 859–873.
2 Ding, B. and Xi, Y. (2004). Stability analysis of generalized predictive control based on Kleinman's controllers. *Science in China, Series F* 47 (4): 458–474.
3 Kleinman, D.L. (1974). Stabilizing a discrete, constant, linear system with application to iterative methods for solving the Riccatti equation. *IEEE Transactions on Automatic Control* 19 (3): 252–254.
4 Xi, Y. and Li, J. (1991). Closed-loop analysis of the generalized predictive control systems. *Control Theory and Applications* 8 (4): 419–424 (in Chinese).
5 Xi, Y. and Zhang, J. (1997). Study on the closed-loop properties of GPC. *Science in China, Series E* 40 (1): 54–63.
6 Xi, Y. (1989). Minimal form of a predictive controller based on the step response model. *International Journal of Control* 49 (1): 57–64.
7 Zhang, J. Study on some theoretical issues of predictive control. PhD thesis, 1997, Shanghai Jiao Tong University (in Chinese).

5

Predictive Control for MIMO Constrained Systems

In the above chapters, both introducing typical predictive control algorithms and investigating the quantitative relationship between the design parameters and the closed-loop performances are all for unconstrained SISO systems. Although predictive control can be adopted in single-loop control to replace the PID controller, its utility is more embodied in controlling constrained multivariable systems. In this chapter, based on the basic algorithms and principles introduced above, we will take the DMC algorithm as an example and introduce the predictive control algorithm for MIMO systems with constraints, which is closer to its real application status.

5.1 Unconstrained DMC for Multivariable Systems

In this section, we first introduce the DMC algorithm for multivariable systems without considering constraints [1]. As introduced in Section 2.1, the single-variable DMC algorithm is based on the following basic principles:

1) Output prediction based on the prediction model using proportion and superposition properties of linear systems;
2) Online rolling optimization based on optimal output tracking and control increment suppression;
3) Error prediction and correction based on real-time measured output.

Obviously these principles can be easily extended to multivariable systems.

Consider a stable MIMO plant with m control inputs and p outputs. Assume that the unit step response of each input u_j to each output y_i has been measured as $a_{ij}(t)$; then the following model vectors can be formed from the sampling values of these step responses:

$$\boldsymbol{a}_{ij} = \left[a_{ij}(1) \ \cdots \ a_{ij}(N)\right]^{\mathrm{T}}, \quad i = 1, \ \ldots, \ p, \quad j = 1, \ \ldots, \ m$$

where N is the model length as presented in Section 2.1. To unify symbols and facilitate programming, here the same model length N is adopted for the step responses between different inputs and outputs. The multivariable DMC algorithm then starts from these model vectors. In the following, similar to Section 2.1, the DMC algorithm for multivariable systems will be introduced based on the predictive control principles.

Predictive Control: Fundamentals and Developments, First Edition. Yugeng Xi and Dewei Li.

(1) Prediction model

For a multivariable linear system, each of its output is affected by several inputs, and the dynamic variation of the output can be superposed by the dynamic changes caused by each input. Firstly, consider the dynamic change of output y_i caused by input u_j. Similar to (2.1), at the sampling time k, the output y_i at future N sampling instants caused by the control increment $\Delta u_j(k)$ can be predicted by

$$\tilde{\boldsymbol{y}}_{i,N1}(k) = \tilde{\boldsymbol{y}}_{i,N0}(k) + \boldsymbol{a}_{ij}\Delta u_j(k) \tag{5.1}$$

where

$$\tilde{\boldsymbol{y}}_{i,N1}(k) = \begin{bmatrix} \tilde{y}_{i,1}(k+1|k) \\ \vdots \\ \tilde{y}_{i,1}(k+N|k) \end{bmatrix}, \quad \tilde{\boldsymbol{y}}_{i,N0}(k) = \begin{bmatrix} \tilde{y}_{i,0}(k+1|k) \\ \vdots \\ \tilde{y}_{i,0}(k+N|k) \end{bmatrix}$$

In addition to using i, j to distinguish the elements of y, u respectively, the meanings of other symbols are the same as that in Section 2.1. The element of $\tilde{\boldsymbol{y}}_{i,N0}(k)$ represents the initial prediction at the sampling time k for output y_i at future N sampling instants if all the control inputs $u_1, \cdots, u_m$ keep unchanged.

Similar to (2.4), if there are M successive control increments of u_j starting from time k, denoted as $\Delta u_j(k), \cdots, \Delta u_j(k+M-1)$, then under these control actions, the prediction of output y_i at future P sampling instants can be given by

$$\tilde{\boldsymbol{y}}_{i,PM}(k) = \tilde{\boldsymbol{y}}_{i,P0}(k) + \boldsymbol{A}_{ij}\Delta \boldsymbol{u}_{j,M}(k) \tag{5.2}$$

where

$$\tilde{\boldsymbol{y}}_{i,PM}(k) = \begin{bmatrix} \tilde{y}_{i,M}(k+1|k) \\ \vdots \\ \tilde{y}_{i,M}(k+P|k) \end{bmatrix}, \quad \tilde{\boldsymbol{y}}_{i,P0}(k) = \begin{bmatrix} \tilde{y}_{i,0}(k+1|k) \\ \vdots \\ \tilde{y}_{i,0}(k+P|k) \end{bmatrix}$$

$$\boldsymbol{A}_{ij} = \begin{bmatrix} a_{ij}(1) & & \boldsymbol{0} \\ \vdots & \ddots & \\ a_{ij}(M) & \cdots & a_{ij}(1) \\ \vdots & & \vdots \\ a_{ij}(P) & \cdots & a_{ij}(P-M+1) \end{bmatrix}, \quad \Delta \boldsymbol{u}_{j,M}(k) = \begin{bmatrix} \Delta u_j(k) \\ \vdots \\ \Delta u_j(k+M-1) \end{bmatrix}$$

Equations (5.1) and (5.2) are prediction models of y_i with respect to single u_j. If y_i is affected by several control inputs, $u_1, \ldots, u_m$, the prediction model can be established by using the superposition property of linear systems as follows.

If each $u_j (j = 1, \ldots, m)$ only has a current control increment $\Delta u_j(k)$, corresponding to (5.1), it follows that

$$\tilde{\boldsymbol{y}}_{i,N1}(k) = \tilde{\boldsymbol{y}}_{i,N0}(k) + \sum_{j=1}^{m} \boldsymbol{a}_{ij}\Delta u_j(k)$$

If, starting from time k, each $u_j(j = 1, \dots, m)$ has M successive control increments $\Delta u_j(k), \dots, \Delta u_j(k+M-1)$, then, corresponding to (5.2), it follows that

$$\tilde{\boldsymbol{y}}_{i,PM}(k) = \tilde{\boldsymbol{y}}_{i,P0}(k) + \sum_{j=1}^{m} \boldsymbol{A}_{ij} \Delta \boldsymbol{u}_{j,M}(k)$$

To simplify the symbols, integrate all y_i in one vector and denote

$$\tilde{\boldsymbol{y}}_{N1}(k) = \begin{bmatrix} \tilde{\boldsymbol{y}}_{1,N1}(k) \\ \vdots \\ \tilde{\boldsymbol{y}}_{p,N1}(k) \end{bmatrix}, \quad \tilde{\boldsymbol{y}}_{N0}(k) = \begin{bmatrix} \tilde{\boldsymbol{y}}_{1,N0}(k) \\ \vdots \\ \tilde{\boldsymbol{y}}_{p,N0}(k) \end{bmatrix}$$

$$\tilde{\boldsymbol{y}}_{PM}(k) = \begin{bmatrix} \tilde{\boldsymbol{y}}_{1,PM}(k) \\ \vdots \\ \tilde{\boldsymbol{y}}_{p,PM}(k) \end{bmatrix}, \quad \tilde{\boldsymbol{y}}_{P0}(k) = \begin{bmatrix} \tilde{\boldsymbol{y}}_{1,P0}(k) \\ \vdots \\ \tilde{\boldsymbol{y}}_{p,P0}(k) \end{bmatrix}$$

$$\overline{\boldsymbol{A}} = \begin{bmatrix} \boldsymbol{a}_{11} & \cdots & \boldsymbol{a}_{1m} \\ \vdots & & \vdots \\ \boldsymbol{a}_{p1} & \cdots & \boldsymbol{a}_{pm} \end{bmatrix}, \quad \boldsymbol{A} = \begin{bmatrix} \boldsymbol{A}_{11} & \cdots & \boldsymbol{A}_{1m} \\ \vdots & & \vdots \\ \boldsymbol{A}_{p1} & \cdots & \boldsymbol{A}_{pm} \end{bmatrix}$$

$$\Delta \boldsymbol{u}(k) = \begin{bmatrix} \Delta u_1(k) \\ \vdots \\ \Delta u_m(k) \end{bmatrix}, \quad \Delta \boldsymbol{u}_M(k) = \begin{bmatrix} \Delta \boldsymbol{u}_{1,M}(k) \\ \vdots \\ \Delta \boldsymbol{u}_{m,M}(k) \end{bmatrix}$$

Then the prediction models of DMC for multivariable systems can generally be given by

$$\tilde{\boldsymbol{y}}_{N1}(k) = \tilde{\boldsymbol{y}}_{N0}(k) + \overline{\boldsymbol{A}} \Delta \boldsymbol{u}(k) \tag{5.3}$$

$$\tilde{\boldsymbol{y}}_{PM}(k) = \tilde{\boldsymbol{y}}_{P0}(k) + \boldsymbol{A} \Delta \boldsymbol{u}_M(k) \tag{5.4}$$

Note that these two prediction models have different time horizons both for control variation and for output prediction, which will be convenient for later direct use.

(2) Rolling optimization

Similar to (2.5) for single variable systems, in rolling optimization of multivariable DMC, m control increment vectors $\Delta \boldsymbol{u}_{1,M}(k), \dots, \Delta \boldsymbol{u}_{m,M}(k)$, each with M step variations, would be determined such that, under their control, p output prediction vectors $\tilde{y}_{1,PM}(k), \dots, \tilde{y}_{1,PM}(k)$, each over future P sampling times, closely track the corresponding desired value vectors $\boldsymbol{w}_1(k), \dots, \boldsymbol{w}_p(k)$; meanwhile the control variations are suppressed. At the sampling time k, the performance index can be written as

$$\min J(k) = \|\boldsymbol{w}(k) - \tilde{\boldsymbol{y}}_{PM}(k)\|_{\boldsymbol{Q}}^2 + \|\Delta \boldsymbol{u}_M(k)\|_{\boldsymbol{R}}^2 \tag{5.5}$$

where

$$w(k)=\begin{bmatrix} w_1(k) \\ \vdots \\ w_p(k) \end{bmatrix},\quad w_i(k)=\begin{bmatrix} w_i(k+1) \\ \vdots \\ w_i(k+P) \end{bmatrix},\quad i=1,\ldots,p$$

$$Q=\text{block diag}\left(Q_1,\ \ldots,\ Q_p\right)$$

$$Q_i=\text{diag}\left[q_i(1),\ \ldots,\ q_i(P)\right],\quad i=1,\ \ldots,\ p$$

$$R=\text{block diag}(R_1,\ \ldots,\ R_m)$$

$$R_j=\text{diag}\left[r_j(1),\ \ldots,\ r_j(M)\right],\quad j=1,\ \ldots,\ m$$

It is obvious that the blocks $Q_1, \ldots, Q_p$ in the error weighting matrix Q correspond to different outputs, while the elements in Q_i correspond to the tracking errors of y_i at different sampling times. Similarly, the blocks $R_1, \ldots, R_m$ in the control weighting matrix R correspond to different control inputs, while the elements in R_j correspond to the suppression on the increments of u_j at different sampling times. Therefore, every element of Q, R in the performance index (5.5) has an intuitive physical meaning, which is very advantageous to parameter tuning.

If constraints are not considered, similar to (2.6), according to the prediction model (5.4), all the control increment vectors optimizing the performance index (5.5) can be calculated by

$$\Delta u_M(k)=\left(A^{\mathrm{T}}QA+R\right)^{-1}A^{\mathrm{T}}Q[w(k)-\tilde{y}_{P0}(k)] \tag{5.6}$$

and the current control increment can be given by

$$\Delta u(k)=D[w(k)-\tilde{y}_{P0}(k)] \tag{5.7}$$

where

$$D=L\left(A^{\mathrm{T}}QA+R\right)^{-1}A^{\mathrm{T}}Q\triangleq\begin{bmatrix} d_{11}^{\mathrm{T}} & \cdots & d_{1p}^{\mathrm{T}} \\ \vdots & & \vdots \\ d_{m1}^{\mathrm{T}} & \cdots & d_{mp}^{\mathrm{T}} \end{bmatrix} \tag{5.8}$$

and the $m \times mM$ dimensional matrix L represents the operation of taking the first, the $(M+1)$th, ..., the $((m-1)M+1)$th rows of the following matrix:

$$L=\begin{bmatrix} 1\ 0\ \cdots\ 0 & & \mathbf{0} \\ & \ddots & \\ \mathbf{0} & & 1\ 0\ \cdots\ 0 \end{bmatrix}$$

Note that here all d_{ji}^{T} ($i=1,\ldots,p, j=1,\ldots,m$) are P-dimensional row vectors. When the step responses are known and the control strategy is fixed, A, Q, R are known, and the elements of D, i.e. all d_{ji}^{T}, can then be off-line calculated using (5.8). When online, m control actions currently to be implemented can be calculated by

$$\begin{aligned} \Delta u_j(k)&=\sum_{i=1}^{p} d_{ji}^{\mathrm{T}}\left[w_i(k)-\tilde{y}_{i,P0}(k)\right] \\ u_j(k)&=u_j(k-1)+\Delta u_j(k),\quad j=1,\ \ldots,\ m \end{aligned} \tag{5.9}$$

(3) Feedback correction

After control implementation at time k, the predicted outputs at the future time $k+1$, i.e. $\tilde{y}_{i,1}(k+1 \mid k)$, $i = 1, \ldots, p$, can be calculated according to the prediction model (5.3). At the next time $k+1$, all the actual outputs $y_i(k+1)$ are measured and compared with corresponding predicted values to construct the prediction error vector

$$\boldsymbol{e}(k+1) = \begin{bmatrix} e_1(k+1) \\ \vdots \\ e_p(k+1) \end{bmatrix} = \begin{bmatrix} y_1(k+1) - \tilde{y}_{1,1}(k+1 \mid k) \\ \vdots \\ y_p(k+1) - \tilde{y}_{p,1}(k+1 \mid k) \end{bmatrix} \tag{5.10}$$

Using this immediate error information, all future prediction errors can be predicted using the heuristic weighting method and then used to compensate the model-based output prediction. After correction, the prediction vector has the form

$$\tilde{\boldsymbol{y}}_{\mathrm{cor}}(k+1) = \tilde{\boldsymbol{y}}_{N1}(k) + \boldsymbol{He}(k+1) \tag{5.11}$$

where

$$\boldsymbol{H} = \begin{bmatrix} \boldsymbol{h}_{11} & \cdots & \boldsymbol{h}_{1p} \\ \vdots & & \vdots \\ \boldsymbol{h}_{p1} & \cdots & \boldsymbol{h}_{pp} \end{bmatrix}, \quad \boldsymbol{h}_{ij} = \begin{bmatrix} h_{ij}(1) \\ \vdots \\ h_{ij}(N) \end{bmatrix}, \quad i,j = 1, \ldots, p$$

$\boldsymbol{H}$ is an error correction matrix constructed from a group of error correction vectors $\boldsymbol{h}_{ij}$. Since the cause of the error is unknown, cross-correction seems meaningless. Thus it is reasonable to let all the cross-correction vectors $\boldsymbol{h}_{ij}(i \neq j)$ be zero, i.e. only the diagonal blocks in $\boldsymbol{H}$ remain, which implies that the predicted output y_i will be corrected only by its own error.

Similar to (2.14), since the time base is moved from k to $k+1$, the corrected prediction vector $\tilde{\boldsymbol{y}}_{\mathrm{cor}}(k+1)$ can construct the initial prediction at time $k+1$ through shifting

$$\tilde{\boldsymbol{y}}_{N0}(k+1) = \boldsymbol{S}_0 \tilde{\boldsymbol{y}}_{\mathrm{cor}}(k+1) \tag{5.12}$$

where

$$\boldsymbol{S}_0 = \begin{bmatrix} \boldsymbol{S} & & \boldsymbol{0} \\ & \ddots & \\ \boldsymbol{0} & & \boldsymbol{S} \end{bmatrix}, \quad \boldsymbol{S} = \begin{bmatrix} 0 & 1 & & 0 \\ \vdots & & \ddots & \ddots \\ & & 0 & 1 \\ 0 & \cdots & 0 & 1 \end{bmatrix}$$

The online computation of the unconstrained multivariable DMC algorithm only concerns three kinds of parameters, including model vectors $\boldsymbol{a}_{ij}$, control vectors $\boldsymbol{d}_{ji}^{\mathrm{T}}$, and correction vectors $\boldsymbol{h}_{ii}$, respectively, where $\boldsymbol{a}_{ij}$ is determined by the step response and the sampling period, $\boldsymbol{h}_{ii}$ can be selected freely, while $\boldsymbol{d}_{ji}^{\mathrm{T}}$ (i.e. $\boldsymbol{D}$) should be off-line calculated by (5.8) in terms of the performance index (5.5). These parameters are stored in the fixed memories for a real-time call. During online control, all the output predictions are firstly initialized by the initially measured actual outputs. At each step, measure the actual

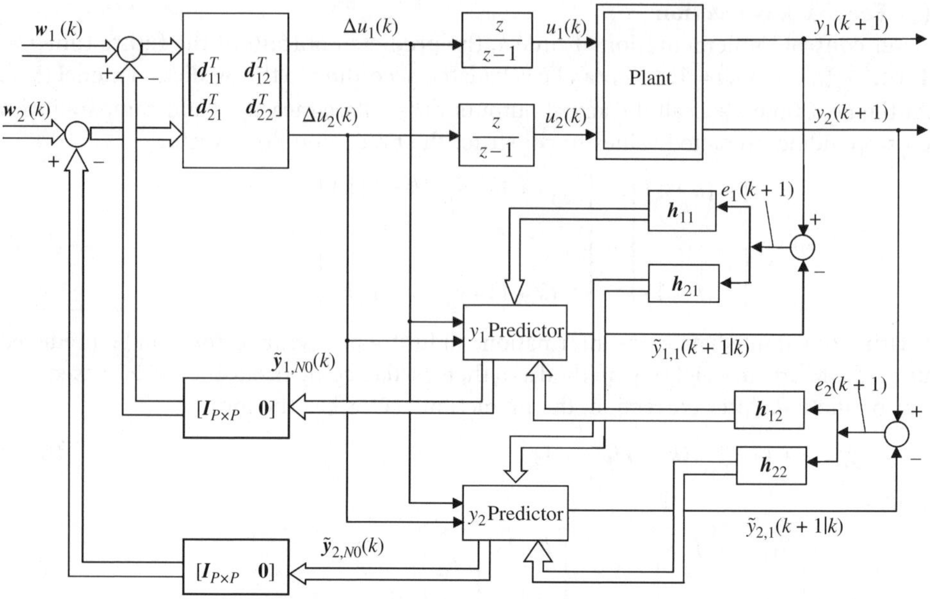

Figure 5.1 Unconstrained multivariable DMC algorithm ($m = p = 2$) [1]. *Source:* Reproduced with permission from Taylor and Francis and Copyright clearance center.

outputs of the plant, then make correction and shifting according to (5.10)–(5.12) and get the initial output predictions at this sampling instant. The current control action can be calculated by (5.9) and then acted to the real plant. In the meantime, the output prediction after implementing current control actions can be calculated by (5.3), among which the output prediction at the next sampling time will be used to compare with the actual outputs at the next time and to construct the output errors (5.10) for error correction (5.11) and shifting (5.12). This process runs repeatedly along with time. The structure of the unconstrained multivariable DMC algorithm for $m = p = 2$ is shown in Figure 5.1 [1].

From the above description and Figure 5.1 it is shown that the extension of the DMC algorithm for SISO systems presented in Section 2.1 to multivariable systems is straightforward. It still adopts the proportion and superposition principles of linear systems. However, in addition to the superposition of control increments at different sampling instants, the superposition of control increments caused by different control variables is also taken into account, i.e. during prediction and optimization, the superposition simply along time is extended to double superposition along both time (different sampling instants) and space (different control variables).

Remark 5.1 Although the DMC algorithm for the multivariable case is in principle not different from that for the single variable case, some specific remarks should be given as follows:

1) For a multivariable plant with m control inputs and p outputs, the step responses of all inputs to all outputs need to be measured. This can be achieved through setting a unit

step to u_j and measuring the corresponding outputs at N sampling instants for all y_i. Such experiments need to be performed one by one for m control inputs, with the obtained step responses having altogether $p \times m$ sets.

2) For set-point control, the control inputs and the outputs should have the following steady-state relationship if the control is stable:

$$\begin{bmatrix} y_{1s} \\ \vdots \\ y_{ps} \end{bmatrix} = \begin{bmatrix} a_{11s} & \cdots & a_{1ms} \\ \vdots & & \vdots \\ a_{p1s} & \cdots & a_{pms} \end{bmatrix} \begin{bmatrix} u_{1s} \\ \vdots \\ u_{ms} \end{bmatrix}$$

where subscript s represents the steady state value of corresponding variable. If the rank of above $p \times m$ matrix is p, there are enough degrees of freedom to control all the plant outputs to arbitrary desired values. However, if $m < p$, the rank of this matrix is always smaller than p. In this case, it is impossible to use less control inputs to make more outputs reach arbitrary desired values, but through coordinating the weighting matrix $\boldsymbol{Q}_i$ in the performance index (5.5), the proximity degrees of different outputs to respective desired values can be artificially specified.

3) If the plant has the same number of inputs and outputs, i.e. $m = p$, then the following case is similar to the single variable case in Section 3.2.2. When constraints are not considered and the optimization strategy is taken as OSOS, i.e. $P = M$, $\boldsymbol{Q} = \boldsymbol{I}$, $\boldsymbol{R} = 0$, then the derived control law is identical to that for $P = M = 1$. To illustrate this, consider (5.6) and also assume that $\boldsymbol{A}$ is invertible; then the optimal solution satisfies

$$\boldsymbol{A}\Delta\boldsymbol{u}_M(k) = \boldsymbol{w}(k) - \tilde{\boldsymbol{y}}_{P0}(k)$$

The corresponding equalities at its first, the $(P+1)$th, …, the $((p-1)P+1)$th rows are given by

$$a_{11}(1)\Delta u_1(k) + \cdots + a_{1m}(1)\Delta u_m(k) = w_1(k+1) - \tilde{y}_{1,0}(k+1 \mid k)$$
$$a_{21}(1)\Delta u_1(k) + \cdots + a_{2m}(1)\Delta u_m(k) = w_2(k+1) - \tilde{y}_{2,0}(k+1 \mid k)$$
$$\vdots$$
$$a_{p1}(1)\Delta u_1(k) + \cdots + a_{pm}(1)\Delta u_m(k) = w_p(k+1) - \tilde{y}_{p,0}(k+1 \mid k)$$

which are independent from $\Delta u_j(k+1)$, $\Delta u_j(k+2)$, etc., i.e. all $\Delta u_j(k)$ can be solved by these p ($p = m$) equalities, which correspond to the optimal solution in the case $P = M = 1$.

4) The above multivariable predictive control algorithm is different from the usual decoupling or decentralized control algorithms. It adopts a global optimization and considers interactive influences between all inputs and outputs in a natural way. All the control inputs (outputs) have the same position in the algorithm no matter how to mark them, so there is no pairing problem in this algorithm.

For the multivariable predictive control algorithm, it is difficult to derive the explicit quantitative description of the closed-loop system and the analytical conclusion reflecting the relationship between design parameters and the closed-loop performances is lacking. As compared with the single variable case, the parameter tuning of the multivariable predictive control algorithm is also more complicated due to the complex interactions between system inputs and outputs. Therefore, the parameter tuning of multivariable DMC is directly based on the physical meaning of these design parameters in the performance index (5.5) and the parameters should be repeatedly updated by combining with simulations. In the following, some rules for parameter tuning of the multivariable DMC are given for reference.

1) For the multivariable DMC algorithm presented above, in order to unify the formulas and simplify the expressions, unified P and M are adopted for optimization horizons of different outputs and control horizons of different inputs, respectively. However, by selecting weighting matrices $\boldsymbol{Q}_i$ and $\boldsymbol{R}_j$, different P_i and M_j can be obtained. For example, let the elements in the weighting matrix $\boldsymbol{Q}_i$ correspond to y_i as $q_i(P_i+1) = \cdots = q_i(P) = 0$; the optimization horizon of y_i can be shortened from P to P_i. Let the elements in the weighting matrix $\boldsymbol{R}_j$ correspond to u_j as $r_j(M_j+1) = \cdots = r_j(M) = 0$; the control horizon of u_j can be shortened to M_j.
2) The optimization horizon P_i corresponding to y_i should cover the main dynamics of step responses of y_i to all control inputs u_j $(j = 1, \ldots, m)$, i.e. P_i must be selected to cross the time delay part or the nonminimum phase reverse part of all $a_{ij}(t)$ and to satisfy

$$\left(\sum_{l=1}^{P_i} a_{ij}(l)q_i(l)\right)a_{ij}(N) > 0,\ j = 1,\ \ldots,\ m$$

which is necessary for obtaining a stable and meaningful control.
3) The control horizon M_j corresponding to input u_j should be selected according to the dynamic complexity of the step responses $a_{ij}(t)$ from u_j to y_i $(i = 1, \ldots, p)$. If all of them exhibit relatively simple dynamics, then u_j does not need to change many times and M_j can be selected to be smaller. Otherwise M_j needs to be larger to increase the degree of freedom for control.
4) During simulation, the elements of $\boldsymbol{Q}_i$, $\boldsymbol{R}_j$ can be adjusted to improve the control performance according to their intuitive physical meanings in the performance index (5.5). If the response of the output y_i is particularly slow, the corresponding element in $\boldsymbol{Q}_i$ can be enlarged, i.e. increase the weighting of the tracking error of this y_i to accelerate its dynamic response. If the variation of the control input u_j is excessive, the corresponding element in $\boldsymbol{R}_j$ can be enlarged to strengthen the suppression on it and to make it change smoothly.
5) It is known through theoretical analysis that if the model is accurate, the selection of $\boldsymbol{h}_{ii}$ does not affect the system dynamics. Therefore, the parameters can be tuned according to the separation principle, i.e. firstly for model accurate case, adjust the design parameters in the performance index (5.5) to get a stable and good dynamic response. Then select $\boldsymbol{h}_{ii}$ to get good disturbance rejection and robustness against model mismatch. Since the output error is corrected independently, $\boldsymbol{h}_{ii}$ can be selected by the correction strategies used in the single variable case, such as equal-value correction (3.20).

5.2 Constrained DMC for Multivariable Systems

The DMC algorithms discussed above, either for the single variable or the multivariable case, do not take the constraints into account. In practical applications, due to physical constraints or for safety considerations, the values of the system outputs, control inputs, and some intermediate variables can only be taken in a certain range, such as that of the valve as an actuator, where its opening can vary only in a certain range. In the level control of a boiler, the level height over certain upper or lower bounds may cause an accident. Therefore, in control implementation, the system inputs and outputs should be restricted to certain ranges according to practical requirements:

$$u_{\min} \le u \le u_{\max} \tag{5.13}$$

$$y_{\min} \le y \le y_{\max} \tag{5.14}$$

In this case, if the optimal control $\boldsymbol{u}(k)$ is still calculated as in the unconstrained case and would be replaced by a corresponding bound value if it destroys the constraint (5.13), the solution is not necessarily optimal. Furthermore, this method can only handle constraints on control inputs u. For output constraints, the condition (5.14) cannot be guaranteed because it must be checked by a posterior test with solved control actions.

In order to get the optimal solution satisfying the constraint conditions, (5.13) and (5.14) must be embedded into the optimization problem. Predictive control has a model prediction function and is thus naturally capable of predicting future constraint conditions and including them in the optimization. Since constraints only affect rolling optimization in predictive control algorithms, in this section, we will focus on two critical issues of constrained optimization in predictive control, i.e. how to formulate the online optimization problem of predictive control if constraints exist and how to efficiently solve the constrained optimization problem. Note that the multivariable DMC algorithm is still taken as an example to illustrate these issues. For other algorithms or cases, the constrained optimization problem can be similarly formulated, but is solved by a proper mathematical programming method according to the problem formulation.

5.2.1 Formulation of the Constrained Optimization Problem in Multivariable DMC

In rolling optimization of multivariable DMC, the optimization problem at each sampling time involves the control increments of all control inputs at future M sampling instants and the predicted values of all system outputs at future P sampling instants. All the control variables should satisfy the constraint condition (5.13), i.e.

$$\begin{aligned} &u_{i,\min} \le u_i(k) = u_i(k-1) + \Delta u_i(k) \le u_{i,\max}, \\ &\qquad\vdots \qquad\qquad\qquad\qquad\qquad\qquad\qquad i = 1, \ldots, m \\ &u_{i,\min} \le u_i(k+M-1) = u_i(k-1) + \cdots + \Delta u_i(k+M-1) \le u_{i,\max} \end{aligned}$$

which can be integrated in the vector form

$$\Delta \boldsymbol{u}_{\min} \le \boldsymbol{B}\Delta \boldsymbol{u}_M(k) \le \Delta \boldsymbol{u}_{\max} \tag{5.15}$$

where

$$\boldsymbol{B} = \text{block diag}\,\underbrace{(\boldsymbol{B}_0,\ \boldsymbol{B}_0,\ \cdots,\ \boldsymbol{B}_0)}_{m\ \text{blocks}},\ \text{with}\ \boldsymbol{B}_0 = \begin{bmatrix} 1 & & \\ 1 & 1 & 0 \\ \vdots & & \ddots \\ 1 & \cdots & 1 \end{bmatrix}_{(M\times M)}$$

$$\Delta\boldsymbol{u}_{\min} = \begin{bmatrix} u_{1,\min} - u_1(k-1) \\ \vdots \\ u_{1,\min} - u_1(k-1) \\ \vdots \\ u_{m,\min} - u_m(k-1) \\ \vdots \\ u_{m,\min} - u_m(k-1) \end{bmatrix}_{(mM\times 1)},\quad \Delta\boldsymbol{u}_{\max} = \begin{bmatrix} u_{1,\max} - u_1(k-1) \\ \vdots \\ u_{1,\max} - u_1(k-1) \\ \vdots \\ u_{m,\max} - u_m(k-1) \\ \vdots \\ u_{m,\max} - u_m(k-1) \end{bmatrix}_{(mM\times 1)}$$

Similarly, the predicted values of system outputs should satisfy the constraint condition (5.14), which can be written in vector form directly according to (5.4):

$$\boldsymbol{y}_{\min} - \tilde{\boldsymbol{y}}_{P0}(k) \le \boldsymbol{A}\Delta\boldsymbol{u}_M(k) \le \boldsymbol{y}_{\max} - \tilde{\boldsymbol{y}}_{P0}(k) \tag{5.16}$$

where

$$\boldsymbol{y}_{\min} = \left[y_{1,\min}\ \cdots\ y_{1,\min}\ \cdots\ y_{p,\min}\ \cdots\ y_{p,\min}\right]^{\mathrm{T}}_{(pP\times 1)}$$

$$\boldsymbol{y}_{\max} = \left[y_{1,\max}\ \cdots\ y_{1,\max}\ \cdots\ y_{p,\max}\ \cdots\ y_{p,\max}\right]^{\mathrm{T}}_{(pP\times 1)}$$

From (5.15) and (5.16) it is known that the constraints both on control inputs and on system outputs can be integrated to the following inequality constraints on control inputs:

$$\boldsymbol{C}\Delta\boldsymbol{u}_M(k) \le \boldsymbol{l} \tag{5.17}$$

where

$$\boldsymbol{C} = \begin{bmatrix} -\boldsymbol{B} \\ \boldsymbol{B} \\ -\boldsymbol{A} \\ \boldsymbol{A} \end{bmatrix},\quad \boldsymbol{l} = \begin{bmatrix} -\Delta\boldsymbol{u}_{\min} \\ \Delta\boldsymbol{u}_{\max} \\ \tilde{\boldsymbol{y}}_{P0}(k) - \boldsymbol{y}_{\min} \\ \boldsymbol{y}_{\max} - \tilde{\boldsymbol{y}}_{P0}(k) \end{bmatrix}$$

$\boldsymbol{C}$, $\boldsymbol{l}$ are all known at the sampling time k. Therefore, at the sampling time k, the online optimization problem with constraints can be formulated as: solve the optimal $\Delta\boldsymbol{u}_M(k)$ with respect to the performance index (5.5) under the inequality constraints (5.17) based on the prediction model (5.4), i.e.

$$\begin{aligned} &\min_{\Delta\boldsymbol{u}_M(k)} J(k) = \|\boldsymbol{w}(k) - \tilde{\boldsymbol{y}}_{PM}(k)\|_{\boldsymbol{Q}}^2 + \|\Delta\boldsymbol{u}_M(k)\|_{\boldsymbol{R}}^2 \\ &\text{s.t.}\ \ \tilde{\boldsymbol{y}}_{PM}(k) = \tilde{\boldsymbol{y}}_{P0}(k) + \boldsymbol{A}\Delta\boldsymbol{u}_M(k) \\ &\qquad \boldsymbol{C}\Delta\boldsymbol{u}_M(k) \le \boldsymbol{l} \end{aligned} \tag{5.18}$$

This kind of optimization problem with a quadratic performance index and linear equality and inequality constraints is usually called Quadratic Programming (QP). In the following, the matrix tearing technique appearing in DMC early applications is firstly introduced to solve this kind of constrained optimization problem. Then the standard QP solving method will be introduced.

5.2.2 Constrained Optimization Algorithm Based on the Matrix Tearing Technique

For the online optimization problem (5.18) with inequality constraints (5.17), it is generally impossible to get its optimal solution in closed form. However, the optimal solution in the unconstrained case is undoubtedly very attractive because it is easy to solve and can be represented analytically in closed form. Normally during control a great number of inequality constraints (5.17) do not work, and only a few inequalities might be destroyed and should be considered. Therefore, a possible way to handle constraints might be as follows: first calculate the unconstrained optimal solution $\Delta \boldsymbol{u}_M(k)$ according to (5.6) and then test (5.17) to see which constraints are destroyed. For destroyed constraints, adjust the corresponding weighting coefficients in $\boldsymbol{Q}$, $\boldsymbol{R}$ according to their physical meanings in the optimization performance index (5.18) to improve the degree of satisfying corresponding constraints. In this way, the optimization problem is solved, tested, and adjusted repeatedly until the results satisfy all the constraints. This method is simple in principle, but the matrix $\boldsymbol{A}^{\mathrm{T}}\boldsymbol{Q}\boldsymbol{A} + \boldsymbol{R}$ should be inverted each time the weighting coefficients are adjusted. For multivariable systems this matrix is always high-dimensional; thus the above method is difficult to meet the requirement for real-time computations.

At the early stage of DMC industrial applications, Prett and Gillette [2] proposed a multivariable constrained optimization method, which fully utilizes the directness and simplicity of the analytical solution in the unconstrained case and reduced the dimension of the matrix to be online inverted through a matrix tearing technique. This method can be used online to solve the constrained optimization problem [2].

Prett and Gillette pointed out that in above constrained optimization problem, the optimization variable $\Delta \boldsymbol{u}_M(k)$ constructs a feasible region in its space, whose boundaries correspond to the critical values of constraint conditions. If the unconstrained optimal solution lies in this region, all the constraint conditions are satisfied and it is also the optimal solution in the constrained case. However, if the unconstrained optimal solution lies outside this region, then it is an infeasible solution for the constrained case because at least some constraints have been violated. In this case, the unconstrained optimal solution must move closer to the feasible region until it falls to the vertex or boundary of the feasible region. Then the previously violated constraint conditions can be critically satisfied and this solution becomes feasible for the constrained optimization, but with the cost of degeneration of optimality due to additional constraints.

According to the above analysis, how to get a simpler online constrained optimization algorithm is now discussed. In the following, we first discuss how to improve the unconstrained optimal solution when it violates some constraint conditions and then give an iterative algorithm for the solution.

(1) Improving the unconstrained optimal solution when it is infeasible for the constrained case

The unconstrained optimal solution can be obtained by neglecting the inequality constraints in the optimization problem (5.18) and given by (5.6):

$$\Delta \boldsymbol{u}_M(k) = (\boldsymbol{A}^{\mathrm{T}}\boldsymbol{Q}\boldsymbol{A} + \boldsymbol{R})^{-1}\boldsymbol{A}^{\mathrm{T}}\boldsymbol{Q}[\boldsymbol{w}(k) - \tilde{\boldsymbol{y}}_{P0}(k)]$$

Then check whether the input and output constraints in (5.17) are violated with this solution. Note that the online optimization in predictive control is performed repeatedly and each time after obtaining the optimal $\Delta \boldsymbol{u}_M(k)$ only the current control $\Delta u(k)$ is implemented. It is therefore unnecessary to check all the constraints over the optimization horizon, i.e. all the inequalities in (5.17). Instead only limited constraints at the sampling instants not far from the current time should be checked. The formulae corresponding to violated constraints are then extracted and set as imposed equalities to make these constraints critically satisfied:

$$\boldsymbol{C}_1 \Delta \boldsymbol{u}_M(k) = \boldsymbol{l}_1 \tag{5.19}$$

where $\boldsymbol{C}_1$ and $\boldsymbol{l}_1$ are composed of the elements of the rows corresponding to the violated constraints in $\boldsymbol{C}$ and $\boldsymbol{l}$, respectively. It is obvious that the row number of $\boldsymbol{C}_1$ or $\boldsymbol{l}_1$ is equal to the number of constraints violated by the unconstrained optimal solution, which is much smaller than the dimension of the matrix to be inverted while calculating $\Delta \boldsymbol{u}_M(k)$.

In order to guide the infeasible solution toward the feasible region, impose the degree of the solution violating the constraints into the optimization problem and revise the performance index (5.18) into

$$\min J'(k) = J(k) + \|\boldsymbol{C}_1 \Delta \boldsymbol{u}_M(k) - \boldsymbol{l}_1\|_{\boldsymbol{S}}^2$$

where $\boldsymbol{S} = s\boldsymbol{I}$, s is a positive number.

With the above performance index and combined with prediction model (5.4), the optimal solution can be given by

$$\Delta \boldsymbol{u}'_M(k) = (\boldsymbol{A}^{\mathrm{T}}\boldsymbol{Q}\boldsymbol{A} + \boldsymbol{R} + \boldsymbol{C}_1^{\mathrm{T}}\boldsymbol{S}\boldsymbol{C}_1)^{-1}[\boldsymbol{A}^{\mathrm{T}}\boldsymbol{Q}(\boldsymbol{w}(k) - \tilde{\boldsymbol{y}}_{P0}(k)) + \boldsymbol{C}_1^{\mathrm{T}}\boldsymbol{S}\boldsymbol{l}_1]$$

Denoting $\boldsymbol{P} = (\boldsymbol{A}^{\mathrm{T}}\boldsymbol{Q}\boldsymbol{A} + \boldsymbol{R})^{-1}$ and using the matrix inverse formula, the matrix to be inverted can be rewritten as

$$(\boldsymbol{P}^{-1} + \boldsymbol{C}_1^{\mathrm{T}}\boldsymbol{S}\boldsymbol{C}_1)^{-1} = \boldsymbol{P} - \boldsymbol{P}\boldsymbol{C}_1^{\mathrm{T}}(\boldsymbol{C}_1\boldsymbol{P}\boldsymbol{C}_1^{\mathrm{T}} + \boldsymbol{S}^{-1})^{-1}\boldsymbol{C}_1\boldsymbol{P} \tag{5.20}$$

and then the above solution can be written as

$$\begin{aligned}
\Delta \boldsymbol{u}'_M(k) &= \boldsymbol{P}\boldsymbol{A}^{\mathrm{T}}\boldsymbol{Q}[w(k) - \tilde{\boldsymbol{y}}_{P0}(k)] - \boldsymbol{P}\boldsymbol{C}_1^{\mathrm{T}}(\boldsymbol{C}_1\boldsymbol{P}\boldsymbol{C}_1^{\mathrm{T}} + \boldsymbol{S}^{-1})^{-1}\boldsymbol{C}_1\boldsymbol{P}\boldsymbol{A}^{\mathrm{T}}\boldsymbol{Q} \\
&\qquad [w(k) - \tilde{\boldsymbol{y}}_{P0}(k)] + \boldsymbol{P}\boldsymbol{C}_1^{\mathrm{T}}\boldsymbol{S}\boldsymbol{l}_1 - \boldsymbol{P}\boldsymbol{C}_1^{\mathrm{T}}(\boldsymbol{C}_1\boldsymbol{P}\boldsymbol{C}_1^{\mathrm{T}} + \boldsymbol{S}^{-1})^{-1}\boldsymbol{C}_1\boldsymbol{P}\boldsymbol{C}_1^{\mathrm{T}}\boldsymbol{S}\boldsymbol{l}_1 \\
&= \Delta \boldsymbol{u}_M(k) - \boldsymbol{P}\boldsymbol{C}_1^{\mathrm{T}}(\boldsymbol{C}_1\boldsymbol{P}\boldsymbol{C}_1^{\mathrm{T}} + \boldsymbol{S}^{-1})^{-1}\boldsymbol{C}_1\Delta \boldsymbol{u}_M(k) + \boldsymbol{P}\boldsymbol{C}_1^{\mathrm{T}}(\boldsymbol{C}_1\boldsymbol{P}\boldsymbol{C}_1^{\mathrm{T}} + \boldsymbol{S}^{-1})^{-1}\boldsymbol{l}_1 \\
&= \Delta \boldsymbol{u}_M(k) - \boldsymbol{P}\boldsymbol{C}_1^{\mathrm{T}}(\boldsymbol{C}_1\boldsymbol{P}\boldsymbol{C}_1^{\mathrm{T}} + \boldsymbol{S}^{-1})^{-1}[\boldsymbol{C}_1\Delta \boldsymbol{u}_M(k) - \boldsymbol{l}_1]
\end{aligned} \tag{5.21}$$

It can be seen that when the critical constraints (5.19) that are violated by the unconstrained optimal solution are imposed on the performance index, the feasibility of the optimal solution can be improved through revising the unconstrained optimal solution $\Delta\boldsymbol{u}_M(k)$ with an additional term. This additional term is related to the degree of the solution violating the critical constraints (5.19). It should be pointed out that $\boldsymbol{C}_1, \boldsymbol{l}_1$ appearing in the additional term should be online determined according to the result of checking the constraint conditions, and thus the matrix $\boldsymbol{C}_1\boldsymbol{P}\boldsymbol{C}_1^{\mathrm{T}} + \boldsymbol{S}^{-1}$ needs to be inverted online. However, the dimension of the matrix to be inverted, i.e. the row number of $\boldsymbol{C}_1$, is only the number of violated constraints and in general is much lower than the dimension of the matrix $\boldsymbol{A}^{\mathrm{T}}\boldsymbol{Q}\boldsymbol{A} + \boldsymbol{R} + \boldsymbol{C}_1^{\mathrm{T}}\boldsymbol{S}\boldsymbol{C}_1$, which should be inverted if $\Delta\boldsymbol{u}'_M(k)$ is directly calculated. Furthermore, since $\boldsymbol{P}$ is off-line calculated in advance, and $\boldsymbol{C}_1\Delta\boldsymbol{u}_M(k) - \boldsymbol{l}_1$ is also calculated during checking whether the constraint conditions are violated, the calculation for the improved control action $\Delta\boldsymbol{u}'_M(k)$, i.e. (5.21), is very simple.

(2) Iteration algorithm for online constrained optimization

During the above calculation, the critical constraints (5.19) are put into the performance index as soft constraints rather than hard constraints; thus for limited s, the constraint conditions given in (5.19) might be still violated by the solution $\Delta\boldsymbol{u}'_M(k)$. Only when $s \to \infty$ does it follow that

$$\Delta\boldsymbol{u}'_M(k) = \Delta\boldsymbol{u}_M(k) - \boldsymbol{P}\boldsymbol{C}_1^{\mathrm{T}}\left(\boldsymbol{C}_1\boldsymbol{P}\boldsymbol{C}_1^{\mathrm{T}}\right)^{-1}[\boldsymbol{C}_1\Delta\boldsymbol{u}_M(k) - \boldsymbol{l}_1]$$

and the constraint conditions (5.19) can be satisfied, i.e.

$$\boldsymbol{C}_1\Delta\boldsymbol{u}'_M(k) = \boldsymbol{C}_1\Delta\boldsymbol{u}_M(k) - \boldsymbol{C}_1\boldsymbol{P}\boldsymbol{C}_1^{\mathrm{T}}\left(\boldsymbol{C}_1\boldsymbol{P}\boldsymbol{C}_1^{\mathrm{T}}\right)^{-1}[\boldsymbol{C}_1\Delta\boldsymbol{u}_M(k) - \boldsymbol{l}_1] = \boldsymbol{l}_1$$

Although the solution obtained by directly setting $s \to \infty$ can guarantee the critical constraints (5.19), some other constraint conditions that formerly satisfy (5.17) might be violated, which will lead to the solving process being out of order. Therefore this problem should be solved iteratively by gradually increasing s.

During iteration, using a similar method to that above, the solution is improved and moved close to the constraint boundaries through checking the constraint conditions and gradually increasing s. To illustrate this procedure, start from the initial values

$$\begin{aligned} &\Delta\boldsymbol{u}_M^0(k) = \boldsymbol{P}_0\boldsymbol{h}_0 \\ &\boldsymbol{P}_0^{-1} = \boldsymbol{A}^{\mathrm{T}}\boldsymbol{Q}\boldsymbol{A} + \boldsymbol{R}, \qquad \boldsymbol{h}_0 = \boldsymbol{A}^{\mathrm{T}}\boldsymbol{Q}(\boldsymbol{w}(k) - \tilde{\boldsymbol{y}}_{P0}(k)) \end{aligned} \tag{5.22}$$

Let the solution obtained from the ith step iteration be

$$\Delta\boldsymbol{u}_M^i(k) = \boldsymbol{P}_i\boldsymbol{h}_i$$

During the $(i+1)$th step iteration, check the constraints and get the critical constraints as

$$\boldsymbol{C}_{i+1}\Delta\boldsymbol{u}_M^{i+1}(k) = \boldsymbol{l}_{i+1}$$

Impose these critical constraints into the performance index and revise it as

$$\min J_{i+1}(k) = J_i(k) + \left\|\boldsymbol{C}_{i+1}\Delta\boldsymbol{u}_M^{i+1}(k) - \boldsymbol{l}_{i+1}\right\|_{\boldsymbol{S}_{i+1}}^2$$

Referring to the deduction of $\Delta \boldsymbol{u}'_M(k)$, the optimal solution can be given by

$$\Delta \boldsymbol{u}_M^{i+1}(k) = \left(\boldsymbol{P}_i^{-1} + \boldsymbol{C}_{i+1}^{\mathrm{T}} \boldsymbol{S}_{i+1} \boldsymbol{C}_{i+1}\right)^{-1} \left(\boldsymbol{h}_i + \boldsymbol{C}_{i+1}^{\mathrm{T}} \boldsymbol{S}_{i+1} \boldsymbol{l}_{i+1}\right)$$

Then the iteration formulae are given by

$$\begin{aligned} &\Delta \boldsymbol{u}_M^{i+1}(k) = \boldsymbol{P}_{i+1} \boldsymbol{h}_{i+1} \\ &\boldsymbol{P}_{i+1}^{-1} = \boldsymbol{P}_i^{-1} + \boldsymbol{C}_{i+1}^{\mathrm{T}} \boldsymbol{S}_{i+1} \boldsymbol{C}_{i+1}, \qquad \boldsymbol{h}_{i+1} = \boldsymbol{h}_i + \boldsymbol{C}_{i+1}^{\mathrm{T}} \boldsymbol{S}_{i+1} \boldsymbol{l}_{i+1} \end{aligned} \tag{5.23}$$

with (5.22) as the initial condition, where s_i in the weighting matrix $\boldsymbol{S}_i = s_i \boldsymbol{I}$ should be gradually increased. It can be seen that at each step of iteration we need to calculate

$$\boldsymbol{P}_{i+1} = \left(\boldsymbol{P}_i^{-1} + \boldsymbol{C}_{i+1}^{\mathrm{T}} \boldsymbol{S}_{i+1} \boldsymbol{C}_{i+1}\right)^{-1} = \boldsymbol{P}_i - \boldsymbol{P}_i \boldsymbol{C}_{i+1}^{\mathrm{T}} \left(\boldsymbol{C}_{i+1} \boldsymbol{P}_i \boldsymbol{C}_{i+1}^{\mathrm{T}} + \boldsymbol{S}_{i+1}^{-1}\right)^{-1} \boldsymbol{C}_{i+1} \boldsymbol{P}_i$$

i.e. the matrix $\boldsymbol{C}_{i+1} \boldsymbol{P}_i \boldsymbol{C}_{i+1}^{\mathrm{T}} + \boldsymbol{S}_{i+1}^{-1}$ needs to be inverted.

This kind of constrained optimization algorithm utilizes the fact that during optimization most constraint conditions can be satisfied and the number of active ones is very few. With the help of the matrix tearing technique the dimension of the matrix to be online inverted is reduced to a minimum such that online inversion of the high-dimensional matrix $\boldsymbol{A}^{\mathrm{T}} \boldsymbol{Q} \boldsymbol{A} + \boldsymbol{R}$ can be avoided, which makes real-time constrained optimization possible. In addition to the boundary constraints, it can also be applied to other cases. For example, in multivariable control, when a fault occurs in a computer output channel and the corresponding control switches to manual with a constant value, the computer cannot calculate the instant control increments according to the original situation. In order to ensure global optimality, an additional constraint of the corresponding control increment equal to zero must be imposed, with which optimal values for other control inputs should be calculated. In this case, the optimal solution under this constraint can be obtained by using (5.23), which guarantees the control optimality with an existing fault, while the continuity and smoothness of the control can also be maintained even if the fault is recovered.

5.2.3 Constrained Optimization Algorithm Based on QP

With constraints, an optimization problem (5.18) with a quadratic performance index and linear equality and inequality constraints should be solved at each time in DMC, which is a standard QP problem. QP is an important branch in mathematic programming and has developed quite maturely with various efficient algorithms and related software. Indeed, the QP method has already been adopted directly to solve the online constrained optimization problem in predictive control and much software for predictive control application also directly take QP as the optimization algorithm. In this section, we briefly introduce how to transfer the online optimization problem of predictive control into a standard QP problem, and then give the basic principle and the algorithm of the active set method [3] for solving the QP problem.

A standard QP problem can be formulated by

$$\begin{aligned} &\min f(\boldsymbol{x}) = \frac{1}{2} \boldsymbol{x}^{\mathrm{T}} \boldsymbol{H} \boldsymbol{x} + \boldsymbol{c}^{\mathrm{T}} \boldsymbol{x} \\ &\text{s.t. } \boldsymbol{B} \boldsymbol{x} \le \boldsymbol{d} \end{aligned} \tag{5.24}$$

where $\boldsymbol{H}$ is a symmetric positive definite matrix and $\boldsymbol{x}$ is the optimization variable. For the online optimization problem in predictive control given by (5.18), after replacing $\tilde{\boldsymbol{y}}_{PM}(k)$ by the prediction model (5.4), the performance index $J(k)$ can be written as

$$\begin{aligned}J(k) &= [\boldsymbol{w}(k)-\tilde{\boldsymbol{y}}_{P0}(k)-\boldsymbol{A}\Delta\boldsymbol{u}_M(k)]^{\mathrm{T}}\boldsymbol{Q}[\boldsymbol{w}(k)-\tilde{\boldsymbol{y}}_{P0}(k)-\boldsymbol{A}\Delta\boldsymbol{u}_M(k)]+\Delta\boldsymbol{u}_M^{\mathrm{T}}(k)\boldsymbol{R}\Delta\boldsymbol{u}_M(k)\\ &= [\boldsymbol{w}(k)-\tilde{\boldsymbol{y}}_{P0}(k)]^{\mathrm{T}}\boldsymbol{Q}[\boldsymbol{w}(k)-\tilde{\boldsymbol{y}}_{P0}(k)]-2[\boldsymbol{w}(k)-\tilde{\boldsymbol{y}}_{P0}(k)]^{\mathrm{T}}\boldsymbol{Q}\boldsymbol{A}\Delta\boldsymbol{u}_M(k)\\ &\quad+\Delta\boldsymbol{u}_M^{\mathrm{T}}(k)\left(\boldsymbol{A}^{\mathrm{T}}\boldsymbol{Q}\boldsymbol{A}+\boldsymbol{R}\right)\Delta\boldsymbol{u}_M(k)\end{aligned}$$

where the first item is known and fixed at sampling time k and is thus independent of the optimization and can be removed from the performance index. Denote

$$\begin{aligned}&\boldsymbol{x}=\Delta\boldsymbol{u}_M(k),\quad \boldsymbol{H}=2\left(\boldsymbol{A}^{\mathrm{T}}\boldsymbol{Q}\boldsymbol{A}+\boldsymbol{R}\right),\\ &\boldsymbol{c}^{\mathrm{T}}=-2(\boldsymbol{w}(k)-\tilde{\boldsymbol{y}}_{P0}(k))^{\mathrm{T}}\boldsymbol{Q}\boldsymbol{A},\quad \boldsymbol{B}=\boldsymbol{C},\quad \boldsymbol{d}=\boldsymbol{l}\end{aligned}$$

Then the online optimization problem in predictive control expressed by (5.18) can be transferred into the standard QP problem (5.24).

In the following how to solve the standard QP problem (5.24) is discussed. Firstly, consider the case only with equality constraints, i.e. the constraint conditions in (5.24) can be written as equality $\boldsymbol{Bx}=\boldsymbol{d}$. Using the Lagrange multiplier method to put the constraint into the performance index, the Lagrange function can be constructed as

$$L(\boldsymbol{x},\boldsymbol{\lambda})=\frac{1}{2}\boldsymbol{x}^{\mathrm{T}}\boldsymbol{H}\boldsymbol{x}+\boldsymbol{c}^{\mathrm{T}}\boldsymbol{x}+\boldsymbol{\lambda}^{\mathrm{T}}(\boldsymbol{B}\boldsymbol{x}-\boldsymbol{d})$$

The equation for optimal solution is given by

$$\begin{aligned}&\boldsymbol{H}\boldsymbol{x}+\boldsymbol{c}+\boldsymbol{B}^{\mathrm{T}}\boldsymbol{\lambda}=0\\ &\boldsymbol{B}\boldsymbol{x}-\boldsymbol{d}=0\end{aligned}$$

where $\boldsymbol{x}$, $\boldsymbol{\lambda}$ can be solved by equation

$$\begin{bmatrix}\boldsymbol{H} & \boldsymbol{B}^{\mathrm{T}}\\ \boldsymbol{B} & \boldsymbol{0}\end{bmatrix}\begin{bmatrix}\boldsymbol{x}\\ \boldsymbol{\lambda}\end{bmatrix}=\begin{bmatrix}-\boldsymbol{c}\\ \boldsymbol{d}\end{bmatrix}\tag{5.25}$$

For QP with inequality constraints, the active set method is often adopted. The basic idea of this method is as follows: any feasible solution $\boldsymbol{x}$ must satisfy all the constraint conditions, among which the equality constraints are called active constraints. All the active constraints form a set, called the active set, denoted by $\boldsymbol{\Omega}(\boldsymbol{x})$. Through solving the QP problem with equality constraints corresponding to those in the active set,

$$\begin{aligned}&\min f(\boldsymbol{x})=\frac{1}{2}\boldsymbol{x}^{\mathrm{T}}\boldsymbol{H}\boldsymbol{x}+\boldsymbol{c}^{\mathrm{T}}\boldsymbol{x}\\ &\text{s.t.}\ \ \boldsymbol{b}_i\boldsymbol{x}=d_i,\quad i\in\Omega(\boldsymbol{x})\end{aligned}\tag{5.26}$$

the feasible solution can be further improved until the optimal solution is obtained, where $\boldsymbol{b}_i$, d_i are the elements in $\boldsymbol{B}$, $\boldsymbol{d}$, respectively, corresponding to the rows of $\boldsymbol{x}$ satisfying equality constraints.

Now start from the arbitrary feasible solution of the QP problem (5.24) $\boldsymbol{x}^*$. Its corresponding active set is denoted as $\Omega^*(\boldsymbol{x}^*)$, i.e.

$$\boldsymbol{b}_i\boldsymbol{x}^*=d_i,\qquad i\in\Omega^*(\boldsymbol{x}^*)$$

Then change $\boldsymbol{x}^*$ to $\boldsymbol{x}^* + \boldsymbol{\delta}$ and from the QP problem with equality constraints

$$\min f(\boldsymbol{x}^* + \boldsymbol{\delta}) = \frac{1}{2}(\boldsymbol{x}^* + \boldsymbol{\delta})^{\mathrm{T}}\boldsymbol{H}(\boldsymbol{x}^* + \boldsymbol{\delta}) + \boldsymbol{c}^{\mathrm{T}}(\boldsymbol{x}^* + \boldsymbol{\delta})$$
$$\text{s.t.} \quad \boldsymbol{b}_i\boldsymbol{\delta} = 0, \qquad i \in \boldsymbol{\Omega}^*(\boldsymbol{x}^*)$$

Since $\boldsymbol{x}^*$ is fixed, the above optimization problem can be rewritten as

$$\begin{aligned} &\min f'(\boldsymbol{\delta}) = \frac{1}{2}\boldsymbol{\delta}^{\mathrm{T}}\boldsymbol{H}\boldsymbol{\delta} + \left(\boldsymbol{x}^{*\mathrm{T}}\boldsymbol{H} + \boldsymbol{c}^{\mathrm{T}}\right)\boldsymbol{\delta} \\ &\text{s.t.} \quad \boldsymbol{b}_i\boldsymbol{\delta} = 0, \qquad i \in \boldsymbol{\Omega}^*(\boldsymbol{x}^*) \end{aligned} \tag{5.27}$$

The optimization problem (5.27) can be similarly solved by using the Lagrange multiplier method and the optimal solution $\boldsymbol{\delta}^*$, $\boldsymbol{\lambda}^*$ can be obtained by solving (5.25) with $\boldsymbol{\delta}$ and $\boldsymbol{H}\boldsymbol{x} + \boldsymbol{c}$ replacing $\boldsymbol{x}$ and $\boldsymbol{c}$, respectively. Whether the solution is optimal and if the active set needs to be adjusted can be judged according to the following rules:

Rule 1. If $\boldsymbol{\delta}^* = 0$ and $\boldsymbol{\lambda}^* \geq 0$, then $\boldsymbol{x}^*$ is the optimal solution of the QP problem (5.24).

Rule 2. If $\boldsymbol{\delta}^* = 0$ but $\lambda_j < 0$, $j \in \Omega^*(\boldsymbol{x}^*)$, then $\boldsymbol{x}^*$ is not the optimal solution. Adjust the active set as $\Omega^*(\boldsymbol{x}^*)/\{j\}$and solve (5.27) again. If there are several j with $\lambda_j < 0$, then select the j corresponding to the most negative λ_j to exit the active set.

Rule 3. If $\boldsymbol{\delta}^* \neq 0$ and $\boldsymbol{x}^* + \boldsymbol{\delta}^*$ is feasible, adjust $\boldsymbol{x}^* = \boldsymbol{x}^* + \boldsymbol{\delta}^*$ and solve (5.27) again.

Rule 4. If $\boldsymbol{\delta}^* \neq 0$ but $\boldsymbol{x}^* + \boldsymbol{\delta}^*$ is infeasible, determine the maximal step length with which the solution starting from $\boldsymbol{x}^*$ along $\boldsymbol{\delta}^*$ direction remains feasible, i.e. $\alpha^* = \max(\alpha > 0 \mid \boldsymbol{x}^* + \alpha\boldsymbol{\delta}^* \text{ feasible})$. In this case there is at least one $l \notin \Omega^*(\boldsymbol{x}^*)$ such that $\boldsymbol{b}_l(\boldsymbol{x}^* + \alpha^*\boldsymbol{\delta}^*) = d_l$; adjust the active set as $\Omega^* \cup \{l\}$ and solve (5.27) again.

Among the above rules, at some step if $\boldsymbol{\delta} \neq 0$, the solution can be further improved under current active constraints, while if $\boldsymbol{\delta} = 0$, it cannot be improved with current active constraints. For the latter case, $\lambda_j < 0$ implies that the corresponding active constraints restrict the improvement of the solution, while $\lambda_j > 0$ implies that the solution cannot be improved even it leaves the corresponding active constraints. Therefore, if all $\lambda_j > 0$, the optimal solution must satisfy all the constraints in the current active set and the current solution cannot be further improved, which means that the current solution is the optimal solution. In the following a simple example is given to illustrate how to use the active set method to solve the constrained optimization problem.

Example 5.1[3] Consider the QP problem (5.24) where

$$\boldsymbol{x} = \begin{bmatrix} x_1 \\ x_2 \end{bmatrix}, \quad \boldsymbol{H} = \begin{bmatrix} 2 & 0 \\ 0 & 2 \end{bmatrix}, \quad \boldsymbol{c} = \begin{bmatrix} -2 \\ -5 \end{bmatrix}$$

The constraint conditions are given by

$$\begin{aligned} -x_1 + 2x_2 &\leq 2 \quad \text{(C1)} \\ x_1 + 2x_2 &\leq 6 \quad \text{(C2)} \\ x_1 - 2x_2 &\leq 2 \quad \text{(C3)} \\ -x_2 &\leq 0 \quad \text{(C4)} \\ -x_1 &\leq 0 \quad \text{(C5)} \end{aligned}$$

Starting from the initial feasible solution $\boldsymbol{x}(0) = [2\ \ 0]^{\mathrm{T}}$, the corresponding active set is $\Omega(\boldsymbol{x}(0)) = \{\mathrm{C3, C4}\}$; solve (5.27) to obtain $\boldsymbol{\delta}(0) = [0\ \ 0]^{\mathrm{T}}$, $\boldsymbol{\lambda}(0) = [-2\ \ -1]^{\mathrm{T}}$.

Since $\boldsymbol{\delta}(0) = 0$, but both $\lambda_1(0) = -2$ and $\lambda_2(0) = -1$ are negative, according to the steps above in Rule 2, keep $\boldsymbol{x}(1) = \boldsymbol{x}(0)$, remove C3 corresponding to λ_1 from $\Omega(\boldsymbol{x}(0))$, and adjust the active set as $\Omega(\boldsymbol{x}(1)) = \{\mathrm{C4}\}$, solve (5.27), and obtain $\boldsymbol{\delta}(1) = [-1\ \ 0]^{\mathrm{T}}$.

Since $\boldsymbol{\delta}(1) \neq 0$ and $\boldsymbol{x}(1) + \boldsymbol{\delta}(1) = [1\ \ 0]^{\mathrm{T}}$ is feasible, according to the steps above in Rule 3, adjust $\boldsymbol{x}$ to $\boldsymbol{x}(2) = [1\ \ 0]^{\mathrm{T}}$, and let $\Omega(\boldsymbol{x}(2)) = \{\mathrm{C4}\}$; then solve (5.27) and get $\boldsymbol{\delta}(2) = [0\ \ 0]^{\mathrm{T}}$, $\boldsymbol{\lambda}(2) = -5$.

Since $\boldsymbol{\delta}(2) = 0$ but $\boldsymbol{\lambda}(2) < 0$, according to the steps above in Rule 2, keep $\boldsymbol{x}(3) = \boldsymbol{x}(2)$, further remove C4 from the active set; the active set then becomes an empty set $\Omega(\boldsymbol{x}(3)) = \{\boldsymbol{\Phi}\}$. Solve (5.27) without constraints to get $\boldsymbol{\delta}(3) = [0\ \ 2.5]^{\mathrm{T}}$.

Since $\boldsymbol{\delta}(3) \neq 0$ and $\boldsymbol{x}(3) + \boldsymbol{\delta}(3) = [1\ \ 2.5]^{\mathrm{T}}$ is infeasible, according to the steps in Rule 4, search along the direction of $\boldsymbol{\delta}(3)$ to get $\alpha^* = 0.6$ and then calculate $\boldsymbol{x}(4) = \boldsymbol{x}(3) + \alpha^*\boldsymbol{\delta}(3) = [1\ \ 1.5]^{\mathrm{T}}$. Note that C1 then becomes an active constraint; adjust the active set as $\Omega(\boldsymbol{x}(4)) = \{\mathrm{C1}\}$ and solve (5.27) to get $\boldsymbol{\delta}(4) = [0.4\ \ 0.2]^{\mathrm{T}}$, $\boldsymbol{\lambda}(4) = 0.8$.

Since $\boldsymbol{\delta}(4) \neq 0$ and $\boldsymbol{x}(4) + \boldsymbol{\delta}(4) = [1.4\ \ 1.7]^{\mathrm{T}}$ is feasible, according to the steps in Rule 3, adjust $\boldsymbol{x}(5) = [1.4\ \ 1.7]^{\mathrm{T}}$. With $\boldsymbol{x}(5)$ and the active set $\Omega(\boldsymbol{x}(5)) = \Omega(\boldsymbol{x}(4)) = \{\mathrm{C1}\}$, solve (5.27) to get $\boldsymbol{\delta}(5) = [0\ \ 0]^{\mathrm{T}}$.

Since $\boldsymbol{\delta}(5) = 0$ and $\lambda(5) = 0.8 > 0$, from Rule 1 it is known that $\boldsymbol{x}(5) = [1.4\ \ 1.7]^{\mathrm{T}}$ is the optimal solution of the QP problem.

The result of each iteration and the relationship with constraints can be illustrated by Figure 5.2. Note that if *C4* instead of *C3* is removed from the active set at the second step, the same final result can be obtained, but the evolution process of optimization is different (see the dotted line in Figure 5.2).

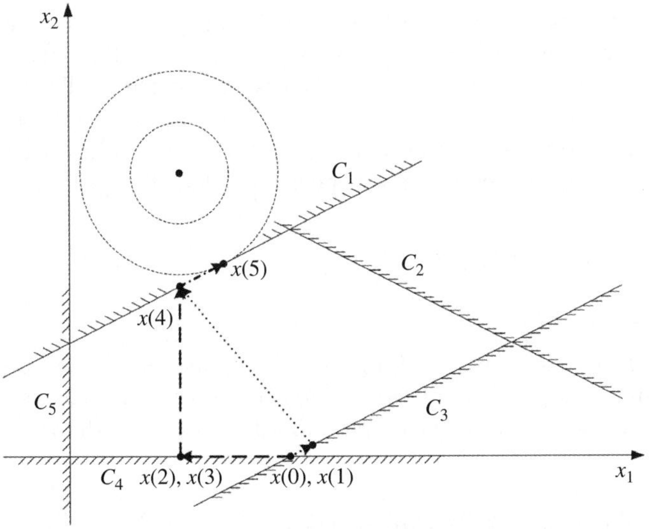

Figure 5.2 The procedure of solving a QP problem using the active set method for Example 5.1 [3]. *Source:* Reproduced with permission from Springer Nature and Copyright clearance center.

Remark 5.2 The QP algorithm using an active constraint set has been developed quite maturely and many efficient computational software have appeared. It is widely applied to solving the online optimization in predictive control applications. However, for predictive control with a non-quadratic performance index, nonlinear constraints, and even a non-linear model, the online optimization in general is not of the QP form. With the corresponding prediction model, constraints, and performance index, it can always be transformed into a mathematical programming problem and solved by corresponding tools. For example, for nonlinear systems with a quadratic performance index and linear constraints, the Sequential Quadratic Programming (SQP) method can be used. Such ability to handle constrained optimization is the main advantage of predictive control as well as the reason for its success in industrial processes. As Morari and Baric [4] pointed out, from a practical point of view, predictive control is the only control technology that is able to handle multivariable systems with constraints in a systematic and transparent manner.

5.3 Decomposition of Online Optimization for Multivariable Predictive Control

In Section 5.1 it was pointed out that tuning the DMC design parameters for a multivariable system is more complicated than that for a single variable system due to the complex interactions between its inputs and outputs. For multivariable systems, the optimization variables are composed of different control inputs, each with a series of control variations along the control horizon, which makes the online optimization problem high-dimensional and the real-time computation burden greatly increased. In this section, some decomposition approaches for online optimization of multivariable predictive control are presented, aiming at decomposing the high-dimensional global optimization problem into lower-dimensional local ones with a lower number of optimization variables, so as to reduce the complexity of online optimization, as well as to simplify the parameter tuning. To highlight the main ideas of different decomposition approaches, only the unconstrained multivariable DMC algorithm in Section 5.1 is taken as an example for illustration. However, the same ideas discussed here are also suitable for the constrained case.

From Section 5.1, the online optimization problem of an unconstrained multivariable DMC algorithm can be given by the performance index (5.5) and the prediction model (5.4):

$$\min_{\Delta \boldsymbol{u}_M(k)} J(k) = \|\boldsymbol{w}(k) - \tilde{\boldsymbol{y}}_{PM}(k)\|_{\boldsymbol{Q}}^2 + \|\Delta \boldsymbol{u}_M(k)\|_{\boldsymbol{R}}^2$$

$$\text{s.t.} \quad \tilde{\boldsymbol{y}}_{PM}(k) = \tilde{\boldsymbol{y}}_{P0}(k) + \boldsymbol{A}\Delta \boldsymbol{u}_M(k)$$

Assume that the plant has the same number of inputs and outputs, i.e. $m = p$. Now decompose the above optimization problem into m subproblems, each as a single variable predictive control problem with input u_i and output y_i. To do that, rewrite the above optimization problem as

$$\begin{aligned} &\min_{\Delta \boldsymbol{u}_M(k)} J(k) = \sum_{i=1}^{m} \left\{ \|\boldsymbol{w}_i(k) - \tilde{\boldsymbol{y}}_{i,PM}(k)\|_{\boldsymbol{Q}_i}^2 + \|\Delta \boldsymbol{u}_{i,M}(k)\|_{\boldsymbol{R}_i}^2 \right\} \\ &\text{s.t.} \quad \tilde{\boldsymbol{y}}_{i,PM}(k) = \tilde{\boldsymbol{y}}_{i,P0}(k) + \sum_{j=1}^{m} \boldsymbol{A}_{ij}\Delta \boldsymbol{u}_{j,M}(k), \quad i = 1, \ldots, m \end{aligned} \tag{5.28}$$

In (5.28) the performance index is decomposed into m indices for single-variable subsystems separately, while the model constraints do not, due to couplings between various inputs and outputs. It is obvious that the critical issue for decomposing the optimization problem is how to handle the input interaction terms. In the following, three strategies to solve this problem will be discussed, which decompose the original optimization problem using different ideas, i.e. global optimization, local optimization with information exchange between subsystems, and local optimization without information exchange between subsystems, respectively.

5.3.1 Hierarchical Predictive Control Based on Decomposition–Coordination

In this subsection, with the help of hierarchical control theory of large-scale systems, a decomposition algorithm for online optimization of predictive control is proposed from the point of view of global optimization [5]. One of the most popular hierarchical control strategies, the goal coordination method [6], is adopted to solve the online optimization problem (5.28). Its main idea is to regard the variable $\tilde{\boldsymbol{y}}_{PM}(k)$ in the interaction constraint in (5.28) as an independent variable, rather than the result caused by $\Delta\boldsymbol{u}_M(k)$. Then the interaction constraint becomes a balance relationship to be pursued between independent variables $\tilde{\boldsymbol{y}}_{PM}(k)$ and $\Delta\boldsymbol{u}_M(k)$. After introducing the coordination variable $\boldsymbol{\lambda}(k)$ to decompose the global optimization problem into subproblems, $\tilde{\boldsymbol{y}}_{i,PM}(k)$ and $\Delta\boldsymbol{u}_{j,\,M}(k)$ of each subsystem can be solved independently if $\boldsymbol{\lambda}(k)$ is given. However, the solved $\tilde{\boldsymbol{y}}_{PM}(k)$ and $\Delta\boldsymbol{u}_M(k)$ do not necessarily satisfy the interaction balance relationship since they are both solved as independent variables. If the balance relationship is not satisfied, the coordination variable $\boldsymbol{\lambda}(k)$ needs to be adjusted and the performance index should be modified. Then solve $\tilde{\boldsymbol{y}}_{i,PM}(k)$ and $\Delta\boldsymbol{u}_{i,\,M}(k)$ again through decomposition. This procedure is repeated iteratively until the interaction balance is reached by $\tilde{\boldsymbol{y}}_{PM}(k)$ and $\Delta\boldsymbol{u}_M(k)$. At that time, $\tilde{\boldsymbol{y}}_{PM}(k)$ satisfies the original interaction constraints and the whole optimization problem is equivalent to the original one; thus the solved $\Delta\boldsymbol{u}_M$ is the optimal solution. More detailed descriptions are given as follows.

According to the strong duality theorem in mathematical programming, the optimal solution of (5.28) is equivalent to the unconstrained optimal solution of the following dual problem:

$$\max_{\boldsymbol{\lambda}(k)}\ \min_{\Delta\boldsymbol{u}_M(k),\tilde{\boldsymbol{y}}_{PM}(k)}\ L(\Delta\boldsymbol{u}_M(k),\ \tilde{\boldsymbol{y}}_{PM}(k),\ \boldsymbol{\lambda}(k)) \tag{5.29}$$

where

$$L(\Delta\boldsymbol{u}_M(k),\ \tilde{\boldsymbol{y}}_{PM}(k),\ \boldsymbol{\lambda}(k)) = J(k) + \sum_{i=1}^{m}\boldsymbol{\lambda}_i^{\mathrm{T}}(k)\left(\tilde{\boldsymbol{y}}_{i,PM}(k) - \tilde{\boldsymbol{y}}_{i,P0}(k) - \sum_{j=1}^{m}\boldsymbol{A}_{ij}\Delta\boldsymbol{u}_{j,M}(k)\right) \tag{5.30}$$

is the Lagrange function of the overall optimization problem, while $\boldsymbol{\lambda}^{\mathrm{T}}(k) = [\boldsymbol{\lambda}_1^{\mathrm{T}}(k)\ \cdots\ \boldsymbol{\lambda}_m^{\mathrm{T}}(k)]$ is the Lagrange multiplier, where $\boldsymbol{\lambda}_i^{\mathrm{T}}(k) = [\lambda_i(k+1)\ \cdots\ \lambda_i(k+P)]$ $(i = 1, \ldots, m)$.

Take $\boldsymbol{\lambda}(k)$ as the coordination factor and let

$$\boldsymbol{\lambda}(k) = \hat{\boldsymbol{\lambda}}(k) = \left[\hat{\boldsymbol{\lambda}}_1^{\mathrm{T}}(k)\ \cdots\ \hat{\boldsymbol{\lambda}}_m^{\mathrm{T}}(k)\right]^{\mathrm{T}}$$

Then (5.29) can be iteratively solved by the following two-level optimization algorithm.

Level 1. Given $\boldsymbol{\lambda}(k)=\hat{\boldsymbol{\lambda}}(k)$, solve $\min_{\Delta \boldsymbol{u}_M(k),\tilde{\boldsymbol{y}}_{PM}(k)} L\left(\Delta \boldsymbol{u}_M(k),\ \tilde{\boldsymbol{y}}_{PM}(k),\ \hat{\boldsymbol{\lambda}}(k)\right)$

Firstly rewrite (5.30) as

$$
\begin{aligned}
&L\left(\Delta \boldsymbol{u}_M(k),\ \tilde{\boldsymbol{y}}_{PM}(k),\ \hat{\boldsymbol{\lambda}}(k)\right)\\
&=\sum_{i=1}^{m}\left\{\left\|\boldsymbol{w}_i(k)-\tilde{\boldsymbol{y}}_{i,PM}(k)\right\|_{\boldsymbol{Q}_i}^2+\left\|\Delta \boldsymbol{u}_{i,M}(k)\right\|_{\boldsymbol{R}_i}^2\right\}\\
&\quad+\sum_{i=1}^{m}\hat{\boldsymbol{\lambda}}_i^{\mathrm{T}}(k)\left(\tilde{\boldsymbol{y}}_{i,PM}(k)-\tilde{\boldsymbol{y}}_{i,P0}(k)-\sum_{j=1}^{m}\boldsymbol{A}_{ij}\Delta \boldsymbol{u}_{j,M}(k)\right)\\
&=\sum_{i=1}^{m}\left\{\left\|\boldsymbol{w}_i(k)-\tilde{\boldsymbol{y}}_{i,PM}(k)\right\|_{\boldsymbol{Q}_i}^2+\left\|\Delta \boldsymbol{u}_{i,M}(k)\right\|_{\boldsymbol{R}_i}^2+\hat{\boldsymbol{\lambda}}_i^{\mathrm{T}}(k)\left(\tilde{\boldsymbol{y}}_{i,PM}(k)-\tilde{\boldsymbol{y}}_{i,P0}(k)\right)\right\}\\
&\quad-\sum_{j=1}^{m}\sum_{i=1}^{m}\left(\hat{\boldsymbol{\lambda}}_j^{\mathrm{T}}(k)\boldsymbol{A}_{ji}\Delta \boldsymbol{u}_{i,M}(k)\right)\\
&=\sum_{i=1}^{m}L_i\left(\Delta \boldsymbol{u}_{i,M}(k),\ \tilde{\boldsymbol{y}}_{i,PM}(k),\ \hat{\boldsymbol{\lambda}}(k)\right)
\end{aligned}
$$

where

$$
\begin{aligned}
L_i\left(\Delta \boldsymbol{u}_{i,M}(k),\ \tilde{\boldsymbol{y}}_{i,PM}(k),\ \hat{\boldsymbol{\lambda}}(k)\right)&=\left\|\boldsymbol{w}_i(k)-\tilde{\boldsymbol{y}}_{i,PM}(k)\right\|_{\boldsymbol{Q}_i}^2+\left\|\Delta \boldsymbol{u}_{i,M}(k)\right\|_{\boldsymbol{R}_i}^2\\
&\quad+\hat{\boldsymbol{\lambda}}_i^{\mathrm{T}}(k)\left(\tilde{\boldsymbol{y}}_{i,PM}(k)-\tilde{\boldsymbol{y}}_{i,P0}(k)\right)-\sum_{j=1}^{m}\left(\hat{\boldsymbol{\lambda}}_j^{\mathrm{T}}(k)\boldsymbol{A}_{ji}\right)\Delta \boldsymbol{u}_{i,M}(k)
\end{aligned}
$$

is only related to $\tilde{\boldsymbol{y}}_{i,PM}(k)$, $\Delta \boldsymbol{u}_{i,M}(k)$, and is independent of the optimization variables in other subsystems. With given $\hat{\boldsymbol{\lambda}}(k)$, the problem of optimizing $L\left(\Delta \boldsymbol{u}_M(k),\ \tilde{\boldsymbol{y}}_{PM}(k),\ \hat{\boldsymbol{\lambda}}(k)\right)$ can be decomposed into m independent subproblems of optimizing $L_i(\Delta \boldsymbol{u}_{i,M}(k),\ \tilde{\boldsymbol{y}}_{i,PM}(k),\ \hat{\boldsymbol{\lambda}}(k))$. For the ith subproblem, it is now an unconstrained optimization problem, and the necessary condition for the extreme value can be given by

$$
\begin{aligned}
\frac{\partial L_i(k)}{\partial \tilde{\boldsymbol{y}}_{i,PM}(k)}&=-2\boldsymbol{Q}_i\left(\boldsymbol{w}_i(k)-\tilde{\boldsymbol{y}}_{i,PM}(k)\right)+\hat{\boldsymbol{\lambda}}_i(k)=0\\
\frac{\partial L_i(k)}{\partial \Delta \boldsymbol{u}_{i,M}(k)}&=2\boldsymbol{R}_i\Delta \boldsymbol{u}_{i,M}(k)-\sum_{j=1}^{m}\left(\boldsymbol{A}_{ji}^{\mathrm{T}}\hat{\boldsymbol{\lambda}}_j(k)\right)=0
\end{aligned}
$$

Then we get

$$
\begin{aligned}
\tilde{\boldsymbol{y}}_{i,PM}^{*}(k)&=\boldsymbol{w}_i(k)-0.5\boldsymbol{Q}_i^{-1}\hat{\boldsymbol{\lambda}}_i(k)\\
\Delta \boldsymbol{u}_{i,M}^{*}(k)&=0.5\boldsymbol{R}_i^{-1}\sum_{j=1}^{m}\left(\boldsymbol{A}_{ji}^{\mathrm{T}}\hat{\boldsymbol{\lambda}}_j(k)\right)
\end{aligned}
\tag{5.31}
$$

Therefore, the optimal solution of the ith subproblem can be obtained by directly calculating (5.31) according to known $\boldsymbol{A}_{ji}$, $\boldsymbol{Q}_i$, $\boldsymbol{R}_i$ and given $\hat{\boldsymbol{\lambda}}(k)$. Optimization problems for m subsystems can be solved in parallel. Note that in order to avoid the appearance of a singular solution, it is always required that the diagonal weighting matrices satisfy $\boldsymbol{Q}>0$, $\boldsymbol{R}>0$.

Level 2. Solve $\max_{\hat{\lambda}(k)} \varphi\left(\hat{\lambda}(k)\right)$ and update the coordination factor $\boldsymbol{\lambda}(k)$, where

$$\varphi\left(\hat{\boldsymbol{\lambda}}(k)\right)=L\left(\Delta\boldsymbol{u}_M^*(k),\ \tilde{\boldsymbol{y}}_{PM}^*(k),\ \hat{\boldsymbol{\lambda}}(k)\right)=\sum_{i=1}^{m}L_i\left(\Delta\boldsymbol{u}_{i,M}^*(k),\ \tilde{\boldsymbol{y}}_{i,PM}^*(k),\ \hat{\boldsymbol{\lambda}}(k)\right)$$

The coordination factor $\boldsymbol{\lambda}(k)$ can be adjusted using the following gradient algorithm:

$$\hat{\boldsymbol{\lambda}}_i^{l+1}(k)=\hat{\boldsymbol{\lambda}}_i^{l}(k)+\alpha(k)\frac{\partial\varphi\left(\hat{\boldsymbol{\lambda}}(k)\right)}{\partial\hat{\boldsymbol{\lambda}}_i^{l}(k)} \tag{5.32}$$

where l is the iteration number and $\alpha(k)$ is the iteration step length. The gradient vector can be obtained by the expression of $\varphi\left(\hat{\boldsymbol{\lambda}}(k)\right)$ combined with (5.30):

$$\frac{\partial\varphi\left(\hat{\boldsymbol{\lambda}}(k)\right)}{\partial\hat{\boldsymbol{\lambda}}_i^{l}(k)}=\tilde{\boldsymbol{y}}_{i,PM}^{l}(k)-\tilde{\boldsymbol{y}}_{i,P0}(k)-\sum_{j=1}^{m}\boldsymbol{A}_{ij}\Delta\boldsymbol{u}_{j,M}^{l}(k) \tag{5.33}$$

Since $\tilde{\boldsymbol{y}}_{i,P0}(k)$ is known, $\tilde{\boldsymbol{y}}_{i,PM}^{l}(k)$ and $\Delta\boldsymbol{u}_{j,M}^{l}(k)$ are calculated at the last iteration and the new $\boldsymbol{\lambda}(k)$ is easily calculated by (5.32) and (5.33). After $\boldsymbol{\lambda}(k)$ is updated, go to level 1 to calculate again. This iteration procedure continues repeatedly until the following condition is satisfied:

$$\left\|\hat{\boldsymbol{\lambda}}_i^{l+1}(k)-\hat{\boldsymbol{\lambda}}_i^{l}(k)\right\|<\varepsilon,\qquad i=1,\ \ldots,\ m \tag{5.34}$$

where ε is a sufficiently small positive number given in advance. Then it is thought that $\boldsymbol{\lambda}(k)$ has reached its optimal value. In the meantime, it is known from (5.33) that $\tilde{\boldsymbol{y}}_{PM}^{l}(k)$ and $\Delta\boldsymbol{u}_M^{l}(k)$ now satisfy the interaction balance relationship; $\Delta\boldsymbol{u}^{l}(k)$, the corresponding current element in $\Delta\boldsymbol{u}_M^{l}(k)$ can then be used to construct the real control action.

The above decomposition–coordination structure of the online optimization of the hierarchical DMC algorithm is shown in Figure 5.3.

Next, consider the relationship of the solution obtained by the decomposition–coordination algorithm and the optimal solution of the original problem. It is easy to prove that the final solution $\Delta\boldsymbol{u}_M^*(k)$ given by the above algorithm is simply the optimal solution of the multivariable DMC (5.6) as long as the iteration procedure is convergent. Since the end condition of iteration (5.34) approximately implies

$$\frac{\partial\varphi\left(\hat{\boldsymbol{\lambda}}(k)\right)}{\partial\hat{\boldsymbol{\lambda}}_i^{l}(k)}=0,\qquad i=1,\ \ldots,\ m$$

It follows from (5.33) that

$$\tilde{\boldsymbol{y}}_{i,PM}^{*}(k)-\tilde{\boldsymbol{y}}_{i,P0}(k)-\sum_{j=1}^{m}\boldsymbol{A}_{ij}\Delta\boldsymbol{u}_{j,M}^{*}(k)=0,\qquad i=1,\ \ldots,\ m$$

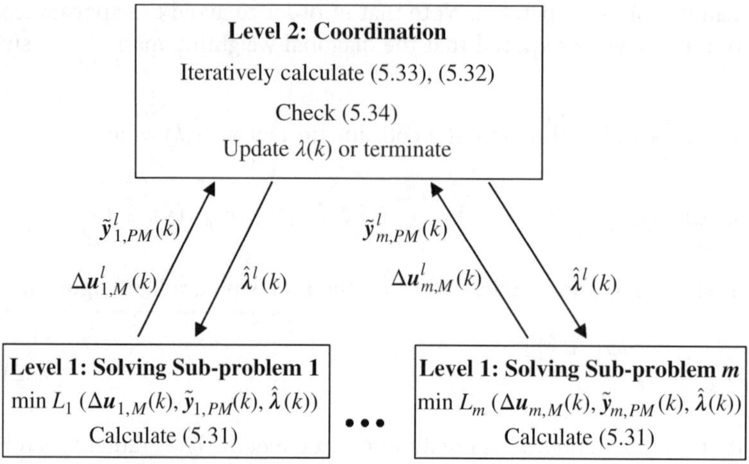

Figure 5.3 Decomposition–coordination structure of the hierarchical DMC algorithm.

Written as a whole gives

$$\tilde{\boldsymbol{y}}^*_{PM}(k) - \tilde{\boldsymbol{y}}_{P0}(k) - \boldsymbol{A}\Delta\boldsymbol{u}^*_M(k) = 0$$

Also write (5.31) as a whole, giving

$$\tilde{\boldsymbol{y}}^*_{PM}(k) = \boldsymbol{w}(k) - 0.5\boldsymbol{Q}^{-1}\hat{\boldsymbol{\lambda}}^*(k)$$
$$\Delta\boldsymbol{u}^*_M(k) = 0.5\boldsymbol{R}^{-1}\boldsymbol{A}^{\mathrm{T}}\hat{\boldsymbol{\lambda}}^*(k)$$

Eliminate $\hat{\boldsymbol{\lambda}}^*(k)$ and $\tilde{\boldsymbol{y}}^*_{PM}(k)$ from above three equations; it follows that

$$\Delta\boldsymbol{u}^*_M(k) = \left(\boldsymbol{A}^{\mathrm{T}}\boldsymbol{Q}\boldsymbol{A} + \boldsymbol{R}\right)^{-1}\boldsymbol{A}^{\mathrm{T}}\boldsymbol{Q}[\boldsymbol{w}(k) - \tilde{\boldsymbol{y}}_{P0}(k)]$$

which is entirely identical to the optimal solution of the multivariable DMC algorithm, i.e. (5.6). It is shown that the above decomposition–coordination algorithm can keep the optimality of the solution, but the analytical optimal solution is obtained by numerical computation iteratively, which does not involve matrix inversion.

The convergence and the convergent speed of the above algorithm depend on the selection of the iteration step length $\alpha(k)$. From (5.32), (5.33), and (5.31) it follows that

$$\begin{aligned}\hat{\boldsymbol{\lambda}}^{l+1}(k) &= \hat{\boldsymbol{\lambda}}^l(k) + \alpha(k)\left[\tilde{\boldsymbol{y}}^l_{PM}(k) - \tilde{\boldsymbol{y}}_{P0}(k) - \boldsymbol{A}\Delta\boldsymbol{u}^l_M(k)\right] \\ &= \hat{\boldsymbol{\lambda}}^l(k) + \alpha(k)\left[\boldsymbol{w}(k) - 0.5\boldsymbol{Q}^{-1}\hat{\boldsymbol{\lambda}}^l(k) - \tilde{\boldsymbol{y}}_{P0}(k) - 0.5\boldsymbol{A}\boldsymbol{R}^{-1}\boldsymbol{A}^{\mathrm{T}}\hat{\boldsymbol{\lambda}}^l(k)\right] \\ &= \left\{\boldsymbol{I} - 0.5\alpha(k)\left(\boldsymbol{Q}^{-1} + \boldsymbol{A}\boldsymbol{R}^{-1}\boldsymbol{A}^{\mathrm{T}}\right)\right\}\hat{\boldsymbol{\lambda}}^l(k) + \alpha(k)[\boldsymbol{w}(k) - \tilde{\boldsymbol{y}}_{P0}(k)]\end{aligned}$$

Denote $\boldsymbol{S} = 0.5(\boldsymbol{Q}^{-1} + \boldsymbol{A}\boldsymbol{R}^{-1}\boldsymbol{A}^{\mathrm{T}})$. The iteration procedure is convergent if

$$\max_i |\lambda_i(\boldsymbol{I} - \alpha(k)\boldsymbol{S})| < 1$$

where $\lambda_i(\cdot)$ represents the ith eigenvalue of a matrix. Thus the range of iteration step length that guarantees the iteration procedure convergent can be theoretically given by

$$0 < \alpha(k) < \frac{2}{\max\limits_{i} \lambda_i(\boldsymbol{S})} \tag{5.35}$$

Since $\boldsymbol{S}$ is a symmetric positive definite matrix, (5.35) indicates that there always exists a suitable iteration step length to make the algorithm convergent. In practice, it is usually possible to find a suitable $\alpha(k)$ by trial.

The above decomposition–coordination method solves the global optimization problem through online iteration. Compared with directly calculating the analytical solution, the high-dimensional matrix inversion can be avoided, which is particularly advantageous when the predictive control law should be adapted online and recalculated. Moreover, after decomposition the subproblems are relatively small scale and can be solved independently. Thus they can be solved in parallel so as to improve the computational efficiency. Compared with the overall problem solving, the requirements on computers and memories are relatively low, while the same optimization result can be obtained. This decomposition–coordination strategy is introduced here for the unconstrained DMC algorithm, but with the same principle it is also applicable for predictive control with constraints or for nonlinear systems.

5.3.2 Distributed Predictive Control

In the decomposition–coordination algorithm presented in the last subsection, the global optimization problem is decomposed into small-scale subproblems and solved independently with local information in level 1. While in level 2, information of other subsystems are used to update the coordination factors, which are then provided to solve the subproblems. From the information point of view, it is essentially a centralized control algorithm with global optimization. With the wide application of computers in the industrial environment and the development of network technology, distributed control has received extensive attention and is widely used. Distributed control refers to the control pattern that distributes the whole control task to multiple controllers and allows the controllers to transfer limited information through communication from each other. In this subsection, we will briefly introduce how to realize online optimization of multivariable predictive control in a distributed way [7].

With the idea of distributed control, the online optimization problem of predictive control described by (5.28) can be distributed to m independent controllers, where the ith controller solves a single variable predictive control subproblem with u_i as the control input and y_i as the output and the corresponding online optimization problem is described by

$$\min_{\Delta \boldsymbol{u}_{i,M}(k)} J_i(k) = \left\| \boldsymbol{w}_i(k) - \tilde{\boldsymbol{y}}_{i,PM}(k) \right\|_{\boldsymbol{Q}_i}^2 + \left\| \Delta \boldsymbol{u}_{i,M}(k) \right\|_{\boldsymbol{R}_i}^2 \tag{5.36}$$

$$\text{s.t.} \quad \tilde{\boldsymbol{y}}_{i,PM}(k) = \tilde{\boldsymbol{y}}_{i,P0}(k) + \sum_{j=1}^{m} \boldsymbol{A}_{ij} \Delta \boldsymbol{u}_{j,M}(k) \tag{5.37}$$

Compared with the global optimization problem (5.28), in the centralized case it can be seen that:

1) The overall optimization problem of multivariable predictive control is replaced by m independent optimization subproblems of single variable predictive control. The global optimization performance index is decomposed into m individual performance indices of subsystems. There is no coordination unit that can have global information. This is a multiperson multiobjective optimization problem.
2) The performance index of each optimization subproblem seems to be only related to its own u_i and y_i. However, if the prediction model (5.37) is substituted into the performance index, it can be found that the optimization of each subproblem is actually related to all control inputs. Thus all the optimization subproblems are interrelated.

This kind of multiperson multiobjective optimization problem can be solved with the help of the concept of Nash optimization in decision theory. Consider a multiobjective decision problem with N decision variables $u_1, \ldots, u_N$, where the decision variable for the ith decision maker is u_i and the performance index is $\min J_i(u_1, \ldots, u_N)$. The Nash optimal solution [8] is referred to a group of decisions $u^* = (u_1^*, \ \ldots, \ u_N^*)$ satisfying

$$J_i(u_1^*, \ldots, u_N^*) \le J_i(u_1^*, \ldots, u_{i-1}^*, u_i, u_{i+1}^*, \ldots, u_N^*), \qquad i = 1, \ \ldots, \ N \tag{5.38}$$

The meaning of Nash optimization can be explained as follows. If the Nash solution is reached, then all the decision makers will never change their own decisions u_i because under this condition all the decision makers have reached an optimum with respect to their own performance indices and further changing u_i will only make J_i worse. Therefore, it is a solution all the decision makers can accept, reflecting the balance between them when seeking their own optimum.

Treat all the distributed controllers as independent decision makers; then according to (5.38), the Nash solution for each controller can be obtained by solving the following optimization problem:

$$\min_{\Delta \boldsymbol{u}_{i,M}(k)} J_i(k) = \left\| \boldsymbol{w}_i(k) - \tilde{\boldsymbol{y}}_{i,PM}(k) \right\|_{\boldsymbol{Q}_i}^2 + \left\| \Delta \boldsymbol{u}_{i,M}(k) \right\|_{\boldsymbol{R}_i}^2$$

$$\text{s.t.} \quad \tilde{\boldsymbol{y}}_{i,PM}(k) = \tilde{\boldsymbol{y}}_{i,P0}(k) + \boldsymbol{A}_{ii} \Delta \boldsymbol{u}_{i,M}(k) + \sum_{\substack{j=1 \\ j \ne i}}^{m} \boldsymbol{A}_{ij} \Delta \boldsymbol{u}_{j,M}^*(k)$$

where $\Delta \boldsymbol{u}_{j,M}^*(k)$ is the Nash optimal solution of other controllers. The solution of the above optimization problem can be given by

$$\Delta \boldsymbol{u}_{i,M}^*(k) = \left(\boldsymbol{A}_{ii}^{\mathrm{T}} \boldsymbol{Q}_i \boldsymbol{A}_{ii} + \boldsymbol{R}_i \right)^{-1} \boldsymbol{A}_{ii}^{\mathrm{T}} \boldsymbol{Q}_i \left(\boldsymbol{w}_i(k) - \tilde{\boldsymbol{y}}_{i,P0}(k) - \sum_{\substack{j=1 \\ j \ne i}}^{m} \boldsymbol{A}_{ij} \Delta \boldsymbol{u}_{j,M}^*(k) \right) \tag{5.39}$$

From (5.39) it can be seen that for the ith controller, in order to calculate the Nash optimal solution $\Delta \boldsymbol{u}_{i,M}^*(k)$ it is necessary to know the Nash optimal solutions of other controllers $\Delta \boldsymbol{u}_{j,M}^*(k), j \ne i$. Therefore, it is a coupled decision problem and can be solved by using prediction and iteration through communication between local controllers. At the lth step of iteration, each controller sends its optimal solution $\Delta\boldsymbol{u}_{j,M}^{l}(k)$ to other controllers through communication. At the $(l+1)$th step, using these obtained $\Delta\boldsymbol{u}_{j,M}^{l}(k)$, the controller i updates its own Nash optimal solution by

$$\Delta\boldsymbol{u}_{i,M}^{l+1}(k) = \left(\boldsymbol{A}_{ii}^{\mathrm{T}}\boldsymbol{Q}_i\boldsymbol{A}_{ii} + \boldsymbol{R}_i\right)^{-1}\boldsymbol{A}_{ii}^{\mathrm{T}}\boldsymbol{Q}_i\left(\boldsymbol{w}_i(k) - \tilde{\boldsymbol{y}}_{i,P0}(k) - \sum_{\substack{j=1\\ j\neq i}}^{m}\boldsymbol{A}_{ij}\Delta\boldsymbol{u}_{j,M}^{l}(k)\right) \tag{5.40}$$

and then sends $\Delta\boldsymbol{u}_{j,M}^{l+1}(k)$ to other controllers This process continues iteratively until the following end condition is satisfied:

$$\left\|\Delta\boldsymbol{u}_{i,M}^{l+1}(k) - \Delta\boldsymbol{u}_{i,M}^{l}(k)\right\| \le \varepsilon, \qquad i = 1, \ldots, m$$

where ε is a given small positive number. The initial solution for iteration can be given arbitrarily, even by setting zero for all, i.e. $\Delta\boldsymbol{u}_{j,M}^{*}(k) = \boldsymbol{0}$, $i = 1,\ldots, m$, which corresponds to the case where each controller solves its own optimal solution independently, without considering the interconnections of other control inputs. During iteration, each controller compares the newly solved optimal solution with the old one, checks whether the end condition of iteration is satisfied, and informs the check result and the newly solved optimal solution to other controllers. If the errors between two successive solutions do not satisfy given precision requirements, the whole system does not reach a Nash equilibrium and thus the local performance index needs to be improved through further iteration. When the iteration is over, each controller will guarantee the performance index minimal. The whole system will reach a Nash equilibrium, which means that all the local controllers achieve the Nash optimum and their own performance indices cannot be further improved. Then the distributed Nash optimization at time k is completed. The information structure of the distributed DMC algorithms is shown in Figure 5.4.

Next consider the convergence condition of the above iterative algorithm. Denote $\boldsymbol{D}_{ii} = \left(\boldsymbol{A}_{ii}^{\mathrm{T}}\boldsymbol{Q}_i\boldsymbol{A}_{ii} + \boldsymbol{R}_i\right)^{-1}\boldsymbol{A}_{ii}^{\mathrm{T}}\boldsymbol{Q}_i$; then (5.40) can be represented by

$$\Delta\boldsymbol{u}_{M}^{l+1}(k) = \boldsymbol{D}_1[\boldsymbol{w}(k) - \tilde{\boldsymbol{y}}_{P0}(k)] + \boldsymbol{D}_0\Delta\boldsymbol{u}_{M}^{l}(k)$$

where

$$\boldsymbol{D}_1 = \begin{bmatrix} \boldsymbol{D}_{11} & & \boldsymbol{0} \\ & \boldsymbol{D}_{22} & \\ & & \ddots & \\ \boldsymbol{0} & & & \boldsymbol{D}_{mm} \end{bmatrix}, \quad \boldsymbol{D}_0 = \begin{bmatrix} \boldsymbol{0} & -\boldsymbol{D}_{11}\boldsymbol{A}_{12} & \cdots & -\boldsymbol{D}_{11}\boldsymbol{A}_{1m} \\ -\boldsymbol{D}_{22}\boldsymbol{A}_{21} & \boldsymbol{0} & & -\boldsymbol{D}_{22}\boldsymbol{A}_{2m} \\ \vdots & & \ddots & \vdots \\ -\boldsymbol{D}_{mm}\boldsymbol{A}_{m1} & -\boldsymbol{D}_{mm}\boldsymbol{A}_{m2} & \cdots & \boldsymbol{0} \end{bmatrix}$$

Then the convergence condition for the iteration is as follows:

$$\max|\lambda(\boldsymbol{D}_0)| < 1$$

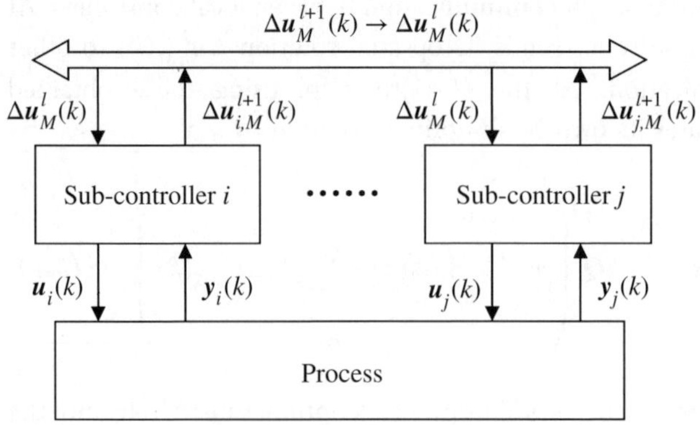

Figure 5.4 Distributed algorithmic architecture of DMC.

and if the iteration process is convergent, the Nash optimal solution can be given by

$$\Delta u_M^*(k) = (I - D_0)^{-1} D_1 [w(k) - \tilde{y}_{P0}(k)]$$

Denote A_1 = block diag($A_{11}, \ldots, A_{mm}$), $A_2 = A - A_1$. It then follows that

$$D_1 = \left(A_1^T Q A_1 + R\right)^{-1} A_1^T Q, \quad D_0 = -D_1 A_2 = -\left(A_1^T Q A_1 + R\right)^{-1} A_1^T Q A_2$$

Furthermore,

$$\begin{aligned}(I - D_0)^{-1} D_1 &= \left[\left(A_1^T Q A_1 + R\right)(I - D_0)\right]^{-1} A_1^T Q \\ &= \left[A_1^T Q A_1 + R + A_1^T Q A_2\right]^{-1} A_1^T Q = \left(A_1^T Q A + R\right)^{-1} A_1^T Q\end{aligned}$$

The Nash optimal solution by distributed control can be written in the form

$$\Delta u_M^*(k) = \left(A_1^T Q A + R\right)^{-1} A_1^T Q [w(k) - \tilde{y}_{P0}(k)] \tag{5.41}$$

which is obviously different from the optimal solution (5.6) by global optimization.

The distributed predictive control presented above focuses on illustrating its principle on information utilization. With the rapid development of communication and network technology, distributed predictive control is widely adopted in practice and has become an important branch of predictive control. The progress of research and applications in this area can be found in [9].

5.3.3 Decentralized Predictive Control

The distributed iterative algorithm in the last section needs to exchange necessary information between local controllers. If the communication between the controllers is not allowed, i.e. for each controller, the control inputs of other subsystems $\Delta u_{j,M}^l(k)$ are unknown when online solving the optimization problem (5.36) and (5.37), then a fully decentralized control structure will be achieved. Decentralized control is an effective control structure developed for controlling large-scale systems with high dimensions. Its basic characteristic is to realize decentralized control using decentralized information, where not only control implementation is decentralized but also the information provided for control is restricted as decentralized local information rather than system global information, i.e. the control law of the ith controller is of the form

$$\boldsymbol{u}_i = \boldsymbol{u}_i(\boldsymbol{I}_i), \qquad i = 1, \ldots, m$$

where $\boldsymbol{I}_i$ is the local information available for the ith controller, i.e. the information of u_i and y_i in the fully decentralized case. This corresponds to using multiple single-loop controllers to control a multivariable system. It is also a special case of distributed control without information exchange between subsystems. With this control structure, each controller only adopts the detected information of its own subsystem to constitute the control action and online communication between subsystems is not required. Obviously, such a fully decentralized control law is the simplest one for implementation. Furthermore, when failure possibly appears in one subsystem, the control of other subsystems can be going on as usual, so the overall reliability of the control system could be improved.

However, the drawback to this decentralized control law is also obvious. Since the inputs and outputs of the multivariable system are highly coupled, each output y_i is affected not only by u_i but also by other u_j $(j \neq i)$. The fully decentralized control law neglects the couplings between the subsystems and will undoubtedly lead to degeneration of the control performance, which has already been seen when controlling a multivariable system using traditional multiple PID loop controllers. To improve the performance of decentralized control systems, it is necessary to properly compensate the lacked information caused by decentralization. Although in the fully decentralized structure such compensation cannot come from other subsystems due to the lack of information exchange, the feedback correction function of predictive control may help to predict and compensate the lacked information. Therefore, when adopting decentralized predictive control to multivariable systems, on the one hand, the computational complexity of overall design and optimization can be reduced by decomposition and the reliability of the control system can be improved, while, on the other hand, the control performance can still be ensured as long as an appropriate feedback correction method is taken to compensate for the lack of relevant information of the subsystem. Through utilizing the advantages both of good control performance from predictive control and of a simple structure from decentralized control, decentralized predictive control can be used as an effective strategy to control multivariable systems. In the following, we introduce the decentralized DMC algorithm for multivariable systems in some detail [10].

In the fully decentralized control structure, for the ith controller, the input and output information of other subsystems is unavailable, and it can only establish a prediction model including input and output information of its own subsystem; i.e. in decentralized control, due to the lack of information, the output prediction can only be given by the following model instead of (5.37):

$$\tilde{\boldsymbol{y}}_{i,PM}(k) = \tilde{\boldsymbol{y}}_{i,P0}(k) + \boldsymbol{A}_{ii}\Delta\boldsymbol{u}_{i,M}(k) \tag{5.42}$$

With the above model and optimization performance index (5.36), the unconstrained optimal solution is given by

$$\Delta\boldsymbol{u}_{i,M}^{*}(k) = \left(\boldsymbol{A}_{ii}^{\mathrm{T}}\boldsymbol{Q}_i\boldsymbol{A}_{ii} + \boldsymbol{R}_i\right)^{-1}\boldsymbol{A}_{ii}^{\mathrm{T}}\boldsymbol{Q}_i\left(\boldsymbol{w}_i(k) - \tilde{\boldsymbol{y}}_{i,P0}(k)\right) \tag{5.43}$$

which is the result of the single-variable predictive control for the ith independent subsystem. However, it is obvious that the prediction model given by (5.42) is inaccurate. Due to the couplings between subsystems, the accurate prediction of y should actually be given by (5.37). This indicates that the prediction model (5.42) is naturally mismatched. Even without disturbance, the actual output is inconsistent with the output calculated by the model (5.42), which is caused by the decentralized control strategy where the relevant information of other subsystems is unavailable and thus not taken into account in the prediction model.

Remark 5.3 The feedback correction in predictive control can compensate for the prediction error caused by model mismatch to a certain extent. According to the feedback correction principle of predictive control, when the control period moves from k to $k+1$, the newly measured output $y_i(k+1)$ should be compared with the model output $\tilde{y}_{i,1}(k+1\,|\,k)$ predicted at k to construct the error $e_i(k+1)$ (see (5.10)), with which the future output predictions are corrected to construct the initial output prediction at $k+1$ through shifting (see (5.11) and (5.12)). This correction can compensate for the influence of unknown factors on output prediction, including model mismatch, disturbance, etc. Since no causal model exists for such a prediction and compensation, the heuristic method is often adopted. In the centralized control case, feedback correction works only when model mismatch or disturbance exists. However, in decentralized control, as mentioned above, the subsystem model (5.42) is naturally mismatched even when the model parameters of a subsystem are accurate and the disturbance does not exist. Therefore, feedback correction plays a more important role in decentralized predictive control than that in centralized case and becomes an indispensable part of adjusting the optimization basis to achieve a better control performance.

In decentralized predictive control, the prediction error includes the couplings between subsystems and its variation is not necessarily slow and smooth. Therefore, a simple weighting correction method as used in centralized control may be unavailable and more effective noncausal prediction methods are required. For example, [10] suggested applying the Holt-Winters approach often used in time series forecasting [11] to error prediction. The main steps of this approach are described as follows. Firstly, construct the error difference according to $e_i(k+1)$ obtained from (5.10) and $e_i(k)$ at the last time:

$$\Delta e_i(k+1) = e_i(k+1) - e_i(k)$$

Then take $\Delta e_i(k+1)$ as the prediction variable. If there is no seasonal factor in its variation, its future prediction can be given by

$$\begin{aligned}
&\Delta \tilde{e}_i(k+1+j\,|\,k+1) = m_i(k+1) + j\gamma_i(k+1), \quad j = 1, \ldots, N \\
&m_i(k+1) = \alpha \Delta e_i(k+1) + (1-\alpha)(m_i(k) + \gamma_i(k)), \quad 0 < \alpha < 1 \\
&\gamma_i(k+1) = \beta(\Delta e_i(k+1) - m_i(k)) + (1-\beta)\gamma_i(k), \quad 0 < \beta < 1
\end{aligned} \tag{5.44}$$

where $m_i(k+1)$ is the base value of $\Delta\tilde{e}_i$, represented by the weighting sum of current $\Delta e_i(k+1)$ and $\Delta\tilde{e}_i(k+1\,|\,k)$ predicted at the last time, $\gamma_i(k+1)$ is the step increment, represented by the weighting sum of the step increment at last time back-calculated by the currently measured actual error and the actual step increment $\gamma_i(k)$ at the last time. The selection for initial values $m_i(0)$, $\gamma_i(0)$ and filtering parameters α, β can be referred to [11].

At the sampling time $k+1$, $m_i(k)$, $\gamma_i(k)$ are known, $\Delta e_i(k+1)$ is constructed by the difference of the actually measured error; then all $\Delta\tilde{e}_i(k+1+j\,|\,k+1)$, $j=1, \ldots, N$ can be calculated according to (5.44). With the definition of Δe_i, the error prediction for future output $\tilde{e}_i(k+1+j\,|\,k+1)$, $j=1, \ldots, N$ can be calculated and used to correct the predictive outputs. After shifting, it follows that

$$\tilde{y}_{i,0}(k+1+j\,|\,k+1) = \tilde{y}_{i,1}(k+1+j\,|\,k) + \tilde{e}_i(k+1+j\,|\,k+1), \quad j=1,\cdots,N$$

After correction using the above method, the initial output prediction $\tilde{\boldsymbol{y}}_{i,N0}(k+1)$ can be obtained and provided to the decentralized controller (5.43) for its use at the sampling time $k+1$.

The decentralized predictive control strategy introduced above decomposes the multivariable system into independent subsystems, aiming at simplifying online computation as well as increasing the reliability of the controlled system. In decentralized predictive control, the local prediction model and local optimization performance index are used, and feedback correction is particularly emphasized to compensate for the lacked information on couplings from other subsystems. Since the feedback correction only uses the error information of its own subsystem, although the prediction error can be compensated to a certain extent by using an appropriate time series forecasting method, the prediction accuracy and the control performance would be in general worse than the other two decomposition strategies.

5.3.4 Comparison of Three Decomposition Algorithms

In this subsection, a brief analysis and comparison on the online optimization of the above three decomposition algorithms for multivariable predictive control are given, where the discussion is still for the unconstrained case because in this case the resultant analytical solutions are intuitive and may be more suitable for comparison.

Consider the online optimization problem of multivariable predictive control represented by (5.4) and (5.5):

$$\min_{\Delta \boldsymbol{u}_M(k)} J(k) = \|\boldsymbol{w}(k) - \tilde{\boldsymbol{y}}_{PM}(k)\|_{\boldsymbol{Q}}^2 + \|\Delta \boldsymbol{u}_M(k)\|_{\boldsymbol{R}}^2$$

$$\text{s.t.}\ \ \tilde{\boldsymbol{y}}_{PM}(k) = \tilde{\boldsymbol{y}}_{P0}(k) + \boldsymbol{A}\Delta \boldsymbol{u}_M(k)$$

The decomposition algorithms for this optimization problem can be divided into the following three typical cases according to different optimization schemes and information structures.

1) **Global optimization with a single objective and centralized control structure**
   This is the hierarchical predictive control algorithm based on decomposition–coordination presented in Section 5.3.1. It uses a single global performance index (5.28) that is the same as (5.4) in the original problem. At the first level, the global optimization problem is decomposed into multiple small-scale optimization subproblems, which can be solved in parallel and thus the computational complexity can be reduced. As to the control structure, if all the tasks of solving the optimization

subproblems at the first level and coordination at the second level are undertaken by different controllers, respectively, then the coordination controller is obviously a global one and has a higher position than the optimization subcontrollers. All the optimization subcontrollers indeed obtain the global information through the coordination factor provided by the coordination controller, so this method has a centralized control structure.

2) **Nash optimization with multiple objectives and a distributed control structure**
   This is the distributed predictive control algorithm presented in Section 5.3.2. It distributes the task of overall optimization to m independent controllers, where each controller has its own performance index (see (5.36)). Therefore, this control algorithm uses multiple objectives of multiple controllers to replace the single global objective. However, due to the interconnections between the models of subproblems, the Nash optimization strategy is adopted to handle the multiobjective optimization. With this control strategy, all the controllers have equal positions, but need to exchange necessary information from each other. It has a distributed control structure.
3) **Decentralized optimization with multiple objectives and a fully decentralized control structure**
   This is the decentralized predictive control algorithm presented in Section 5.3.3. Similar to the distributed predictive control, it distributes the overall optimization task to independent controllers and replaces the original single global objective by multiple objectives of multiple controllers. However, due to the fully decentralized structure, both the objective of each controller and the model it uses are independent from each other. Therefore, the multiobjective optimization problem is actually decomposed into multiple independent optimization problems, each with a single objective. Since information exchange between subsystems is not allowed, both the control law and model prediction can only use the input and output information of its own subsystem. The prediction error caused by lacked coupling information needs to be compensated through feedback correction, which is also based on its own output errors.

The optimal control laws obtained by these three decomposition algorithms are compared in the following.

1) Decomposition–coordination algorithm. It has been proved that the optimal control law converges to the optimal solution of the original optimization problem (5.6):

$$\Delta \boldsymbol{u}_M^*(k) = \left(\boldsymbol{A}^{\mathrm{T}}\boldsymbol{Q}\boldsymbol{A} + \boldsymbol{R}\right)^{-1}\boldsymbol{A}^{\mathrm{T}}\boldsymbol{Q}[\boldsymbol{w}(k) - \tilde{\boldsymbol{y}}_{P0}(k)]$$

2) Distributed algorithm. The Nash optimal solution is given by (5.41)

$$\Delta \boldsymbol{u}_M^*(k) = \left(\boldsymbol{A}_1^{\mathrm{T}}\boldsymbol{Q}\boldsymbol{A} + \boldsymbol{R}\right)^{-1}\boldsymbol{A}_1^{\mathrm{T}}\boldsymbol{Q}[\boldsymbol{w}(k) - \tilde{\boldsymbol{y}}_{P0}(k)]$$

3) Fully decentralized algorithm. According to (5.43) and the definition of $\boldsymbol{A}_1$ it follows that

$$\Delta \boldsymbol{u}_M^*(k) = \left(\boldsymbol{A}_1^{\mathrm{T}}\boldsymbol{Q}\boldsymbol{A}_1 + \boldsymbol{R}\right)^{-1}\boldsymbol{A}_1^{\mathrm{T}}\boldsymbol{Q}[\boldsymbol{w}(k) - \tilde{\boldsymbol{y}}_{P0}(k)]$$

where matrix $\boldsymbol{A}_1$ is the main diagonal block part of the matrix $\boldsymbol{A}$, corresponding to the responses of individual subsystems. These formulae reflect the difference between optimization ideas in three algorithms. From the deduction of the optimal control laws it can be seen that, in the fully decentralized algorithm, the control law only contains $\boldsymbol{A}_1$ because the couplings between subsystems are neglected both in the model prediction and in the optimization, while in the distributed algorithm both $\boldsymbol{A}_1$ and $\boldsymbol{A}$ appear in the control law because during optimization of the subproblem, the couplings between the subsystems are treated as known quantities, retained in model prediction ($\boldsymbol{A}$ appears) but not involved in optimization ($\boldsymbol{A}_1$ appears).

These formulae also reflect the degree of optimal control laws of three decomposition algorithms close to the optimal solution of the original multivariable predictive control, where the decomposition–coordination algorithm can get the optimal solution of the original problem while the fully decentralized algorithm has the largest deviation from the optimal solution although the deviation can be reduced to a certain extent through feedback correction.

Example 5.2 Comparison of three decomposition algorithms for multivariable predictive control

Consider a linear system with three inputs and three outputs:

$$\boldsymbol{G}(s)=\begin{bmatrix}\dfrac{1}{100s+1} & \dfrac{1}{100s+1} & \dfrac{1}{200s+1}\\ \dfrac{-1.25}{50s+1} & \dfrac{3.75}{50s+1} & \dfrac{1}{50s+1}\\ \dfrac{-2}{200s+1} & \dfrac{2}{200s+1} & \dfrac{3.5}{100s+1}\end{bmatrix}$$

Divide it into three subsystems and set the sampling period $T = 20$ sec onds. Use the hierarchical, distributed, and decentralized predictive control algorithms presented in Sections 5.3.1 to 5.3.3, respectively. For each subsystem, select $P = 5, M = 3, \boldsymbol{Q} = \boldsymbol{I}, \boldsymbol{R} = 0.5\boldsymbol{I}$. In order to focus on the influence of different decomposition algorithms on the control performance, the feedback correction is not considered here. The simulation results of the input and output curves using three decomposition algorithms are shown in Figure 5.5a, b, and c, respectively. Take the performance index with the running time $T_{\text{sim}} = 300$ seconds for comparison; it follows that $J_a = 1.8279, J_b = 1.8853, J_c = 2.3917$. This shows that the performance of the hierarchical algorithm is better than that of the distributed algorithm and the latter is better than the decentralized algorithm. Note that here the performance of the decentralized algorithm directly results from decentralized optimization and without feedback correction, so the performance degeneration is obvious.

In order to highlight the differences and relationships of different decomposition algorithms, the above discussion is only for the unconstrained case where an analytical solution is available. However, the main ideas of theses algorithms are also suitable for systems with input and output constraints. In this case the constraints can be put into the optimization subproblems, and the coupling terms in the constraints can be handled using the similar idea of handling the model coupling terms. The online optimization subproblems in general need to be solved iteratively, but the online computation burden can be greatly reduced because the scale of the optimization subproblem has been reduced through decomposition. Therefore, decomposition is more meaningful when dealing with constrained predictive control problems.

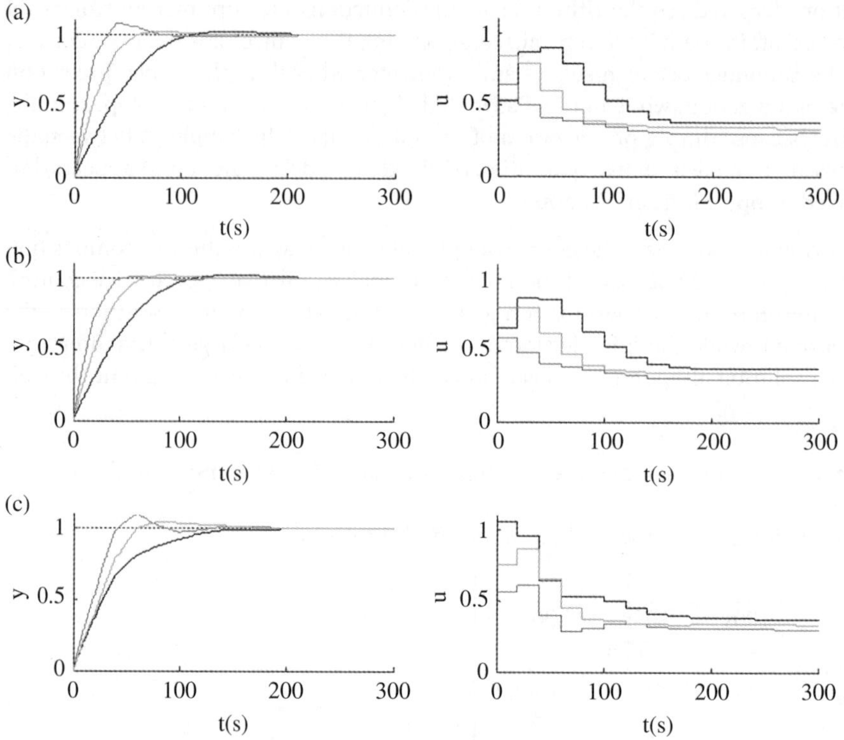

Figure 5.5 Comparison of three decomposition algorithms: (a) hierarchical algorithm, (b) distributed algorithm, (c) decentralized algorithm.

5.4 Summary

In this chapter, taking the DMC algorithm as an example, the predictive control algorithm for multivariable systems is introduced, which is a straightforward extension of the predictive control algorithm for single variable systems. Compared with the single variable algorithm, the main difference is that the prediction of future system output should consider not only the variation of the control input at different time instances but also the influence from different control inputs. Therefore, the output response is the double superposition of the control inputs along both time and space.

In practical applications, constraints on system inputs, states, and outputs often exist. It is well recognized that predictive control is able to explicitly handle the constraints in online optimization. In many cases, the online optimization problem of constrained multivariable predictive control can be formulated as a QP problem with a quadratic performance index and linear inequality constraints. It needs to be solved iteratively by a matrix tearing technique or a standard QP algorithm. For more general cases, for example for nonlinear systems/constraints or a nonquadratic performance index, the online optimization problem with constraints can be similarly formulated, but the solving method should be properly selected according to the optimization formulation.

In order to simplify the online computation of multivariable predictive control algorithms, three decomposition strategies for its online optimization are presented: centralized control structure with global optimization based on decomposition–coordination, distributed control structure with a Nash optimization solved by information exchange and iteration, and a fully decentralized control structure with independent optimization using feedback correction for coupling compensation. These three kinds of decomposition algorithms correspond to different optimization ideas and information structures, having different degrees close to the original optimal solution. Since they all decompose the online computation task of the overall optimization problem into small-scale subproblems, which can be solved in parallel, the computational complexity of solving the online optimization problem can be greatly reduced.

References

1 Xi, Y. (1989). New design method for discrete time multi-variable predictive controllers. *International Journal of Control* 49 (1): 45–56.
2 Prett, D.M. and Gillete, R.D. Optimization and constrained multivariable control of a catalytic creaking unit, *Proceedings of the 1980 Joint American Control Conference*, San Francisco, CA (13–15 August 1980), WP5-C.
3 Nocedal, J. and Wright, S.J. (2006). *Numerical Optimization*, 2e. Springer.
4 Morari, M. and Baric, M. (2006). Recent developments in the control of constrained hybrid systems. *Computers and Chemical Engineering* 30: 1619–1631.
5 Zhang, Z., Xi, Y., and Xu, X. (1988). Hierarchical predictive control for large scale industrial systems. In: *Automatic Control Tenth Triennial World Congress of IFAC*, vol. 7 (ed. R. Isermann), 91–96. Oxford: Pergamon Press.
6 Singh, M.G. and Titli, A. (1978). *Systems: Decomposition, Optimization and Control*. New York: Pergamon Press.
7 Du, X.N., Xi, Y.G., and Li, S.Y. (2001). Distributed model predictive control for large scale systems. *Proceedings of ACC* (4): 3142–3143.
8 Singh, M.G. (1981). *Decentralized Control*. Amsterdam: North-Holland.
9 Christofides, P.D., Scattolini, R., de la Pena, D.M., and Liu, J. (2013). Distributed model predictive control: a tutorial review and future research directions. *Computers and Chemical Engineering* 51: 21–41.
10 Xu, X., Xi, Y., and Zhang, Z. (1988). Decentralized predictive control (DPC) of large scale systems. *Information and Decision Technologies* 14: 307–322.
11 Chartfield, C. (1984). *The Analysis of Time Series: An Introduction*, 3e. London: Chapman and Hall.

6

Synthesis of Stable Predictive Controllers

Since the 1970s, predictive control has been widely used in the industrial fields because of its applicability to complex industrial processes. At the same time, its theoretical research has also received extensive attention by both industrial and academic communities. Through making a historical view on the progress of the predictive control theory, it could be found that it has roughly gone through two stages [1].

During the period of the 1980s to the early 1990s, the theoretical research on predictive control was primarily motivated by the practical requirements of industrial applications. At this stage, various predictive control algorithms such as MAC, DMC, and GPC were successively proposed and applied to industrial processes. The industrial community particularly asked for theoretical guidance on parameter tuning when applying these algorithms to practice. To meet this requirement, the theoretical study often started from existing predictive control algorithms and focused on exploring the quantitative relationship between the design parameters and the closed-loop performances such as stability, dynamic response, etc. as discussed in Chapters 3 and 4. We call the predictive control theory at this stage the (classical) quantitative analysis theory of predictive control. However, the quantitative analysis is only possible for the unconstrained case where an explicit analytical solution for online optimization is available. Meanwhile, since the relationship between the design parameters and the system performances is not direct and concise, even for unconstrained SISO predictive control systems, only limited results have been achieved by using some special techniques. Therefore, by the middle of the 1990s, the quantitative analysis theory gradually faded out from the study of predictive control.

In view of the essential difficulties of the research on quantitative analysis of predictive control systems, in the middle 1990s, the academic community turned the thinking of theoretical research on predictive control from "analyzing the stability of the existing algorithms" to "synthesizing new algorithms with guaranteed stability. This started a new stage of research on predictive control theory which we call (modern) qualitative synthesis theory of predictive control. It is obvious that this study was motivated by the requirements of the academic community on exploring the theoretical essence of predictive control after reviewing and introspecting existing predictive control theory and research methods. It is not limited to investigating the existing predictive control algorithms, but tries to synthesize new algorithms from a theoretical viewpoint to guarantee closed-loop stability and control performance. During this period of research on

Predictive Control: Fundamentals and Developments, First Edition. Yugeng Xi and Dewei Li.

predictive control theory, with optimal control theory as the most important theoretical reference, the Lyapunov stability analysis as the basic method to guarantee control performance, the invariant set, linear matrix inequality (LMI), etc., as fundamental tools, and the control performance guarantee under the rolling optimization as the research focus, rich research contents have been constituted, showing academic profundity and methodological innovation. For more than two decades, the qualitative synthesis theory of predictive control has become the mainstream of theoretical research in the predictive control field.

In this and the next chapter, through interpreting the basic ideas in the research of qualitative synthesis theory of predictive control, some representative research results of synthesizing stable and robust predictive controllers will be introduced, which provide readers with basic knowledge for further research.

6.1 Fundamental Philosophy of the Qualitative Synthesis Theory of Predictive Control

In the 1990s, the theory of model predictive control (MPC) was rapidly developed. Particularly, optimal control theory and Lyapunov approach were utilized in the study of MPC and led to a number of important results. Mayne et al. [2] deeply revealed the relationship between MPC and optimal control, and summarized the fundamental ideas and conditions of stability analysis and synthesis of MPC. In the following, we will give a brief illustration for that according to [2].

6.1.1 Relationships between MPC and Optimal Control

As pointed out by Mayne et al. [2], MPC is not a new control method. It essentially solves a standard optimal control problem except that it adopts a limited time horizon instead of the infinite horizon usually employed in optimal control. The main difference between these two control methods is that MPC solves the optimal control problem online for the current state of the plant, while optimal control determines off-line an optimal feedback control law for all states.

It is well known that the optimal control problem can be solved either by the dynamic programming approach or by the maximum value principle. The dynamic programming provides sufficient conditions for the optimality and results in a constructive process for determining the optimal feedback controller $\boldsymbol{u}(k) = \boldsymbol{K}(\boldsymbol{x}(k))$. The obtained control law is applicable to any time instant. However, this approach needs to solve a Hamilton–Jacobi–Bellman (HJB) differential or difference equation, which is usually a hard job, while the maximum value principle provides the necessary condition of optimality and determines the optimal open-loop control sequence $\boldsymbol{u}^0(\boldsymbol{x}(0))$ for a given initial state $\boldsymbol{x}(0)$. The optimization problem is solved by the mathematical programming method, which is less difficult than dynamic programming, but the solution depends on the initial state $\boldsymbol{x}(0)$ and is not universal for all time instants.

In order to make full use of the advantages of the above two approaches, note that for the infinite horizon optimization problem the obtained optimal solutions using these two approaches are the same. At any time k, if an open-loop infinite horizon optimization problem with $\boldsymbol{x}(k)$ as the initial state is solved by mathematical programming and the optimal control sequence $\boldsymbol{u}^0(\boldsymbol{x}(k))$ is obtained, then its first element, i.e. the current control $\boldsymbol{u}(k) = \boldsymbol{u}^0(0, \boldsymbol{x}(k))$, should be the same as that obtained by the optimal feedback control law $\boldsymbol{u}(k) = \boldsymbol{K}(\boldsymbol{x}(k))$ solved by dynamic programming. The link $\boldsymbol{K}(\boldsymbol{x}(k)) = \boldsymbol{u}^0(0, \boldsymbol{x}(k))$ implies that the optimal feedback control law can be step-by-step constituted by the optimal controls obtained from successively solving an open-loop optimal control problem with initial state $\boldsymbol{x}(k)$ at each time k. Therefore, predictive control obtains the optimal feedback control law through repeatedly solving the open-loop optimal control problem in rolling style. In this sense, predictive control also solves the optimal control problem. It differs from traditional optimal control merely in its implementation.

Remark 6.1 [2] also pointed out that the above link can be found in some previous literature, such as the principle of optimality in [3]. Lee and Markus also stated in [4] (p. 423): "One technique for obtaining a feedback controller synthesis from knowledge of open loop controllers is to measure the current control process state and then compute very rapidly for the open loop control function. The first portion of this function is then used during a short time interval, after which a new measurement of the process state is made and a new open loop control function is computed for this new measurement. The procedure is then repeated." This is consistent with predictive control. Thus predictive control can be regarded as a technique developed from optimal control.

For the traditional infinite horizon optimal control, it is recognized [2, 5] that optimality does not imply stability. However, under certain conditions (i.e. stabilizability and detectability), the infinite horizon optimal controller can guarantee the stability of the closed-loop system and the optimal value function is commonly chosen as an appropriate Lyapunov function. Due to the relationship between predictive control and optimal control, it would be appropriate to borrow the idea and the method in infinite horizon optimal control to design a predictive controller with guaranteed stability. However, for practical use, it is not realistic to solve an open-loop optimal control problem with an infinite horizon at each time instant. If predictive control adopts a finite prediction horizon instead of an infinite one in online optimization, the guaranteed stability may be lost. This makes the predictive control differ from traditional optimal control and brings new challenges.

Indeed, before the appearance of predictive control, many efforts had been made on the study of guaranteed stability of adopting receding horizon control with finite horizon optimization to implement infinite horizon LQ control. For example, Kwon and Pearson [6] suggested that a zero terminal constraint $\boldsymbol{x}(k+N) = \boldsymbol{0}$ should be added in receding horizon control. This leads to a feedback control law solved by a Riccati-type iteration over a finite time interval which could guarantee the stability of the closed-loop system. However, this study was only for unconstrained linear systems.

The above discussion about the relationship between the optimal control and predictive control inspires us to make the following two points:

1) From an implementation point of view, the open-loop optimization problem solved by predictive control each time adopts a finite prediction horizon rather than an infinite one. Thus the control law solved by predictive control is actually not the same as that solved by infinite horizon optimization. In order to analyze the closed-loop

stability of predictive control by means of the infinite horizon optimal control theory, the open-loop optimization problem in predictive control should be modified such that it adopts a finite optimization horizon but can approximate the infinite horizon optimization.

2) In optimal control theory, the optimal value function is often chosen as a Lyapunov function for stability analysis. This idea can be used for reference in stability analysis of predictive control. However, because of the use of finite horizon optimization and the rolling style implementation, the optimization problems of predictive control at successive time instants are independent from each other. This makes the corresponding performances (i.e. value functions) have no relevance and are difficult to compare, which is the main difficulty of the stability analysis for predictive control and needs to be handled by some special methods.

In the following two subsections, these two points are discussed in some detail.

6.1.2 Infinite Horizon Approximation of Online Open-Loop Finite Horizon Optimization

Consider the nonlinear system described by the following difference equation:

$$\begin{aligned} \boldsymbol{x}(k+1) &= \boldsymbol{f}(\boldsymbol{x}(k), \boldsymbol{u}(k)) \\ \boldsymbol{y}(k) &= \boldsymbol{g}(\boldsymbol{x}(k)) \end{aligned} \tag{6.1}$$

where $\boldsymbol{f}(.)$ is a nonlinear function of the measurable system state $\boldsymbol{x}(k)$ and control input $\boldsymbol{u}(k)$, and $\boldsymbol{g}(.)$ is the system output defined as a nonlinear function of the system state $\boldsymbol{x}(k)$, $\boldsymbol{x}(k) \in \mathrm{R}^n$, $\boldsymbol{u}(k) \in \mathrm{R}^m$, and $\boldsymbol{y}(k) \in \mathrm{R}^p$. Without loss of generality, assume that $\boldsymbol{f}(.)$ and $\boldsymbol{g}(.)$ satisfy $\boldsymbol{f}(\mathbf{0}, \mathbf{0}) = \mathbf{0}$ and $g(\mathbf{0}) = \mathbf{0}$, i.e. the origin is an equilibrium point of system (6.1). In addition, the control inputs and the system states should satisfy the following constraints:

$$\boldsymbol{u}(k) \in \boldsymbol{\Omega}_u \tag{6.2}$$

$$\boldsymbol{x}(k) \in \boldsymbol{\Omega}_x \tag{6.3}$$

where $\boldsymbol{\Omega}_u$ ($\mathbf{0} \in \boldsymbol{\Omega}_u$) is a convex, compact subset on R^m and $\boldsymbol{\Omega}_x$ ($\mathbf{0} \in \boldsymbol{\Omega}_x$) is a convex, closed subset on R^n.

At each time instant k, the infinite horizon optimization problem starting from $\boldsymbol{x}(k)$ can be described by

$$\begin{aligned} \text{IHO}: \quad & \min_{\boldsymbol{u}(k+i|k)\ i\geq 0} J_\infty(k) = \sum_{i=0}^{\infty} l(\boldsymbol{x}(k+i\,|\,k), \boldsymbol{u}(k+i\,|\,k)) \\ \text{s.t.}\ & \boldsymbol{x}(k+i+1\,|\,k) = \boldsymbol{f}(\boldsymbol{x}(k+i\,|\,k), \boldsymbol{u}(k+i\,|\,k)),\ i = 0, 1, \ldots \\ & \boldsymbol{x}(k+i\,|\,k) \in \boldsymbol{\Omega}_x,\ \boldsymbol{u}(k+i\,|\,k) \in \boldsymbol{\Omega}_u,\ i = 0, 1, \ldots \\ & \boldsymbol{x}(k\,|\,k) = \boldsymbol{x}(k) \end{aligned} \tag{6.4}$$

where $l(\cdot, \cdot) \geq 0$ is the nonlinear performance index and $l(\mathbf{0}, \mathbf{0}) = 0$.

Since solving infinite number of control inputs is impossible for real implementation, predictive control adopts a finite horizon optimization instead of an infinite one. At time k, the finite horizon optimization problem starting from $\boldsymbol{x}(k)$ can be described by

$$\begin{aligned}
\text{FHO}: \quad & \min_{u(k+i|k),0\le i\le N-1} J_N(k)=\sum_{i=0}^{N-1} l(\boldsymbol{x}(k+i\,|\,k),\ \boldsymbol{u}(k+i\,|\,k)) \\
\text{s.t.}\ & \boldsymbol{x}(k+i+1\,|\,k)=\boldsymbol{f}(\boldsymbol{x}(k+i\,|\,k),\ \boldsymbol{u}(k+i\,|\,k)),\ i=0,\ \ldots,\ N-1 \\
& \boldsymbol{u}(k+i\,|\,k)\in \boldsymbol{\Omega}_u,\ i=0,\ \ldots,\ N-1 \\
& \boldsymbol{x}(k+i\,|\,k)\in \boldsymbol{\Omega}_x,\ i=1,\ \ldots,\ N \\
& \boldsymbol{x}(k\,|\,k)=\boldsymbol{x}(k)
\end{aligned} \tag{6.5}$$

Compare the optimization problems (6.4) and (6.5). It is easy to find that the performance indices of these two optimization problems are related by

$$J_\infty(k)=J_N(k)+J_{N,\infty}(k) \tag{6.6}$$

where

$$J_{N,\infty}(k)=\sum_{i=N}^{\infty} l(\boldsymbol{x}(k+i\,|\,k),\ \boldsymbol{u}(k+i\,|\,k))$$

As mentioned above, in order to analyze and synthesize predictive control systems by means of mature infinite horizon optimal control theory, the online optimization problem in predictive control should be modified to approximate the infinite horizon optimization. The above Eq. (6.6) indicates that when using the FHO problem (6.5) to approximate the IHO problem (6.4), the performance index (6.5) needs to be modified to compensate for the infinite horizon part $J_{N,\infty}(k)$ after the prediction horizon. Note that $J_{N,\infty}(k)$ is determined by the terminal state $\boldsymbol{x}(k+N\,|\,k)$ and future control inputs $\boldsymbol{u}(k+N+i\,|\,k)$ $(i\ge 0)$, and is unknown before the optimization problem is solved. However, it can be approximated by using different strategies to impose specific conditions on the terminal state $\boldsymbol{x}(k+N\,|\,k)$ and on future control inputs $\boldsymbol{u}(k+N+i)$ $(i\ge 0)$. In the following, three commonly used strategies to approximate $J_{N,\infty}(k)$ are briefly described.

1) **Zero terminal constraint (also terminal equality constraint)**
   In the FHO problem (6.5), if the condition $\boldsymbol{x}(k+N\,|\,k)=\boldsymbol{0}$ is imposed, then with $\boldsymbol{u}(k+i\,|\,k)\equiv\boldsymbol{0}$, $i\ge N$, we can get $J_{N,\infty}(k)=0$ according to (6.1) and $\boldsymbol{f}(\boldsymbol{0},\boldsymbol{0})=\boldsymbol{0}$, $l(\boldsymbol{0},\boldsymbol{0})=\boldsymbol{0}$. Therefore, the FHO problem (6.5) with the zero terminal constraint $\boldsymbol{x}(k+N\,|\,k)=\boldsymbol{0}$ can be employed to approximate the IHO problem (6.4). In some literature, the zero terminal constraint is also relaxed into the terminal equality constraint.
2) **Terminal cost function**
   Regard $J_{N,\infty}(k)$ in (6.6) as the performance index of the optimal control starting from time $k+N$, where it should be a function of the corresponding initial state $\boldsymbol{x}(k+N\,|\,k)$. The concrete form of this function is generally unknown. However, if a known function $F(\boldsymbol{x}(k+N\,|\,k))$ can be selected as its upper bound, then the FHO problem (6.5) with $F(\boldsymbol{x}(k+N\,|\,k))$ as a terminal cost function added in the performance index can approximate the IHO problem (6.4).
3) **Terminal constraint set**
   For the FHO problem (6.5), impose the condition $\boldsymbol{x}(k+N\,|\,k)\in\boldsymbol{X}_f$, where $\boldsymbol{X}_f$ is called the terminal constraint set, and assume that if the system state enters $\boldsymbol{X}_f$ some simple feedback control law is adopted to stabilize the system. Then an upper

bound of the performance index after the system state enters the terminal constraint set can be obtained. Thus, the FHP problem (6.5) can be modified to approximate the IHO problem (6.4).

Figure 6.1 shows the above three strategies of using the finite horizon optimization to approximate the infinite horizon optimization.

The combination of the above three strategies, especially the terminal cost function and the terminal constraint set, makes great progress in the research on synthesizing stable predictive controllers, resulting in a number of significant theoretical results. In 2000, Mayne et al. [2] summed up the previous research results and proposed that the terminal cost function, the terminal constraint set, and the local controller are three ingredients of stabilizing predictive controllers, which provides the fundamental way to synthesize predictive controllers with guaranteed stability.

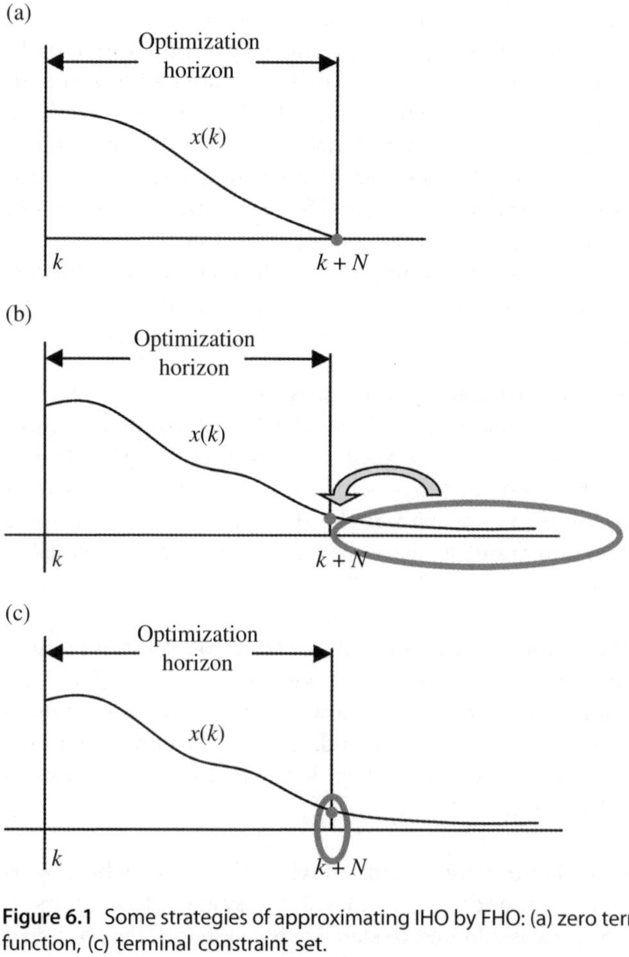

Figure 6.1 Some strategies of approximating IHO by FHO: (a) zero terminal constraint, (b) terminal cost function, (c) terminal constraint set.

6.1.3 Recursive Feasibility in Rolling Optimization

After being modified by the terminal cost function and the terminal constraint set, the online optimization problem of predictive control at time k can be formulated as

$$
\begin{aligned}
\text{PC:} \quad & \min_{\boldsymbol{u}(k+i|k),0\le i\le N-1} J_N(k)=\sum_{i=0}^{N-1} l(\boldsymbol{x}(k+i\mid k),\ \boldsymbol{u}(k+i\mid k))+F(\boldsymbol{x}(k+N\mid k))\\
\text{s.t.}\quad & \boldsymbol{x}(k+i+1\mid k)=\boldsymbol{f}(\boldsymbol{x}(k+i\mid k),\ \boldsymbol{u}(k+i\mid k)),\quad i=0,\ldots,N-1\\
& \boldsymbol{u}(k+i\mid k)\in\boldsymbol{\Omega}_u,\quad i=0,\ldots,N-1\\
& \boldsymbol{x}(k+i\mid k)\in\boldsymbol{\Omega}_x,\quad i=1,\ldots,N\\
& \boldsymbol{x}(k+N\mid k)\in\boldsymbol{X}_f\\
& \boldsymbol{x}(k\mid k)=\boldsymbol{x}(k)
\end{aligned}
\tag{6.7}
$$

At time k, the optimal control sequence $\boldsymbol{U}^*(k)=\{\boldsymbol{u}^*(k\mid k),\ldots,\boldsymbol{u}^*(k+N-1\mid k)\}$ can be obtained by solving the above optimization problem, which results in the optimal state trajectory $\boldsymbol{X}^*(k)=\{\boldsymbol{x}^*(k+1\mid k),\ldots,\boldsymbol{x}^*(k+N\mid k)\}$. The optimal value of the performance index $J_N^*(k)$ can be accordingly calculated. Because of the rolling optimization character of predictive control, at time $k+1$ the above optimization problem (6.7) is repeatedly solved and we can obtain the optimal control sequence $\boldsymbol{U}^*(k+1)=\{\boldsymbol{u}^*(k+1\mid k+1),\ldots,\boldsymbol{u}^*(k+N\mid k+1)\}$ and the optimal state trajectory $\boldsymbol{X}^*(k+1)=\{\boldsymbol{x}^*(k+2\mid k+1),\ldots,\boldsymbol{x}^*(k+N+1\mid k+1)\}$ as well as the optimal value of the performance index $J_N^*(k+1)$ at time $k+1$.

In infinite horizon optimal control, the optimal value function is often taken as a Lyapunov function for stability analysis and synthesis. If the same idea is used for predictive control, i.e. to take the optimal value of the performance index $J_N^*(k)$ as a Lyapunov function and try to make it decreasing or at least nonincreasing with k, some challenging problems would be faced. The first one is that the optimal value $J_N^*(k)$ of the performance index in predictive control is in general impossible to be expressed in a closed form of system states, which is however often required for stability analysis. This is because of the absence of explicit control laws (such as state feedback laws) for the open-loop optimization problem (6.7) when state or input constraints exist. As a result, in most cases $J_N^*(k)$ can only be expressed by the optimal control sequence $\boldsymbol{U}^*(k)$ and the optimal state trajectory $\boldsymbol{X}^*(k)$, both of which are calculated numerically. The second one is that, although $J_N^*(k)$ can be expressed by the optimal $\boldsymbol{U}^*(k)$ and $\boldsymbol{X}^*(k)$, it is difficult to compare $J_N^*(k)$ and $J_N^*(k+1)$. The reason is that the optimal $\boldsymbol{U}^*(k)$ and $\boldsymbol{U}^*(k+1)$ are solved by the optimization problem (6.7) at time k and $k+1$, respectively, while these two optimization problems are independent from each other. Thus there is no relationship between $\boldsymbol{U}^*(k)$ and $\boldsymbol{U}^*(k+1)$ and nor between $J_N^*(k)$ and $J_N^*(k+1)$. This indicates that the Lyapunov stability analysis of predictive control is different from that in traditional infinite horizon optimal control. The difficulty is essentially caused by the rolling style implementation of finite horizon optimization in predictive control.

For such a kind of optimal control problem implemented in a rolling style, an ingenious idea was proposed to overcome the above-mentioned difficulty when Keerthi and Gilbert analyzed the stability of the receding horizon control systems in [7]. This idea was later on also used by Scokaert and Clarke in analyzing the stability of the GPC system (such as in [8]) and is now widely adopted in synthesizing stable predictive control systems. The key issue is to introduce a candidate solution for the optimization problem at time $k+1$: $\boldsymbol{U}(k+1) = \{\boldsymbol{u}(k+1 \mid k+1), \ldots, \boldsymbol{u}(k+N \mid k+1)\}$, which can be used as a bridge between $\boldsymbol{U}^*(k)$ and $\boldsymbol{U}^*(k+1)$. Denote the value of the performance index at time $k+1$ with $\boldsymbol{U}(k+1)$ as $J_N(k+1)$. The candidate solution $\boldsymbol{U}(k+1)$ should be so selected that the resultant $J_N(k+1)$ is comparable with both $J_N^*(k)$ and $J_N^*(k+1)$.

On the one hand, $\boldsymbol{U}(k+1) = \{\boldsymbol{u}(k+1 \mid k+1), \ldots, \boldsymbol{u}(k+N \mid k+1)\}$ should be constructed closely related to $\boldsymbol{U}^*(k)$. For example, the first $N-1$ elements of $\boldsymbol{U}(k+1)$ can be set by shifting the elements in $\boldsymbol{U}^*(k)$, i.e. $\boldsymbol{u}(k+i \mid k+1) = \boldsymbol{u}^*(k+i \mid k)$, $i = 1, \ldots, N-1$, while the last element $\boldsymbol{u}(k+N \mid k+1)$ can be additionally constructed by some specific strategies (see Figure 6.2). With this setting, it is obvious that except for the first element $\boldsymbol{u}^*(k \mid k)$ in $\boldsymbol{U}^*(k)$ and the last element $\boldsymbol{u}(k+N \mid k+1)$ in $\boldsymbol{U}(k+1)$, all other elements of $\boldsymbol{U}^*(k)$ and $\boldsymbol{U}(k+1)$ are identical. Thus they will result in the same state scenario until time $k+N$, which makes it easy to compare $J_N^*(k)$ and $J_N(k+1)$. Furthermore, if $\boldsymbol{u}(k+N \mid k+1)$ is properly selected, it is possible to make $J_N(k+1) \le J_N^*(k)$.

On the other hand, $\boldsymbol{U}(k+1)$ should be a feasible solution of the optimization problem at time $k+1$. This means that $\boldsymbol{U}(k+1)$ and the resultant system states should satisfy all the input and state constraints (including the terminal set constraint) in the optimization problem at time $k+1$. For example, with the above selected $\boldsymbol{U}(k+1)$, most constraints are already satisfied due to the same controls and states caused by $\boldsymbol{U}(k+1)$ and $\boldsymbol{U}^*(k)$. Only a few new input and state constraints caused by $\boldsymbol{u}(k+N \mid k+1)$ should be judged.

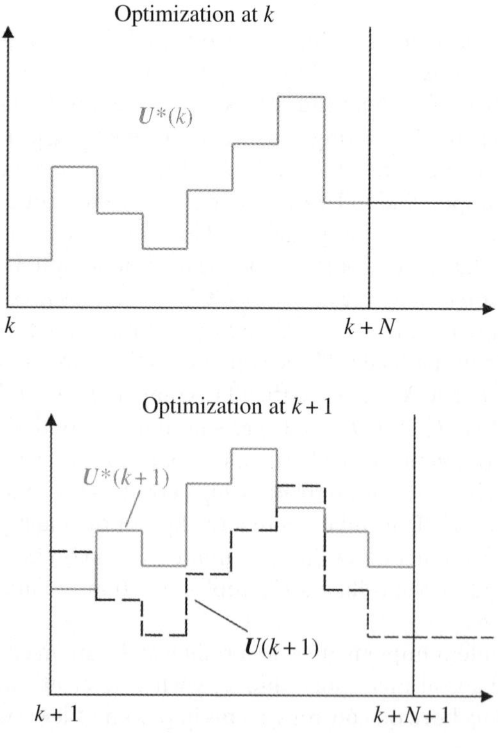

Figure 6.2 Optimal solution $\boldsymbol{U}^*(k)$, $\boldsymbol{U}^*(k+1)$ and the candidate solution $\boldsymbol{U}(k+1)$.

What is most important is that, as long as $\boldsymbol{U}(k+1)$ is a feasible solution at time $k+1$, $J_N^*(k+1) \le J_N(k+1)$ always holds no matter how much we know about the optimal solution $\boldsymbol{U}^*(k+1)$ at time $k+1$.

As discussed above, if $\boldsymbol{U}(k+1)$ is selected such that both $J_N(k+1) \le J_N^*(k)$ and $J_N^*(k+1) \le J_N(k+1)$ hold, i.e. $J_N^*(k+1) \le J_N(k+1) \le J_N^*(k)$, and the equality holds if and only if $\boldsymbol{x}(k) = \boldsymbol{0}$ and $\boldsymbol{u}(k \mid k) = \boldsymbol{0}$, then $J_N^*(k)$ can be taken as a Lyapunov function to guarantee that the predictive control system is stable. The above idea constructs the most fundamental and most commonly used method for stability synthesis of predictive control systems. Note that the constitution of the candidate solution $\boldsymbol{U}(k+1)$ plays a critical role. It should be constructed from the optimal solution $\boldsymbol{U}^*(k)$ and guarantee $J_N(k+1) \le J_N^*(k)$. Meanwhile, it must be feasible for the optimization problem at time $k+1$ to guarantee $J_N^*(k+1) \le J_N(k+1)$. If a feasible solution $\boldsymbol{U}(k+1)$ of the optimization problem at time $k+1$ can be constructed from the optimal (also feasible) solution $\boldsymbol{U}^*(k)$ of the optimization problem at time k, the predictive controller is called recursively feasible. Therefore the recursive feasibility is always the precondition for synthesizing stable predictive controllers.

6.1.4 Preliminary Knowledge

Before the following discussion, some preliminary knowledge of the invariant set and LMI will be briefly introduced, which are commonly used in the qualitative synthesis theory of predictive control.

(1) Invariant set [9]

The theory of the invariant set plays an important role in control system analysis and design. It is a fundamental tool used to analyze and synthesize constrained predictive control systems. Referring to [9], some basic concepts and conclusions on invariant sets are briefly introduced, which will be used in later sections.

Definition 6.1 [9] The set $\boldsymbol{\Omega} \subset \mathrm{R}^n$ is said to be positively invariant for a system of the form

$$\boldsymbol{x}(k+1) = \boldsymbol{f}(\boldsymbol{x}(k))$$

if for all $\boldsymbol{x}(0) \in \boldsymbol{\Omega}$ the solution $\boldsymbol{x}(k) \in \boldsymbol{\Omega}$ for $k > 0$.

From this definition it is obvious that for a dynamic system, a subset $\boldsymbol{\Omega}$ of the state space is a positively invariant set if $\boldsymbol{\Omega}$ contains the system state at some time and will then also contain it in the future.

Definition 6.2 [9] The set $\boldsymbol{\Omega} \subset \mathrm{R}^n$ is said to be controlled invariant for the system (6.1) if there exists a feedback control law $\boldsymbol{u}(k) = \boldsymbol{h}(\boldsymbol{x}(k))$ or $\boldsymbol{u}(k) = \boldsymbol{w}(\boldsymbol{y}(k))$ that assures the existence and uniqueness of the solution on R^+ and it is such that $\boldsymbol{\Omega}$ is positively invariant for the closed-loop system.

According to the above definitions, if at some time the system state is in the controlled invariant set $\boldsymbol{\Omega}$, then it will remain in $\boldsymbol{\Omega}$ under the corresponding feedback control law. This property is very helpful for analyzing the future system dynamics and designing proper control laws. Note that in the definition the term of $\boldsymbol{\Omega}$ positively invariant is referred to the closed-loop system $\boldsymbol{x}(k+1) = \boldsymbol{f}(\boldsymbol{x}(k), \boldsymbol{h}(\boldsymbol{x}(k)))$ or $\boldsymbol{x}(k+1) = \boldsymbol{f}(\boldsymbol{x}(k), \boldsymbol{w}(\boldsymbol{g}(\boldsymbol{x}(k))))$. For the original system (6.1), the same set $\boldsymbol{\Omega}$ is controlled invariant, which means that it could become an invariant set by proper feedback control.

Remark 6.2 For the system (6.1) with state constraint (6.3) and input constraint (6.2), if a controlled invariant set $\boldsymbol{\Omega}$ from Definition 6.2 is a subset of the admissible state domain $\boldsymbol{\Omega}_x$, and the feedback control $\boldsymbol{u}(k) = \boldsymbol{h}(\boldsymbol{x}(k))$ for the state in $\boldsymbol{\Omega}$ belongs to the admissible input domain $\boldsymbol{\Omega}_u$, i.e. $\boldsymbol{x}(k) \in \boldsymbol{\Omega}$ if $\boldsymbol{x}(0) \in \boldsymbol{\Omega}$, $\boldsymbol{\Omega} \subseteq \boldsymbol{\Omega}_x$, and $\boldsymbol{u}(k) \in \boldsymbol{\Omega}_u$, $\forall \boldsymbol{x}(k) \in \boldsymbol{\Omega}$, then any state starting from the invariant set $\boldsymbol{\Omega}$ and controlled by $\boldsymbol{u}(k) = \boldsymbol{h}(\boldsymbol{x}(k))$ will remain in $\boldsymbol{\Omega}$ and the state constraint (6.3) and input constraint (6.2) will not be violated. This makes the invariant set widely exploited, especially for systems with constraints.

To construct an invariant set, the concept of the contractive set is introduced as follows.

Definition 6.3 [9] A compact set $\boldsymbol{\Omega} \subset \mathrm{R}^n$ is contractive for a discrete-time system

$$\boldsymbol{x}(k+1) = \boldsymbol{f}(\boldsymbol{x}(k), \boldsymbol{u}(k))$$

if there exists a control function $\boldsymbol{u}(\boldsymbol{x})$ and a positive $\lambda < 1$, such that if $\boldsymbol{x}(k) \in \boldsymbol{\Omega}$ then $\boldsymbol{x}(k+1) \in \lambda\boldsymbol{\Omega}$.

As stated in [9], an invariant set $\boldsymbol{\Omega}$ is contractive if it is invariant and whenever the state is on the boundary, the control can "push it towards the interior." We can therefore construct an invariant set by constructing a contractive set.

Two important families of positively invariant sets, or controlled invariant sets, and their associated controllers are particularly successful in solving control engineering problems, i.e. the ellipsoidal sets and the polyhedral sets.

The ellipsoidal set is most commonly used as an invariant set and is defined by

$$\boldsymbol{\Omega} = \{\boldsymbol{x} \in \mathrm{R}^n \mid \boldsymbol{x}^{\mathrm{T}}\boldsymbol{Q}^{-1}\boldsymbol{x} \le 1\} \tag{6.8}$$

where the matrix $\boldsymbol{Q}$ is a symmetric positive-definite matrix and the use of the inverse of $\boldsymbol{Q}$ in (6.8) is to accord with the notations in the following sections.

For a linear time-invariant system

$$\boldsymbol{x}(k+1) = \boldsymbol{A}\boldsymbol{x}(k)$$

a positively invariant set $\boldsymbol{\Omega} = \{\boldsymbol{x} \in \mathrm{R}^n \mid \boldsymbol{x}^{\mathrm{T}}\boldsymbol{Q}^{-1}\boldsymbol{x} \le 1\}$ can be constructed by the boundary contraction condition:

$$\boldsymbol{A}^{\mathrm{T}}\boldsymbol{Q}^{-1}\boldsymbol{A} - \boldsymbol{Q}^{-1} \le -\boldsymbol{P}$$

where $\boldsymbol{Q}$, $\boldsymbol{P}$ are symmetric positive-definite matrices.

This is obvious from the fact that if the system state $\boldsymbol{x}(k) \in \boldsymbol{\Omega}$, i.e. $\boldsymbol{x}^{\mathrm{T}}(k)\boldsymbol{Q}^{-1}\boldsymbol{x}(k) \leq 1$, then $\boldsymbol{x}^{\mathrm{T}}(k+1)\boldsymbol{Q}^{-1}\boldsymbol{x}(k+1) = \boldsymbol{x}^{\mathrm{T}}(k)\boldsymbol{A}^{\mathrm{T}}\boldsymbol{Q}^{-1}\boldsymbol{A}\boldsymbol{x}(k) < \boldsymbol{x}^{\mathrm{T}}(k)\boldsymbol{Q}^{-1}\boldsymbol{x}(k) \leq 1$, i.e. $\boldsymbol{x}(k+1) \in \boldsymbol{\Omega}$, and thus $\boldsymbol{\Omega}$ is a positively invariant set of the considered system.

For a linear time-invariant control system with constraints (6.2) and (6.3)

$$\boldsymbol{x}(k+1) = \boldsymbol{A}\boldsymbol{x}(k) + \boldsymbol{B}\boldsymbol{u}(k),\ \boldsymbol{u}(k) \in \boldsymbol{\Omega}_u,\ \boldsymbol{x}(k) \in \boldsymbol{\Omega}_x \tag{6.9}$$

the set $\boldsymbol{\Omega} = \{\boldsymbol{x} \in \mathrm{R}^n \mid \boldsymbol{x}^{\mathrm{T}}\boldsymbol{Q}^{-1}\boldsymbol{x} \leq 1\}$ is a controlled invariant set of this system with the feedback control law $\boldsymbol{u}(k) = \boldsymbol{K}\boldsymbol{x}(k)$ if the symmetric positive-definite matrix $\boldsymbol{Q}$ satisfies

$$(\boldsymbol{A}+\boldsymbol{B}\boldsymbol{K})^{\mathrm{T}}\boldsymbol{Q}^{-1}(\boldsymbol{A}+\boldsymbol{B}\boldsymbol{K}) - \boldsymbol{Q}^{-1} \leq -\boldsymbol{P} \tag{6.10}$$

and $\boldsymbol{\Omega} \subseteq \boldsymbol{\Omega}_x$; also $\boldsymbol{u}(k) = \boldsymbol{K}\boldsymbol{x}(k) \in \boldsymbol{\Omega}_u$ if $\boldsymbol{x}(k) \in \boldsymbol{\Omega}$, where $\boldsymbol{P}$ is a symmetric positive definite matrix.

This can be similarly proved. If the system state $\boldsymbol{x}(k) \in \boldsymbol{\Omega}$, i.e. $\boldsymbol{x}^{\mathrm{T}}(k)\boldsymbol{Q}^{-1}\boldsymbol{x}(k) \leq 1$, then under feedback control $\boldsymbol{u}(k) = \boldsymbol{K}\boldsymbol{x}(k)$, it follows that

$$\boldsymbol{x}^{\mathrm{T}}(k+1)\boldsymbol{Q}^{-1}\boldsymbol{x}(k+1) = \boldsymbol{x}^{\mathrm{T}}(k)(\boldsymbol{A}+\boldsymbol{B}\boldsymbol{K})^{\mathrm{T}}\boldsymbol{Q}^{-1}(\boldsymbol{A}+\boldsymbol{B}\boldsymbol{K})\boldsymbol{x}(k) < \boldsymbol{x}^{\mathrm{T}}(k)\boldsymbol{Q}^{-1}\boldsymbol{x}(k) \leq 1$$

i.e. $\boldsymbol{x}(k+1) \in \boldsymbol{\Omega}$. This means that once the system state is in the invariant set $\boldsymbol{\Omega}$, then it will remain in $\boldsymbol{\Omega}$ under feedback control $\boldsymbol{u}(k) = \boldsymbol{K}\boldsymbol{x}(k)$. Furthermore, the conditions on $\boldsymbol{\Omega}$ related to admissible domains $\boldsymbol{\Omega}_x$ and $\boldsymbol{\Omega}_u$ ensure that the constraints are satisfied as long as the system state is in the invariant set $\boldsymbol{\Omega}$.

Remark 6.3 To simplify the description, in the following we sometimes use the terminology "invariant set" instead of "positively invariant set" and "controlled invariant set" given strictly by Definitions 6.1 and 6.2 if it is not difficult to identify which kind of invariance it refers to according to the system description.

Compared with the ellipsoidal set, the polyhedral set may be more useful for predictive control because most of practical constraints are given as linear inequalities. A polyhedral set can be formulated as

$$\boldsymbol{\Lambda} = \{\boldsymbol{x} \mid \boldsymbol{F}\boldsymbol{x} \leq \bar{\mathbf{1}}\}$$

where $\boldsymbol{F}$ is an $r \times n$ matrix and $\bar{\mathbf{1}}$ is an r-dimensional column vector with all entries 1. For a polyhedral invariant set of a system like $\boldsymbol{x}(k+1) = \boldsymbol{A}\boldsymbol{x}(k)$, the following theorem is given.

Theorem 6.1 [9] The following conditions are equivalent.

1) The polyhedral compact set $\boldsymbol{\Lambda}$ is positively invariant (contractive) for $\boldsymbol{x}(k+1) = \boldsymbol{A}\boldsymbol{x}(k)$.
2) There exists a positive $\lambda \leq 1$ $(\lambda < 1)$ and an $r \times r$ non-negative matrix $\boldsymbol{H}$ such that

$$\boldsymbol{H}\bar{\mathbf{1}} \leq \lambda\bar{\mathbf{1}},\quad \boldsymbol{H}\boldsymbol{F} = \boldsymbol{F}\boldsymbol{A}$$

3) There exists a positive $\lambda \leq 1$ $(\lambda < 1)$ and an $s \times s$ non-negative matrix $\boldsymbol{P}$ such that

$$\bar{\mathbf{1}}^{\mathrm{T}}\boldsymbol{P} \leq \lambda\bar{\mathbf{1}}^{\mathrm{T}},\quad \boldsymbol{X}\boldsymbol{P} = \boldsymbol{A}\boldsymbol{X}$$

where $\boldsymbol{X}$ is an $n \times s$ matrix where each column is a vertex of the polyhedral set $\boldsymbol{\Lambda}$.

As indicated by [9], the above theorem shows that if a polyhedral set Λ is given (i.e. $\boldsymbol{F}$ or $\boldsymbol{X}$ are fixed), checking its positive invariance is a linear programming problem, while if $\boldsymbol{F}$ or $\boldsymbol{X}$ are to be determined, the conditions become bilinear. Therefore, constructing a polyhedral invariant set Λ is in principle harder than computing an ellipsoidal invariant set.

For a constrained autonomous linear system $\boldsymbol{x}(k+1) = \boldsymbol{A}\boldsymbol{x}(k)$ with $\boldsymbol{x}(k) \in \boldsymbol{\Omega}_x$, we call the initial state $\boldsymbol{x}(0)$ an admissible state if $\boldsymbol{x}(0) \in \boldsymbol{\Omega}_x$; then $\boldsymbol{x}(i) \in \boldsymbol{\Omega}_x$, $i > 0$. Define the maximal admissible set $\boldsymbol{S} = \{\boldsymbol{x}(0) \mid \boldsymbol{A}^i\boldsymbol{x}(0) \in \boldsymbol{\Omega}_x, \forall\, i \geq 0\}$. It is obvious that $\boldsymbol{S} \subseteq \boldsymbol{\Omega}_x$. If $\boldsymbol{\Omega}_x$ is a set of polyhedral constraints, $\boldsymbol{S}$ is also a polyhedral set. Note that there exist algorithms in the literature to calculate the maximal admissible set, so we can obtain $\boldsymbol{S}$ as a polyhedral invariant set of the system. For the constrained linear control system (6.9), the fixed feedback control law $\boldsymbol{u}(k) = \boldsymbol{K}\boldsymbol{x}(k)$ can be firstly adopted to convert it into an autonomous system $\boldsymbol{x}(k+1) = \boldsymbol{A}_K\boldsymbol{x}(k)$ with $\boldsymbol{A}_K = \boldsymbol{A} + \boldsymbol{B}\boldsymbol{K}$. However, its maximal admissible set $\boldsymbol{S}$ can be calculated with these algorithms only if $\boldsymbol{K}$ is fixed, because it depends on the feedback gain $\boldsymbol{K}$.

The concept of an invariant set is fundamental and very important for theoretical research of predictive control. More details on the invariant set can be found in the literatures, such as [9].

(2) Linear Inequality Matrix (LMI) [10, 11]

The LMI is a matrix inequality with the following form:

$$\boldsymbol{F}(\boldsymbol{x}) = \boldsymbol{F}_0 + \sum_{i=1}^{l} x_i \boldsymbol{F}_i > \boldsymbol{0}$$

where $x_1, x_2, \ldots, x_l$ are the variables, $\boldsymbol{F}_i \in \mathrm{R}^n$, $i = 0, \ldots, l$ are given symmetric matrices, and $\boldsymbol{F}(\boldsymbol{x}) > \boldsymbol{0}$ means that $\boldsymbol{F}(\boldsymbol{x})$ is positive-definite. For multiple LMIs, they can be taken as the elements of a block diagonal matrix to form a single LMI.

In the study of control theory, matrix inequalities are usually converted into LMIs by Schur complements. For symmetric matrices $\boldsymbol{Q}(\boldsymbol{x})$, $\boldsymbol{R}(\boldsymbol{x})$ and matrix $\boldsymbol{S}(\boldsymbol{x})$, all depending affinely on $\boldsymbol{x}$, the LMI

$$\begin{bmatrix} \boldsymbol{Q}(\boldsymbol{x}) & \boldsymbol{S}(\boldsymbol{x}) \\ \boldsymbol{S}^{\mathrm{T}}(\boldsymbol{x}) & \boldsymbol{R}(\boldsymbol{x}) \end{bmatrix} > \boldsymbol{0} \tag{6.11a}$$

is equivalent to the matrix inequalities

$$\boldsymbol{R}(\boldsymbol{x}) > \boldsymbol{0}, \quad \boldsymbol{Q}(\boldsymbol{x}) - \boldsymbol{S}(\boldsymbol{x})\boldsymbol{R}^{-1}(\boldsymbol{x})\boldsymbol{S}^{\mathrm{T}}(\boldsymbol{x}) > \boldsymbol{0} \tag{6.11b}$$

or

$$\boldsymbol{Q}(\boldsymbol{x}) > \boldsymbol{0}, \quad \boldsymbol{R}(\boldsymbol{x}) - \boldsymbol{S}^{\mathrm{T}}(\boldsymbol{x})\boldsymbol{Q}^{-1}(\boldsymbol{x})\boldsymbol{S}(\boldsymbol{x}) > \boldsymbol{0} \tag{6.11c}$$

As a result, it is easy to show that the matrix inequality

$$\boldsymbol{Q}(\boldsymbol{x}) - \boldsymbol{S}_1(\boldsymbol{x})\boldsymbol{R}_1^{-1}(\boldsymbol{x})\boldsymbol{S}_1^{\mathrm{T}}(\boldsymbol{x}) - \cdots - \boldsymbol{S}_m(\boldsymbol{x})\boldsymbol{R}_m^{-1}(\boldsymbol{x})\boldsymbol{S}_m^{\mathrm{T}}(\boldsymbol{x}) > \boldsymbol{0} \tag{6.11d}$$

can equivalently be converted to the LMI form

$$\begin{bmatrix} \boldsymbol{Q}(\boldsymbol{x}) & \boldsymbol{S}_1(\boldsymbol{x}) & \cdots & \boldsymbol{S}_m(\boldsymbol{x}) \\ \boldsymbol{S}_1^{\mathrm{T}}(\boldsymbol{x}) & \boldsymbol{R}_1(\boldsymbol{x}) & \cdots & \boldsymbol{0} \\ \vdots & \vdots & \ddots & \vdots \\ \boldsymbol{S}_m^{\mathrm{T}}(\boldsymbol{x}) & \boldsymbol{0} & \cdots & \boldsymbol{R}_m(\boldsymbol{x}) \end{bmatrix} > \boldsymbol{0} \tag{6.11e}$$

where symmetric matrices $\boldsymbol{Q}(\boldsymbol{x})$, $\boldsymbol{R}_i(\boldsymbol{x})$ and matrices $\boldsymbol{S}_i(\boldsymbol{x})$, $i = 1, \ldots, m$ depend affinely on $\boldsymbol{x}$.

By the Schur complement, if the state $\boldsymbol{x}(k)$ is in the ellipsoidal set $\boldsymbol{\Omega}$ given by (6.8), the following LMI holds:

$$\begin{bmatrix} 1 & \boldsymbol{x}^{\mathrm{T}}(k) \\ \boldsymbol{x}(k) & \boldsymbol{Q} \end{bmatrix} \geq 0 \tag{6.12}$$

Then, for the linear time-invariant control system (6.9), if the ellipsoidal set $\boldsymbol{\Omega}$ given by (6.8) is controlled invariant under the feedback control law $\boldsymbol{u}(k) = \boldsymbol{K}\boldsymbol{x}(k)$, we can discuss the condition to ensure that $\boldsymbol{u}(k) = \boldsymbol{K}\boldsymbol{x}(k) \in \boldsymbol{\Omega}_u$ if $\boldsymbol{x}(k) \in \boldsymbol{\Omega}$. Two cases are discussed.

1) The constraint of $\boldsymbol{u} \in \boldsymbol{\Omega}_u$ is specified as

$$\|\boldsymbol{u}(k+i)\|_2 \leq u_{\max}, \quad i = 0, 1, \ldots$$

Denote $\boldsymbol{K} = \boldsymbol{Y}\boldsymbol{Q}^{-1}$. For $\boldsymbol{x}(k+i) \in \boldsymbol{\Omega}$, i.e. $\boldsymbol{x}^{\mathrm{T}}(k+i)\boldsymbol{Q}^{-1}\boldsymbol{x}(k+i) \leq 1$, it follows that

$$\begin{aligned} \|\boldsymbol{u}(k+i)\|_2^2 &= \left\|\boldsymbol{Y}\boldsymbol{Q}^{-1/2}\boldsymbol{Q}^{-1/2}\boldsymbol{x}(k+i)\right\|_2^2 \\ &\leq \left\|\boldsymbol{Y}\boldsymbol{Q}^{-1/2}\right\|_2^2 \left\|\boldsymbol{Q}^{-1/2}\boldsymbol{x}(k+i)\right\|_2^2 \leq \left\|\boldsymbol{Y}\boldsymbol{Q}^{-1/2}\right\|_2^2 \quad i = 0, 1, \ldots \end{aligned}$$

Thus the constraint $\boldsymbol{u} \in \boldsymbol{\Omega}_u$ can be ensured by the following condition:

$$\left\|\boldsymbol{Y}\boldsymbol{Q}^{-1/2}\right\|_2^2 = \boldsymbol{Q}^{-1/2}\boldsymbol{Y}^T\boldsymbol{Y}\boldsymbol{Q}^{-1/2} \leq u_{\max}^2 \boldsymbol{I}$$

where $\boldsymbol{I}$ is the identity matrix with proper dimension. By left and right multiplying $\boldsymbol{Q}^{1/2}$, it follows that

$$\boldsymbol{Q} - \boldsymbol{Y}^{\mathrm{T}} \frac{1}{u_{\max}^2} \boldsymbol{Y} \geq \boldsymbol{0}$$

which can be converted into LMI by the Schur complement

$$\begin{bmatrix} u_{\max}^2 \boldsymbol{I} & \boldsymbol{Y} \\ \boldsymbol{Y}^{\mathrm{T}} & \boldsymbol{Q} \end{bmatrix} \geq \boldsymbol{0} \tag{6.13}$$

2) The constraint $\boldsymbol{u} \in \boldsymbol{\Omega}_u$ is specified as the peak bound of the individual input component, i.e.

$$\left|\boldsymbol{u}_j(k+i)\right| \leq u_{j,\max}, \quad i = 0, 1, \ldots, \quad j = 1, \ldots, m$$

A sufficient condition can be similarly derived to ensure that $\boldsymbol{u}(k) = \boldsymbol{K}\boldsymbol{x}(k) \in \boldsymbol{\Omega}_u$ if $\boldsymbol{x}(k) \in \boldsymbol{\Omega}$. Let $\boldsymbol{h}_j^{\mathrm{T}}$ be jth row of an $m \times m$ identity matrix and rewrite the above condition as

$$\left|\boldsymbol{h}_j^{\mathrm{T}}\boldsymbol{u}(k+i)\right|^2=\left|\boldsymbol{h}_j^{\mathrm{T}}\boldsymbol{Y}\boldsymbol{Q}^{-1}\boldsymbol{x}(k+i)\right|^2\le\left\|\boldsymbol{h}_j^{\mathrm{T}}\boldsymbol{Y}\boldsymbol{Q}^{-1/2}\right\|_2^2\left\|\boldsymbol{Q}^{-1/2}\boldsymbol{x}(k+i)\right\|_2^2$$
$$\le\left\|\boldsymbol{h}_j^{\mathrm{T}}\boldsymbol{Y}\boldsymbol{Q}^{-1/2}\right\|_2^2=\left(\boldsymbol{Y}\boldsymbol{Q}^{-1}\boldsymbol{Y}^{\mathrm{T}}\right)_{jj}\le u_{j,\max}^2,\quad i=0,\ 1,\ \dots,\ j=1,\ \dots,\ m$$

which can be written as

$$\boldsymbol{h}_j^{\mathrm{T}}\boldsymbol{Y}\boldsymbol{Q}^{-1}\boldsymbol{Y}^{\mathrm{T}}\boldsymbol{h}_j\le u_{j,\max}^2,\quad j=1,\ \dots,\ m$$

or in LMI form as

$$\begin{bmatrix} u_{j,\max}^2 & \boldsymbol{h}_j^{\mathrm{T}}\boldsymbol{Y} \\ \boldsymbol{Y}^{\mathrm{T}}\boldsymbol{h}_j & \boldsymbol{Q}\end{bmatrix}\ge\boldsymbol{0},\quad j=1,\ \dots,\ m$$

This condition can also be written in a compact form with a symmetric matrix $\boldsymbol{X}$ [11]:

$$\begin{bmatrix} \boldsymbol{X} & \boldsymbol{Y} \\ \boldsymbol{Y}^{\mathrm{T}} & \boldsymbol{Q}\end{bmatrix}\ge\boldsymbol{0},\ \text{where}\ \boldsymbol{X}_{jj}\le u_{j,\max}^2,\quad j=1,\ \dots,\ m \tag{6.14}$$

Next consider the output constraints for the linear time-invariant control system (6.9) with output $\boldsymbol{y}(k)=\boldsymbol{C}\boldsymbol{x}(k)$. Two cases are discussed as follows.

1) The output constraint is specified as

$$\|\boldsymbol{y}(k+i)\|_2\le y_{\max},\quad i=1,\ 2,\ \dots$$

With $\boldsymbol{x}(k+i)\in\boldsymbol{\Omega}$, $i=0, 1, \dots$, it follows that

$$\|\boldsymbol{y}(k+i+1)\|_2^2=\left\|\boldsymbol{C}\left(\boldsymbol{A}+\boldsymbol{B}\boldsymbol{Y}\boldsymbol{Q}^{-1}\right)\boldsymbol{x}(k+i)\right\|_2^2$$
$$\le\left\|\boldsymbol{C}(\boldsymbol{A}\boldsymbol{Q}+\boldsymbol{B}\boldsymbol{Y})\boldsymbol{Q}^{-1/2}\right\|_2^2\left\|\boldsymbol{Q}^{-1/2}\boldsymbol{x}(k+i)\right\|_2^2\le\left\|\boldsymbol{C}(\boldsymbol{A}\boldsymbol{Q}+\boldsymbol{B}\boldsymbol{Y})\boldsymbol{Q}^{-1/2}\right\|_2^2,\quad i=0,\ 1,\ \dots$$

Thus the output constraints are satisfied if the following condition holds:

$$\boldsymbol{Q}^{-1/2}(\boldsymbol{A}\boldsymbol{Q}+\boldsymbol{B}\boldsymbol{Y})^{\mathrm{T}}\boldsymbol{C}^{\mathrm{T}}\boldsymbol{C}(\boldsymbol{A}\boldsymbol{Q}+\boldsymbol{B}\boldsymbol{Y})\boldsymbol{Q}^{-1/2}\le y_{\max}^2\boldsymbol{I}$$

By left and right multiplying $\boldsymbol{Q}^{1/2}$, we get

$$\boldsymbol{Q}-(\boldsymbol{A}\boldsymbol{Q}+\boldsymbol{B}\boldsymbol{Y})^{\mathrm{T}}\boldsymbol{C}^{\mathrm{T}}\left(y_{\max}^2\boldsymbol{I}\right)^{-1}\boldsymbol{C}(\boldsymbol{A}\boldsymbol{Q}+\boldsymbol{B}\boldsymbol{Y})\ge\boldsymbol{0}$$

which can be converted into the LMI form

$$\begin{bmatrix} \boldsymbol{Q} & (\boldsymbol{A}\boldsymbol{Q}+\boldsymbol{B}\boldsymbol{Y})^{\mathrm{T}}\boldsymbol{C}^{\mathrm{T}} \\ \boldsymbol{C}(\boldsymbol{A}\boldsymbol{Q}+\boldsymbol{B}\boldsymbol{Y}) & y_{\max}^2\boldsymbol{I}\end{bmatrix}\ge\boldsymbol{0} \tag{6.15}$$

2) The output constraint is specified as the peak bound of the individual output component, i.e.

$$\left|y_j(k+i)\right|\le y_{j,\max},\quad i=1,\ 2,\ \dots,\ j=1,\ \dots,\ r$$

Let $\boldsymbol{h}_j^{\mathrm{T}}$ be the jth row of an $r \times r$ identity matrix. With $\boldsymbol{x}(k+i) \in \boldsymbol{\Omega}, i = 0, 1, \ldots$, it follows that

$$\begin{aligned}
\left|y_j(k+i+1)\right|^2 &= \left|\boldsymbol{h}_j^{\mathrm{T}}\boldsymbol{y}(k+i+1)\right|^2 = \left|\boldsymbol{h}_j^{\mathrm{T}}\boldsymbol{C}\left(\boldsymbol{A}+\boldsymbol{BYQ}^{-1}\right)\boldsymbol{x}(k+i)\right|^2 \\
&\le \left\|\boldsymbol{h}_j^{\mathrm{T}}\boldsymbol{C}(\boldsymbol{AQ}+\boldsymbol{BY})\boldsymbol{Q}^{-1/2}\right\|_2^2 \left\|\boldsymbol{Q}^{-1/2}\boldsymbol{x}(k+i)\right\|_2^2 \\
&\le \left\|\boldsymbol{h}_j^{\mathrm{T}}\boldsymbol{C}(\boldsymbol{AQ}+\boldsymbol{BY})\boldsymbol{Q}^{-1/2}\right\|_2^2, \qquad i = 0, 1, \ldots
\end{aligned}$$

Thus the output constraints hold if the following condition is satisfied:

$$\boldsymbol{h}_j^{\mathrm{T}}\boldsymbol{C}(\boldsymbol{AQ}+\boldsymbol{BY})\boldsymbol{Q}^{-1}(\boldsymbol{AQ}+\boldsymbol{BY})^{\mathrm{T}}\boldsymbol{C}^{\mathrm{T}}\boldsymbol{h}_j \le y_{j,\max}^2$$

which can be converted into the LMI form as follows:

$$\begin{bmatrix} y_{j,\max}^2 & \boldsymbol{h}_j^{\mathrm{T}}\boldsymbol{C}(\boldsymbol{AQ}+\boldsymbol{BY}) \\ (\boldsymbol{AQ}+\boldsymbol{BY})^{\mathrm{T}}\boldsymbol{C}^{\mathrm{T}}\boldsymbol{h}_j & \boldsymbol{Q} \end{bmatrix} \ge \boldsymbol{0}, \quad j = 1, \ldots, r$$

Similarly this condition can also be written in a compact form with a symmetric matrix $\boldsymbol{Z}$:

$$\begin{bmatrix} \boldsymbol{Z} & \boldsymbol{C}(\boldsymbol{AQ}+\boldsymbol{BY}) \\ (\boldsymbol{AQ}+\boldsymbol{BY})^{\mathrm{T}}\boldsymbol{C}^{\mathrm{T}} & \boldsymbol{Q} \end{bmatrix} \ge \boldsymbol{0}, \text{ where } Z_{jj} \le y_{j,\max}^2, \; j = 1, \ldots, r \tag{6.16}$$

Other kinds of constraints can be converted into LMI forms in a similar way. Since there are several efficient toolboxes to solve LMIs (such as LMI-Tools in MATLAB), the online optimization problems of predictive control can also be solved by converting the original problem into the LMI form.

6.2 Synthesis of Stable Predictive Controllers

Based on the basic ideas used in analysis and synthesis of MPC presented in the last section, the theoretical research of MPC has been developed rapidly. A great number of MPC synthesis methods with guaranteed stability were proposed to solve different problems by utilizing different techniques. In this section, we will introduce some typical methods to further explain how the basic ideas mentioned above are specifically implemented and adopted in synthesizing stable MPC controllers.

6.2.1 Predictive Control with Zero Terminal Constraints

The zero terminal constraint was early adopted in the study of the infinite horizon LQ control problem to obtain a stabilizing controller [6, 7]. This method was also applied to the design of the stable GPC controller, for example with zero terminal constraints on the output tracking errors [8].

At time k the optimization problem with zero terminal constraints can be formulated as

$$\begin{aligned}
\text{PC1:} \quad & \min_{u(k+i|k),0\le i\le N-1} J_N(k)=\sum_{i=0}^{N-1} l(\boldsymbol{x}(k+i\mid k),\ \boldsymbol{u}(k+i\mid k)) \\
\text{s.t.}\ & \boldsymbol{x}(k+i+1\mid k)=\boldsymbol{f}(\boldsymbol{x}(k+i\mid k),\ \boldsymbol{u}(k+i\mid k)),\ \ i=0,\ \ldots,\ N-1 \\
& \boldsymbol{u}(k+i\mid k)\in\boldsymbol{\Omega}_u,\ \ i=0,\ldots,\ N-1 \\
& \boldsymbol{x}(k+i\mid k)\in\boldsymbol{\Omega}_x,\ \ i=1,\ldots,\ N \\
& \boldsymbol{x}(k+N\mid k)=\boldsymbol{0} \\
& \boldsymbol{x}(k\mid k)=\boldsymbol{x}(k)
\end{aligned} \tag{6.17}$$

If Problem (6.17) with system state $\boldsymbol{x}(k)$ is feasible at time k, assume that its optimal solution is $\boldsymbol{U}^*(k)=\{\boldsymbol{u}^*(k\mid k),\ \ldots,\ \boldsymbol{u}^*(k+N-1\mid k)\}$, with which the optimal state trajectory is given by $\boldsymbol{X}^*(k)=\{\boldsymbol{x}^*(k+1\mid k),\ \ldots,\ \boldsymbol{x}^*(k+N\mid k)\}$ and the corresponding optimal value of performance index is

$$J_N^*(k)=\sum_{i=0}^{N-1} l(\boldsymbol{x}^*(k+i\mid k),\ \boldsymbol{u}^*(k+i\mid k))$$

It is obvious that the elements of $\boldsymbol{U}^*(k)$ and $\boldsymbol{X}^*(k)$ satisfy all the constraints in Problem (6.17), including $\boldsymbol{x}^*(k+N\mid k)=\boldsymbol{0}$.

At time $k+1$, construct a solution as

$$\boldsymbol{U}(k+1)=\{\boldsymbol{u}(k+1\mid k+1),\ \ldots,\ \boldsymbol{u}(k+N-1\mid k+1),\ \boldsymbol{0}\},$$

where $\boldsymbol{u}(k+i\mid k+1)=\boldsymbol{u}^*(k+i\mid k)$, $i=1,\ \ldots,\ N-1$. It is obvious that $\boldsymbol{U}(k+1)$ will result in the system states

$$\boldsymbol{X}(k+1)=\{\boldsymbol{x}(k+2\mid k+1),\ \ldots,\ \boldsymbol{x}(k+N+1\mid k+1)\}$$

where $\boldsymbol{x}(k+i+1\mid k+1)=\boldsymbol{x}^*(k+i+1\mid k)$, $i=1,\ \ldots,\ N-1$. and

$$\boldsymbol{x}(k+N+1\mid k+1)=\boldsymbol{f}(\boldsymbol{x}(k+N\mid k+1),\ \boldsymbol{u}(k+N\mid k+1))=\boldsymbol{f}(\boldsymbol{x}^*(k+N\mid k),\ \boldsymbol{0})=\boldsymbol{0}$$

due to $\boldsymbol{u}(k+N\mid k+1)=\boldsymbol{0}$, $\boldsymbol{x}^*(k+N\mid k)=\boldsymbol{0}$, and $\boldsymbol{f}(\boldsymbol{0},\ \boldsymbol{0})=\boldsymbol{0}$. Therefore the elements of $\boldsymbol{U}(k+1)$ and $\boldsymbol{X}(k+1)$ satisfy all the constraints of Problem (6.17) at time $k+1$. Thus $\boldsymbol{U}(k+1)$ is a feasible solution of Problem (6.17) and there must exist the optimal solution $\boldsymbol{U}^*(k+1)$ with the value of performance index $J_N^*(k+1)$ satisfying

$$J_N^*(k+1)\le J_N(k+1)$$

On the other hand, the value of the performance index with respect to $\boldsymbol{U}(k+1)$ is given by

$$\begin{aligned}
J_N(k+1) &= \sum_{i=0}^{N-1} l(\boldsymbol{x}(k+i+1\mid k+1),\ \boldsymbol{u}(k+i+1\mid k+1)) \\
&= \sum_{i=0}^{N-2} l(\boldsymbol{x}^*(k+i+1\mid k),\ \boldsymbol{u}^*(k+i+1\mid k))+l(\boldsymbol{x}(k+N\mid k+1),\ \boldsymbol{u}(k+N\mid k+1)) \\
&= J_N^*(k)-l(\boldsymbol{x}^*(k\mid k),\ \boldsymbol{u}^*(k\mid k))\le J_N^*(k).
\end{aligned}$$

Then it follows that

$$J_N^*(k+1) \le J_N^*(k)$$

where the equality holds only if $\boldsymbol{x}(k) = 0$, $\boldsymbol{u}^*(k \mid k) = 0$. Hence, the optimal value function $J_N^*(k)$ can be used as a Lyapunov function to guarantee the closed-loop stability of the predictive control system by using the Lyapunov approach.

Although the above approach of synthesizing predictive control systems with zero terminal constraints is quite simple, it is conservative when using it to approximate the future inputs and outputs in an infinite time horizon. Meanwhile, the zero terminal constraint condition is too strict to be ensured. Especially for systems with disturbance, it is difficult to guarantee the terminal state to be zero, which often results in infeasibility of the optimization problem. In practice, it is often needed to relax the zero terminal constraints, which leads to the synthesis approaches using a terminal cost function or/and a terminal set constraint.

6.2.2 Predictive Control with Terminal Cost Functions

The method using a terminal cost function is an extension of the method with a zero terminal constraint. From Problem (6.17) it can be seen that a zero terminal constraint is used to impose the constraint $\boldsymbol{x}(k+N \mid k) = \boldsymbol{0}$ into the FHO Problem (6.5). This additional constraint is equivalent to putting an infinitely large weighting on the terminal state in the performance index. With this understanding, the strategy of the terminal cost function is to relax the weighting matrix on the terminal state from infinitely large to a finite value so as to reduce the conservativeness of the zero terminal constraint. This method is initially proposed for predictive control of unconstrained linear systems and is then extended to the constrained and nonlinear systems.

(1) The unconstrained linear system

Consider the unconstrained linear system

$$\boldsymbol{x}(k+1) = \boldsymbol{A}\boldsymbol{x}(k) + \boldsymbol{B}\boldsymbol{u}(k) \tag{6.18}$$

where $\boldsymbol{x}(k) \in \mathrm{R}^n$, $\boldsymbol{u}(k) \in \mathrm{R}^m$. The optimization problem of predictive control with a terminal cost function is formulated as

$$\begin{aligned}
\text{PC2}: \quad & \min_{\boldsymbol{u}(k+i|k),\,0\le i\le N-1} J_N(k) = \sum_{i=0}^{N-1}\left[\|\boldsymbol{x}(k+i\mid k)\|_{\boldsymbol{Q}}^2 + \|\boldsymbol{u}(k+i\mid k)\|_{\boldsymbol{R}}^2\right] + F(\boldsymbol{x}(k+N\mid k)) \\
& \text{s.t.}\ \ \boldsymbol{x}(k+i+1\mid k) = \boldsymbol{A}\boldsymbol{x}(k+i\mid k) + \boldsymbol{B}\boldsymbol{u}(k+i\mid k), \quad i = 0, \ldots, N-1 \\
& \qquad \boldsymbol{x}(k\mid k) = \boldsymbol{x}(k)
\end{aligned} \tag{6.19}$$

where the weighting matrix $\boldsymbol{Q} > \boldsymbol{0}$, $\boldsymbol{R} > \boldsymbol{0}$. It is obvious from (6.6) that to make the cost function of Problem (6.19) close to that with an infinite horizon, the terminal cost function $F(\boldsymbol{x}(k+N \mid k))$ should approximate the remaining part of the infinite cost function after the prediction horizon in (6.6), i.e.

$$J_{N,\infty}(k) = \sum_{i=N}^{\infty}\left[\|\boldsymbol{x}(k+i\mid k)\|_{\boldsymbol{Q}}^2 + \|\boldsymbol{u}(k+i\mid k)\|_{\boldsymbol{R}}^2\right]$$

Therefore the selection of $F(\boldsymbol{x}(k+N\mid k))$ is closely related to the future control inputs $\boldsymbol{u}(k+i\mid k)$, $i \ge N$.

In [12], control inputs after time $k+N$ are assumed to be zeros, i.e. $\boldsymbol{u}(k+i\mid k) = \boldsymbol{0}$, $i \ge N$. Then the system model (6.18) after time $k+N$ can be written as follows:

$$\boldsymbol{x}(k+i+1\mid k) = \boldsymbol{A}\boldsymbol{x}(k+i\mid k), \quad i \ge N$$

It follows that

$$J_{N,\infty}(k) = \sum_{i=N}^{\infty}\|\boldsymbol{x}(k+i\mid k)\|_{\boldsymbol{Q}}^2 = \sum_{i=0}^{\infty}\boldsymbol{x}^{\mathrm{T}}(k+N\mid k)\left(\boldsymbol{A}^{\mathrm{T}}\right)^i\boldsymbol{Q}\boldsymbol{A}^i\boldsymbol{x}(k+N\mid k)$$

Zheng [12] suggested that the terminal cost function could be chosen:

$$F(\boldsymbol{x}(k+N\mid k)) = \|\boldsymbol{x}(k+N\mid k)\|_{\boldsymbol{P}}^2, \quad \boldsymbol{P} = \sum_{i=0}^{\infty}\left(\boldsymbol{A}^{\mathrm{T}}\right)^i\bar{\boldsymbol{Q}}\boldsymbol{A}^i \tag{6.20}$$

to approximate $J_{N,\infty}(k)$, where $\bar{\boldsymbol{Q}} \ge \boldsymbol{Q}$. With this terminal cost function, in the following it can be shown that if Problem (6.19) is feasible at $k = 0$, i.e. $J_N^*(0)$ is bounded, then the closed-loop system controlled by solving Problem (6.19) is asymptotically stable.

Assume that the optimal control inputs obtained by solving Problem (6.19) at time k and the resultant system states are $\boldsymbol{U}^*(k)$ and $\boldsymbol{X}^*(k)$, respectively. At time $k+1$, construct a solution of Problem (6.19) as $\boldsymbol{U}(k+1) = \{\boldsymbol{u}^*(k+1\mid k), \ldots, \boldsymbol{u}^*(k+N-1\mid k), \boldsymbol{0}\}$. It follows that

$$J_N^*(k) = \sum_{i=0}^{N-1}\left[\|\boldsymbol{x}^*(k+i\mid k)\|_{\boldsymbol{Q}}^2 + \|\boldsymbol{u}^*(k+i\mid k)\|_{\boldsymbol{R}}^2\right] + \|\boldsymbol{x}^*(k+N\mid k)\|_{\boldsymbol{P}}^2$$

$$\begin{aligned} J_N(k+1) &= \sum_{i=1}^{N}\left[\|\boldsymbol{x}(k+i\mid k+1)\|_{\boldsymbol{Q}}^2 + \|\boldsymbol{u}(k+i\mid k+1)\|_{\boldsymbol{R}}^2\right] + \|\boldsymbol{x}(k+N+1\mid k+1)\|_{\boldsymbol{P}}^2 \\ &= \sum_{i=1}^{N-1}\left[\|\boldsymbol{x}^*(k+i\mid k)\|_{\boldsymbol{Q}}^2 + \|\boldsymbol{u}^*(k+i\mid k)\|_{\boldsymbol{R}}^2\right] + \|\boldsymbol{x}^*(k+N\mid k)\|_{\boldsymbol{Q}+\boldsymbol{A}^{\mathrm{T}}\boldsymbol{P}\boldsymbol{A}}^2 \end{aligned}$$

and

$$J_N^*(k) - J_N(k+1) = \|\boldsymbol{x}^*(k\mid k)\|_{\boldsymbol{Q}}^2 + \|\boldsymbol{u}^*(k\mid k)\|_{\boldsymbol{R}}^2 + \|\boldsymbol{x}^*(k+N\mid k)\|_{\boldsymbol{P}-\boldsymbol{Q}-\boldsymbol{A}^{\mathrm{T}}\boldsymbol{P}\boldsymbol{A}}^2 \ge 0.$$

From $J_N^*(0) < \infty$, we know that $J_N(1) < \infty$. Similarly, $J_N(k+1) < \infty$ can be deduced by iteration, i.e. $\boldsymbol{U}(k+1)$ is a feasible solution for Problem (6.19) at time $k+1$. Then it follows that $J_N^*(k+1) \le J_N^*(k)$, where the equality holds only if all $\boldsymbol{u}$ and $\boldsymbol{x}$ are zeros, which means that the closed-loop system is asymptotically stable.

In [13], the control inputs after time $k+N$ are assumed to be an LQ control law starting from the initial state $\boldsymbol{x}(k+N\mid k)$, i.e. $\boldsymbol{u}(k+i) = \boldsymbol{H}\boldsymbol{x}(k+i)$ $(i \ge N)$. Then, according to the optimal control theory, the optimal value of $J_{N,\infty}(k)$ is a quadric function of $\boldsymbol{x}(k+N\mid k)$, which can be taken as a terminal cost function to approximate $J_{N,\infty}(k)$ with the form

$$F(\boldsymbol{x}(k+N \mid k)) = \|\boldsymbol{x}(k+N \mid k)\|_{\boldsymbol{P}}^2 \tag{6.21}$$

where $\boldsymbol{P} > \boldsymbol{0}$ and for $\boldsymbol{H} \in \mathrm{R}^{m \times n}$ satisfies

$$(\boldsymbol{A}+\boldsymbol{BH})^{\mathrm{T}}\boldsymbol{P}(\boldsymbol{A}+\boldsymbol{BH}) + \boldsymbol{Q} + \boldsymbol{H}^{\mathrm{T}}\boldsymbol{RH} < \boldsymbol{P} \tag{6.22}$$

For this case, also assume that the optimal control inputs and the corresponding states at time k are $\boldsymbol{U}^*(k)$ and $\boldsymbol{X}^*(k)$. The corresponding optimal value of the performance index is

$$J_N^*(k) = \sum_{i=0}^{N-1}\left[\|\boldsymbol{x}^*(k+i \mid k)\|_{\boldsymbol{Q}}^2 + \|\boldsymbol{u}^*(k+i \mid k)\|_{\boldsymbol{R}}^2\right] + \|\boldsymbol{x}^*(k+N \mid k)\|_{\boldsymbol{P}}^2$$

At time $k+1$, construct $\boldsymbol{U}(k+1) = \{\boldsymbol{u}^*(k+1 \mid k), \ldots, \boldsymbol{u}^*(k+N-1 \mid k), \boldsymbol{Hx}^*(k+N \mid k)\}$. The resultant performance index is

$$\begin{aligned}
J_N(k+1) &= \sum_{i=0}^{N-1}\left[\|\boldsymbol{x}(k+i+1 \mid k+1)\|_{\boldsymbol{Q}}^2 + \|\boldsymbol{u}(k+i+1 \mid k+1)\|_{\boldsymbol{R}}^2\right] + \|\boldsymbol{x}(k+N+1 \mid k+1)\|_{\boldsymbol{P}}^2 \\
&= \sum_{i=0}^{N-2}\left[\|\boldsymbol{x}^*(k+i+1 \mid k)\|_{\boldsymbol{Q}}^2 + \|\boldsymbol{u}^*(k+i+1 \mid k)\|_{\boldsymbol{R}}^2\right] + \|\boldsymbol{x}^*(k+N \mid k)\|_{\boldsymbol{Q}}^2 \\
&\quad + \|\boldsymbol{Hx}^*(k+N \mid k)\|_{\boldsymbol{R}}^2 + \|(\boldsymbol{A}+\boldsymbol{BH})\boldsymbol{x}^*(k+N \mid k)\|_{\boldsymbol{P}}^2
\end{aligned}$$

The comparison between $J_N^*(k)$ and $J_N(k+1)$ can yield

$$\begin{aligned}
J_N(k+1) &= J_N^*(k) + \|\boldsymbol{x}^*(k+N \mid k)\|_{\boldsymbol{Q}}^2 + \|\boldsymbol{Hx}^*(k+N \mid k)\|_{\boldsymbol{R}}^2 - \|\boldsymbol{x}^*(k \mid k)\|_{\boldsymbol{Q}}^2 \\
&\quad - \|\boldsymbol{u}^*(k \mid k)\|_{\boldsymbol{R}}^2 + \|(\boldsymbol{A}+\boldsymbol{BH})\boldsymbol{x}^*(k+N \mid k)\|_{\boldsymbol{P}}^2 - \|\boldsymbol{x}^*(k+N \mid k)\|_{\boldsymbol{P}}^2 \\
&= J_N^*(k) - \|\boldsymbol{x}^*(k \mid k)\|_{\boldsymbol{Q}}^2 - \|\boldsymbol{u}^*(k \mid k)\|_{\boldsymbol{R}}^2 + \|\boldsymbol{x}^*(k+N \mid k)\|_{\boldsymbol{Q}+\boldsymbol{H}^{\mathrm{T}}\boldsymbol{RH}+(\boldsymbol{A}+\boldsymbol{BH})^{\mathrm{T}}\boldsymbol{P}(\boldsymbol{A}+\boldsymbol{BH})-\boldsymbol{P}}^2 \\
&\le J_N^*(k)
\end{aligned}$$

Meanwhile, since there is no constraint, the constructed solution is always feasible. Therefore $J_N^*(k+1) \le J_N(k+1)$ and then

$$J_N^*(k+1) \le J_N^*(k)$$

i.e. $J_N^*(k)$ can be used as a Lyapunov function to guarantee the closed-loop stability.

Remark 6.4 In the above two cases, the terminal cost function is designed by directly approximating the value of the objective function after the finite prediction horizon. By adding the terminal cost function, the optimization problem of predictive control with a finite prediction horizon is actually transformed into an infinite horizon optimization problem but with restricted input structures after the finite prediction horizon, i.e. $\boldsymbol{u}(k+i) = \boldsymbol{0}$, $i \ge N$ for the former case and $\boldsymbol{u}(k+i) = \boldsymbol{Hx}(k+i)$, $i \ge N$ for the latter case. It is obvious that these control strategies are not strictly in accord with the optimal control law of the infinite horizon optimization, but are only an approximation, but the closed-loop stability of predictive control can be guaranteed. Furthermore, the stability proof also shows that selection of the terminal cost function is closely related to the adopted control strategy after the finite prediction horizon, although it does not explicitly appear in the terminal cost function.

(2) The constrained nonlinear system
Consider a nonlinear system

$$\boldsymbol{x}(k+1)=\boldsymbol{f}(\boldsymbol{x}(k),\ \boldsymbol{u}(k),\ k) \tag{6.23}$$

where $\boldsymbol{x}(k) \in \mathrm{R}^n$, $\boldsymbol{u}(k) \in \mathrm{R}^m$. The system is subjected to the state and input constraints

$$\boldsymbol{x} \in \boldsymbol{\Omega}_x, \quad \boldsymbol{u} \in \boldsymbol{\Omega}_u \tag{6.24}$$

At time k the online optimization problem of predictive control with a terminal cost function can be formulated as

$$\begin{aligned}
\text{PC3:} \quad & \min_{\boldsymbol{u}(k+i|k),0\le i\le N-1} J_N(k)=\sum_{i=0}^{N-1} l(\boldsymbol{x}(k+i\mid k),\ \boldsymbol{u}(k+i\mid k))+F(\boldsymbol{x}(k+N\mid k)) \\
\text{s.t.} \quad & \boldsymbol{x}(k+i+1\mid k)=\boldsymbol{f}(\boldsymbol{x}(k+i\mid k),\ \boldsymbol{u}(k+i\mid k),\ k+i),\quad i=0,\ \dots,\ N-1 \\
& \boldsymbol{u}(k+i\mid k)\in\boldsymbol{\Omega}_u,\quad i=0,\ \dots,\ N-1 \\
& \boldsymbol{x}(k+i\mid k)\in\boldsymbol{\Omega}_x,\quad i=1,\ \dots,\ N \\
& \boldsymbol{x}(k\mid k)=\boldsymbol{x}(k)
\end{aligned} \tag{6.25}$$

For this category of problem, De Nicolao et al. [14] suggested firstly to find the feasible region of the feedback control law which exponentially stabilizes the system and then to determine the terminal cost function. For system (6.23), the linearized system within a neighbor region of the equilibrium point $\boldsymbol{x} = \mathbf{0}$, $\boldsymbol{u} = \mathbf{0}$ is given by

$$\boldsymbol{x}(k+1)=\boldsymbol{A}(k)\boldsymbol{x}(k)+\boldsymbol{B}(k)\boldsymbol{u}(k) \tag{6.26}$$

where $\boldsymbol{A}(k) = \partial \boldsymbol{f}(\boldsymbol{x},\ \boldsymbol{u},\ k)/\partial \boldsymbol{x}|_{\boldsymbol{x}=0,\ \boldsymbol{u}=0}$ and $\boldsymbol{B}(k)=\partial \boldsymbol{f}(\boldsymbol{x},\ \boldsymbol{u},\ k)/\partial \boldsymbol{u}|_{\boldsymbol{x}=0,\ \boldsymbol{u}=0}$.

Assume that there is a feedback gain matrix $\boldsymbol{H}(k)$ to make $\boldsymbol{A}(k) + \boldsymbol{B}(k)\boldsymbol{H}(k)$ exponentially stable. Then, under certain conditions, we can get the closed-loop nonlinear system by this feedback gain matrix with the origin as its exponentially stable equilibrium point:

$$\boldsymbol{x}(k+1)=\boldsymbol{f}(\boldsymbol{x}(k),\ \boldsymbol{H}(k)\boldsymbol{x}(k),\ k) \tag{6.27}$$

Define $\boldsymbol{X}_H(t)$ as such a set that if $\boldsymbol{x}(t) \in \boldsymbol{X}_H(t)$, then starting from $\boldsymbol{x}(t)$ and adopting the feedback control law $\boldsymbol{u}(t+i) = \boldsymbol{H}(t+i)\boldsymbol{x}(t+i)$ for $i \ge 0$ to system (6.23), the resultant state trajectory $\boldsymbol{x}_C(t+i,\ \boldsymbol{x}(t))$ and the input sequence $\boldsymbol{H}(t+i)\boldsymbol{x}_C(t+i,\ \boldsymbol{x}(t))$ for $i \ge 0$ will satisfy the constraint (6.24) and the closed-loop system (6.27) is exponentially stable.

For Problem (6.25) at time k, denote the terminal state as $\boldsymbol{x}_N \triangleq \boldsymbol{x}(k+N\mid k)$. De Nicolao et al. [14] suggested that the following function could be utilized as the terminal cost function of Problem (6.25):

$$F(\boldsymbol{x}_N)=\begin{cases} \sum_{i=N}^{\infty} l(\boldsymbol{x}_C(k+i,\boldsymbol{x}_N),\ \boldsymbol{H}(k+i)\boldsymbol{x}_C(k+i,\boldsymbol{x}_N),\ k+i), & \boldsymbol{x}_N\in\boldsymbol{X}_H(k+N) \\ \infty, & \text{otherwise} \end{cases} \tag{6.28}$$

which is the cost from $k+N$ to infinity incurred by adopting the control law $\boldsymbol{u}(\cdot) = \boldsymbol{H}(\cdot)\boldsymbol{x}(\cdot)$ to system (6.23) after the optimization horizon $k+N$.

Remark 6.5 The use of the terminal cost function (6.28) implies that the control sequence $\boldsymbol{U}_N(k) = \{\boldsymbol{u}(k\,|\,k), \ldots, \boldsymbol{u}(k+N-1\,|\,k)\}$ is admissible for Problem (6.25) only if the input and state constraints in (6.25) and the constraint $\boldsymbol{x}_N \in \boldsymbol{X}_{\boldsymbol{H}}(k+N)$ are simultaneously satisfied when $\boldsymbol{U}_N(k)$ is applied to the system (6.23). This implicitly includes a terminal set constraint, i.e. using the control sequence $\boldsymbol{U}_N(k)$ to steer the system state to the terminal set $\boldsymbol{X}_{\boldsymbol{H}}(k+N)$, before adopting the linear control law $\boldsymbol{u}(\cdot) = \boldsymbol{H}(\cdot)\boldsymbol{x}(\cdot)$.

For the stability of the closed-loop system, it is easy to show the decreasing property of the optimal value function with time. To do that, extend the length of the prediction horizon at time k from N to $N+1$ and let

$$\boldsymbol{U}_{N+1}(k) = \left\{\boldsymbol{u}^*(k\,|\,k),\ \ldots,\ \boldsymbol{u}^*(k+N-1\,|\,k),\ \boldsymbol{H}(k+N)\boldsymbol{x}_N^*\right\}$$

It is obvious that the added input $\boldsymbol{u}(k+N\,|\,k) = \boldsymbol{H}(k+N)\boldsymbol{x}_N^*$ and the resultant state $\boldsymbol{x}_{N+1} = \boldsymbol{f}\left(\boldsymbol{x}_N^*,\ \boldsymbol{H}(k+N)\boldsymbol{x}_N^*,\ k\right)$ satisfy the constraints for Problem (6.25) with the prediction horizon $N+1$, and $\boldsymbol{x}_{N+1} \in \boldsymbol{X}_{\boldsymbol{H}}(N+1)$. Therefore $\boldsymbol{U}_{N+1}(k)$ is a feasible solution of Problem (6.25) at time k with the prediction horizon $N+1$. In fact, $J_{N+1}(k) = J_N^*(k)$.

At time $k+1$, select the last N elements of $\boldsymbol{U}_{N+1}(k)$ to construct a solution to Problem (6.25) with the prediction horizon N:

$$\boldsymbol{U}_N(k+1) = \left\{\boldsymbol{u}^*(k+1\,|\,k),\ \ldots,\ \boldsymbol{u}^*(k+N-1\,|\,k),\ \boldsymbol{H}(k+N)\boldsymbol{x}_N^*\right\}$$

It is obviously feasible for this optimization problem and thus it follows that

$$J_N^*(k+1) \le J_N(k+1) = J_{N+1}(k) - l(\boldsymbol{x}^*(k\,|\,k),\ \boldsymbol{u}^*(k\,|\,k)) \le J_{N+1}(k) = J_N^*(k)$$

where the equality holds only if $\boldsymbol{x} = \boldsymbol{0}$, $\boldsymbol{u} = \boldsymbol{0}$; thus the closed-loop system is exponentially stable.

Note that the infinite summation in the terminal cost function (6.28) can in principle be approximated by the sum of limited but enough items. The system states and the control inputs appearing in these items can be calculated starting from $\boldsymbol{x}_N$ and using (6.27) recursively. In addition, if the linearized system (6.26) is time-invariant and Problem (6.25) adopts a quadratic function such as the following objective function,

$$J_N(k) = \sum_{i=0}^{N-1}\left[\|\boldsymbol{x}(k+i\,|\,k)\|_{\boldsymbol{Q}}^2 + \|\boldsymbol{u}(k+i\,|\,k)\|_{\boldsymbol{R}}^2\right] + F(\boldsymbol{x}(k+N\,|\,k))$$

we can firstly derive the LQ feedback gain of the linearized system

$$\boldsymbol{H} = -\left(\boldsymbol{R} + \boldsymbol{B}^{\mathrm{T}}\boldsymbol{P}\boldsymbol{B}\right)^{-1}\boldsymbol{B}^{\mathrm{T}}\boldsymbol{P}\boldsymbol{A}$$

where $\boldsymbol{P}$ is the unique positive-definite solution of the following Riccati equation:

$$\boldsymbol{P} = \boldsymbol{A}^{\mathrm{T}}\boldsymbol{P}\boldsymbol{A} + \boldsymbol{Q} - \boldsymbol{A}^{\mathrm{T}}\boldsymbol{P}\boldsymbol{B}\left(\boldsymbol{R} + \boldsymbol{B}^{\mathrm{T}}\boldsymbol{P}\boldsymbol{B}\right)^{-1}\boldsymbol{B}^{\mathrm{T}}\boldsymbol{P}\boldsymbol{A}$$

and then design the terminal cost function as

$$F(\boldsymbol{x}_N) = \sum_{i=N}^{M-1}\left[\|\boldsymbol{x}(k+i\,|\,k)\|_{\boldsymbol{Q}}^2 + \|\boldsymbol{H}\boldsymbol{x}(k+i\,|\,k)\|_{\boldsymbol{R}}^2\right] + \|\boldsymbol{x}(k+M\,|\,k)\|_{\boldsymbol{P}}^2$$

where $M > N$. This means that in the optimization problem (6.25) at time k, it is assumed that the LQ optimal control law $\boldsymbol{u}(\cdot) = \boldsymbol{Hx}(\cdot)$ for the linearized system is implemented after the time $k + N$, so as to construct the above terminal cost function $F(\boldsymbol{x}_N)$ to approximate the optimal control with an infinite horizon. Note that in order to avoid handling infinite items, the value function of the standard LQ control with the initial state $\boldsymbol{x}(k + M \mid k)$, i.e. the last term in $F(\boldsymbol{x}_N)$, is used to replace the costs after time $k + M$ because the system state will be steered into a sufficiently small neighborhood of the origin and its dynamic behavior will be very close to linear. Thus the part of the optimization after time $k + M$ can be regarded as a standard LQ problem.

From the above discussion it can be seen that since the terminal cost function is used to approximate the cost of the optimization problem after time $k + N$, the control strategy after the finite prediction horizon N should be considered. Although this control strategy will not actually be implemented, it is implied in the expression of the terminal cost function and has an essential impact on the closed-loop stability. This idea will be shown more straightforwardly and obviously in the following section on terminal set constraints.

6.2.3 Predictive Control with Terminal Set Constraints

The terminal set constraint is another kind of extension of the zero terminal constraint, which directly relaxes the condition $\boldsymbol{x}(k + N \mid k) = \boldsymbol{0}$ in the online optimization problem (6.17) into $\boldsymbol{x}(k + N \mid k) \in \boldsymbol{X}_f$, where $\boldsymbol{X}_f$ is called the terminal set. In general, driving a system state into a set is easier than driving it to a fixed point. Therefore the terminal set constraint is less conservative than the zero terminal constraint. This approach is proposed by Michalska and Mayne [15] using the dual-mode control strategy. They adopted two control modes to predictive control for a class of constrained nonlinear systems, i.e. a sequence of free control inputs and a local feedback control law, where the sequence of free control inputs within a finite prediction horizon firstly steers the system state into a neighbor set $\boldsymbol{X}_f$ of the origin. A local linear feedback control law was then adopted to stabilize the system after the state entered the terminal set. The principle of dual-mode control is shown in Figure 6.3.

The dual-mode control strategy with terminal set constraints has been widely used in the theoretical research of predictive control. Furthermore, the terminal set constraint is often combined with the terminal cost function for a better approximation of the infinite horizon optimal control. In the following, taking the linear constrained system with the quadratic performance index as an example, the predictive control strategy with a terminal set constraint and terminal cost function is introduced [13].

Consider the linear system

$$\boldsymbol{x}(k+1) = \boldsymbol{Ax}(k) + \boldsymbol{Bu}(k) \tag{6.29}$$

where $\boldsymbol{x}(k) \in \mathrm{R}^n$, $\boldsymbol{u}(k) \in \mathrm{R}^m$, and the input and state constraints are given by

$$\begin{aligned} &\left|u_j\right| \le u_{j,\max}, \quad j = 1, \ldots, m \\ &\left|(\boldsymbol{Gx})_j\right| \le g_{j,\max}, \quad j = 1, \ldots, p \end{aligned} \tag{6.30}$$

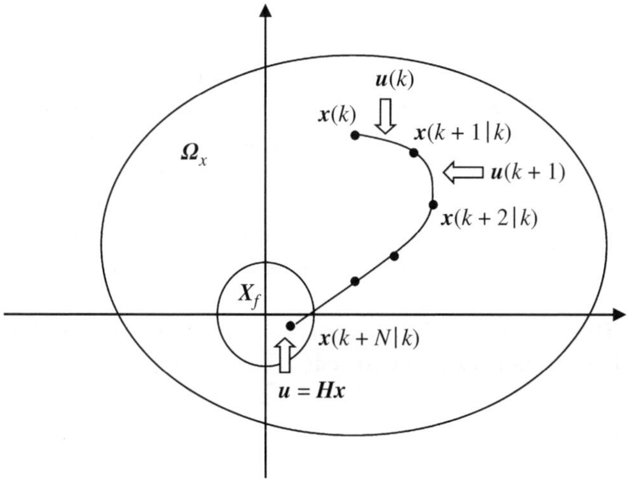

Figure 6.3 Principles of dual-mode control [15]. *Source:* Reproduced with permission from H. Michalska and D.Q. Mayne of IEEE.

where $\boldsymbol{G} \in \mathrm{R}^{p \times n}$. The optimization problem of predictive control with the terminal set constraint and the terminal cost function can be formulated as

$$\begin{aligned}
\text{PC4:} \quad & \min_{\boldsymbol{u}(k+i|k), 0 \le i \le N-1} J_N(k) = \sum_{i=0}^{N-1} \left[\|\boldsymbol{x}(k+i\mid k)\|_{\boldsymbol{Q}}^2 + \|\boldsymbol{u}(k+i\mid k)\|_{\boldsymbol{R}}^2 \right] + \|\boldsymbol{x}(k+N\mid k)\|_{\boldsymbol{P}}^2 \\
\text{s.t.} \quad & \boldsymbol{x}(k+i+1\mid k) = \boldsymbol{A}\boldsymbol{x}(k+i\mid k) + \boldsymbol{B}\boldsymbol{u}(k+i\mid k), \quad i = 0, \ldots, N-1 \\
& \left| (\boldsymbol{u}(k+i\mid k))_j \right| \le u_{j,\max}, \quad i = 0, \ldots, N-1, \quad j = 1, \ldots, m \\
& \left| (\boldsymbol{G}\boldsymbol{x}(k+i\mid k))_j \right| \le g_{j,\max}, \quad i = 0, \ldots, N, \quad j = 1, \ldots, p \\
& \boldsymbol{x}(k+N\mid k) \in \boldsymbol{\Omega} = \left\{ \boldsymbol{x} \in \mathrm{R}^n \mid \boldsymbol{x}^{\mathrm{T}} \boldsymbol{S}^{-1} \boldsymbol{x} \le 1 \right\} \\
& \boldsymbol{x}(k\mid k) = \boldsymbol{x}(k)
\end{aligned} \tag{6.31}$$

For the above optimization problem, the dual-mode control strategy is adopted, i.e. firstly to use the free control inputs $\boldsymbol{u}(k+i \mid k)$, $i = 0, \ldots, N-1$ to steer the system state into the terminal set $\boldsymbol{\Omega}$ and then to use the local feedback control law $\boldsymbol{u} = \boldsymbol{H}\boldsymbol{x}$ to stabilize the system. In the following, how to guarantee that the closed-loop system is asymptotically stable is discussed.

Assume that Problem (6.31) has the optimal solution $\boldsymbol{U}^*(k)$ at time k, which results in corresponding system states $\boldsymbol{X}^*(k)$ and the optimal value of performance index

$$J_N^*(k) = \sum_{i=0}^{N-1} \left[\|\boldsymbol{x}^*(k+i\mid k)\|_{\boldsymbol{Q}}^2 + \|\boldsymbol{u}^*(k+i\mid k)\|_{\boldsymbol{R}}^2 \right] + \|\boldsymbol{x}^*(k+N\mid k)\|_{\boldsymbol{P}}^2$$

At time $k+1$, construct $\boldsymbol{U}(k+1) = \{\boldsymbol{u}^*(k+1 \mid k), \ldots, \boldsymbol{u}^*(k+N-1 \mid k), \boldsymbol{H}\boldsymbol{x}^*(k+N \mid k)\}$. According to the idea of stability analysis for predictive control systems discussed in

Section 6.1.3, we explore conditions to ensure that $\boldsymbol{U}(k+1)$ is feasible at time $k+1$ and $J_N(k+1) < J_N^*(k)$.

Firstly, since $\boldsymbol{U}^*(k)$ and $\boldsymbol{X}^*(k)$ are the optimal solution of Problem (6.31) at time k and the resultant state trajectory, respectively, the items $\boldsymbol{u}^*(k+1 \mid k), \ldots, \boldsymbol{u}^*(k+N-1 \mid k)$ in $\boldsymbol{U}(k+1)$ and the corresponding states $\boldsymbol{x}^*(k+1 \mid k), \ldots, \boldsymbol{x}^*(k+N \mid k)$ must satisfy the constraints in Problem (6.31), and meanwhile $\boldsymbol{x}^*(k+N \mid k) \in \boldsymbol{\Omega}$. Thus it is only necessary to consider whether the last item of $\boldsymbol{U}(k+1)$, i.e. $\boldsymbol{Hx}^*(k+N \mid k)$, satisfies the input constraint and whether the resultant state $\boldsymbol{x}(k+N+1 \mid k+1)$ satisfies the state constraint and belongs to set $\boldsymbol{\Omega}$. If so, it can be concluded that $\boldsymbol{U}(k+1)$ is feasible for Problem (6.31) at time $k+1$.

Let $\boldsymbol{S} = \boldsymbol{P}^{-1}$, $\boldsymbol{\Omega} = \{\boldsymbol{x} \in \mathrm{R}^n \mid \boldsymbol{x}^{\mathrm{T}}\boldsymbol{Px} \le 1\}$, and $\boldsymbol{H} = \boldsymbol{YS}^{-1}$. From the LMI (6.14), when $\boldsymbol{x}^*(k+N \mid k) \in \boldsymbol{\Omega}$, the following input constraints can be ensured:

$$\left|(\boldsymbol{Hx}^*(k+N \mid k))_j\right| \le u_{j,\max}, \quad j = 1, \ldots, m$$

if there exists a symmetric matrix $\boldsymbol{X}$ such that

$$\begin{bmatrix} \boldsymbol{X} & \boldsymbol{Y} \\ \boldsymbol{Y}^{\mathrm{T}} & \boldsymbol{S} \end{bmatrix} \ge \boldsymbol{0} \tag{6.32}$$

where $\boldsymbol{X}_{jj} \le u_{j,\max}^2$, $j = 1, \ldots, m$.

According to $\boldsymbol{x}(k+N+1 \mid k+1) = (\boldsymbol{A} + \boldsymbol{BH})\boldsymbol{x}^*(k+N \mid k)$ and $\boldsymbol{x}^*(k+N \mid k) \in \boldsymbol{\Omega}$, similarly referring to (6.16), the following state constraints can be ensured:

$$\left|(\boldsymbol{Gx}(k+N+1 \mid k+1))_j\right| \le g_{j,\max}, \quad j = 1, \ldots, p$$

if there exists a symmetric matrix $\boldsymbol{Z}$ such that

$$\begin{bmatrix} \boldsymbol{Z} & \boldsymbol{G}(\boldsymbol{AS} + \boldsymbol{BY}) \\ (\boldsymbol{AS} + \boldsymbol{BY})^{\mathrm{T}}\boldsymbol{G}^{\mathrm{T}} & \boldsymbol{S} \end{bmatrix} \ge \boldsymbol{0} \tag{6.33}$$

where $\boldsymbol{Z}_{jj} \le g_{j,\max}^2$, $j = 1, \ldots, p$.

On the other hand, the state $\boldsymbol{x}(k+N+1 \mid k+1)$ can be ensured within the set $\boldsymbol{\Omega}$ by the condition that the terminal set $\boldsymbol{\Omega}$ is an invariant set under the feedback control law $\boldsymbol{u} = \boldsymbol{Hx}$ (see (6.10)), i.e.

$$(\boldsymbol{A} + \boldsymbol{BH})^{\mathrm{T}}\boldsymbol{P}(\boldsymbol{A} + \boldsymbol{BH}) - \boldsymbol{P} < \boldsymbol{0} \tag{6.34}$$

From the above analysis, if conditions (6.32) to (6.34) hold, then $\boldsymbol{U}(k+1)$ is a feasible solution of Problem (6.31) at time $k+1$ and the corresponding cost value is given by

$$\begin{aligned} J_N(k+1) &= \sum_{i=0}^{N-2}\left[\|\boldsymbol{x}^*(k+i+1 \mid k)\|_{\boldsymbol{Q}}^2 + \|\boldsymbol{u}^*(k+i+1 \mid k)\|_{\boldsymbol{R}}^2\right] + \|\boldsymbol{x}^*(k+N \mid k)\|_{\boldsymbol{Q}}^2 \\ &\quad + \|\boldsymbol{Hx}^*(k+N \mid k)\|_{\boldsymbol{R}}^2 + \|(\boldsymbol{A} + \boldsymbol{BH})\boldsymbol{x}^*(k+N \mid k)\|_{\boldsymbol{P}}^2 \\ &= J_N^*(k) - \|\boldsymbol{x}^*(k \mid k)\|_{\boldsymbol{Q}}^2 - \|\boldsymbol{u}^*(k \mid k)\|_{\boldsymbol{R}}^2 + \|\boldsymbol{x}^*(k+N \mid k)\|_{\boldsymbol{Q}+\boldsymbol{H}^{\mathrm{T}}\boldsymbol{RH}+(\boldsymbol{A}+\boldsymbol{BH})^{\mathrm{T}}\boldsymbol{P}(\boldsymbol{A}+\boldsymbol{BH})-\boldsymbol{P}}^2 \end{aligned}$$

Therefore, $J_N(k+1) < J_N^*(k)$ and thus $J_N^*(k+1) < J_N^*(k)$ if the following condition with respect to the terminal cost function holds:

$$\boldsymbol{Q} + \boldsymbol{H}^{\mathrm{T}}\boldsymbol{R}\boldsymbol{H} + (\boldsymbol{A} + \boldsymbol{B}\boldsymbol{H})^{\mathrm{T}}\boldsymbol{P}(\boldsymbol{A} + \boldsymbol{B}\boldsymbol{H}) - \boldsymbol{P} < \boldsymbol{0} \tag{6.35}$$

Note that condition (6.34) is covered by condition (6.35) and it can therefore be concluded that if the terminal set and the terminal cost function are designed to satisfy conditions (6.32), (6.33), (6.35), and Problem (6.31) is feasible at time $k = 0$, the closed-loop predictive control system controlled by solving Problem (6.31) is asymptotically stable.

In order to obtain matrices $\boldsymbol{P}$ and $\boldsymbol{H}$, note that $\boldsymbol{S} = \boldsymbol{P}^{-1}$ and $\boldsymbol{H} = \boldsymbol{Y}\boldsymbol{S}^{-1}$ and rewrite (6.35) as

$$\boldsymbol{S} - \boldsymbol{S}^{\mathrm{T}}\boldsymbol{Q}\boldsymbol{S} - \boldsymbol{Y}^{\mathrm{T}}\boldsymbol{R}\boldsymbol{Y} - (\boldsymbol{A}\boldsymbol{S} + \boldsymbol{B}\boldsymbol{Y})^{\mathrm{T}}\boldsymbol{S}^{-1}(\boldsymbol{A}\boldsymbol{S} + \boldsymbol{B}\boldsymbol{Y}) > \boldsymbol{0}$$

By the Schur complement (6.11d) and (6.11e), its LMIs can be given by

$$\begin{bmatrix} \boldsymbol{S} & (\boldsymbol{A}\boldsymbol{S} + \boldsymbol{B}\boldsymbol{Y})^{\mathrm{T}} & (\boldsymbol{Q}^{1/2}\boldsymbol{S})^{\mathrm{T}} & (\boldsymbol{R}^{1/2}\boldsymbol{Y})^{\mathrm{T}} \\ \boldsymbol{A}\boldsymbol{S} + \boldsymbol{B}\boldsymbol{Y} & \boldsymbol{S} & \boldsymbol{0} & \boldsymbol{0} \\ \boldsymbol{Q}^{1/2}\boldsymbol{S} & \boldsymbol{0} & \boldsymbol{I} & \boldsymbol{0} \\ \boldsymbol{R}^{1/2}\boldsymbol{Y} & \boldsymbol{0} & \boldsymbol{0} & \boldsymbol{I} \end{bmatrix} > \boldsymbol{0} \tag{6.36}$$

Thus, if the matrices $\boldsymbol{S}$ and $\boldsymbol{Y}$ in LIMs (6.32), (6.33), and (6.36) are solved, $\boldsymbol{P}$ and $\boldsymbol{H}$ can be obtained immediately.

Remark 6.6 At each time k, the predictive controller solves the optimization problem PC4 using dual-mode control where a sequence of free control inputs $\boldsymbol{u}(k+i \mid k)$, $i = 0, \ldots, N-1$ steer the system state into the terminal set $\boldsymbol{\Omega}$ and then a local feedback control law $\boldsymbol{u} = \boldsymbol{H}\boldsymbol{x}$ is adopted. Due to the rolling style of predictive control, only the optimal $\boldsymbol{u}^*(k \mid k)$ is actually implemented and the remaining parts of future controls are not put into use. This might lead to the conclusion that the predictive controller has nothing to do with the feedback control law $\boldsymbol{u} = \boldsymbol{H}\boldsymbol{x}$. However, this is not the case because the terminal set $\boldsymbol{\Omega}$ as well as the terminal cost function in the optimization problem PC4 are both directly related to matrix $\boldsymbol{P}$, while during its solving the feedback control gain $\boldsymbol{H}$ should be properly selected to satisfy the condition (6.35) with respect to $\boldsymbol{P}$ such that the closed-loop stability can be guaranteed. This means that the predictive control law $\boldsymbol{u}^*(k)$ is implicitly affected by the feedback control law $\boldsymbol{u} = \boldsymbol{H}\boldsymbol{x}$ in the terminal constraint set, although the latter is not actually put into use.

The above design method for stable predictive controllers with a terminal cost function and terminal set constraint makes use of the dual mode control strategy. It is easy to find that both the terminal cost function and the terminal set constraint are closely related to the control strategy adopted after the system state enters into the terminal set. In the dual mode control discussed above, the optimization horizon, i.e. the number of free control variables, is fixed as N. However, using a variable optimization horizon has also been studied in some literature. For example, in the pioneering literature on the dual-mode control method [15], the optimization horizon is designed to be decreased gradually until the system state enters into the terminal set. Once the system state is in the terminal set, the feedback control law will substitute the receding horizon optimization to control the system.

6.3 General Stability Conditions of Predictive Control and Suboptimality Analysis

6.3.1 General Stability Conditions of Predictive Control

By introducing the synthesis methods for stable predictive controllers discussed in the last section, it is shown that the terminal cost function, terminal constraint set, and local stabilizing controller are three ingredients for synthesizing stable predictive controllers. In 2000, Mayne et al. [2] surveyed the related research results and gave the basic conditions for guaranteed stability of predictive control systems in a general form.

(1) General stability condition

Consider the online optimization problem of predictive control in general form (6.7):

$$\begin{aligned}
\text{PC:} \quad & \min_{\boldsymbol{u}(k+i|k),0\le i\le N-1} J_N(k)=\sum_{i=0}^{N-1} l(\boldsymbol{x}(k+i\mid k),\ \boldsymbol{u}(k+i\mid k))+F(\boldsymbol{x}(k+N\mid k)) \\
& \text{s.t.}\ \ \boldsymbol{x}(k+i+1\mid k)=\boldsymbol{f}(\boldsymbol{x}(k+i\mid k),\ \boldsymbol{u}(k+i\mid k)),\ \ i=0,\ \dots,\ N-1 \\
& \qquad \boldsymbol{u}(k+i\mid k)\in\boldsymbol{\Omega}_u,\ \ i=0,\ \dots,\ N-1 \\
& \qquad \boldsymbol{x}(k+i\mid k)\in\boldsymbol{\Omega}_x,\ \ i=1,\ \dots,\ N \\
& \qquad \boldsymbol{x}(k+N\mid k)\in\boldsymbol{X}_f,\quad \boldsymbol{x}(k\mid k)=\boldsymbol{x}(k)
\end{aligned} \tag{6.37}$$

where the terminal set is $\boldsymbol{X}_f$, the terminal cost function is $F(\boldsymbol{x}(k+N))$, and the local control law is $\boldsymbol{u}=\boldsymbol{H}(\boldsymbol{x})$. Assume that if the system state enters into the terminal set $\boldsymbol{X}_f$, the local control law will be adopted to stabilize the system. Then the following conditions will ensure that the closed-loop system is asymptotically stable:

$$\begin{aligned}
&\text{A1:}\ \ \boldsymbol{X}_f\subset\boldsymbol{\Omega}_x,\ \boldsymbol{X}_f\ \text{closed},\ \boldsymbol{0}\in\boldsymbol{X}_f \\
&\text{A2:}\ \ \boldsymbol{H}(\boldsymbol{x})\in\boldsymbol{\Omega}_u,\ \ \forall\boldsymbol{x}\in\boldsymbol{X}_f \\
&\text{A3:}\ \ \boldsymbol{f}(\boldsymbol{x},\ \boldsymbol{H}(\boldsymbol{x}))\in\boldsymbol{X}_f,\ \ \forall\boldsymbol{x}\in\boldsymbol{X}_f \\
&\text{A4:}\ \ F(\boldsymbol{f}(\boldsymbol{x},\ \boldsymbol{H}(\boldsymbol{x})))-F(\boldsymbol{x})+l(\boldsymbol{x},\ \boldsymbol{H}(\boldsymbol{x}))\le 0,\ \ \forall\boldsymbol{x}\in\boldsymbol{X}_f
\end{aligned} \tag{6.38}$$

In the above four conditions, A1 means that the states in the terminal set satisfy state constraints. A2 means that the local control law in the terminal set satisfies input constraints. A3 means that the terminal set is an invariant set under the local control law $\boldsymbol{u}=\boldsymbol{H}(\boldsymbol{x})$. A4 means that the value of the terminal cost function decreases under the local control law. Why these four conditions can ensure the closed-loop predictive control system stable is briefly explained as follows.

At time k, assume Problem (6.37) is feasible. The optimal control sequence is solved as $\boldsymbol{U}^*(k)=\{\boldsymbol{u}^*(k\mid k),\ \dots,\ \boldsymbol{u}^*(k+N-1\mid k)\}$, with which the corresponding state trajectory $\boldsymbol{X}^*(k)=\{\boldsymbol{x}^*(k+1\mid k),\cdots,\boldsymbol{x}^*(k+N\mid k)\}$ can be calculated, and the optimal value of the performance index $J_N^*(k)$ can be obtained. It follows that

$$
\begin{aligned}
&\boldsymbol{u}^*(k+i \mid k) \in \boldsymbol{\Omega}_u, \quad i=0, \ldots, N-1 \\
&\boldsymbol{x}^*(k+i \mid k) \in \boldsymbol{\Omega}_x, \quad i=1, \ldots, N \\
&\boldsymbol{x}^*(k+N \mid k) \in \boldsymbol{X}_f \\
&J_N^*(k) = \sum_{i=0}^{N-1} l(\boldsymbol{x}^*(k+i \mid k), \boldsymbol{u}^*(k+i \mid k)) + F(\boldsymbol{x}^*(k+N \mid k))
\end{aligned}
\tag{6.39}
$$

At time $k+1$, construct the control sequence $\boldsymbol{U}(k+1)$, whose first $N-1$ elements are obtained by forward-shifting the elements in $\boldsymbol{U}^*(k)$, i.e. $\boldsymbol{u}(k+i \mid k+1) = \boldsymbol{u}^*(k+i \mid k)$, $i = 1, \ldots, N-1$, and the last one is chosen as $\boldsymbol{u}(k+N \mid k+1) = \boldsymbol{H}(\boldsymbol{x}^*(k+N \mid k))$. That is,

$$
\boldsymbol{U}(k+1) = \{\boldsymbol{u}^*(k+1 \mid k), \ \ldots, \ \boldsymbol{u}^*(k+N-1 \mid k), \ \boldsymbol{H}(\boldsymbol{x}^*(k+N \mid k))\}
$$

The corresponding state trajectory is given by

$$
\boldsymbol{X}(k+1) = \{\boldsymbol{x}^*(k+2 \mid k), \ \ldots, \ \boldsymbol{x}^*(k+N \mid k), \boldsymbol{f}(\boldsymbol{x}^*(k+N \mid k), \ \boldsymbol{H}(\boldsymbol{x}^*(k+N \mid k)))\}
$$

Since Problem (6.37) is feasible at time k, all the constraints in (6.39) are satisfied by the elements of $\boldsymbol{U}^*(k)$ and $\boldsymbol{X}^*(k)$, and combined with the conditions A1 to A3, it can be concluded that $\boldsymbol{U}(k+1)$ and $\boldsymbol{X}(k+1)$ satisfy the constraints of Problem (6.37) at time $k+1$. That is, $\boldsymbol{U}(k+1)$ is a feasible solution at time $k+1$. Since according to the optimal control theory the value of the performance index $J_N(k+1)$ corresponding to any feasible $\boldsymbol{U}(k+1)$ must be larger than or equal to the optimal value $J_N^*(k+1)$, together with the condition A4 it follows that

$$
\begin{aligned}
&J_N^*(k+1) \le J_N(k+1) \\
&= J_N^*(k) - l(\boldsymbol{x}(k \mid k), \ \boldsymbol{u}^*(k \mid k)) - F(\boldsymbol{x}^*(k+N \mid k)) \\
&\quad + F(\boldsymbol{f}(\boldsymbol{x}^*(k+N \mid k), \ \boldsymbol{H}(\boldsymbol{x}^*(k+N \mid k)))) + l(\boldsymbol{x}^*(k+N \mid k), \ \boldsymbol{H}(\boldsymbol{x}^*(k+N \mid k))) \\
&\quad \le J_N^*(k) - l(\boldsymbol{x}(k \mid k), \ \boldsymbol{u}^*(k \mid k))
\end{aligned}
$$

Therefore, if $J_N^*(k)$ is chosen as a Lyapunov function, the value of $J_N^*(k)$ will be descending as long as $\boldsymbol{x}(k)$ and $\boldsymbol{u}^*(k \mid k)$ are not simultaneously zero. This can guarantee that the closed-loop system is asymptotically stable.

Based on above analysis, the monotonicity property of the optimal value function is further discussed, which is closely related to the stability of the closed-loop predictive control system.

(2) The monotonicity property of the optimal value function and the relationship of it to the closed-loop stability

This property was observed and adopted by Chen and Shaw in studying the receding horizon control [16]. For a predictive controller with an optimization horizon N, solve Problem (6.37) starting from state $\boldsymbol{x}$ and denote the optimal value function as $J_N^*(\boldsymbol{x})$, which only depends on $\boldsymbol{x}$. If the state $\boldsymbol{x}$ should be more specialized by time such as $\boldsymbol{x}(k)$ at time k, then $J_N^*(\boldsymbol{x})$ can be accordingly expressed as $J_N^*(\boldsymbol{x}(k))$ and so on.

According to the principle of optimality, at time k, for all $\boldsymbol{x}(k) \in \boldsymbol{X}_N$, it follows that

$$
J_N^*(\boldsymbol{x}(k)) = J_{N-1}^*(\boldsymbol{x}^*(k+1 \mid k)) + l(\boldsymbol{x}(k \mid k), \ \boldsymbol{u}^*(k \mid k))
$$

where $\boldsymbol{X}_N$ is the set of states that can be controlled by predictive control with a fixed horizon N. At time $k+1$, obtain $J_N^*(\boldsymbol{x}(k+1))$ by solving Problem (6.37) with an optimization horizon N starting from $\boldsymbol{x}(k+1) = \boldsymbol{x}^*(k+1 \mid k)$. Then it follows that

$$J_N^*(\boldsymbol{x}(k+1)) - J_N^*(\boldsymbol{x}(k)) = J_N^*(\boldsymbol{x}(k+1)) - J_{N-1}^*(\boldsymbol{x}(k+1)) - l(\boldsymbol{x}(k \mid k),\ \boldsymbol{u}^*(k \mid k))$$

From the above it is obvious that if for all states $\boldsymbol{x} \in \boldsymbol{X}_{N-1}$ the predictive controller can guarantee

$$J_N^*(\boldsymbol{x}) \le J_{N-1}^*(\boldsymbol{x}) \tag{6.40}$$

then it follows that

$$J_N^*(\boldsymbol{x}(k+1)) \le J_N^*(\boldsymbol{x}(k)) - l(\boldsymbol{x}(k \mid k),\ \boldsymbol{u}^*(k \mid k))$$

and the optimal value function $J_N^*(\boldsymbol{x}(k))$ can be chosen as a Lyapunov function to guarantee the closed-loop stability.

Equation (6.40) means that the optimal performance index of predictive control with an optimization horizon N is not larger than that with an optimization horizon $N-1$. This is called the monotonicity property [2]. The approach of synthesizing a predictive control system with guaranteed stability based on (6.40) is different from the direct method [2] based on the above conditions A1 to A4. Although the derivation of the monotonicity condition (6.40) is direct and natural, as an indirect method it does not indicate how condition (6.40) can be satisfied.

Remark 6.7 Intuitively, compared with $J_{N-1}(\boldsymbol{x}(k))$, $J_N(\boldsymbol{x}(k))$ has an additional non-negative term $l(\boldsymbol{x}(k+N-1 \mid k), \boldsymbol{u}(k+N-1 \mid k))$ and thus seems impossible to be smaller than $J_{N-1}(\boldsymbol{x}(k))$ if $\boldsymbol{U}_N$ is composed of the same control input sequence $\boldsymbol{U}_{N-1}$ and an additional $\boldsymbol{u}(k+N-1)$. However, it is worth noting that the inequality (6.40) compares the optimal performance indices corresponding to the optimal control sequences solved by two different optimization problems with the optimization horizons $N-1$ and N, respectively. This means that it is possible to select proper optimization strategies such that the resultant optimal performance indices $J_N^*(\cdot)$ and $J_{N-1}^*(\cdot)$ satisfy the condition (6.40). In fact, this condition is closely related to the direct method. In the following, we show that for Problem (6.37) with the same control strategy assumed above, if the conditions A1 to A4 are satisfied, then condition (6.40) can be derived.

At time k, through solving Problem (6.37), the optimal performance index can be given by

$$J_N^*(k) = \sum_{i=0}^{N-1} l(\boldsymbol{x}^*(k+i \mid k),\ \boldsymbol{u}^*(k+i \mid k)) + F(\boldsymbol{x}^*(k+N \mid k))$$

Also at time k, if the optimization horizon is extended to $N+1$, according to the control strategy for solving Problem (6.37), let

$$\boldsymbol{U}_{N+1}(k) = \{\boldsymbol{u}^*(k \mid k),\ \boldsymbol{u}^*(k+1 \mid k),\ \dots,\ \boldsymbol{u}^*(k+N-1 \mid k),\ \boldsymbol{H}(\boldsymbol{x}^*(k+N \mid k))\}$$

Conditions A1 to A3 ensure that the above solution is feasible for Problem (6.37) with the optimization horizon $N+1$. Denote the corresponding performance index as $J_{N+1}(k)$. With condition A4 we get

$$\begin{aligned} J_{N+1}(k) &= \sum_{i=0}^{N-1} l(\boldsymbol{x}^*(k+i\,|\,k),\ \boldsymbol{u}^*(k+i\,|\,k)) + l(\boldsymbol{x}^*(k+N\,|\,k),\ \boldsymbol{H}(\boldsymbol{x}^*(k+N\,|\,k))) \\ &\quad + F(\boldsymbol{f}(\boldsymbol{x}^*(k+N\,|\,k),\ \boldsymbol{H}(\boldsymbol{x}^*(k+N\,|\,k)))) \\ &= J_N^*(k) - F(\boldsymbol{x}^*(k+N\,|\,k)) + l(\boldsymbol{x}^*(k+N\,|\,k),\ \boldsymbol{H}(\boldsymbol{x}^*(k+N\,|\,k))) \\ &\quad + F(\boldsymbol{f}(\boldsymbol{x}^*(k+N\,|\,k),\ \boldsymbol{H}(\boldsymbol{x}^*(k+N\,|\,k)))) \le J_N^*(k) \end{aligned}$$

Note that $\boldsymbol{U}_{N+1}(k)$ is not the optimal solution. Denote the optimal performance index obtained by solving Problem (6.37) with the optimization horizon $N+1$ at time k as $J_{N+1}^*(k)$. It follows that

$$J_{N+1}^*(k) - J_N^*(k) \le J_{N+1}(k) - J_N^*(k) \le 0$$

Therefore, with the control strategy for solving Problem (6.37) described above, if the conditions A1 to A4 of the direct method are satisfied, the condition of the monotonicity can also be satisfied and the closed-loop stability is ensured.

From the derivation of $J_{N+1}(k)$ it can be seen that starting from the same system state $\boldsymbol{x}(k)$, although the performance index of the optimization problem with the optimization horizon $N+1$ has one positive term more than that with the optimization horizon N, the degree of freedom for control is however increased. Therefore, if the control strategy is properly selected, even though $J_{N+1}(k)$ is not the optimal performance index, it may still be smaller than $J_N^*(k)$.

6.3.2 Suboptimality Analysis of Predictive Control

For the nonlinear system (6.1), with the initial state $\boldsymbol{x}(0)$, consider the following infinite horizon optimal control problem:

$$\begin{aligned} \text{IHC}: \quad & \min_{\boldsymbol{u}(k+i|k)} J_\infty(k) = \sum_{i=0}^{\infty} l(\boldsymbol{x}(k+i\,|\,k),\ \boldsymbol{u}(k+i\,|\,k)) \\ \text{s.t.}\quad & \boldsymbol{x}(k+i+1\,|\,k) = \boldsymbol{f}(\boldsymbol{x}(k+i\,|\,k),\ \boldsymbol{u}(k+i\,|\,k)),\quad i=0,\ 1,\ \ldots \\ & \boldsymbol{u}(k+i\,|\,k) \in \boldsymbol{\Omega}_u,\quad i=0,\ 1,\ \ldots \\ & \boldsymbol{x}(k+i\,|\,k) \in \boldsymbol{\Omega}_x,\quad i=1,\ 2,\ \ldots \end{aligned} \tag{6.41}$$

Assume that this optimization problem can be solved and the corresponding global optimal value of the performance index is denoted by $J_\infty^*(\boldsymbol{x}(0))$. In real applications, predictive control is adopted to implement optimal control and Problem (6.41) is substituted by online solving finite horizon optimization problems in rolling style. At each time k, a finite horizon optimization problem (6.37) starting from $\boldsymbol{x}(k)$ is solved and in the obtained optimal control sequence $\boldsymbol{U}^*(k)$ only the first element, $\boldsymbol{u}^*(k\,|\,k)$, i.e. the current control action, is actually implemented. Thus the performance index actually achieved by predictive control should be

$$J_\infty^P(\boldsymbol{x}(0)) = \sum_{k=0}^{\infty} l(\boldsymbol{x}(k\,|\,k),\ \boldsymbol{u}^*(k\,|\,k)) \tag{6.42}$$

This means that the global performance index achieved by predictive control is composed of the values of the performance index obtained by solving Problem (6.37) at each time and accumulated from $k = 0$ to infinity, where the performance index at each time is calculated by the actual system state $\boldsymbol{x}(k)$ and control input $\boldsymbol{u}^*(k \mid k)$. It is obvious that the global performance index (6.42) achieved by predictive control is in general not the same as the global optimal value $J_\infty^*(\boldsymbol{x}(0))$. To evaluate the suboptimality, assume that the predictive control system satisfies conditions A1 to A4 and is asymptotically stable. Then it follows that

$$J_N^*(k) - J_N^*(k+1) \geq l(\boldsymbol{x}(k \mid k),\ \boldsymbol{u}^*(k \mid k)) \tag{6.43}$$

Accumulating (6.43) from $k = 0$ to $k = \infty$, it follows that

$$J_N^*(0) - J_N^*(\infty) \geq \sum_{k=0}^{\infty} l(\boldsymbol{x}(k \mid k),\ \boldsymbol{u}^*(k \mid k)) = J_\infty^P(\boldsymbol{x}(0))$$

Since conditions A1 to A4 are satisfied, the closed-loop system is asymptotically stable, i.e. $J_N^*(\infty) = 0$. Considering that $J_\infty^*(\boldsymbol{x}(0))$ is the global optimal value, we can obtain

$$J_\infty^*(\boldsymbol{x}(0)) \leq J_\infty^P(\boldsymbol{x}(0)) \leq J_N^*(0) \tag{6.44}$$

This equation indicates that the global performance achieved by the predictive controller with finite horizon optimization is not optimal and its upper bound $J_N^*(0)$ is the optimal performance index given by solving the finite horizon optimization problem (6.37) at time $k = 0$.

By applying (6.43) step by step, we further obtain

$$J_N^*(0) \geq J_N^*(1) + l(\boldsymbol{x}(0 \mid 0),\ \boldsymbol{u}^*(0 \mid 0)) \geq \cdots \geq J_N^*(k) + \sum_{i=0}^{N-1} l(\boldsymbol{x}(k \mid k),\ \boldsymbol{u}^*(k \mid k))$$

$$\geq \cdots \geq \sum_{k=0}^{\infty} l(\boldsymbol{x}(k \mid k),\ \boldsymbol{u}^*(k \mid k)) = J_\infty^P(\boldsymbol{x}(0))$$

which yield the following conclusions on the suboptimality of the global performance of predictive control systems:

1) In predictive control, the finite horizon optimization problem (6.37) approximates the infinite horizon optimal control problem through introducing a terminal set constraint and terminal cost function. However, adopting the fixed local feedback control law after the prediction horizon restricts the degree of freedom of the infinite horizon optimization and will result in suboptimality of the control performance.
2) With the optimization horizon moving forward, resolving the optimization problem (6.37) implies that a new control degree of freedom is released compared with the problem solved at the previous time. Therefore the global control performance can be improved over time.
3) At each time k, the suboptimality of the global control performance of predictive control systems can be evaluated by summing the two parts: the actual value of the performance index caused by previous control and the optimal value of the performance index solved by the current finite horizon optimization problem (6.37). The upper bound of the suboptimality of the global control performance decreases over time.

6.4 Summary

In this chapter, the basic ideas used in the research of the qualitative synthesis theory of predictive control are introduced with emphasis on synthesizing stable predictive controllers, aiming at providing the readers with fundamental knowledge for theoretical research on predictive control.

The maturity of predictive control theory is marked by the qualitative synthesis theory developed since the 1990s with rapidly growing theoretical results on synthesizing stable and robust predictive controllers. Different from the classical quantitative analysis theory as described in Chapter 4, this kind of research puts predictive control in the framework of optimal control. It is not limited to analyzing existing predictive control algorithms, but synthesizes various new algorithms in accordance with the specific characters of predictive control from the point of view of a performance guarantee. These studies open up a new research system in predictive control and result in many new theoretical results with profound ideas and innovative methods.

The qualitative synthesis theory focuses on how to synthesize predictive control algorithms with guaranteed stability with respect to the rolling style of optimization in predictive control. In this chapter, the basic ideas of qualitative synthesis theory and the relationship of predictive control with optimal control are firstly introduced. Taking the classical infinite horizon optimal control as reference, the main difficulties of the stability guarantee for predictive control are analyzed, and how these difficulties can be solved are described in detail. Some representative methods for synthesizing stable predictive controllers are introduced for illustration. Based on these synthesis methods, three ingredients, i.e. terminal cost function, terminal constraint set, and local stabilizing controller, and the general conditions for synthesizing stable predictive controllers are introduced. The suboptimality of predictive control is analyzed. It is also indicated that the suboptimality of the global control performance of predictive control would be continuously improved over time.

References

1 Xi, Y. and Li, D. (2008). Fundamental philosophy and status of qualitative synthesis of model predictive control. *Acta Automatica Sinica* 34 (10): 1225–1234. (in Chinese).

2 Mayne, D.Q., Rawling, J.B., Rao, C.V. et al. (2000). Constrained model predictive control: stability and optimality. *Automatica* 36 (6): 789–814.

3 Bellman, R. (1957). *Dynamic Programming*. Princeton: Princeton University Press.

4 Lee, E.B. and Markus, L. (1967). *Foundations of Optimal Control Theory*. New York: Wiley.

5 Kalman, R.E. (1960). Contributions to the theory of optimal control. *Boletin Sociedad Matematica Mexicana* 5: 102–119.

6 Kwon, W.H. and Pearson, A.E. (1978). On feedback stabilization of time-varying discrete linear systems. *IEEE Transactions on Automatic Control* 23 (3): 479–481.

7 Keerthi, S.S. and Gilbert, E.G. (1988). Optimal infinite-horizon feedback laws for a general class of constrained discrete-time systems: stability and moving-horizon approximations. *Journal of Optimization Theory and Applications* 57 (2): 265–293.

8 Scokaert, P.O.M. and Clarke, D.W. (1994). Stabilising properties of constrained predictive control. *IEE Proceedings D (Control Theory and Applications)* 141 (5): 295–304.

9 Blanchini, F. (1999). Set invariance in control. *Automatica* 35 (11): 1747–1767.

10 Boyd, S., Ghaoui, L.E., Feron, E. et al. (1994). *Linear Matrix Inequalities in System and Control Theory*. Philadelphia: Society for Industrial and Applied Mathematics.

11 Kothare, M.V., Balakrishnan, V., and Morari, M. (1996). Robust constrained model predictive control using linear matrix inequalities. *Automatica* 32 (10): 1361–1379.

12 Zheng, A. (1997). Stability of model predictive control with time-varying weights. *Computers and Chemical Engineering* 21 (12): 1389–1393.

13 Lee, J.W., Kwon, W.H., and Choi, J.H. (1998). On stability of constrained receding horizon control with finite terminal weighting matrix. *Automatica* 34 (12): 1607–1612.

14 De Nicolao, J., Magni, L., and Scattolini, R. (1998). Stabilizing receding horizon control of nonlinear time-varying systems. *IEEE Transactions on Automatic Control* 43 (7): 1030–1036.

15 Michalska, H. and Mayne, D.Q. (1993). Robust receding horizon control of constrained nonlinear systems. *IEEE Transactions on Automatic Control* 38 (11): 1623–1633.

16 Chen, C.C. and Shaw, L. (1982). On receding horizon feedback control. *Automatica* 18 (3): 349–352.

7

Synthesis of Robust Model Predictive Control

The methods for synthesizing stable predictive controllers introduced in the last chapter are with respect to systems with an accurate model and without disturbance. However, in real applications, model uncertainty or unknown disturbance always exists, which may affect the control performance of the predictive control system, and may even make the closed-loop system unstable. Over the last two decades, how to design a predictive controller with guaranteed stability for uncertain systems has become an important topic of the qualitative synthesis theory of predictive control. Many novel methods were proposed and important results have been obtained. In this chapter, we will introduce the basic philosophy of synthesizing robust model predictive controllers (RMPCs) for systems with some typical categories of uncertainties. According to most of the literature, the uncertain systems discussed here are classified into two main categories: systems subjected to model uncertainty, particularly with polytopic type uncertainties, and systems subjected to external disturbance.

7.1 Robust Predictive Control for Systems with Polytopic Uncertainties

7.1.1 Synthesis of RMPC Based on Ellipsoidal Invariant Sets

Many uncertain systems can be described by the following linear time varying (LTV) system with polytopic uncertainty:

$$\begin{aligned} &\boldsymbol{x}(k+1)=\boldsymbol{A}(k)\boldsymbol{x}(k)+\boldsymbol{B}(k)\boldsymbol{u}(k)\\ &\boldsymbol{y}(k)=\boldsymbol{C}\boldsymbol{x}(k)\\ &[\boldsymbol{A}(k)\ \ \boldsymbol{B}(k)]\in\boldsymbol{\Pi} \end{aligned} \tag{7.1}$$

where $\boldsymbol{A}(k)\in \mathrm{R}^{n\times n}$, $\boldsymbol{B}(k)\in \mathrm{R}^{n\times m}$, and

$$\boldsymbol{\Pi}=\mathrm{Co}\{[\boldsymbol{A}_1\ \ \boldsymbol{B}_1],\ \ldots,\ [\boldsymbol{A}_L\ \ \boldsymbol{B}_L]\} \tag{7.2}$$

Predictive Control: Fundamentals and Developments, First Edition. Yugeng Xi and Dewei Li.

where Co means a convex hull, i.e. for any time-varying $\boldsymbol{A}(k)$ and $\boldsymbol{B}(k)$, $[\boldsymbol{A}(k)\quad \boldsymbol{B}(k)] \in \boldsymbol{\Pi}$ means that there exists a group of non-negative parameters $\lambda_l(k)$, $l = 1, \ldots, L$ satisfying

$$[\boldsymbol{A}(k)\ \ \boldsymbol{B}(k)] = \sum_{l=1}^{L} \lambda_l(k)[\boldsymbol{A}_l\ \ \boldsymbol{B}_l], \quad \sum_{l=1}^{L} \lambda_l(k) = 1,\ \ \lambda_l(k) \ge 0,\ \ l = 1, \ldots, L \tag{7.3}$$

The uncertainty described by (7.2) and (7.3) is called polytopic uncertainty, which can be formulated by a convex hull with L given vertices $[\boldsymbol{A}_l\quad \boldsymbol{B}_l]$. The real system parameters $[\boldsymbol{A}(k)\quad \boldsymbol{B}(k)]$ will always be in the convex hull and λ_l are the convex combination parameters, either time-varying or time-invariant.

The system model with polytopic uncertainty can describe both nonlinear systems which are linearized at several equilibrium points and some linear systems with time-varying parameters. For example, the linear parameter varying (LPV) system can be formulated by (7.2) and (7.3). Therefore, much RMPC literature focuses on this kind of uncertain system.

In addition, consider the constraints on system inputs and outputs:

$$\|\boldsymbol{u}(k)\|_2 \le u_{\max},\ \ k = 0,\ 1, \ldots \tag{7.4}$$

$$\|\boldsymbol{y}(k)\|_2 \le y_{\max},\ \ k = 0,\ 1, \ldots \tag{7.5}$$

Note that in order to highleight the basic idea of RMPC synthesis, (7.4) and (7.5) are taken as examples of system constraints. For other kinds of constraint, such as the peak bounds on individual components of system input or output, as discussed in Section 6.1.4, the results derived here only need to be adjusted accordingly.

For this kind of uncertain system described by (7.1) to (7.3), and with input and output constraints (7.4) and (7.5), referring to the linear robust control theory, the following infinite horizon "min-max" problem is proposed at time k:

$$\min_{\boldsymbol{u}(k+i|k), i\ge 0}\ \ \max_{[\boldsymbol{A}(k+i)\ \boldsymbol{B}(k+i)] \subset \boldsymbol{\Pi}, i \ge 0} J_\infty(k)$$

$$\begin{aligned}
& J_\infty(k) = \sum_{i=0}^{\infty} \left[\boldsymbol{x}^{\mathrm{T}}(k+i\,|\,k)\mathcal{Q}\boldsymbol{x}(k+i\,|\,k) + \boldsymbol{u}^{\mathrm{T}}(k+i\,|\,k)\mathcal{R}\boldsymbol{u}(k+i\,|\,k)\right] \\
\text{s.t.}\ \ & \boldsymbol{x}(k+i+1\,|\,k) = \boldsymbol{A}(k+i)\boldsymbol{x}(k+i\,|\,k) + \boldsymbol{B}(k+i)\boldsymbol{u}(k+i\,|\,k),\ \ k,\ i \ge 0 \\
& \boldsymbol{y}(k+i\,|\,k) = \boldsymbol{C}\boldsymbol{x}(k+i\,|\,k),\ \ k,\ i \ge 0 \\
& \|\boldsymbol{u}(k+i\,|\,k)\|_2 \le u_{\max},\ \ k,\ i \ge 0 \\
& \|\boldsymbol{y}(k+i\,|\,k)\|_2 \le y_{\max},\ \ k \ge 0,\ \ i \ge 1 \\
& \boldsymbol{x}(k\,|\,k) = \boldsymbol{x}(k)
\end{aligned} \tag{7.6}$$

Different from the optimization problem in previous chapters, the min-max problem (7.6) results from the uncertainty of the system model. This means that the RMPC will optimize the future performance index with an infinite horizon so as to achieve an optimal solution for the worst case in all possible model variations in $\boldsymbol{\Pi}$.

The min-max problem (7.6) is a reasonable formulation of RMPC for an uncertain system (7.1) to (7.5), but is difficult to solve. Firstly, the objective of the minimization in (7.6) results from the maximization procedure and is heavily affected by system uncertainty, while the objective of the usual optimization method is explicitly given by analytical functions or variables. Secondly, the infinite terms in the objective function make the optimization computationally intractable. A specific strategy should be adopted to make the optimization problem solvable. In the following, referring to [1], we introduce the synthesis method of RMPC for this kind of polytopic uncertain system, explaining how to use an effective strategy to overcome these difficulties and how to implement it using LMIs and ellipsoidal invariant sets.

(1) Strategy and key ideas

1) Introducing an upper bound for the maximization problem

Consider the quadratic function $V(i, k) = \boldsymbol{x}(k+i \mid k)^{\mathrm{T}} \boldsymbol{P} \boldsymbol{x}(k+i \mid k)$, where $\boldsymbol{P}$ is a symmetric positive definite matrix. At time k, for all $\boldsymbol{x}(k+i \mid k)$ and $\boldsymbol{u}(k+i \mid k)$ satisfying (7.1) to (7.3), impose the following condition on the corresponding function V:

$$V(i+1, k) - V(i, k) \le -[\boldsymbol{x}^{\mathrm{T}}(k+i \mid k)\mathcal{Q}\boldsymbol{x}(k+i \mid k) + \boldsymbol{u}^{\mathrm{T}}(k+i \mid k)\mathcal{R}\boldsymbol{u}(k+i \mid k)]$$
$$\forall [\boldsymbol{A}(k+i) \;\; \boldsymbol{B}(k+i)] \in \boldsymbol{\Pi}, \;\; i \ge 0 \tag{7.7}$$

This condition implies that the closed-loop system is stable. Therefore $\boldsymbol{x}(\infty \mid k) = 0$ and thus $V(\infty, k) = 0$. Summing (7.7) from $i = 0$ to $i = \infty$, it follows that

$$-V(0, k) \le -J_{\infty}(k)$$

Then define an upper bound $\gamma > 0$ on the robust performance index

$$\max_{[\boldsymbol{A}(k+i) \;\; \boldsymbol{B}(k+i)] \subset \boldsymbol{\Pi}, \, i \ge 0} J_{\infty}(k) \le V(0, k) \le \gamma \tag{7.8}$$

In this way the min-max problem (7.6) can be converted into a minimization problem with a simple linear objective γ.

2) Adopting a single feedback control law for the infinite horizon

In order to handle the infinite number of control actions in problem (7.6), introduce the ellipsoid controlled invariant set

$$\boldsymbol{\Omega} = \{\boldsymbol{x} \in \mathrm{R}^n \mid \boldsymbol{x}^{\mathrm{T}} \boldsymbol{Q}^{-1} \boldsymbol{x} \le 1\}, \;\; \boldsymbol{Q} > \boldsymbol{0} \tag{7.9}$$

with a corresponding state feedback control law

$$\boldsymbol{u}(k+i \mid k) = \boldsymbol{F}\boldsymbol{x}(k+i \mid k), \;\; i \ge 0 \tag{7.10}$$

Control invariance means that if the system state $\boldsymbol{x}(k \mid k) = \boldsymbol{x}(k) \in \boldsymbol{\Omega}$, then under the control law (7.10), $\boldsymbol{x}(k+1 \mid k) \in \boldsymbol{\Omega}$. This can be ensured by

$$[(\boldsymbol{A}(k) + \boldsymbol{B}(k)\boldsymbol{F})\boldsymbol{x}(k)]^{\mathrm{T}} \boldsymbol{Q}^{-1} [(\boldsymbol{A}(k) + \boldsymbol{B}(k)\boldsymbol{F})\boldsymbol{x}(k)] \le \boldsymbol{x}^{\mathrm{T}}(k) \boldsymbol{Q}^{-1} \boldsymbol{x}(k)$$

and furthermore

$$(\boldsymbol{A}(k) + \boldsymbol{B}(k)\boldsymbol{F})^{\mathrm{T}} \boldsymbol{Q}^{-1} (\boldsymbol{A}(k) + \boldsymbol{B}(k)\boldsymbol{F}) \le \boldsymbol{Q}^{-1} \tag{7.11}$$

Use of the above state feedback control law and the corresponding ellipsoid controlled invariant set make it possible to parametrize the control actions to a tractable form as well as to keep the system state within the invariant set.

(2) Constraints formulated in LMIs

Using the above key ideas, the min-max problem (7.6) can be converted into an optimization problem of minimizing the upper bound γ by using the state feedback control law (7.10) with the following constraints: the upper bound constraint (7.8), controlled invariant set conditions (7.9) and (7.11), and input/output constraints (7.4) and (7.5), where γ, $\boldsymbol{P}$, $\boldsymbol{Q}$, $\boldsymbol{F}$ should be determined. Note that the quadratic form of $\boldsymbol{x}^{\mathrm{T}}\boldsymbol{P}\boldsymbol{x}$ in the upper bound constraint (7.8) is in accordance with the ellipsoid set representation in (7.9). Use $\boldsymbol{P} = \gamma\boldsymbol{Q}^{-1}$ to reduce the complexity. Then, with the help of the Schur Complement of LMIs (6.11) and the properties of the convex hull (7.3), the details of synthesizing the RMPC can be given as follows.

1) The upper bound of the maximum of the objective function (7.8). From (7.8) it follows that

$$\max_{[\boldsymbol{A}(k+i)\ \boldsymbol{B}(k+i)]\subset\boldsymbol{\Pi},\, i\geq 0} J_\infty(k) \leq V(0,\ k) = \boldsymbol{x}^{\mathrm{T}}(k\,|\,k)\boldsymbol{P}\boldsymbol{x}(k\,|\,k) \leq \gamma$$

With $\boldsymbol{P} = \gamma\boldsymbol{Q}^{-1}$ and $\boldsymbol{x}(k\,|\,k) = \boldsymbol{x}(k)$ the above inequality is exactly the invariant set condition (7.9) and can be rewritten in LMI form as

$$\begin{bmatrix} 1 & \boldsymbol{x}^{\mathrm{T}}(k) \\ \boldsymbol{x}(k) & \boldsymbol{Q} \end{bmatrix} \geq \boldsymbol{0} \tag{7.12}$$

2) Imposing $V(i, k)$ descending and satisfying (7.7). With $V(i, k) = \boldsymbol{x}(k+i\,|\,k)^{\mathrm{T}}\boldsymbol{P}\boldsymbol{x}(k+i\,|\,k)$ and (7.1), (7.10), the condition (7.7) can be expressed as

$$\forall[\boldsymbol{A}(k+i)\quad \boldsymbol{B}(k+i)] \in \boldsymbol{\Pi},\quad i\geq 0$$

$$\boldsymbol{x}^{\mathrm{T}}(k+i\,|\,k)\left[(\boldsymbol{A}(k+i)+\boldsymbol{B}(k+i)\boldsymbol{F})^{\mathrm{T}}\boldsymbol{P}(\boldsymbol{A}(k+i)+\boldsymbol{B}(k+i)\boldsymbol{F})-\boldsymbol{P}\right]\boldsymbol{x}(k+i\,|\,k)$$
$$\leq -\left[\boldsymbol{x}^{\mathrm{T}}(k+i\,|\,k)\left(\mathcal{Q}+\boldsymbol{F}^{T}\mathcal{R}\boldsymbol{F}\right)\boldsymbol{x}(k+i\,|\,k)\right]$$

This inequality can be ensured by the following matrix inequality:

$$(\boldsymbol{A}(k+i)+\boldsymbol{B}(k+i)\boldsymbol{F})^{\mathrm{T}}\boldsymbol{P}(\boldsymbol{A}(k+i)+\boldsymbol{B}(k+i)\boldsymbol{F})-\boldsymbol{P} \leq -\left(\mathcal{Q}+\boldsymbol{F}^{\mathrm{T}}\mathcal{R}\boldsymbol{F}\right) \tag{7.13}$$

With $\boldsymbol{P} = \gamma\boldsymbol{Q}^{-1}$, left- and right-multiplying $\boldsymbol{Q}$ on both sides will yield

$$\gamma\left[\boldsymbol{Q}(\boldsymbol{A}(k+i)+\boldsymbol{B}(k+i)\boldsymbol{F})^{\mathrm{T}}\boldsymbol{Q}^{-1}(\boldsymbol{A}(k+i)+\boldsymbol{B}(k+i)\boldsymbol{F})\boldsymbol{Q}-\boldsymbol{Q}\right] \leq -\left(\boldsymbol{Q}\mathcal{Q}\boldsymbol{Q}+\boldsymbol{Q}\boldsymbol{F}^{\mathrm{T}}\mathcal{R}\boldsymbol{F}\boldsymbol{Q}\right)$$

Let $\boldsymbol{Y} = \boldsymbol{F}\boldsymbol{Q}$ and rewrite the above matrix inequality as

$$\boldsymbol{Q}-(\boldsymbol{A}(k+i)\boldsymbol{Q}+\boldsymbol{B}(k+i)\boldsymbol{Y})^{\mathrm{T}}\boldsymbol{Q}^{-1}(\boldsymbol{A}(k+i)\boldsymbol{Q}+\boldsymbol{B}(k+i)\boldsymbol{Y})-\gamma^{-1}\left(\boldsymbol{Q}\mathcal{Q}\boldsymbol{Q}+\boldsymbol{Y}^{\mathrm{T}}\mathcal{R}\boldsymbol{Y}\right)\geq\boldsymbol{0}$$

According to (6.11d) and (6.11e), this condition can be expressed in the following LMI form:

$$\begin{bmatrix} \boldsymbol{Q} & * & * & * \\ \boldsymbol{A}(k+i)\boldsymbol{Q}+\boldsymbol{B}(k+i)\boldsymbol{Y} & \boldsymbol{Q} & * & * \\ \mathcal{Q}^{1/2}\boldsymbol{Q} & \boldsymbol{0} & \gamma\boldsymbol{I} & * \\ \mathcal{R}^{1/2}\boldsymbol{Y} & \boldsymbol{0} & \boldsymbol{0} & \gamma\boldsymbol{I} \end{bmatrix} \geq \boldsymbol{0}$$

where the symbol $*$ is used to induce a symmetric structure of the matrix for simplicity.

Since the uncertain parameters $\boldsymbol{A}(k+i)$ and $\boldsymbol{B}(k+i)$ satisfy the polytopic condition (7.3) and can be represented by a linear combination of parameters of the vertices $[\boldsymbol{A}_l \quad \boldsymbol{B}_l]$ in the polytopic model, the above condition is satisfied if the following group of LMIs holds:

$$\begin{bmatrix} \boldsymbol{Q} & * & * & * \\ \boldsymbol{A}_l\boldsymbol{Q}+\boldsymbol{B}_l\boldsymbol{Y} & \boldsymbol{Q} & * & * \\ \mathcal{Q}^{1/2}\boldsymbol{Q} & \boldsymbol{0} & \gamma\boldsymbol{I} & * \\ \mathcal{R}^{1/2}\boldsymbol{Y} & \boldsymbol{0} & \boldsymbol{0} & \gamma\boldsymbol{I} \end{bmatrix} \geq \boldsymbol{0}, \quad l=1,\ \dots,\ L \tag{7.14}$$

3) $\boldsymbol{x}(k)$ in the ellipsoid set $\boldsymbol{\Omega} = \{\boldsymbol{x} \in \mathrm{R}^n \mid \boldsymbol{x}^{\mathrm{T}}\boldsymbol{Q}^{-1}\boldsymbol{x} \leq 1\}$. This is also described by (7.12) because the condition $\boldsymbol{x}^{\mathrm{T}}(k)\boldsymbol{Q}^{-1}\boldsymbol{x}(k) \leq 1$ is identical to $\boldsymbol{x}^{\mathrm{T}}(k)\boldsymbol{P}\boldsymbol{x}(k) \leq \gamma$ under the selection $\boldsymbol{P} = \gamma\boldsymbol{Q}^{-1}$ and $\gamma > 0$.
4) $\boldsymbol{\Omega}$ is an invariant set of system (7.1) under the feedback control law (7.10), i.e. (7.11):

$$(\boldsymbol{A}(k)+\boldsymbol{B}(k)\boldsymbol{F})^{\mathrm{T}}\boldsymbol{Q}^{-1}(\boldsymbol{A}(k)+\boldsymbol{B}(k)\boldsymbol{F}) \leq \boldsymbol{Q}^{-1}$$

It is obviously covered by the $V(i, k)$ descending condition (7.13) with $\boldsymbol{P} = \gamma\boldsymbol{Q}^{-1}$ and $\gamma > 0$. Thus the corresponding LMI has already been included in (7.14).
5) Constraints on control inputs (7.4). Referring to (6.13) in Section 6.1.4, it can be given by

$$\begin{bmatrix} u_{\max}^2\boldsymbol{I} & \boldsymbol{Y} \\ \boldsymbol{Y}^{\mathrm{T}} & \boldsymbol{Q} \end{bmatrix} \geq \boldsymbol{0} \tag{7.15}$$

6) Constraints on system outputs (7.5). According to (7.5) and referring to (6.15) in Section 6.1.4, it follows that

$$\begin{bmatrix} \boldsymbol{Q} & (\boldsymbol{A}_l\boldsymbol{Q}+\boldsymbol{B}_l\boldsymbol{Y})^{\mathrm{T}}\boldsymbol{C}^{\mathrm{T}} \\ \boldsymbol{C}(\boldsymbol{A}_l\boldsymbol{Q}+\boldsymbol{B}_l\boldsymbol{Y}) & y_{\max}^2\boldsymbol{I} \end{bmatrix} \geq \boldsymbol{0}, \quad l=1,\ \dots,\ L \tag{7.16}$$

(3) RMPC algorithm with robust stability

Summing the above derivations, at time k, solving the min-max problem (7.6) can be converted into solving the following optimization problem:

$$\begin{aligned} &\min_{\gamma,\boldsymbol{Q},\boldsymbol{Y}} \ \gamma \\ &\text{s.t.} \ \ (7.12),\ (7.14),\ (7.15),\ (7.16) \end{aligned} \tag{7.17}$$

After $\boldsymbol{Q}$ and $\boldsymbol{Y}$ are solved, we can obtain $\boldsymbol{F} = \boldsymbol{Y}\boldsymbol{Q}^{-1}$ and $\boldsymbol{u}(k) = \boldsymbol{F}\boldsymbol{x}(k)$, i.e. the current control action at time k. Note that $\{\gamma, \boldsymbol{Q}, \boldsymbol{Y}\}$ are solved at time k, and strictly they should be denoted by $\{\gamma(k), \boldsymbol{Q}(k), \boldsymbol{Y}(k)\}$.

Now the RMPC algorithm based on ellipsoidal invariant set can be given as folllows.

Algorithm 7.1 RMPC Algorithm Based on the Ellipsoidal Invariant Set [1]

Step 1. Solve the optimization problem (7.17) according to the current system state $\boldsymbol{x}(k)$. Obtain the optimal γ, $\boldsymbol{Q}$, $\boldsymbol{Y}$.

Step 2. Calculate $\boldsymbol{F} = \boldsymbol{Y}\boldsymbol{Q}^{-1}$ according to the optimization solution at Step 1 and then obtain $\boldsymbol{u}(k) = \boldsymbol{F}\boldsymbol{x}(k)$.

Step 3. Implement $\boldsymbol{u}(k)$ on the controlled plant and then return to Step 1 the next time.

For Algorithm 7.1, the following theorem on the stability of the closed-loop system can be given.

Theorem 7.1 For the system (7.1) to (7.5) controlled by RMPC that online solves the optimization problem (7.17), if the problem (7.17) is feasible for the current state $\boldsymbol{x}(k)$ at time k, then the proposed RMPC is recursively feasible and the closed-loop system is asymptotically stable.

Proof: At time k, starting from $\boldsymbol{x}(k)$, the optimization problem (7.17) is feasible. Denote the optimal solution as $\{\gamma^*(k), \boldsymbol{Q}^*(k), \boldsymbol{Y}^*(k)\}$. Then the optimal control at time k is given by $\boldsymbol{u}^*(k) = \boldsymbol{F}^*(k)\boldsymbol{x}(k)$ with $\boldsymbol{F}^*(k) = \boldsymbol{Y}^*(k)(\boldsymbol{Q}^*(k))^{-1}$ and $\gamma^*(k) \geq V(0, k)$ according to (7.8).

Using (7.7), $V(1, k) < V(0, k)$ for all $[\boldsymbol{A}(k) \quad \boldsymbol{B}(k)] \in \boldsymbol{\Pi}$ if $\boldsymbol{x}(k) \neq \boldsymbol{0}$. At time $k+1$, get the actual $\boldsymbol{x}(k+1)$ through measurement, which is indeed one of the possible $\boldsymbol{x}(k+1 \mid k)$ for a specific $[\boldsymbol{A}(k) \quad \boldsymbol{B}(k)] \in \boldsymbol{\Pi}$, which satisfies the above inequality, i.e. $V(\boldsymbol{x}(k+1), k) \triangleq \boldsymbol{x}(k+1)^{\mathrm{T}}\boldsymbol{P}^*(k)\boldsymbol{x}(k+1) < V(0, k)$. Note that $V(\boldsymbol{x}(k+1), k)$ is fixed for definite $\boldsymbol{x}(k+1)$, so we can define a constant $a = V(\boldsymbol{x}(k+1), k)/\gamma^*(k) < 1$ if $\boldsymbol{x}(k) \neq \boldsymbol{0}$.

For the optimization problem (7.17) at time $k+1$, let $\{a\gamma^*(k), a\boldsymbol{Q}^*(k), a\boldsymbol{Y}^*(k)\}$ be a candidate solution. It satisfies the condition (7.14) if both sides of the inequalities are multiplied by a. It is also easy to verify that the constructed solution satisfies conditions (7.15) and (7.16). Meanwhile, we have $\boldsymbol{x}^T(k+1)(a\boldsymbol{Q}^*(k))^{-1}\boldsymbol{x}(k+1) = a^{-1}(\gamma^*(k))^{-1}\boldsymbol{x}^T(k+1)\boldsymbol{P}^*(k)\boldsymbol{x}(k+1) = 1$ which implies that the condition (7.12) is also satisfied. Hence, the problem (7.17) is feasible at time $k+1$. This argument can be continued for times $k+2, k+3, \ldots$, so it can be concluded that the proposed RMPC is recursively feasible.

In addition, with the above constructed solution, the value of the objective function at time $k+1$ is $a\gamma^*(k)$, which is smaller than $\gamma^*(k)$ due to $a < 1$. Denote $\gamma^*(k+1)$ as the optimal value of the objective function of Problem (7.17) at time $k+1$; then $\gamma^*(k+1) \leq a\gamma^*(k) < \gamma^*(k)$. Thus the optimal value function $\gamma^*(k)$ is strictly decreasing if $\boldsymbol{x}(k) \neq \boldsymbol{0}$ and can be chosen as a Lyapunov function. The closed-loop system is asymptotically stable.

Remark 7.1 The strictly decreasing condition (7.7) of $V(i, k)$ implies the contractive property of the domain of system states. It is imposed to solve the optimization problem at time k, where $V(i, k)$ is defined for all predicted future system states controlled by the optimal control law at time k. However, the strictly decreasing property of $V(i, k)$ does not mean that it can be chosen as a global Lyapumov function for guaranteed stability. Due to the rolling style of RMPC, at time $k+1$, the condition (7.7) will be in the form of $V(i, k+1)$ rather than $V(i, k)$. Accordingly $V(i, k+1)$ is defined for system states controlled by the optimal control law at time $k+1$. This means that $V(i, k)$ is only a local Lyapumov function for the system dynamic evolution under control actions determined at time k, and the condition (7.7) of $V(i, k)$ is actually only valid at time k, such as $V(1, k) < V(0, k)$. Nevertheless, we can make full use of this local contraction information by introducing $a = V(\boldsymbol{x}(k+1), k))/\gamma^*(k) < 1$ as the actual contraction ratio of the system state domain from time k to $k+1$, and contracting the initial invariant set for the optimization problem at time $k+1$ accordingly. Since such a contraction constant $a < 1$ exists at each control period, it is possible to step by step bring this local contraction property into the global process to steer the system state to the origin. This idea motivates construction of a contractive candidate solution $\{a\gamma^*(k), a\boldsymbol{Q}^*(k), a\boldsymbol{Y}^*(k)\}$ in the proof of Theorem 7.1. It is an appropriate bridge to link the optimal value functions at two successive times. On the one hand, it provides a better control performance $a\gamma^*(k)$ directly comparable with the optimal $\gamma^*(k)$ at time k. On the other hand, note that at time $k+1$ the system state $\boldsymbol{x}(k+1)$ has been steered to a contractive set $\bar{\boldsymbol{\Omega}} = \{\boldsymbol{x} \in \mathrm{R}^n \mid \boldsymbol{x}^{\mathrm{T}}(\boldsymbol{Q}^*(k))^{-1}\boldsymbol{x} \le a,\ a < 1\}$, which is included in $\boldsymbol{\Omega}$, and the same feedback gain $\boldsymbol{F}(k+1) = a\boldsymbol{Y}^*(k)(a\boldsymbol{Q}^*(k))^{-1} = \boldsymbol{F}^*(k)$ is adopted, thus making the feasibility of this candidate solution at time $k+1$ easy to prove. In this way the asymptotical stability of the closed-loop system can be proved.

The above proposed RMPC should solve the optimization problem (7.17) online by using LMIs toolboxes. The online computational burden is determined by the number of optimization variables and the number of rows of LMIs. It is easy to see that both of these two numbers will increase very quickly with the increase of the system dimensions and the number of vertices.

7.1.2 Improved RMPC with Parameter-Dependent Lyapunov Functions

During the RMPC design in the last section, each time when solving the optimization problem (7.6) a Lyapunov function $V(\boldsymbol{x}) = \boldsymbol{x}^{\mathrm{T}}\boldsymbol{P}\boldsymbol{x}$ with fixed $\boldsymbol{P}$ is adopted and the strictly descending condition of $V(\boldsymbol{x})$ is imposed for all admissible model parameters. The use of this unique Lyapunov function for all possible models with parameters in the convex hull makes the problem solving easy but is conservative. To improve the control performance, Cuzzola et al. [2] and Mao [3] proposed that multiple Lyapunov functions could be used to design RMPC. When solving the optimization problem (7.6) at a fixed time, Lyapunov functions are defined for all the models corresponding to the vertices of the polytope and then a Lyapunov function is constructed by weighting these Lyapunov functions with the model combination coefficients λ_l. This kind of Lyapunov function is called a parameter-dependent Lyapunov function. In the following, the main idea of synthesizing RMPC based on parameter-dependent Lyapunov functions proposed in [2] and [3] is briefly introduced.

Consider the LTV system with polytopic uncertainty and constraints, (7.1) to (7.5). To solve the optimization problem (7.6) at time k, instead of the single Lyapunov function $V(\boldsymbol{x}) = \boldsymbol{x}^{\mathrm{T}}\boldsymbol{P}\boldsymbol{x}$ with fixed $\boldsymbol{P}$, as presented in Section 7.1.1, firstly define the Lyapunov functions

$$V_l(i,k) = \boldsymbol{x}^{\mathrm{T}}(k+i \mid k)\boldsymbol{P}_l\boldsymbol{x}(k+i \mid k), \quad l = 1, \ldots, L$$

for each linear time-invariant system

$$\boldsymbol{x}(k+1) = \boldsymbol{A}_l\boldsymbol{x}(k) + \boldsymbol{B}_l\boldsymbol{u}(k), \quad l = 1, \ldots, L$$

where $[\boldsymbol{A}_l \quad \boldsymbol{B}_l]$ represents a vertex of the polytope $\boldsymbol{\Pi}$. Then construct the following Lyapunov function:

$$V(i,k) = \sum_{l=1}^{L} \lambda_l(k+i)V_l(i,k) = \boldsymbol{x}^{\mathrm{T}}(k+i \mid k)\boldsymbol{P}(i,k)\boldsymbol{x}(k+i \mid k) \tag{7.18}$$

with

$$\boldsymbol{P}(i,k)=\sum_{l=1}^{L}\lambda_l(k+i)\boldsymbol{P}_l \tag{7.19}$$

where $\lambda_l(k+i)$ for $k\geq 0$, $i\geq 0$ represents the combination coefficient of $[\boldsymbol{A}_l \quad \boldsymbol{B}_l]$ in $[\boldsymbol{A}(k+i)\ \boldsymbol{B}(k+i)]$ (see (7.3)). It is obvious that $\boldsymbol{P}(i,k)$ is time-varying along with the combination coefficients and can handle uncertainty more finely to reduce conservatism.

Along the same way as presented in Section 7.1.1, for all $\boldsymbol{x}(k+i\mid k)$ and $\boldsymbol{u}(k+i\mid k)$ satisfying (7.1) to (7.3), impose the following condition on the corresponding function $V(i,k)$:

$$\begin{aligned}V(i+1,k)-V(i,k)<-[\boldsymbol{x}^{\mathrm{T}}(k+i\mid k)\mathcal{Q}\boldsymbol{x}(k+i\mid k)+\boldsymbol{u}^{\mathrm{T}}(k+i\mid k)\mathcal{R}\boldsymbol{u}(k+i\mid k)]\\ \forall[\boldsymbol{A}(k+i)\ \ \boldsymbol{B}(k+i)]\in\boldsymbol{\Pi},\ \ i\geq 0\end{aligned} \tag{7.20}$$

By summing (7.20) from $i=0$ to ∞, we can similarly obtain

$$\max_{[\boldsymbol{A}(k+i)\boldsymbol{B}(k+i)]\subset\boldsymbol{\Pi},i\geq 0} J_\infty(k)\leq V(0,k)=\boldsymbol{x}^{\mathrm{T}}(k\mid k)\boldsymbol{P}(0,k)\boldsymbol{x}(k\mid k)\leq\gamma$$

According to (7.19), this inequality holds if

$$\boldsymbol{x}^{\mathrm{T}}(k\mid k)\boldsymbol{P}_l\boldsymbol{x}(k\mid k)\leq\gamma,\quad l=1,\ \dots,\ L$$

Let $\boldsymbol{P}_l=\gamma\boldsymbol{Q}_l^{-1}$ and noting that $\boldsymbol{x}(k\mid k)=\boldsymbol{x}(k)$, the above inequalities can be represented by

$$\begin{bmatrix}1 & \boldsymbol{x}^{\mathrm{T}}(k)\\ \boldsymbol{x}(k) & \boldsymbol{Q}_l\end{bmatrix}\geq\boldsymbol{0},\quad l=1,\ \dots,\ L \tag{7.21}$$

Still assume the state feedback control law $\boldsymbol{u}(k+i\mid k)=\boldsymbol{F}\boldsymbol{x}(k+i\mid k)$, $i\geq 0$, with a fixed feedback gain $\boldsymbol{F}$; the condition (7.20) holds if

$$\begin{aligned}&(\boldsymbol{A}(k+i)+\boldsymbol{B}(k+i)\boldsymbol{F})^{\mathrm{T}}\boldsymbol{P}(i+1,k)(\boldsymbol{A}(k+i)+\boldsymbol{B}(k+i)\boldsymbol{F})-\boldsymbol{P}(i,k)\\ &<-(\mathcal{Q}+\boldsymbol{F}^T\mathcal{R}\boldsymbol{F}),\qquad \forall[\boldsymbol{A}(k+i)\ \ \boldsymbol{B}(k+i)]\in\boldsymbol{\Pi},\ \ i\geq 0\end{aligned} \tag{7.22}$$

With the help of the Schur complement, it can be written as the following matrix inequality:

$$\begin{bmatrix}\boldsymbol{P}(i,k) & * & * & *\\ \boldsymbol{P}(i+1,k)(\boldsymbol{A}(k+i)+\boldsymbol{B}(k+i)\boldsymbol{F}) & \boldsymbol{P}(i+1,k) & * & *\\ \mathcal{Q}^{1/2} & \boldsymbol{0} & \boldsymbol{I} & *\\ \mathcal{R}^{1/2}\boldsymbol{F} & \boldsymbol{0} & \boldsymbol{0} & \boldsymbol{I}\end{bmatrix}>\boldsymbol{0},\quad \forall[\boldsymbol{A}(k+i)\ \ \boldsymbol{B}(k+i)]\in\boldsymbol{\Pi},\ \ i\geq 0 \tag{7.23}$$

where the symbol $*$ in the matrix expression is used to induce a symmetric structure.

In order to distinguish the combination coefficients λ at time $k+i$ and $k+i+1$, use different subscript l and j, respectively, to mark them, i.e.

$$[\boldsymbol{A}(k+i)\ \ \boldsymbol{B}(k+i)] = \sum_{l=1}^{L} \lambda_l(k+i)[\boldsymbol{A}_l\ \ \boldsymbol{B}_l], \quad \boldsymbol{P}(i,k) = \sum_{l=1}^{L} \lambda_l(k+i)\boldsymbol{P}_l$$

$$\boldsymbol{P}(i+1,k) = \sum_{j=1}^{L} \lambda_j(k+i+1)\boldsymbol{P}_j$$

Then, according to (7.3), it is easy to show that the matrix inequality (7.23) holds if

$$\begin{bmatrix} \boldsymbol{P}_l & * & * & * \\ \boldsymbol{P}_j(\boldsymbol{A}_l + \boldsymbol{B}_l\boldsymbol{F}) & \boldsymbol{P}_j & * & * \\ \mathcal{Q}^{1/2} & \boldsymbol{0} & \boldsymbol{I} & * \\ \mathcal{R}^{1/2}\boldsymbol{F} & \boldsymbol{0} & \boldsymbol{0} & \boldsymbol{I} \end{bmatrix} > \boldsymbol{0}, \quad l = 1, \ldots, L, \quad j = 1, \ldots, L$$

Let $\boldsymbol{P}_l = \gamma\boldsymbol{Q}_l^{-1}$ and $\boldsymbol{P}_j = \gamma\boldsymbol{Q}_j^{-1}$, this can be rewritten as

$$\begin{bmatrix} \gamma\boldsymbol{Q}_l^{-1} & * & * & * \\ \gamma\boldsymbol{Q}_j^{-1}(\boldsymbol{A}_l + \boldsymbol{B}_l\boldsymbol{F}) & \gamma\boldsymbol{Q}_j^{-1} & * & * \\ \mathcal{Q}^{1/2} & \boldsymbol{0} & \boldsymbol{I} & * \\ \mathcal{R}^{1/2}\boldsymbol{F} & \boldsymbol{0} & \boldsymbol{0} & \boldsymbol{I} \end{bmatrix} > \boldsymbol{0}, \quad l = 1, \ldots, L, \quad j = 1, \ldots, L \tag{7.24}$$

By multiplying diag $(\gamma^{-1/2}\boldsymbol{Q}_l, \gamma^{-1/2}\boldsymbol{Q}_j, \gamma^{1/2}\boldsymbol{I}, \gamma^{1/2}\boldsymbol{I})$ on both sides of (7.24), the following matrix inequalities can be equivalently obtained:

$$\begin{bmatrix} \boldsymbol{Q}_l & * & * & * \\ (\boldsymbol{A}_l + \boldsymbol{B}_l\boldsymbol{F})\boldsymbol{Q}_l & \boldsymbol{Q}_j & * & * \\ \mathcal{Q}^{1/2}\boldsymbol{Q}_l & \boldsymbol{0} & \gamma\boldsymbol{I} & * \\ \mathcal{R}^{1/2}\boldsymbol{F}\boldsymbol{Q}_l & \boldsymbol{0} & \boldsymbol{0} & \gamma\boldsymbol{I} \end{bmatrix} > \boldsymbol{0}, \quad l = 1, \ldots, L, \quad j = 1, \ldots, L$$

If all $\boldsymbol{Q}_i$ are the same and denoted by $\boldsymbol{Q}$, this is the case of using a single Lyapunov function discussed in Section 7.1.1. By letting $\boldsymbol{Y} = \boldsymbol{FQ}$, these matrix inequalities can be easily written in the form of LMIs with γ, $\boldsymbol{Q}$, $\boldsymbol{Y}$ to be determined (see (7.14)). The feedback control gain $\boldsymbol{F}$ can then be obtained by $\boldsymbol{F} = \boldsymbol{YQ}^{-1}$ with solved $\boldsymbol{Y}$, $\boldsymbol{Q}$. However, in the case of adopting multiple Lyapunov functions, in the above matrix inequalities the feedback control gain $\boldsymbol{F}$ appears in the product terms $\boldsymbol{FQ}_l$ for all $\boldsymbol{Q}_l$, $l = 1, \ldots, L$. By similarly defining $\boldsymbol{Y}_l = \boldsymbol{FQ}_l$ these matrix inequalities might be rewritten into LMIs with γ, $\boldsymbol{Q}_l$, $\boldsymbol{Y}_l$, $l = 1, \ldots, L$ to be determined. However, with the solved pairs $\{\boldsymbol{Y}_l, \boldsymbol{Q}_l\}$, $l = 1, \ldots, L$, it is impossible to find a feedback control gain $\boldsymbol{F}$ satisfying all the L matrix equations $\boldsymbol{Y}_l = \boldsymbol{FQ}_l$.

In order to get an available feedback gain $\boldsymbol{F}$, when defining a new matrix variable while converting the matrix inequality into LMI, it should be ensured that $\boldsymbol{F}$ could be uniquely solved by it. In spite of different $\boldsymbol{Q}_l$, refer to the single $\boldsymbol{Q}$ case and define $\boldsymbol{F} = \boldsymbol{YG}^{-1}$ with $\boldsymbol{G}$ an invertible matrix to be determined. Multiplying diag $(\gamma^{-1/2}\boldsymbol{G}^{\mathrm{T}}, \gamma^{-1/2}\boldsymbol{Q}_j, \gamma^{1/2}\boldsymbol{I}, \gamma^{1/2}\boldsymbol{I})$ and diag $(\gamma^{-1/2}\boldsymbol{G}, \gamma^{-1/2}\boldsymbol{Q}_j, \gamma^{1/2}\boldsymbol{I}, \gamma^{1/2}\boldsymbol{I})$ on the left and right side of (7.24), respectively, we get

$$\begin{bmatrix} G^{\mathrm{T}}Q_l^{-1}G & * & * & * \\ (A_l+B_lF)G & Q_j & * & * \\ \mathcal{Q}^{1/2}G & 0 & \gamma I & * \\ \mathcal{R}^{1/2}FG & 0 & 0 & \gamma I \end{bmatrix} > 0, \quad l=1,\ldots,L,\ j=1,\ldots,L$$

Using Y instead of FG, then the only nonlinear term in the matrix is $G^{\mathrm{T}}Q_l^{-1}G$. Note that with matrix inequality $(G-Q_l)^{\mathrm{T}}Q_l^{-1}(G-Q_l)\geq 0$, it follows that $G^{\mathrm{T}}Q_l^{-1}G\geq G+G^{\mathrm{T}}-Q_l$. Therefore the above matrix inequalities are satisfied if the following LMIs hold:

$$\begin{bmatrix} G+G^{\mathrm{T}}-Q_l & * & * & * \\ A_lG+B_lY & Q_j & * & * \\ \mathcal{Q}^{1/2}G & 0 & \gamma I & * \\ \mathcal{R}^{1/2}Y & 0 & 0 & \gamma I \end{bmatrix} > 0, \quad l=1,\ldots,L,\ j=1,\ldots,L \tag{7.25}$$

Thus, the descending condition (7.20) of the parameter-dependent Lyapunov function can be guaranteed by LMIs (7.25) if there exist L symmetric matrices Q_l, $l=1,\ldots,L$, a pair of matrices $\{Y, G\}$, and a positive γ satisfying (7.25).

Then consider the constraints on control inputs (7.4). Note that the feedback control gain F is now $F = YG^{-1}$, and at time k the input constraint (7.4) can be specified as

$$\begin{aligned} \|u(k+i\mid k)\|_2^2 &= \|Fx(k+i\mid k)\|_2^2 \\ &= x^{\mathrm{T}}(k+i\mid k)\left(G^{-1}\right)^{\mathrm{T}}Y^{\mathrm{T}}YG^{-1}x(k+i\mid k)\leq u_{\max}^2, \quad i\geq 0 \end{aligned}$$

Due to $x^{\mathrm{T}}(k+i\mid k)Q^{-1}(i,k)x(k+i\mid k)\leq 1$, this condition can be ensured by

$$Y^{\mathrm{T}}Y/u_{\max}^2\leq G^{\mathrm{T}}Q^{-1}(i,k)G, \quad i\geq 0$$

Since $Q^{-1}(i,k)=\sum_{l=1}^{L}\lambda_l(k+i)Q_l^{-1}$, and considering $G+G^{\mathrm{T}}-Q_l\leq G^{\mathrm{T}}Q_l^{-1}G$, the above condition holds if the following condition holds:

$$Y^{\mathrm{T}}Y/u_{\max}^2\leq G+G^{\mathrm{T}}-Q_l, \quad l=1,\ldots,L$$

which can be written in LMI form as

$$\begin{bmatrix} u_{\max}^2 I & Y \\ Y^{\mathrm{T}} & G+G^{\mathrm{T}}-Q_l \end{bmatrix}\geq 0, \quad l=1,\ldots,L \tag{7.26}$$

Finally, consider the constraints on system outputs (7.5), i.e. at time k:

$$\|y(k+i+1\mid k)\|_2^2=\left\|C\left(A(k+i)+B(k+i)YG^{-1}\right)x(k+i\mid k)\right\|_2^2\leq y_{\max}^2, \quad i\geq 0$$

Since at time k, $x^{\mathrm{T}}(k+i\mid k)Q^{-1}(i,k)x(k+i\mid k)\leq 1$, this constraint can be satisfied if the following condition holds:

$$(A(k+i)G+B(k+i)Y)^{\mathrm{T}}C^{\mathrm{T}}\left(y_{\max}^2 I\right)^{-1}C(A(k+i)G+B(k+i)Y)\leq G^{\mathrm{T}}Q^{-1}(i,k)G$$

which can be written in the form of the following matrix inequality:

$$\begin{bmatrix} \boldsymbol{G}^{\mathrm{T}}\boldsymbol{Q}^{-1}(i,k)\boldsymbol{G} & (\boldsymbol{A}(k+i)\boldsymbol{G}+\boldsymbol{B}(k+i)\boldsymbol{Y})^{\mathrm{T}}\boldsymbol{C}^{\mathrm{T}} \\ \boldsymbol{C}(\boldsymbol{A}(k+i)\boldsymbol{G}+\boldsymbol{B}(k+i)\boldsymbol{Y}) & y_{\max}^2\boldsymbol{I} \end{bmatrix} \geq \boldsymbol{0}, \qquad i \geq 0$$

Due to (7.3) and

$$\boldsymbol{Q}^{-1}(i,k) = \sum_{l=1}^{L} \lambda_l(k+i)\boldsymbol{Q}_l^{-1}$$

also consider $\boldsymbol{G}^{\mathrm{T}}\boldsymbol{Q}_l^{-1}\boldsymbol{G} \geq \boldsymbol{G}+\boldsymbol{G}^{\mathrm{T}}-\boldsymbol{Q}_l$. The above matrix inequalities can be ensured by the following LMIs:

$$\begin{bmatrix} \boldsymbol{G}+\boldsymbol{G}^{\mathrm{T}}-\boldsymbol{Q}_l & (\boldsymbol{A}_l\boldsymbol{G}+\boldsymbol{B}_l\boldsymbol{Y})^{\mathrm{T}}\boldsymbol{C}^{\mathrm{T}} \\ \boldsymbol{C}(\boldsymbol{A}_l\boldsymbol{G}+\boldsymbol{B}_l\boldsymbol{Y}) & y_{\max}^2\boldsymbol{I} \end{bmatrix} \geq \boldsymbol{0}, \qquad l = 1, \ldots, L \tag{7.27}$$

According to the above discussion, the RMPC synthesis with parameter-dependent Lyapunov functions can be formulated as solving the following optimization problem at each time k:

$$\begin{aligned} &\min_{\gamma,\boldsymbol{Q}_l,\boldsymbol{Y},\boldsymbol{G}} \gamma \\ &\text{s.t. } (7.21), (7.25), (7.26), (7.27) \end{aligned} \tag{7.28}$$

With $\boldsymbol{Y}$, $\boldsymbol{G}$ solved, the feedback gain at time k can be obtained by $\boldsymbol{F}(k) = \boldsymbol{Y}\boldsymbol{G}^{-1}$ and the control action at time k can then be calculated by $\boldsymbol{u}(k) = \boldsymbol{F}(k)\boldsymbol{x}(k)$ and implemented.

Compared with the design in Section 7.1.1, it is easy to show that Problem (7.17) is a special case of Problem (7.28) if we select a single Lyapunov function with $L = 1$ and set $\boldsymbol{G} = \boldsymbol{Q} = \boldsymbol{Q}_l$, with which the constraints in (7.28) are exactly reduced into those given in (7.17). The use of parameter-dependent Lyapunov functions brings a greater degree of freedom to RMPC design and a better control performance could be achieved. Of course, the less conservative result is obtained at the cost of increasing the number of optimization variables and more rows of LMIs, which may cause the online computational burden to be heavier than that in Section 7.1.1.

7.1.3 Synthesis of RMPC with Dual-Mode Control

It is easy to find that the RMPC designs in Section 7.1.1 and 7.1.2 adopt a single feedback control law to approximate the future control strategy for the infinite horizon optimization problem at time k. In order to improve the control performance, in addition to introducing parameter-dependent Lyapunov functions, as presented in the last section, considering more reasonable future control strategies is also an attractive way. Some researchers explored bringing dual-mode control [4] (see Section 6.2.3) into the RMPC design, such as Wan and Kothare [5] and Ding et al. [6]. In this section, we will briefly introduce this RMPC design method.

Consider the same RMPC problem in Section 7.1.1. At each time, an optimization problem (7.6) should be solved, with respect to the model (7.1) with polytopic uncertainties (7.2) and (7.3) and input and output constraints (7.4) and (7.5). According to the dual-mode control, at time k, the future control inputs are assumed to be

$$\boldsymbol{u}(k+i\,|\,k)=\begin{cases}\boldsymbol{u}(k+i\,|\,k), & 0\le i<N\\ \boldsymbol{F}\boldsymbol{x}(k+i\,|\,k), & i\ge N\end{cases}\tag{7.29}$$

where N is the length of the control horizon, which is the same as the optimization horizon. With this strategy, N free control inputs are firstly adopted to steer the system state into the terminal set and $\boldsymbol{F}$ is the gain of the state feedback control law in the terminal set.

Consider the min-max performance index in the optimization problem (7.6) at time k:

$$\min_{\boldsymbol{u}(k+i|k)}\ \max_{[\boldsymbol{A}(k+i)\ \boldsymbol{B}(k+i)]\subset\boldsymbol{\Pi}} J_\infty(k)=\sum_{i=0}^{\infty}\left[\boldsymbol{x}(k+i\,|\,k)^{\mathrm{T}}\mathcal{Q}\boldsymbol{x}(k+i\,|\,k)+\boldsymbol{u}(k+i\,|\,k)^{\mathrm{T}}\mathcal{R}\boldsymbol{u}(k+i\,|\,k)\right]$$

Since two different control modes are adopted, $J_\infty(k)$ can accordingly be divided into two parts: the input and state items before and after the state entering the terminal set, respectively, i.e.

$$J_1(k)=\sum_{i=0}^{N-1}\left[\boldsymbol{x}^{\mathrm{T}}(k+i\,|\,k)\mathcal{Q}\boldsymbol{x}(k+i\,|\,k)+\boldsymbol{u}^{\mathrm{T}}(k+i\,|\,k)\mathcal{R}\boldsymbol{u}(k+i\,|\,k)\right]$$

$$J_2(k)=\sum_{i=N}^{\infty}\left[\boldsymbol{x}^{\mathrm{T}}(k+i\,|\,k)\mathcal{Q}\boldsymbol{x}(k+i\,|\,k)+\boldsymbol{u}^{\mathrm{T}}(k+i\,|\,k)\mathcal{R}\boldsymbol{u}(k+i\,|\,k)\right]$$

Firstly we derive the upper bounds of the maximum of the above two objective functions under polytopic uncertainty, respectively.

For the first part, using the system model (7.1) starting from $\boldsymbol{x}(k\,|\,k)=\boldsymbol{x}(k)$, it is easy to get the system states controlled by the N free control inputs as shown in (7.29):

$$\begin{bmatrix}\boldsymbol{x}(k+1\,|\,k)\\ \vdots\\ \boldsymbol{x}(k+N\,|\,k)\end{bmatrix}=\begin{bmatrix}\boldsymbol{A}(k)\\ \vdots\\ \boldsymbol{A}(k+N-1)\cdots\boldsymbol{A}(k+1)\boldsymbol{A}(k)\end{bmatrix}\boldsymbol{x}(k)+$$

$$\begin{bmatrix}\boldsymbol{B}(k) & 0 & \cdots & 0\\ \vdots & \ddots & \ddots & \vdots\\ \vdots & & \ddots & 0\\ \boldsymbol{A}(k+N-1)\cdots\boldsymbol{A}(k+1)\boldsymbol{B}(k) & \cdots & \cdots & \boldsymbol{B}(k+N-1)\end{bmatrix}\times\begin{bmatrix}\boldsymbol{u}(k\,|\,k)\\ \vdots\\ \boldsymbol{u}(k+N-1\,|\,k)\end{bmatrix}$$

or in the matrix form

$$\begin{bmatrix}\boldsymbol{X}(k)\\ \boldsymbol{x}(k+N\,|\,k)\end{bmatrix}=\begin{bmatrix}\tilde{\boldsymbol{A}}(k)\\ \tilde{\boldsymbol{A}}_N(k)\end{bmatrix}\boldsymbol{x}(k)+\begin{bmatrix}\tilde{\boldsymbol{B}}(k)\\ \tilde{\boldsymbol{B}}_N(k)\end{bmatrix}\boldsymbol{U}(k)\tag{7.30}$$

where

$$\begin{aligned}\boldsymbol{X}(k)&=[\boldsymbol{x}^{\mathrm{T}}(k+1\,|\,k)\ \ \boldsymbol{x}^{\mathrm{T}}(k+2\,|\,k)\ \cdots\ \boldsymbol{x}^{\mathrm{T}}(k+N-1\,|\,k)]^{\mathrm{T}}\\ \boldsymbol{U}(k)&=[\boldsymbol{u}^{\mathrm{T}}(k\,|\,k)\ \ \boldsymbol{u}^{\mathrm{T}}(k+1\,|\,k)\ \cdots\ \boldsymbol{u}^{\mathrm{T}}(k+N-1\,|\,k)]^{\mathrm{T}}\end{aligned}\tag{7.31}$$

Note that all the coefficients of matrices $\tilde{\boldsymbol{A}}(k)$, $\tilde{\boldsymbol{B}}(k)$ in (7.30) are composed of $\boldsymbol{A}(k+i)$, $\boldsymbol{B}(k+i)$, which belong to the convex hull $\boldsymbol{\Pi}$ described by (7.2) and (7.3). If the combination parameters $\lambda_l(k+i)$, $l = 1, \ldots, L$, are time-varying, the system uncertainty will be expanded with time. We use time-related subscripts to distinguish time-varying combination parameters and denote the combination parameters at time $k+i$ as $\lambda_l(k+i) \triangleq \lambda_{l_i}$, $l = 1, \ldots, L$, $i = 0, 1, \ldots$, $l_i = 1, \ldots, L$. Then $[\boldsymbol{A}(k+i) \;\; \boldsymbol{B}(k+i)] = \sum_{l=1}^{L} \lambda_l(k+i)[\boldsymbol{A}_l \;\; \boldsymbol{B}_l] \triangleq \sum_{l_i=1}^{L} \lambda_{l_i}[\boldsymbol{A}_{l_i} \;\; \boldsymbol{B}_{l_i}]$.

To illustrate how the system uncertainty expands with time, start from the fixed $\boldsymbol{x}(k)$ at time k. For $\boldsymbol{x}(k+1 \mid k)$, uncertainty is only caused by $[\boldsymbol{A}(k) \;\; \boldsymbol{B}(k)]$ and can be described by the polytope with L vertices $[\boldsymbol{A}_{l_0} \;\; \boldsymbol{B}_{l_0}]$, $l_0 = 1, \ldots, L$. For $\boldsymbol{x}(k+2 \mid k)$, with the same procedure for each possible $\boldsymbol{x}(k+1 \mid k)$, uncertainty will be expanded due to $[\boldsymbol{A}(k+1) \;\; \boldsymbol{B}(k+1)]$, which can be described by the same type of polytope with L vertices $[\boldsymbol{A}_{l_1} \;\; \boldsymbol{B}_{l_1}]$, $l_1 = 1, \ldots, L$, but with independent combination parameters. Therefore the uncertainty of $\boldsymbol{x}(k+2 \mid k)$ should be described by a composite polytope with L^2 vertices $[\boldsymbol{A}_{l_0} \; \boldsymbol{B}_{l_0} \; \boldsymbol{A}_{l_1} \; \boldsymbol{B}_{l_1}]$, $l_0 = 1, \ldots, L$, $l_1 = 1, \ldots, L$. For example, the term $\boldsymbol{A}(k+1)\boldsymbol{A}(k)\boldsymbol{x}(k)$ in $\boldsymbol{x}(k+2 \mid k)$ can be expressed by

$$\left(\sum_{l_1=1}^{L}\sum_{l_0=1}^{L} \lambda_{l_1}\lambda_{l_0}\boldsymbol{A}_{l_1}\boldsymbol{A}_{l_0}\right)\boldsymbol{x}(k)$$

With the above understanding, a convex hull $\tilde{\boldsymbol{\Pi}}$ with $\tilde{L} = L^{N-1}$ vertices can now be used to describe the polytopic uncertainties of $\left[\tilde{\boldsymbol{A}}(k) \;\; \tilde{\boldsymbol{B}}(k)\right]$ corresponding to $\boldsymbol{x}(k+1 \mid k)$ to $\boldsymbol{x}(k+N-1 \mid k)$, i.e.

$$\tilde{\boldsymbol{\Pi}} = \mathrm{Co}\left\{\left[\tilde{\boldsymbol{A}}_1 \;\; \tilde{\boldsymbol{B}}_1\right], \ldots, \left[\tilde{\boldsymbol{A}}_{\tilde{L}} \;\; \tilde{\boldsymbol{B}}_{\tilde{L}}\right]\right\} \tag{7.32}$$

$$\left[\tilde{\boldsymbol{A}}(k) \;\; \tilde{\boldsymbol{B}}(k)\right] = \sum_{\tilde{l}=1}^{\tilde{L}} \xi_{\tilde{l}}(k)\left[\tilde{\boldsymbol{A}}_{\tilde{l}} \;\; \tilde{\boldsymbol{B}}_{\tilde{l}}\right], \quad \sum_{\tilde{l}=1}^{\tilde{L}} \xi_{\tilde{l}}(k) = 1, \quad \xi_{\tilde{l}}(k) \geq 0, \quad \tilde{l} = 1, \ldots, \tilde{L} \tag{7.33}$$

where $\left[\tilde{\boldsymbol{A}}_{\tilde{l}} \;\; \tilde{\boldsymbol{B}}_{\tilde{l}}\right]$ is the $\tilde{l}$th vertex of the polytope $\tilde{\boldsymbol{\Pi}}$. Similarly, a convex hull $\tilde{\boldsymbol{\Pi}}_N$ with $\tilde{L}_N = L^N$ vertices can be used to describe the polytopic uncertainties of $\left[\tilde{\boldsymbol{A}}_N(k) \; \tilde{\boldsymbol{B}}_N(k)\right]$ corresponding to the terminal state $\boldsymbol{x}(k+N \mid k)$:

$$\tilde{\boldsymbol{\Pi}}_N = \mathrm{Co}\left\{\left[\tilde{\boldsymbol{A}}_{N,1}(k) \;\; \tilde{\boldsymbol{B}}_{N,1}(k)\right], \ldots, \left[\tilde{\boldsymbol{A}}_{N,\tilde{L}_N}(k) \;\; \tilde{\boldsymbol{B}}_{N,\tilde{L}_N}(k)\right]\right\} \tag{7.34}$$

$$\left[\tilde{\boldsymbol{A}}_N(k) \; \tilde{\boldsymbol{B}}_N(k)\right] = \sum_{\tilde{l}_N=1}^{\tilde{L}_N} \xi_{\tilde{l}_N}(k)\left[\tilde{\boldsymbol{A}}_{N,\tilde{l}_N}\tilde{\boldsymbol{B}}_{N,\tilde{l}_N}\right], \quad \sum_{\tilde{l}_N=1}^{\tilde{L}_N} \xi_{\tilde{l}_N}(k) = 1, \quad \xi_{\tilde{l}_N}(k) \geq 0, \quad \tilde{l}_N = 1, \ldots, \tilde{L}_N \tag{7.35}$$

Now we can define an upper bound for $J_1(k)$. According to (7.30), it follows that

$$\max_{[\boldsymbol{A}(k+i) \;\; \boldsymbol{B}(k+i)] \subset \boldsymbol{\Pi}} J_1(k) = \sum_{i=0}^{N-1}\left[\boldsymbol{x}(k+i \mid k)^{\mathrm{T}}\mathcal{Q}\boldsymbol{x}(k+i \mid k) + \boldsymbol{u}(k+i \mid k)^{\mathrm{T}}\mathcal{R}\boldsymbol{u}(k+i \mid k)\right]$$

$$= \max_{\left[\tilde{\boldsymbol{A}}(k) \;\; \tilde{\boldsymbol{B}}(k)\right] \subset \tilde{\boldsymbol{\Pi}}} \|\boldsymbol{x}(k \mid k)\|_{\mathcal{Q}}^2 + \left\|\tilde{\boldsymbol{A}}(k)\boldsymbol{x}(k) + \tilde{\boldsymbol{B}}(k)\boldsymbol{U}(k)\right\|_{\tilde{\mathcal{Q}}}^2 + \|\boldsymbol{U}(k)\|_{\tilde{\mathcal{R}}}^2$$

where $\tilde{\mathcal{Q}}$, $\tilde{\mathcal{R}}$ are the diagonal matrices, respectively, composed of $\mathcal{Q}$, $\mathcal{R}$ with proper dimensions. Remove the fixed term $\|\boldsymbol{x}(k \mid k)\|_{\mathcal{Q}}^2$ and define an upper bound γ_1 for the remaining part. It follows that

$$\left\|\tilde{\boldsymbol{A}}(k)\boldsymbol{x}(k)+\tilde{\boldsymbol{B}}(k)\boldsymbol{U}(k)\right\|_{\tilde{\mathcal{Q}}}^2+\|\boldsymbol{U}(k)\|_{\tilde{\mathcal{R}}}^2 \le \gamma_1, \quad \left[\tilde{\boldsymbol{A}}(k)\;\; \tilde{\boldsymbol{B}}(k)\right] \subset \tilde{\boldsymbol{\Pi}}$$

Due to the polytopic characters of the system parameters (7.32) and (7.33), it can be ensured by

$$\begin{bmatrix} \tilde{\mathcal{Q}}^{-1} & 0 & \tilde{\boldsymbol{A}}_{\tilde{l}}\boldsymbol{x}(k)+\tilde{\boldsymbol{B}}_{\tilde{l}}\boldsymbol{U}(k) \\ 0 & \tilde{\mathcal{R}}^{-1} & \boldsymbol{U}(k) \\ \boldsymbol{x}^{\mathrm{T}}(k)\tilde{\boldsymbol{A}}_{\tilde{l}}^{T}+\boldsymbol{U}(k)^{\mathrm{T}}\tilde{\boldsymbol{B}}_{\tilde{l}}^{\mathrm{T}} & \boldsymbol{U}^{\mathrm{T}}(k) & \gamma_1 \end{bmatrix} \ge 0, \quad \tilde{l}=1, \dots, \tilde{L} \tag{7.36}$$

The input and output constriants (7.4) and (7.5) can be expressed by

$$\|\boldsymbol{u}(k+i \mid k)\|_2^2 = \|\boldsymbol{L}_{u,i+1}\boldsymbol{U}(k)\|_2^2 \le u_{\max}^2, \quad i=0, \dots, N-1$$

$$\begin{aligned}\|\boldsymbol{y}(k+i \mid k)\|_2^2 &= \|\boldsymbol{C}\boldsymbol{x}(k+i \mid k)\|_2^2 = \|\boldsymbol{C}\boldsymbol{L}_{x,i}\boldsymbol{X}(k)\|_2^2 \\ &= \left\|\boldsymbol{C}\boldsymbol{L}_{x,i}\left(\tilde{\boldsymbol{A}}(k)\boldsymbol{x}(k)+\tilde{\boldsymbol{B}}(k)\boldsymbol{U}(k)\right)\right\|_2^2 \le y_{\max}^2, \quad i=1, \dots, N-1, \quad \left[\tilde{\boldsymbol{A}}(k)\;\; \tilde{\boldsymbol{B}}(k)\right] \in \tilde{\boldsymbol{\Pi}}\end{aligned}$$

where $\boldsymbol{L}_{u,\, i+1}(\boldsymbol{L}_{x,\, i})$ is a proper matrix that selects $\boldsymbol{u}(k+i \mid k)(\boldsymbol{x}(k+i \mid k))$ from $\boldsymbol{U}(k)(\boldsymbol{X}(k))$ in (7.31). These two conditions can be ensured if the following LMIs are satisfied:

$$\begin{bmatrix} u_{\max}^2 & \boldsymbol{U}^{\mathrm{T}}(k)\boldsymbol{L}_{u,i+1}^{\mathrm{T}} \\ \boldsymbol{L}_{u,i+1}\boldsymbol{U}(k) & \boldsymbol{I} \end{bmatrix} \ge \boldsymbol{0}, \quad i=0, \dots, N-1 \tag{7.37}$$

$$\begin{bmatrix} y_{\max}^2 & \left(\tilde{\boldsymbol{A}}_{\tilde{l}}\boldsymbol{x}(k)+\tilde{\boldsymbol{B}}_{\tilde{l}}\boldsymbol{U}(k)\right)^{\mathrm{T}}\boldsymbol{L}_{x,i}^{\mathrm{T}}\boldsymbol{C}^{\mathrm{T}} \\ \boldsymbol{C}\boldsymbol{L}_{x,i}\left(\tilde{\boldsymbol{A}}_{\tilde{l}}\boldsymbol{x}(k)+\tilde{\boldsymbol{B}}_{\tilde{l}}\boldsymbol{U}(k)\right) & \boldsymbol{I} \end{bmatrix} \ge \boldsymbol{0}, \quad i=1, \dots, N-1, \quad \tilde{l}=1, \dots, \tilde{L} \tag{7.38}$$

For the second part, the optimization problem with performance index $J_2(k)$ is almost the same as that discussed in Section 7.1.1, with the only difference being that the initial state $\boldsymbol{x}(k \mid k)$ is replaced by $\boldsymbol{x}(k+N \mid k)$. We can similarly define a quadratic function $V(\boldsymbol{x})=\boldsymbol{x}^{\mathrm{T}}\boldsymbol{P}\boldsymbol{x}$ to obtain an upper bound γ for the maximization problem, and define a controlled invariant set $\boldsymbol{\Omega}=\{\boldsymbol{x} \in \mathrm{R}^n \mid \boldsymbol{x}^{\mathrm{T}}\boldsymbol{Q}^{-1}\boldsymbol{x} \le 1\}$ with a corresponding feedback control law $\boldsymbol{u}=\boldsymbol{F}\boldsymbol{x}$ for all $k+i$, $i \ge N$. $\boldsymbol{\Omega}$ is just the terminal set that the system state should be steered to by the first N free control actions. Let $\boldsymbol{P}=\gamma\boldsymbol{Q}^{-1}$ and $\boldsymbol{F}=\boldsymbol{Y}\boldsymbol{Q}^{-1}$; the following conditions similar to (7.12) and (7.14) to (7.16) can be obtained:

$$\begin{bmatrix} 1 & * \\ \tilde{\boldsymbol{A}}_{N,\tilde{l}_N}\boldsymbol{x}(k)+\tilde{\boldsymbol{B}}_{N,\tilde{l}_N}\boldsymbol{U}(k) & \boldsymbol{Q} \end{bmatrix} \ge \boldsymbol{0}, \quad \tilde{l}_N=1, \dots, \tilde{L}_N \tag{7.39}$$

$$\begin{bmatrix} \boldsymbol{Q} & * & * & * \\ \boldsymbol{A}_l\boldsymbol{Q}+\boldsymbol{B}_l\boldsymbol{Y} & \boldsymbol{Q} & * & * \\ \mathcal{Q}^{1/2}\boldsymbol{Q} & \boldsymbol{0} & \gamma\boldsymbol{I} & * \\ \mathcal{R}^{1/2}\boldsymbol{Y} & \boldsymbol{0} & \boldsymbol{0} & \gamma\boldsymbol{I} \end{bmatrix} \geq \boldsymbol{0}, \quad l=1, \ldots, L \tag{7.40}$$

$$\begin{bmatrix} u_{\max}^2\boldsymbol{I} & \boldsymbol{Y} \\ \boldsymbol{Y}^{\mathrm{T}} & \boldsymbol{Q} \end{bmatrix} \geq \boldsymbol{0} \tag{7.41}$$

$$\begin{bmatrix} \boldsymbol{Q} & * \\ \boldsymbol{C}(\boldsymbol{A}_l\boldsymbol{Q}+\boldsymbol{B}_l\boldsymbol{Y}) & y_{\max}^2\boldsymbol{I} \end{bmatrix} \geq \boldsymbol{0}, \quad l=1, \ldots, L \tag{7.42}$$

Then the min-max problem (7.6) to be solved at time k can be converted into the following optimization problem:

$$\begin{aligned} &\min_{\gamma,\gamma_1,\boldsymbol{Q},\boldsymbol{Y},\boldsymbol{U}(k)} \gamma+\gamma_1 \\ &\text{s.t. } (7.36) \text{ to } (7.42) \end{aligned} \tag{7.43}$$

Remark 7.2 The optimization problem (7.43) in RMPC is straightforward to derive from the dual-mode control and is parallel to that in Section 7.1.1. However, as indicated by Pluymer et al. in [7], with this kind of design, the closed-loop stability cannot be guaranteed because there exist difficulties with recursive feasibility. For control policy with free control inputs, assume that Problem (7.43) is feasible at time k. A feasible solution $\boldsymbol{U}(k+1)$ at time $k+1$ is often constructed by shifting forward the last $N-1$ components in $\boldsymbol{U}^*(k)$ and setting the last one, $\boldsymbol{u}(k+N\mid k+1) = \boldsymbol{F}\boldsymbol{x}^*(k+N\mid k)$. This has been succesfully used for stable MPC design as in Section 6.2.3, but is unavailable here. Note that condition (7.39) with (7.30) means the terminal state $\boldsymbol{x}^*(k+N\mid k)$ belonging to the invariant set $\boldsymbol{\Omega}$, i.e. $\boldsymbol{x}^*(k+N\mid k)\in\boldsymbol{\Omega}$. For recursive feasibility, $\boldsymbol{x}(k+N+1\mid k+1)\in\boldsymbol{\Omega}$ should be proved when using $\boldsymbol{U}(k+1)$ at time $k+1$. Furthermore, as a candidate solution, $\boldsymbol{U}(k+1)$ must be fixed. Consider the last component $\boldsymbol{u}(k+N\mid k+1)$ in $\boldsymbol{U}(k+1)$. If it is set as $\boldsymbol{F}\boldsymbol{x}^*(k+N\mid k)$, due to the inviant set condition implied in (7.40), i.e. $\boldsymbol{A}_l\boldsymbol{x}^*(k+N\mid k)+\boldsymbol{B}_l\boldsymbol{F}\boldsymbol{x}^*(k+N\mid k)\in\boldsymbol{\Omega}$, $\forall\boldsymbol{x}^*(k+N\mid k)\in\boldsymbol{\Omega}$, we can ensure $\boldsymbol{x}(k+N+1\mid k+1)\in\boldsymbol{\Omega}$. However, this solution is not fixed due to uncertain $\boldsymbol{x}^*(k+N\mid k)$. If a fixed $\boldsymbol{x}^f(k+N\mid k)$ among possible $\boldsymbol{x}^*(k+N\mid k)$ is selected such that $\boldsymbol{u}^f(k+N\mid k+1) = \boldsymbol{F}\boldsymbol{x}^f(k+N\mid k)$ can be fixed, the condition $\boldsymbol{x}(k+N+1\mid k+1)\in\boldsymbol{\Omega}$, i.e. $\boldsymbol{A}_l\boldsymbol{x}^*(k+N\mid k)+\boldsymbol{B}_l\boldsymbol{F}\boldsymbol{x}^f(k+N\mid k)\in\boldsymbol{\Omega}$, $\forall\boldsymbol{x}^*(k+N\mid k)\in\boldsymbol{\Omega}$ cannot be ensured because it is impossible to keep all $\boldsymbol{x}^*(k+N\mid k)\in\boldsymbol{\Omega}$ still in $\boldsymbol{\Omega}$ by the unique $\boldsymbol{u}^f(k+N\mid k+1)$. Therefore the recursive feasibility cannot be guaranteed by this kind of $\boldsymbol{U}(k+1)$.

Pluymer et al. thus proposed a control policy based on a tree structure spanned by free control inputs according to the character of polytopic uncertainty [7]. With this structure and the system model (7.1) to (7.3), the system state at any future time is uncertain but can be directly described by a polytope with corresponding state vertices (see Figure 7.1), where at each time a node means that a possible state vertex is driven by a corresponding input. In this way, the uncertainty expansion of system inputs and states could be explicitly described in accordance with the model uncertainty.

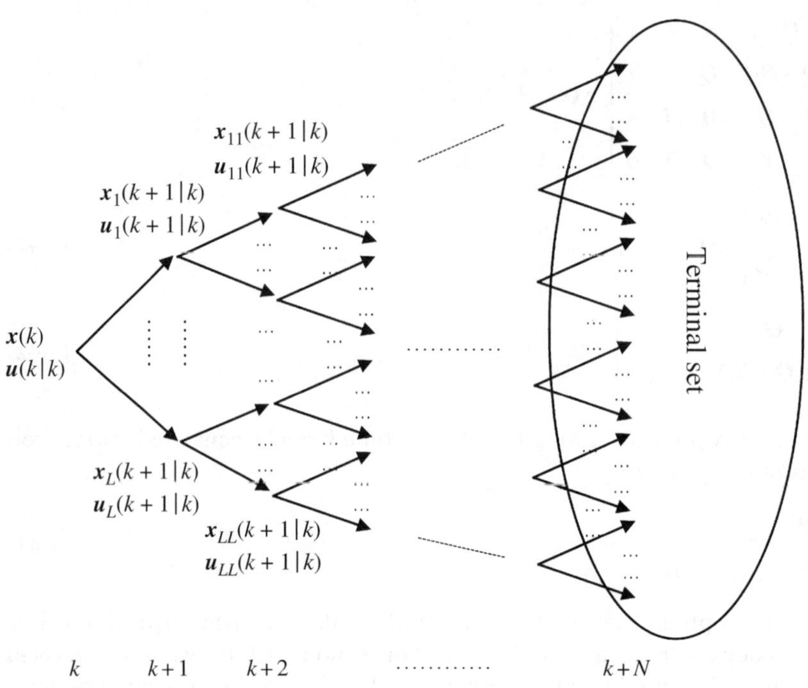

Figure 7.1 State trajectory in a tree structure with polytopic uncertainty.

In Figure 7.1, at time k, the system state $\boldsymbol{x}(k)$ is known. According to the polytopic uncertainty of system model (7.1) to (7.3), the state under the control of $\boldsymbol{u}(k \mid k)$ can be given by

$$\boldsymbol{x}(k+1 \mid k) = \boldsymbol{A}(k)\boldsymbol{x}(k) + \boldsymbol{B}(k)\boldsymbol{u}(k \mid k) = \sum_{l_0=1}^{L}(\lambda_{l_0}\boldsymbol{A}_{l_0})\boldsymbol{x}(k) + \sum_{l_0=1}^{L}(\lambda_{l_0}\boldsymbol{B}_{l_0})\boldsymbol{u}(k \mid k)$$

$$= \sum_{l_0=1}^{L}\lambda_{l_0}(\boldsymbol{A}_{l_0}\boldsymbol{x}(k) + \boldsymbol{B}_{l_0}\boldsymbol{u}(k \mid k)) \triangleq \sum_{l_0=1}^{L}\lambda_{l_0}\boldsymbol{x}_{l_0}(k+1 \mid k),$$

$$\sum_{l_0=1}^{L}\lambda_{l_0} = 1, \quad \lambda_{l_0} \geq 0, \; l_0 = 1, \; \ldots, \; L$$

which indicates that the system state $\boldsymbol{x}(k+1 \mid k)$ must be the convex combination of L state vertices $\boldsymbol{x}_{l_0}(k+1 \mid k)$, $l_0 = 1, \ldots, L$ (the nodes at time $k+1$ in Figure 7.1), each corresponding to the fixed model with the vertex parameter $[\boldsymbol{A}_{l_0} \;\; \boldsymbol{B}_{l_0}]$, respectively.

For the sake of simplification, in the following we use the symbol $l_0, \ldots, l_m = *$ to represent $l_0 = *, \cdots, l_m = *$, and $\sum_{l_m \cdots l_0 = 1}^{L}(\cdot)$ to represent $\sum_{l_m=1}^{L} \cdots \sum_{l_0=1}^{L}(\cdot)$. For the system state $\boldsymbol{x}(k+2 \mid k)$, in terms of L state vertices $\boldsymbol{x}_{l_0}(k+1 \mid k)$, define different inputs $\boldsymbol{u}_{l_0}(k+1 \mid k)$ for each state vertex and let

$$\boldsymbol{u}(k+1 \mid k) = \sum_{l_0=1}^{L}\lambda_{l_0}\boldsymbol{u}_{l_0}(k+1 \mid k)$$

Similarly, under the control of $\boldsymbol{u}_{l_0}(k+1 \mid k)$, each state vertex $\boldsymbol{x}_{l_0}(k+1 \mid k)$ will evolve to L state vertices $\boldsymbol{x}_{l_1 l_0}(k+2 \mid k)$, l_0, $l_1 = 1, \ldots, L$. We have

$$
\begin{aligned}
\boldsymbol{x}(k+2\,|\,k) &= \boldsymbol{A}(k+1)\boldsymbol{x}(k+1\,|\,k) + \boldsymbol{B}(k+1)\boldsymbol{u}(k+1\,|\,k) \\
&= \sum_{l_1=1}^{L}(\lambda_{l_1}\boldsymbol{A}_{l_1})\sum_{l_0=1}^{L}(\lambda_{l_0}\boldsymbol{x}_{l_0}(k+1\,|\,k)) + \sum_{l_1=1}^{L}(\lambda_{l_1}\boldsymbol{B}_{l_1})\sum_{l_0=1}^{L}(\lambda_{l_0}\boldsymbol{u}_{l_0}(k+1\,|\,k)) \\
&= \sum_{l_1=1}^{L}\sum_{l_0=1}^{L}\lambda_{l_1}\lambda_{l_0}(\boldsymbol{A}_{l_1}\boldsymbol{x}_{l_0}(k+1\,|\,k) + \boldsymbol{B}_{l_1}\boldsymbol{u}_{l_0}(k+1\,|\,k)) \triangleq \sum_{l_1 l_0=1}^{L}\lambda_{l_1}\lambda_{l_0}\boldsymbol{x}_{l_1 l_0}(k+2\,|\,k), \\
&\quad \sum_{l_1 l_0=1}^{L}\lambda_{l_1}\lambda_{l_0} = 1,\ \ \lambda_{l_1},\lambda_{l_0} \ge 0,\ \ l_1, l_0 = 1,\ \ldots,\ L
\end{aligned}
$$

where the system state $\boldsymbol{x}(k+2\,|\,k)$ must be the convex combination of L^2 state vertices $\boldsymbol{x}_{l_1 l_0}(k+2\,|\,k)$, l_1, $l_0 = 1, \ldots, L$ (i.e. the nodes at time $k+2$ in Figure 7.1), where each corresponds to the fixed model with parameters of the vertex $[\boldsymbol{A}_{l_0}\ \boldsymbol{B}_{l_0}\ \boldsymbol{A}_{l_1}\ \boldsymbol{B}_{l_1}]$, respectively.

This procedure is continued until time $k+N$. As the result, the uncertainty of the system state $\boldsymbol{x}(k+i\,|\,k)$ can still be described by polytope but the number of state vertices expands with time, as shown by the tree in Figure 7.1. Note that here a control policy of using different control inputs corresponding to each expanded state vertex is adopted, so the control inputs and the states in (7.31) are given by

$$
\begin{aligned}
\boldsymbol{u}(k+i\,|\,k) &= \sum_{l_{i-1}\cdots l_0=1}^{L}\left(\prod_{h=0}^{i-1}\lambda_{l_h}\right)\boldsymbol{u}_{l_{i-1}\cdots l_0}(k+i\,|\,k),\ \ i=1,\ \ldots,\ N-1 \\
\boldsymbol{x}(k+i\,|\,k) &= \sum_{l_{i-1}\cdots l_0=1}^{L}\left(\prod_{h=0}^{i-1}\lambda_{l_h}\right)\boldsymbol{x}_{l_{i-1}\cdots l_0}(k+i\,|\,k),\ \ i=1,\ \ldots,\ N \\
&\sum_{l_{i-1}\cdots l_0=1}^{L}\left(\prod_{h=0}^{i-1}\lambda_{l_h}\right) = 1,\ \ \lambda_{l_{i-1}},\ldots,\lambda_{l_0} \ge 0,\ \ \ l_0,\ \ldots,\ l_{i-1} = 1,\ \ldots,\ L
\end{aligned}
\tag{7.44}
$$

where

$$
\begin{aligned}
&\boldsymbol{x}_{l_0}(k+1\,|\,k) = \boldsymbol{A}_{l_0}\boldsymbol{x}(k) + \boldsymbol{B}_{l_0}\boldsymbol{u}(k\,|\,k) \\
&\boldsymbol{x}_{l_i\cdots l_0}(k+i+1\,|\,k) = \boldsymbol{A}_{l_i}\boldsymbol{x}_{l_{i-1}\cdots l_0}(k+i\,|\,k) + \boldsymbol{B}_{l_i}\boldsymbol{u}_{l_{i-1}\cdots l_0}(k+i\,|\,k) \\
&\quad = \boldsymbol{A}_{l_i}\cdots\boldsymbol{A}_{l_0}\boldsymbol{x}(k) + \boldsymbol{A}_{l_i}\cdots\boldsymbol{A}_{l_1}\boldsymbol{B}_{l_0}\boldsymbol{u}(k\,|\,k) + \cdots + \boldsymbol{B}_{l_i}\boldsymbol{u}_{l_{i-1}\cdots l_0}(k+i\,|\,k), \\
&\qquad\quad l_0,\ \ldots,\ l_i = 1,\ldots,L,\ \ i=\ 1,\ \ldots,\ N-1
\end{aligned}
\tag{7.45}
$$

Each time the control input is not a single one but is represented in a more detailed form corresponding to the tree structure

$$
\boldsymbol{U}(k) = \begin{bmatrix} \bar{\boldsymbol{u}}(k\,|\,k) \\ \vdots \\ \bar{\boldsymbol{u}}(k+i\,|\,k) \\ \vdots \\ \bar{\boldsymbol{u}}(k+N-1\,|\,k) \end{bmatrix} \ \text{with}\ \bar{\boldsymbol{u}}(k+i\,|\,k) = \begin{bmatrix} \boldsymbol{u}_{\underbrace{1\cdots 1}_{i}}(k+i\,|\,k) \\ \vdots \\ \boldsymbol{u}_{\underbrace{L\cdots L}_{i}}(k+i\,|\,k) \end{bmatrix},\ i=1,\ \ldots,\ N-1
\tag{7.46}
$$

Then the state evolution (7.30) under the new control input $\boldsymbol{U}(k)$ with $1 + L + \cdots + L^{N-1}$ components and new matrices $\tilde{\boldsymbol{A}}$, $\tilde{\boldsymbol{B}}$, $\tilde{\boldsymbol{A}}_N$, $\tilde{\boldsymbol{B}}_N$ with proper dimensions can be similarly derived. According to the new form of $\boldsymbol{U}(k)$ and new matrices, the optimization problem with constraint conditions similar to (7.43) and (7.36) to (7.42) can be established.

For the new formulated optimization problem, assume that it has an optimal solution $\boldsymbol{U}^*(k)$ at time k. Let $l_i(k)$ be the subscript l_i defined at time k. Due to vertices expansion in Figure 7.1, $\bar{\boldsymbol{u}}(k+i\,|\,k+1)$ has L^{i-1} components $\boldsymbol{u}_{l_{i-2}(k+1)\cdots l_0(k+1)}(k+i\,|\,k+1)$ while $\boldsymbol{u}^*(k+i\,|\,k)$ has L^i components $\boldsymbol{u}^*_{l_{i-1}(k)\cdots l_0(k)}(k+i\,|\,k)$. Obviously, for $\boldsymbol{U}(k)$ with this new structure, during proof of recursive feasibility, the candidate solution $\boldsymbol{U}(k+1)$ at time $k+1$ cannot be constructed by using a shifting operation to set $\bar{\boldsymbol{u}}(k+i\,|\,k+1) = \bar{\boldsymbol{u}}^*(k+i\,|\,k)$ as usual. However, as shown in Figure 7.1, $\boldsymbol{u}(k+1\,|\,k+1)$ corresponds to L $\boldsymbol{u}^*_{l_0(k)}(k+1\,|\,k)$ with different $l_0(k)$, and each $\boldsymbol{u}_{l_{i-2}(k+1)\cdots l_0(k+1)}(k+i\,|\,k+1)$ corresponds to L $\boldsymbol{u}^*_{l_{i-1}(k)\cdots l_1(k)}(k+i\,|\,k)$ resulting from $\boldsymbol{u}^*_{l_{i-1}(k)\cdots l_1(k)l_0(k)}(k+i\,|\,k)$ with different $l_0(k)$. Therefore each component in $\bar{\boldsymbol{u}}(k+i\,|\,k+1)$, i.e. $\boldsymbol{u}_{l_{i-2}(k+1)\cdots l_0(k+1)}(k+i\,|\,k+1)$, can be set by the weighted sum of $\boldsymbol{u}^*_{l_{i-1}(k)\cdots l_1(k)l_0(k)}(k+i\,|\,k)$, i.e.

$$\boldsymbol{u}(k+1\,|\,k+1) = \sum_{l_0(k)=1}^{L} \lambda_{l_0(k)} \boldsymbol{u}^*_{l_0(k)}(k+1\,|\,k)$$

$$\boldsymbol{u}_{l_{i-2}(k+1)\cdots l_0(k+1)}(k+i\,|\,k+1) = \sum_{l_0(k)=1}^{L} \lambda_{l_0(k)} \boldsymbol{u}^*_{l_{i-1}(k)\cdots l_1(k)l_0(k)}(k+i\,|\,k), \quad i = 2, \ldots, N-1 \tag{7.47a}$$

Similarly, set

$$\boldsymbol{u}_{l_{N-2}(k+1)\cdots l_0(k+1)}(k+N\,|\,k+1) = \boldsymbol{F}^*(k) \sum_{l_0(k)=1}^{L} \lambda_{l_0(k)} \boldsymbol{x}^*_{l_{N-1}(k)\cdots l_1(k)l_0(k)}(k+N\,|\,k) \tag{7.47b}$$

Since all $\lambda_{l_0(k)}$ are fixed at time k, the components of $\bar{\boldsymbol{u}}$ given by (7.47) are also fixed and the constructed $\boldsymbol{U}(k+1)$ is available. Furthermore, with this setting and (7.44), it is easy to prove that

$$\boldsymbol{u}(k+1\,|\,k+1) = \sum_{l_0(k)=1}^{L} \lambda_{l_0(k)} \boldsymbol{u}^*_{l_0(k)}(k+1\,|\,k) = \boldsymbol{u}^*(k+1\,|\,k)$$

$$\begin{aligned}
\boldsymbol{u}(k+1+i\,|\,k+1) &= \sum_{l_{i-1}(k+1)\cdots l_0(k+1)=1}^{L} \left(\prod_{h=0}^{i-1} \lambda_{l_h(k+1)}\right) \boldsymbol{u}_{l_{i-1}(k+1)\cdots l_0(k+1)}(k+1+i\,|\,k+1) \\
&= \sum_{l_i(k)\cdots l_1(k)=1}^{L} \left(\prod_{h=1}^{i} \lambda_{l_h(k)}\right) \sum_{l_0(k)=1}^{L} \lambda_{l_0(k)} \boldsymbol{u}^*_{l_i(k)\cdots l_1(k)l_0(k)}(k+1+i\,|\,k) \\
&= \sum_{l_i(k)\cdots l_0(k)=1}^{L} \left(\prod_{h=0}^{i} \lambda_{l_h(k)}\right) \boldsymbol{u}^*_{l_i(k)\cdots l_1(k)l_0(k)}(k+1+i\,|\,k) = \boldsymbol{u}^*(k+1+i\,|\,k), \quad i = 1, \ldots, N-2
\end{aligned}$$

$$\begin{aligned}
\boldsymbol{u}(k+N \mid k+1) &= \sum_{l_{N-2}(k+1)\cdots l_0(k+1)=1}^{L} \left(\prod_{h=0}^{N-2} \lambda_{l_h(k+1)}\right) \boldsymbol{u}_{l_{N-2}(k+1)\cdots l_0(k+1)}(k+N \mid k+1) \\
&= \sum_{l_{N-1}(k)\cdots l_1(k)=1}^{L} \left(\prod_{h=1}^{N-1} \lambda_{l_h(k)}\right) \boldsymbol{F}^*(k) \sum_{l_0(k)=1}^{L} \lambda_{l_0(k)} \boldsymbol{x}^*_{l_{N-1}(k)\cdots l_1(k) l_0(k)}(k+N \mid k) \\
&= \boldsymbol{F}^*(k) \sum_{l_{N-1}(k)\cdots l_0(k)=1}^{L} \left(\prod_{h=0}^{N-1} \lambda_{l_h(k)}\right) \boldsymbol{x}^*_{l_{N-1}(k)\cdots l_1(k) l_0(k)}(k+N \mid k) = \boldsymbol{F}^*(k)\boldsymbol{x}^*(k+N \mid k)
\end{aligned}$$

This indicates that the use of $\boldsymbol{U}(k)$ with new components in (7.46), on the one hand, makes the constructed $\boldsymbol{U}(k+1)$ available at time $k+1$ and, on the other hand, implies that the control inputs $\boldsymbol{u}(k+i \mid k+1)$ have the above relationship with $\boldsymbol{u}^*(k+i \mid k)$ and $\boldsymbol{x}^*(k+N \mid k)$, which is just what we originally expect to guarantee on the feasibility of the optimization problem (7.43) at time $k+1$. For more details, refer to [8].

The design in this section uses the dual-mode control concept. With additional free control inputs this kind of RMPC can improve the control performance and enlarge the initial feasible region, but at the cost of bringing a heavy online computational burden and difficuty on recursive feasibility, particularly when the combination parameters describing uncertainty are time varying.

7.1.4 Synthesis of RMPC with Multistep Control Sets

The RMPC described in the last section improves the control performance through adding free control inputs before employing a fixed-state feedback control law, which brings more degrees of freedom for control but also increases the complexity to guarantee the recursive feasibility because of uncertainty expansion. This contradiction was resolved by introducing an open-loop tree structure for state and control expansion. From another perspective, note that the state feedback law is often helpful to ensure recursive feasibility, such as its use in the terminal set. Li and Xi [9] proposed the concept of the multistep control set and used it to design a sequence of feedback control laws instead of free control inputs used in the dual-mode framework. It seems an efficient way to guarantee recursive feasibility and thus is helpful to RMPC synthesis. In this section, we will briefly introduce the synthesis method of RMPC using multistep control sets.

(1) Multistep control set

Definition 7.1 For system (7.1) to (7.5), if all the system states in set $\boldsymbol{S}_0 = \{\boldsymbol{x} \mid \boldsymbol{x}^{\mathrm{T}} \boldsymbol{Q}_0^{-1} \boldsymbol{x} \le 1\}$ can be steered into an invariant set step by step under the control actions given by a sequence of admissible feedback control laws, then set $\boldsymbol{S}_0$ is called as a multistep control set of system (7.1) to (7.5). Meanwhile, if the length of the sequence of admissible feedback control laws is $s-1$, then set $\boldsymbol{S}_0$ is called an s-step control set.

From the definition, it is easy to see that any controlled invariant set is a one-step control set. Furthermore, it is also easy to prove that a set is an s-step control set if all the system states in it can be steered into an $(s-1)$-step control set by an admissible feedback control law. With this property, in the following we discuss how to design an s-step ellipsoidal control set for the constrained polytopic uncertain system (7.1) to (7.5).

An s-step control set can be designed by a series of ellipsoidal sets $\boldsymbol{S}_i = \{\boldsymbol{x} \mid \boldsymbol{x}^{\mathrm{T}} \boldsymbol{Q}_i^{-1} \boldsymbol{x} \le 1\}$, $i = 0, \cdots, s-1$. When the system state $\boldsymbol{x}$ is in the ellipsoidal set $\boldsymbol{S}_{i-1}$, one can use an admissible feedback control law $\boldsymbol{F}_{i-1}\boldsymbol{x}$ to steer the system state $\boldsymbol{x}$ into the ellipsoidal set $\boldsymbol{S}_i$, while the last ellipsoidal set $\boldsymbol{S}_{s-1} = \{\boldsymbol{x} \mid \boldsymbol{x}^{\mathrm{T}} \boldsymbol{Q}_{s-1}^{-1} \boldsymbol{x} \le 1\}$ is a controlled invariant set with the feedback control law $\boldsymbol{F}_{s-1}\boldsymbol{x}$ in usual sense. Let $\boldsymbol{F}_i = \boldsymbol{Y}_i \boldsymbol{Q}_i^{-1}$. Conditions for $\boldsymbol{S}_0$ as an s-step control set can be given by

$$
\begin{aligned}
&\begin{bmatrix} \boldsymbol{Q}_{i-1} & * \\ \boldsymbol{A}_l \boldsymbol{Q}_{i-1} + \boldsymbol{B}_l \boldsymbol{Y}_{i-1} & \boldsymbol{Q}_i \end{bmatrix} > 0, \quad i = 1, \ldots, s-1, \quad l = 1, \ldots, L \\
&\begin{bmatrix} \boldsymbol{Q}_{s-1} & * \\ \boldsymbol{A}_l \boldsymbol{Q}_{s-1} + \boldsymbol{B}_l \boldsymbol{Y}_{s-1} & \boldsymbol{Q}_{s-1} \end{bmatrix} > 0, \quad l = 1, \ldots, L
\end{aligned}
\tag{7.48}
$$

with

$$
\begin{aligned}
&\begin{bmatrix} u_{\max}^2 \boldsymbol{I} & \boldsymbol{Y}_i \\ \boldsymbol{Y}_i^{\mathrm{T}} & \boldsymbol{Q}_i \end{bmatrix} \ge \boldsymbol{0}, \quad i = 0, \ldots, s-1 \\
&\begin{bmatrix} \boldsymbol{Q}_i & (\boldsymbol{A}_l \boldsymbol{Q}_i + \boldsymbol{B}_l \boldsymbol{Y}_i)^{\mathrm{T}} \boldsymbol{C}^{\mathrm{T}} \\ \boldsymbol{C}(\boldsymbol{A}_l \boldsymbol{Q}_i + \boldsymbol{B}_l \boldsymbol{Y}_i) & y_{\max}^2 \boldsymbol{I} \end{bmatrix} \ge \boldsymbol{0}, \quad i = 0, \ldots, s-1, \quad l = 1, \ldots, L
\end{aligned}
\tag{7.49}
$$

where conditions (7.48) ensure that the states in $\boldsymbol{S}_{i-1}$ enter into set $\boldsymbol{S}_i$ at the next time instant under the feedback control law $\boldsymbol{u} = \boldsymbol{F}_{i-1}\boldsymbol{x}$ and meanwhile the set $\boldsymbol{S}_{s-1}$ is a controlled invariant set with the control law $\boldsymbol{u} = \boldsymbol{F}_{s-1}\boldsymbol{x}$. In addition, conditions (7.49) are used to ensure the input and state constraints, which are similar to (7.15), (7.16) in Section 7.1.1.

According to Definition 7.1 and the conditions (7.48) and (7.49), it is easy to show the following properties of a multistep control set.

Property 1 For the uncertain system (7.1) to (7.5), an s-step control set is also an $(s+i)$-step control set $(i \ge 0)$.

Property 2 If $\boldsymbol{S}_0$ is an s-step control set of the uncertain system (7.1) to (7.5) with corresponding series of ellipsoidal sets $\boldsymbol{S}_i$, $i = 1, \ldots, s-1$, then each $\boldsymbol{S}_i$ is also an s-step control set of system (7.1) to (7.5).

Property 3 For system (7.1) to (7.5), if there exist n s-step control sets $\boldsymbol{S}_{0,h}$, $h = 1, \ldots, n$, whose corresponding $\boldsymbol{Q}_{i,h}$, $\boldsymbol{Y}_{i,h}$, $i = 0, \ldots, s-1$, $h = 1, \ldots, n$ satisfy (7.48) and (7.49), then their convex combination $\hat{\boldsymbol{S}}_0 = \{\boldsymbol{x} \mid \boldsymbol{x}^{\mathrm{T}} \hat{\boldsymbol{Q}}_0^{-1} \boldsymbol{x} \le 1\}$ is also an s-step control set, where the convex combination refers to $\hat{\boldsymbol{Q}}_i \triangleq \sum_{h=1}^{n} \lambda_h \boldsymbol{Q}_{i,h}$, $\hat{\boldsymbol{Y}}_i \triangleq \sum_{h=1}^{n} \lambda_h \boldsymbol{Y}_{i,h}$ in corresponding $\hat{\boldsymbol{S}}_i$, $i = 0, \ldots, s-1$ and $\lambda_h \ge 0$, $h = 1, \ldots, n$, $\sum_{h=1}^{n} \lambda_h \le 1$.

(2) RMPC synthesis with a multistep control set

Now for the constrained uncertain system (7.1) to (7.5), consider the online optimization problem (7.6) at time k. Since the current system state $\boldsymbol{x}(k)$ is determined, its quadratic term can be firstly removed from $J(k)$. Then rewrite the performance index in the optimization problem (7.6) as

$$J(k) = \boldsymbol{u}^{\mathrm{T}}(k \mid k)\mathcal{R}\boldsymbol{u}(k \mid k) + \sum_{i=1}^{\infty}\left[\boldsymbol{x}^{\mathrm{T}}(k+i \mid k)\mathcal{Q}\boldsymbol{x}(k+i \mid k) + \boldsymbol{u}^{\mathrm{T}}(k+i \mid k)\mathcal{R}\boldsymbol{u}(k+i \mid k)\right]$$

Using the concept of a multistep control set, the following control strategy is adopted:

$$\pi:\ \boldsymbol{u}(k+i \mid k) = \begin{cases} \boldsymbol{u}(k \mid k), & i=0 \\ \boldsymbol{F}_i\boldsymbol{x}(k+i \mid k), & N>i>0 \\ \boldsymbol{F}_N\boldsymbol{x}(k+i \mid k), & i \ge N \end{cases} \tag{7.50}$$

which implies that a series of ellipsoidal sets $S_i = \left\{\boldsymbol{x} \mid \boldsymbol{x}^{\mathrm{T}}\boldsymbol{Q}_i^{-1}\boldsymbol{x} \le 1\right\}$, $i = 1, \ldots, N$ is available to construct a multistep control set. At first, use a free control action $\boldsymbol{u}(k)$ to steer the system state $\boldsymbol{x}(k)$ into the ellipsoidal set S_1. Then for $i = 1, \ldots, N-1$, step by step use the feedback control law $\boldsymbol{F}_i\boldsymbol{x}$ to steer the system state from the ellipsoidal set S_i to S_{i+1}. The last ellipsoidal set S_N is a controlled invariant set where the feedback control law $\boldsymbol{F}_N\boldsymbol{x}$ is adopted. The control strategy π is to mix the quasi-min-max RMPC in [10] and the multistep control set, where the multistep control set is used as the terminal set of the quasi-min-max RMPC.

With control strategy π, choose quadratic functions $V(i, k) = \boldsymbol{x}^{\mathrm{T}}(k+i \mid k)\boldsymbol{P}_i\boldsymbol{x}(k+i \mid k)$, $i = 1, \ldots, N-1$ for feedback control laws with gain $\boldsymbol{F}_i$. Let $\boldsymbol{P}_i = \boldsymbol{P}_N$ when $i \ge N$. Similar to that in Section 7.1.1, the optimization problem to be solved at time k is derived as follows.

1) Imposing $V(i, k)$ descending and satisfying the following conditions:

$$\begin{aligned} V(i+1,k) - V(i,k) &= \|(\boldsymbol{A}(k+i) + \boldsymbol{B}(k+i)\boldsymbol{F}_i)\boldsymbol{x}(k+i \mid k)\|_{\boldsymbol{P}_{i+1}}^2 - \|\boldsymbol{x}(k+i \mid k)\|_{\boldsymbol{P}_i}^2 \\ &\le -\left(\|\boldsymbol{x}(k+i \mid k)\|_{\mathcal{Q}}^2 + \|\boldsymbol{u}(k+i \mid k)\|_{\mathcal{R}}^2\right),\ \forall[\boldsymbol{A}(k+i)\ \ \boldsymbol{B}(k+i)] \in \boldsymbol{\Pi},\ i \ge 1 \end{aligned} \tag{7.51}$$

2) Obtaining the upper bound of the objective function. Summing Eq. (7.51) from $i = 1$ to $i = \infty$, it follows that

$$J(k) \le \|\boldsymbol{u}(k \mid k)\|_{\mathcal{R}}^2 + V(1,k) \tag{7.52}$$

Define the upper bounds for both $\|\boldsymbol{u}(k \mid k)\|_{\mathcal{R}}^2$ and $V(1, k)$:

$$\|\boldsymbol{u}(k \mid k)\|_{\mathcal{R}}^2 \le \gamma_0 \tag{7.53}$$

$$V(1,k) = \|\boldsymbol{A}(k)\boldsymbol{x}(k) + \boldsymbol{B}(k)\boldsymbol{u}(k \mid k)\|_{\boldsymbol{P}_1}^2 \le \gamma,\ [\boldsymbol{A}(k)\ \ \boldsymbol{B}(k)] \in \boldsymbol{\Pi} \tag{7.54}$$

Then it follows that

$$\max_{[\boldsymbol{A}(k+i)\ \boldsymbol{B}(k+i)] \subset \boldsymbol{\Pi}, i \ge 0} J(k) \le \gamma_0 + \gamma$$

3) Using $\boldsymbol{u}(k \mid k)$ to steer the state $\boldsymbol{x}(k)$ into the ellipsoidal set $S_1 = \{\boldsymbol{x} \mid \boldsymbol{x}^{\mathrm{T}} \boldsymbol{Q}_1^{-1} \boldsymbol{x} \le 1\}$:

$$\|\boldsymbol{A}(k)\boldsymbol{x}(k) + \boldsymbol{B}(k)\boldsymbol{u}(k \mid k)\|_{\boldsymbol{Q}_1^{-1}}^2 \le 1, \quad [\boldsymbol{A}(k) \;\; \boldsymbol{B}(k)] \in \boldsymbol{\Pi} \tag{7.55}$$

$$\|\boldsymbol{u}(k \mid k)\|_2 \le u_{\max} \tag{7.56}$$

$$\|\boldsymbol{y}(k+1 \mid k)\|_2 = \|\boldsymbol{C}(\boldsymbol{A}(k)\boldsymbol{x}(k) + \boldsymbol{B}(k)\boldsymbol{u}(k \mid k))\|_2 \le y_{\max}, \quad [\boldsymbol{A}(k) \;\; \boldsymbol{B}(k)] \in \boldsymbol{\Pi} \tag{7.57}$$

4) Steering the system state into the invariant set S_N by the N-step control set:

$$\begin{aligned} &\|(\boldsymbol{A}(k+i) + \boldsymbol{B}(k+i)\boldsymbol{F}_i)\boldsymbol{x}(k+i \mid k)\|_{\boldsymbol{Q}_{i+1}^{-1}}^2 \le 1, \\ &\qquad [\boldsymbol{A}(k+i) \;\; \boldsymbol{B}(k+i)] \in \boldsymbol{\Pi}, \;\; i = 1, \ldots, N-1 \end{aligned} \tag{7.58}$$

$$\begin{aligned} &\|(\boldsymbol{A}(k+i) + \boldsymbol{B}(k+i)\boldsymbol{F}_N)\boldsymbol{x}(k+i \mid k)\|_{\boldsymbol{Q}_N^{-1}}^2 \le 1, \\ &\qquad [\boldsymbol{A}(k+i) \;\; \boldsymbol{B}(k+i)] \in \boldsymbol{\Pi}, \;\; i \ge N \end{aligned} \tag{7.59}$$

$$\|\boldsymbol{u}(k+i \mid k)\|_2 = \|\boldsymbol{F}_i \boldsymbol{x}(k+i \mid k)\|_2 \le u_{\max}, \qquad i \ge 1 \tag{7.60}$$

$$\begin{aligned} &\|\boldsymbol{y}(k+i+1 \mid k)\|_2 = \|\boldsymbol{C}(\boldsymbol{A}(k+i) + \boldsymbol{B}(k+i)\boldsymbol{F}_i)\boldsymbol{x}(k+i \mid k)\|_2 \le y_{\max}, \\ &\qquad [\boldsymbol{A}(k+i) \;\; \boldsymbol{B}(k+i)] \in \boldsymbol{\Pi}, \;\; i \ge 1 \end{aligned} \tag{7.61}$$

Now all of the above conditions will be converted into LMI forms. Let $\boldsymbol{F}_i = \boldsymbol{Y}_i \boldsymbol{Q}_i^{-1}$, $\boldsymbol{P}_i = \gamma \boldsymbol{Q}_i^{-1}$, $i = 1, \ldots, N$. Then the bound condition (7.54) is equivalent to the ellipsoidal set condition (7.55), and the multistep control set conditions (7.58) and (7.59) could be covered by the condition (7.51) together with (7.55). Therefore we only need to derive LMIs for other conditions. Similar to (7.14) but with a different $\boldsymbol{P}_i$, the $V(i, k)$ descending condition (7.51) can be written as

$$\begin{bmatrix} \boldsymbol{Q}_i & * & * & * \\ \boldsymbol{A}_l \boldsymbol{Q}_i + \boldsymbol{B}_l \boldsymbol{Y}_i & \boldsymbol{Q}_{i+1} & * & * \\ \mathcal{Q}^{1/2} \boldsymbol{Q}_i & \boldsymbol{0} & \gamma \boldsymbol{I} & * \\ \mathcal{R}^{1/2} \boldsymbol{Y}_i & \boldsymbol{0} & \boldsymbol{0} & \gamma \boldsymbol{I} \end{bmatrix} \ge \boldsymbol{0}, \quad i = 1, \ldots, N-1, \;\; l = 1, \ldots, L \tag{7.62}$$

$$\begin{bmatrix} \boldsymbol{Q}_N & * & * & * \\ \boldsymbol{A}_l \boldsymbol{Q}_N + \boldsymbol{B}_l \boldsymbol{Y}_N & \boldsymbol{Q}_N & * & * \\ \mathcal{Q}^{1/2} \boldsymbol{Q}_N & \boldsymbol{0} & \gamma \boldsymbol{I} & * \\ \mathcal{R}^{1/2} \boldsymbol{Y}_N & \boldsymbol{0} & \boldsymbol{0} & \gamma \boldsymbol{I} \end{bmatrix} \ge \boldsymbol{0}, \quad l = 1, \ldots, L \tag{7.63}$$

The bound condition (7.53) will be

$$\begin{bmatrix} \gamma_0 & \boldsymbol{u}^{\mathrm{T}}(k \mid k) \\ \boldsymbol{u}(k \mid k) & \mathcal{R}^{-1} \end{bmatrix} \ge \boldsymbol{0} \tag{7.64}$$

The condition (7.55) of $\boldsymbol{x}(k)$ steered into the ellipsoidal set S_1 can be written as

$$\begin{bmatrix} 1 & * \\ \boldsymbol{A}_l \boldsymbol{x}(k) + \boldsymbol{B}_l \boldsymbol{u}(k \mid k) & \boldsymbol{Q}_1 \end{bmatrix} \ge \boldsymbol{0}, \quad l = 1, \ldots, L \tag{7.65}$$

The input/output constraints at the first step (7.56) and (7.57) can be guaranteed by

$$\begin{bmatrix} u_{\max}^2 & \boldsymbol{u}^{\mathrm{T}}(k \mid k) \\ \boldsymbol{u}(k \mid k) & \boldsymbol{I} \end{bmatrix} \ge \boldsymbol{0} \tag{7.66}$$

$$\begin{bmatrix} y_{\max}^2 \boldsymbol{I} & * \\ \boldsymbol{C}(\boldsymbol{A}_l \boldsymbol{x}(k) + \boldsymbol{B}_l \boldsymbol{u}(k \mid k)) & \boldsymbol{I} \end{bmatrix} \geq \boldsymbol{0}, \quad l = 1, \ldots, L \tag{7.67}$$

Referring to (6.13) and (6.15), the input/output constraints when $\boldsymbol{x}$ enters the multistep control set, i.e. (7.60) and (7.61), can be guaranteed by

$$\begin{bmatrix} u_{\max}^2 \boldsymbol{I} & \boldsymbol{Y}_i \\ \boldsymbol{Y}_i^{\mathrm{T}} & \boldsymbol{Q}_i \end{bmatrix} \geq \boldsymbol{0}, \quad i = 1, \ldots, N \tag{7.68}$$

$$\begin{bmatrix} y_{\max}^2 \boldsymbol{I} & * \\ (\boldsymbol{A}_l \boldsymbol{Q}_i + \boldsymbol{B}_l \boldsymbol{Y}_i)^{\mathrm{T}} \boldsymbol{C}^{\mathrm{T}} & \boldsymbol{Q}_i \end{bmatrix} \geq \boldsymbol{0}, \quad i = 1, \cdots, N, \; l = 1, \cdots, L \tag{7.69}$$

Now the RMPC algorithm based on the multistep control set can be given.

(3) Algorithm and robust stability

Algorithm 7.2 RMPC algorithm based on the multistep control set [9]

Step 1. Solve the following optimization problem according to the current system state $\boldsymbol{x}(k)$. Obtain the optimal $\boldsymbol{u}(k)$:

$$\begin{aligned} &\min_{\gamma_0, \gamma, \boldsymbol{u}(k), \boldsymbol{Q}_i, \boldsymbol{Y}_i} \quad \gamma_0 + \gamma \\ &\text{s.t.} \quad (7.62) \text{ to } (7.69) \end{aligned} \tag{7.70}$$

Step 2. Implement $\boldsymbol{u}(k)$ on the controlled plant and return to Step 1 at the next time.

The following theorem is given for the stability of the above algorithm.

Theorem 7.2 For the system (7.1) to (7.5) controlled by RMPC with online solving the optimization problem (7.70), if the problem (7.70) is feasible for the current state $\boldsymbol{x}(k)$ at time k, then the proposed RMPC is recursively feasible and the closed-loop system is asymptotically stable.

Proof: At time k, starting from $\boldsymbol{x}(k)$, the problem (7.70) is feasible. Denote the optimal solution as $\{\gamma_0^*(k), \gamma^*(k), \boldsymbol{u}^*(k), \boldsymbol{Q}_1^*(k), \boldsymbol{Y}_1^*(k), \ldots, \boldsymbol{Q}_{N-1}^*(k), \boldsymbol{Y}_{N-1}^*(k), \boldsymbol{Q}_N^*(k), \boldsymbol{Y}_N^*(k)\}$. Denote $\boldsymbol{F}_i^*(k) = \boldsymbol{Y}_i^*(k)\left(\boldsymbol{Q}_i^*(k)\right)^{-1}$, $i = 1, \ldots, N$. It is known that $\boldsymbol{x}(k+1 \mid k) = \boldsymbol{A}(k)\boldsymbol{x}(k) + \boldsymbol{B}(k)\boldsymbol{u}^*(k)$, $\boldsymbol{u}(k+1 \mid k) = \boldsymbol{F}_1^*(k)\boldsymbol{x}(k+1 \mid k)$, $\boldsymbol{x}(k+2 \mid k) = \boldsymbol{A}(k+1)\boldsymbol{x}(k+1 \mid k) + \boldsymbol{B}(k+1)\boldsymbol{u}(k+1 \mid k)$, and $V(2, k) < V(1, k) < \gamma^*(k)$ if $\boldsymbol{x}(k) \neq \boldsymbol{0}$.

At time $k + 1$, obtain the actual system state $\boldsymbol{x}(k+1)$ through measurement; then both $\boldsymbol{u}(k+1) = \boldsymbol{F}_1^*(k)\boldsymbol{x}(k+1)$ and $V(k+1) = \boldsymbol{x}(k+1)^{\mathrm{T}}\left(\gamma^*(k)\left(\boldsymbol{Q}_1^*(k)\right)^{-1}\right)\boldsymbol{x}(k+1)$ are fixed. However, the predicted $\boldsymbol{x}(k+2 \mid k) = \left(\boldsymbol{A}(k+1) + \boldsymbol{B}(k+1)\boldsymbol{F}_1^*(k)\right)\boldsymbol{x}(k+1)$ is still uncertain. Define $V(k+2) = \mathrm{Sup}\left\{V(2,k) = \boldsymbol{x}(k+2 \mid k)^{\mathrm{T}}\left(\gamma^*(k)\left(\boldsymbol{Q}_2^*(k)\right)^{-1}\right)\boldsymbol{x}(k+2 \mid k) \mid [\boldsymbol{A}(k+1)\ \boldsymbol{B}(k+1) \in \boldsymbol{\Pi}]\right\}$ which is unknown but fixed. It is obvious that $V(k+2) \leq V(k+1) < \gamma^*(k)$ if $\boldsymbol{x}(k) \neq \boldsymbol{0}$. Then a constant $a = V(k+2)/\gamma^*(k) < 1$ can be defined.

For the optimization problem (7.70) at $k+1$, let the candidate solution $\{\|F_1^*(k)x(k+1)\|_{\mathcal{R}}^2,\ V(k+2), F_1^*(k)x(k+1), aQ_2^*(k), aY_2^*(k), \ldots, aQ_N^*(k), aY_N^*(k), aQ_N^*$ $(k), aY_N^*(k)\}$ It is easy to verify that this solution satisfies conditions (7.62) to (7.64) and (7.68) and(7.69) at time $k+1$. For (7.65), according to the expressions of $x(k+2\,|k)$ and $V(k+2)$, it follows that $V(k+2)\geq\ x(k+1)^{\mathrm{T}}(A(k+1)+B(k+1)F_1^*(k))^{\mathrm{T}}$ $\left(\gamma^*(k)\left(Q_2^*(k)\right)^{-1}\right)(A(k+1)+B(k+1)F_1^*(k))x(k+1)$ for $[A(k+1)\quad B(k+1)]\in\boldsymbol{\Pi}$. Note that $V(k+2)=a\gamma^*(k)$, which is exactly the condition (7.65) at time $k+1$, i.e.

$$\begin{bmatrix} 1 & * \\ A_l x(k+1)+B_l F_1^*(k)x(k+1) & aQ_2^*(k)\end{bmatrix}\geq \mathbf{0},\quad l=1,\ldots,L$$

The conditions (7.66) and (7.67) with the candidate solution at time $k+1$ have the forms

$$\begin{bmatrix} u_{\max}^2 & * \\ F_1^*(k)x(k+1) & I\end{bmatrix}\geq \mathbf{0}$$

$$\begin{bmatrix} y_{\max}^2 I & * \\ C(A_l+B_lF_1^*(k))x(k+1) & I\end{bmatrix}\geq \mathbf{0},\quad l=1,\ldots,L$$

They can be ensured by using $x(k+1)^{\mathrm{T}}\left(Q_1^*(k)\right)^{-1}x(k+1)<1$ and (7.68) and (7.69) at time k, respectively. Thus this candidate solution is proved feasible at time $k+1$.

Denote the optimal γ_0, γ solved at time $k+1$ as $\gamma_0^*(k+1), \gamma^*(k+1)$; then we have $\gamma_0^*(k+1)+\gamma^*(k+1)\leq \|F_1^*(k)x(k+1)\|_{\mathcal{R}}^2+V(k+2)$. On the other hand, from (7.51) it follows that $\|F_1^*(k)x(k+1)\|_{\mathcal{R}}^2+V(k+2)\leq V(k+1)<\gamma^*(k)$, so $\gamma_0^*(k+1)+\gamma^*(k+1)<\gamma_0^*(k)+\gamma^*(k)$. The optimal value function $\gamma_0^*(k)+\gamma^*(k)$ is strictly decreasing if $x(k)\neq\mathbf{0}$ and can be chosen as a Lyapunov function. The closed-loop system is asymptotically stable.

Remark 7.3 By introducing a free control input and a sequence of feedback control laws before the terminal invariant set, the control strategy π in (7.50) brings more degrees of freedom to the RMPC, which is helpful to improve the control performance and enlarge the feasible region of the RMPC in Section 7.1.1. However, as compared with the control strategy (7.29) in Section 7.3.3 where a series of free control inputs are introduced before the terminal set, it can be seen that the control law (7.50) is a special case of (7.29) with the optimization horizon $N+1$. Therefore the feedback RMPC based on the multistep control set presented here is not superior to the RMPC with a series of free control inputs in terms of improving the control performance and enlarging the feasible region. However, as mentioned above, the RMPC using a series of free control inputs is difficult to guarantee recursive feasibility and would lead to a heavy computational burden due to uncertain expansion. However, the feedback RMPC can easily handle the recursive feasibility problem, as shown in the proof of Theorem 7.2. This might be the greatest advantage of using the RMPC based on the multistep control set.

Remark 7.4 The RMPC control strategy (7.50) introduces an additional free control input $\boldsymbol{u}(k\mid k)$ before the system state enters the multistep control set because more free control inputs are helpful for achieving a better control performance. However, only one free control input can be adopted otherwise the same problem on recursive feasibility as indicated by Remark 7.2 will occur. In general, with h free control inputs, no matter whether or not feedback controls follow them, when constructing a candidate solution at time $k+1$, the last free control input cannot be set by shifting and is generally constructed by state feedback, i.e. $\boldsymbol{u}(k+h\mid k+1)=\boldsymbol{F}_1^*\boldsymbol{x}(k+h\mid k)$. If $h=1$, $\boldsymbol{u}(k+1)=\boldsymbol{F}_1^*\boldsymbol{x}(k+1)$ is fixed because $\boldsymbol{x}(k+1)$ is available at time $k+1$. Otherwise $\boldsymbol{u}(k+h\mid k+1)$ is uncertain because of uncertain $\boldsymbol{x}(k+h\mid k)$ and thus the condidate solution is unavailable. If a fixed $\boldsymbol{x}^f(k+h\mid k)$ is assumed among all possible states $\boldsymbol{x}(k+h\mid k)$, then $\boldsymbol{u}^f(k+h)=\boldsymbol{F}_1^*\boldsymbol{x}^f(k+h\mid k)$ in the candidate solution can be fixed. However, as this fixed $\boldsymbol{u}^f(k+h)$ cannot steer all the possible states $\boldsymbol{x}(k+h\mid k)$ into the followed multistep control set or terminal invariant set, recursive feasibility cannot be guaranteed. To avoid that, the control strategy used here only adopts one free control input at the beginning.

7.2 Robust Predictive Control for Systems with Disturbances

In the last section, we focused on synthesizing robust predictive controllers for systems with parameter uncertainty. In real applications, in addition to such unmodeled dynamics, external disturbance always exists and affects the controlled system away from satisfactory performance. How to deal with external disturbance is thus also a hot topic in RMPC study and applications. In this section, two catalogs of synthesis approaches for such types of RMPC are introduced.

7.2.1 Synthesis with Disturbance Invariant Sets

Consider the following linear time-invariant system with external disturbance

$$\boldsymbol{x}(k+1)=\boldsymbol{A}\boldsymbol{x}(k)+\boldsymbol{B}\boldsymbol{u}(k)+\boldsymbol{\omega}(k) \tag{7.71}$$

where $\boldsymbol{A}\in\mathrm{R}^{n\times n}$, $\boldsymbol{B}\in\mathrm{R}^{n\times m}$, and $\boldsymbol{\omega}(k)\in\mathrm{R}^n$ represent the external disturbance. For the controlled system, the state and input constraints are given in general forms as

$$\boldsymbol{x}(k)\in\boldsymbol{\Omega}_x \tag{7.72}$$

$$\boldsymbol{u}(k)\in\boldsymbol{\Omega}_u \tag{7.73}$$

where $\boldsymbol{\Omega}_x$ and $\boldsymbol{\Omega}_u$ are the admissible sets of states and inputs, respectively. The notations of Minkowski set operations are also given as follows, which will be used later. For two sets $\boldsymbol{A}$, $\boldsymbol{B}$, set addition $\boldsymbol{A}\oplus\boldsymbol{B}\triangleq\{a+b\mid a\in\boldsymbol{A}, b\in\boldsymbol{B}\}$, and set subtraction $\boldsymbol{A}\ominus\boldsymbol{B}\triangleq\{a\mid a\oplus\boldsymbol{B}\subseteq\boldsymbol{A}\}$.

If the disturbance is unbounded and varies arbitrarily, it is impossible to guarantee the constraints satisfied. In practical applications, the disturbance with a too large amplitude is often handled by some other measures. It is therefore generally assumed that the disturbance in (7.71) is bounded as

$$\boldsymbol{\omega}(k)\in\boldsymbol{\Omega}\subset\mathrm{R}^n,\quad \boldsymbol{\Omega}=\left\{\boldsymbol{\omega}(k)\;\middle|\;\|\boldsymbol{\omega}(k)\|_2\le\bar{\omega}\right\} \tag{7.74}$$

For system (7.71) to (7.74), the appearance of the disturbance makes the constraint handling much more complicated. If the approaches for RMPC synthesis in Section 7.1 are adopted, the influence caused by disturbance will be time varying, so the conditions that ensure the satisfaction of the input and state constraints are also time varying. In order to simplify the disturbance handling, Mayne et al. [11] proposed an approach to synthesize RMPC by using disturbance invariant sets. In the following, the main idea of this approach is introduced.

(1) Disturbance invariant set

Firstly, similar to the definition of the invariant set given in Chapter 6, the following definition on the disturbance invariant set is given.

Definition 7.2 For system $\boldsymbol{x}(k+1) = \boldsymbol{A}\boldsymbol{x}(k) + \boldsymbol{\omega}(k)$, $\boldsymbol{Z}$ is a disturbance invariant set if $\boldsymbol{x}(k) \in \boldsymbol{Z}$; then $\boldsymbol{x}(k+1) = \boldsymbol{A}\boldsymbol{x}(k) + \boldsymbol{\omega}(k) \in \boldsymbol{Z}$ for all $\boldsymbol{\omega}(k) \in \boldsymbol{\Omega}$.

It is obvious that if the system $\boldsymbol{x}(k+1) = \boldsymbol{A}\boldsymbol{x}(k)$ is unstable, no disturbance invariant set exists. For a control system given by $\boldsymbol{x}(k+1) = \boldsymbol{A}\boldsymbol{x}(k) + \boldsymbol{B}\boldsymbol{u}(k) + \boldsymbol{\omega}(k)$, a feedback control law $\boldsymbol{u}(k) = \boldsymbol{K}\boldsymbol{x}(k)$, often taken as the optimal controller of the unconstrained LQ problem with $(\boldsymbol{A}, \boldsymbol{B}, \mathcal{Q}, \mathcal{R})$, can be used to make $\boldsymbol{x}(k+1) = \boldsymbol{A}_K\boldsymbol{x}(k)$ with $\boldsymbol{A}_K = \boldsymbol{A} + \boldsymbol{B}\boldsymbol{K}$ stable. The disturbance invariant set $\boldsymbol{Z}$ designed for the resultant system $\boldsymbol{x}(k+1) = \boldsymbol{A}_K\boldsymbol{x}(k) + \boldsymbol{\omega}(k)$ makes $\boldsymbol{x}(k+1) \in \boldsymbol{Z}$ for all $\boldsymbol{\omega}(k) \in \boldsymbol{\Omega}$ when $\boldsymbol{x}(k) \in \boldsymbol{Z}$ and $\boldsymbol{u}(k) = \boldsymbol{K}\boldsymbol{x}(k)$.

(2) RMPC synthesis by dividing the uncertain system into nominal and disturbed subsystems

Like the invariant set, the disturbance invariant set can keep the system state in it always within it by using the corresponding feedback control law even when the system is disturbed. This is convenient when handling constraints. Since system (7.71) is linear and of an additive character, [11] proposed an RMPC synthesis approach by dividing the system (7.71) into two subsystems:

$$\bar{\boldsymbol{x}}(k+1) = \boldsymbol{A}\bar{\boldsymbol{x}}(k) + \boldsymbol{B}\bar{\boldsymbol{u}}(k) \tag{7.75a}$$

$$\hat{\boldsymbol{x}}(k+1) = \boldsymbol{A}\hat{\boldsymbol{x}}(k) + \boldsymbol{B}\hat{\boldsymbol{u}}(k) + \boldsymbol{\omega}(k) \tag{7.75b}$$

$$\boldsymbol{x}(k) = \bar{\boldsymbol{x}}(k) + \hat{\boldsymbol{x}}(k) \tag{7.75c}$$

where subsystem (7.75a) is the nominal system of the original system (7.71). The basic idea to synthesize RMPC is as follows. Firstly, as mentioned above, use the feedback control law $\hat{\boldsymbol{u}}(k) = \boldsymbol{K}\hat{\boldsymbol{x}}(k)$ for the subsystem (7.75b) to get $\hat{\boldsymbol{x}}(k+1) = \boldsymbol{A}_K\hat{\boldsymbol{x}}(k) + \boldsymbol{\omega}(k)$ and find a disturbance invariant set $\boldsymbol{Z}$ for it. That is, under the control of $\hat{\boldsymbol{u}}(k)$, the state component $\hat{\boldsymbol{x}}(k) \in \boldsymbol{Z}$ will be kept in $\boldsymbol{Z}$ when the subsystem is disturbed by $\boldsymbol{\omega}(k) \in \boldsymbol{\Omega}$, i.e. $\boldsymbol{A}_K\boldsymbol{Z} \oplus \boldsymbol{\Omega} \subseteq \boldsymbol{Z}$. Then, for the nominal system (7.75a), use the conventional predictive control approach and optimize the nominal control component $\bar{\boldsymbol{u}}(k)$ in order to steer the nominal state $\bar{\boldsymbol{x}}(k)$ to the origin under tighter constraints, which take the influence of the disturbance into account. The real control input will be composed of $\boldsymbol{u}(k) = \bar{\boldsymbol{u}}(k) + \boldsymbol{K}\hat{\boldsymbol{x}}(k)$. With this basic idea, when designing an RMPC using the disturbance invariant set, the admissible input set can be simply divided into two parts: one for designing the disturbance invariant set while the other for stabilizing the nominal system. After the disturbance has been separated, the nominal system will be stabilized by usual predictive control synthesis methods, which simplifies the design.

Synthesis of the disturbed subsystem

For subsystem (7.75b), a disturbance invariant set $\boldsymbol{Z}$ would be found such that with the feedback control law $\hat{\boldsymbol{u}}(k)=\boldsymbol{K}\hat{\boldsymbol{x}}(k)$, $\hat{\boldsymbol{x}}(k+1)=\boldsymbol{A}_K\hat{\boldsymbol{x}}(k)+\boldsymbol{\omega}(k)\in\boldsymbol{Z}$ if $\hat{\boldsymbol{x}}(k)\in\boldsymbol{Z}$. The disturbance invariant set $\boldsymbol{Z}$ should be as small as possible to reduce conservativeness. Its form should also be suitable for optimization in the RMPC design. It has been suggested [11] that an algorithm be used to compute a polytopic, disturbance invariant, outer approximation of the minimal disturbance set, which can be taken as the disturbance invariant set $\boldsymbol{Z}$ here. If the nominal state reaches zero, the system state will always be in set $\boldsymbol{Z}$. Therefore the disturbance invariant set $\boldsymbol{Z}$ serves as the "origin" for the original system with a bounded but unknown disturbance.

Synthesis of the nominal subsystem

Now turn to control of the nominal subsystem (7.75a). It is a linear time-invariant system without uncertainty and disturbance, and thus can be easily handled by conventional predictive control approaches as presented in Chapter 6. The dual-mode control strategy with N free control actions $\{\bar{\boldsymbol{u}}(k),\ldots,\bar{\boldsymbol{u}}(k+N-1)\}$ followed by the feedback control law $\bar{\boldsymbol{u}}(k)=\boldsymbol{K}\bar{\boldsymbol{x}}(k)$ is adopted. At time k, the cost function of the online optimization problem is given by

$$\min_{\bar{\boldsymbol{u}}(k+i)} J_N(k)=\sum_{i=0}^{N-1}\left[\bar{\boldsymbol{x}}^{\mathrm{T}}(k+i)\mathcal{Q}\bar{\boldsymbol{x}}(k+i)+\bar{\boldsymbol{u}}^{\mathrm{T}}(k+i)\mathcal{R}\bar{\boldsymbol{u}}(k+i)\right]+\bar{\boldsymbol{x}}^{\mathrm{T}}(k+N)\boldsymbol{P}\bar{\boldsymbol{x}}(k+N) \tag{7.76}$$

where the last term in (7.76) is the terminal cost function. This optimization problem is very similar to Problem PC4, (6.31) in Section 6.2.3, and could be solved in the same way. However, some important issues different from conventional predictive control should be pointed out.

1) Tighter state and input constraints. Since the optimization is only for the nominal subsystem (7.75a) with state $\bar{\boldsymbol{x}}(k)$ and control input $\bar{\boldsymbol{u}}(k)$, the constraints (7.72) and (7.73) for the real system state $\boldsymbol{x}(k)$ and input $\boldsymbol{u}(k)$ cannot be directly used here. According to (7.75c), the admissible set $\boldsymbol{\Omega}_x$ is shared by $\bar{\boldsymbol{x}}(k)$ and $\hat{\boldsymbol{x}}(k)$. Therefore the admissible set of $\bar{\boldsymbol{x}}(k)$ should be constructed by subtracting the admissible set of $\hat{\boldsymbol{x}}(k)$ from $\boldsymbol{\Omega}_x$. A similar consideration is also for constructing the admissible set of $\bar{\boldsymbol{u}}(k)$ as well as for setting the constraint on the terminal set $\boldsymbol{X}_f$ used in dual-mode control. Therefore, we have tighter state, input, and terminal constraints as follows:

$$\bar{\boldsymbol{x}}(k+i)\in\boldsymbol{\Omega}_x\ominus\boldsymbol{Z},\qquad i=0,1,\ldots,N-1 \tag{7.77}$$

$$\bar{\boldsymbol{u}}(k+i)\in\boldsymbol{\Omega}_u\ominus\boldsymbol{K}\boldsymbol{Z},\qquad i=0,1,\ldots,N-1 \tag{7.78}$$

$$\bar{\boldsymbol{x}}(k+N)\in\boldsymbol{X}_f\subset\boldsymbol{\Omega}_x\ominus\boldsymbol{Z} \tag{7.79}$$

2) Initial state $\bar{\boldsymbol{x}}(k)$ as an optimization variable. When solving the optimization problem at time k, the real system state $\boldsymbol{x}(k)$ is assumed available and is used as the initial state of the optimization in conventional predictive control. However, the optimization here is with respect to $\bar{\boldsymbol{x}}$ and $\bar{\boldsymbol{u}}$. Nothing is known about $\bar{\boldsymbol{x}}(k)$ and thus $\bar{\boldsymbol{x}}(k)$ cannot be set as the initial state in the optimization, but would be taken as an additional optimization variable and handled as a parameter of the control law. It can vary

but would be constrained by the following initial condition with the known system state $\boldsymbol{x}(k)$:

$$\boldsymbol{x}(k) \in \bar{\boldsymbol{x}}(k) \oplus \boldsymbol{Z} \tag{7.80}$$

Thus, the optimization problem at time k can be formulated as

$$\begin{aligned} \min_{\bar{\boldsymbol{x}}(k), \boldsymbol{U}(k)} \ & J_N(k) = \sum_{i=0}^{N-1} (\|\bar{\boldsymbol{x}}(k+i)\|_{\mathcal{Q}}^2 + \|\bar{\boldsymbol{u}}(k+i)\|_{\mathcal{R}}^2) + \|\bar{\boldsymbol{x}}(k+N)\|_{\boldsymbol{P}}^2 \\ \text{s.t.} \ & \bar{\boldsymbol{x}}(k+i+1) = \boldsymbol{A}\bar{\boldsymbol{x}}(k+i) + \boldsymbol{B}\bar{\boldsymbol{u}}(k+i), \quad i = 0, \ldots, N-1 \\ & (7.77) \text{ to } (7.80) \end{aligned} \tag{7.81}$$

where $\boldsymbol{U}(k) \triangleq (\bar{\boldsymbol{u}}(k), \ \ldots, \ \bar{\boldsymbol{u}}(k+N-1))$. the terminal set $\boldsymbol{X}_f$ and the weighting matrix $\boldsymbol{P}$ in the terminal cost function should satisfy the following conditions:

$$\text{(C1)} \boldsymbol{A}_K \boldsymbol{X}_f \subset \boldsymbol{X}_f, \ \ \boldsymbol{X}_f \subset \boldsymbol{\Omega}_x \ominus \boldsymbol{Z}, \ \ \boldsymbol{K}\boldsymbol{X}_f \subset \boldsymbol{\Omega}_u \ominus \boldsymbol{K}\boldsymbol{Z} \tag{7.82a}$$

$$\text{(C2)} (\boldsymbol{A}_K \bar{\boldsymbol{x}})^{\mathrm{T}} \boldsymbol{P} \boldsymbol{A}_K \bar{\boldsymbol{x}} + \bar{\boldsymbol{x}}^{\mathrm{T}} \mathcal{Q} \bar{\boldsymbol{x}} + (\boldsymbol{K}\bar{\boldsymbol{x}})^{\mathrm{T}} \mathcal{R} \boldsymbol{K} \bar{\boldsymbol{x}} \le \bar{\boldsymbol{x}}^{\mathrm{T}} \boldsymbol{P} \bar{\boldsymbol{x}}, \ \ \forall \bar{\boldsymbol{x}} \in \boldsymbol{X}_f \tag{7.82b}$$

where the first condition is to ensure that the terminal set $\boldsymbol{X}_f$ is a controlled invariant set for the nominal system (7.75a) with state component $\bar{\boldsymbol{x}}$ and feedback control law $\bar{\boldsymbol{u}}(k) = \boldsymbol{K}\bar{\boldsymbol{x}}(k)$ when $\bar{\boldsymbol{x}}$ enters $\boldsymbol{X}_f$, while the second condition is to ensure that the value of the objective function is nonincreasing within the terminal set; see condition A4 in (6.38).

Note that the disturbance invariant set $\boldsymbol{Z}$, the terminal set $\boldsymbol{X}_f$, and set operations can be calculated off-line. Thus the online optimization problem (7.81) is simple and straightforward.

After the optimal $\bar{\boldsymbol{x}}^*(k)$ and $\boldsymbol{U}^*(k)$ are solved, the following control law will be applied to the real system:

$$\kappa_N^*(k) = \bar{\boldsymbol{u}}^*(k) + \boldsymbol{K}(\boldsymbol{x}(k) - \bar{\boldsymbol{x}}^*(k)) \tag{7.83}$$

The associated optimal state sequence $\boldsymbol{X}^*(k) \triangleq \{\bar{\boldsymbol{x}}^*(k), \ \ldots, \ \bar{\boldsymbol{x}}^*(k+N)\}$ as well as the optimal value of the cost function $J_N^*(k)$ can also be obtained.

Remark 7.5 It should be pointed out that although the optimization problem (7.81) is formulated only for states and inputs of the nominal system rather than for those of the real system, the constraints on the real states and inputs can still be ensured because the disturbance components are already considered in the terminal set and the state and input constraints by using the disturbance invariant set. It is easy to verify $\boldsymbol{x}(k) \in \boldsymbol{\Omega}_x$ by $\boldsymbol{x}(k) = \bar{\boldsymbol{x}}(k) + \hat{\boldsymbol{x}}(k)$ with $\bar{\boldsymbol{x}}(k) \in \boldsymbol{\Omega}_x \ominus \boldsymbol{Z}$ (see (7.77) and (7.82a)) and $\hat{\boldsymbol{x}}(k) \in \boldsymbol{Z}$, and also to verify $\boldsymbol{u}(k) \in \boldsymbol{\Omega}_u$ by $\boldsymbol{u}(k) = \bar{\boldsymbol{u}}(k) + \boldsymbol{K}(\boldsymbol{x}(k) - \bar{\boldsymbol{x}}(k))$ with $\bar{\boldsymbol{u}}(k) \in \boldsymbol{\Omega}_u \ominus \boldsymbol{K}\boldsymbol{Z}$ (see (7.78) and (7.82a)) and $\hat{\boldsymbol{u}}(k) = \boldsymbol{K}\hat{\boldsymbol{x}}(k) \in \boldsymbol{K}\boldsymbol{Z}$.

(3) Theoretical properties of the RMPC controller

For the optimization problem (7.81), the optimal solution and the control law depend on the system state $\boldsymbol{x}(k)$. The following important results were given in [11] after investigating the theoretical properties of the solution and the controlled system:

1) For all $\boldsymbol{x}(k) \in \boldsymbol{Z}$, $J^*(\boldsymbol{x}) = 0$, $\bar{\boldsymbol{x}}^*(\boldsymbol{x}) = 0$, $\boldsymbol{U}^*(\boldsymbol{x}) = \{\boldsymbol{0}, \ldots, \boldsymbol{0}\}$, $\boldsymbol{X}^*(\boldsymbol{x}) = \{\boldsymbol{0}, \ldots, \boldsymbol{0}\}$, and the control law $\boldsymbol{\kappa}_N^*(\boldsymbol{x}) = \boldsymbol{K}\boldsymbol{x}$.
2) Let $\boldsymbol{X}_N$ be the set of $\boldsymbol{x}(k)$ for which there always exists $\bar{\boldsymbol{x}}(k)$ satisfying (7.77) to (7.80). Suppose that $\boldsymbol{x}(k) \in \boldsymbol{X}_N$ so that $(\bar{\boldsymbol{x}}^*(k), \boldsymbol{U}^*(k))$ exists and is feasible for Problem (7.81) at time k. Then, for all $\boldsymbol{x}(k+1) \in \boldsymbol{A}\boldsymbol{x} + \boldsymbol{B}\boldsymbol{\kappa}_N^*(\boldsymbol{x}) \oplus \boldsymbol{\Omega}$, $(\bar{\boldsymbol{x}}^*(k+1),\ \bar{\boldsymbol{U}}(k+1))$ with $\bar{\boldsymbol{U}}(k+1) \triangleq \{\bar{\boldsymbol{u}}^*(k+1), \ldots, \bar{\boldsymbol{u}}^*(k+N-1), \boldsymbol{K}\bar{\boldsymbol{x}}^*(k+N)\}$ is feasible at time $k+1$ and $J_N^*(k+1) \le J_N^*(k) - \left(\|\bar{\boldsymbol{x}}^*(k)\|_Q^2 + \|\bar{\boldsymbol{u}}^*(k)\|_R^2\right)$.
3) The set $\boldsymbol{Z}$ is robustly exponentially stable for the uncertain controlled system $\boldsymbol{x}(k+1) = \boldsymbol{A}\boldsymbol{x}(k) + \boldsymbol{B}\boldsymbol{\kappa}_N^*(k) + \boldsymbol{\omega}(k)$, $\boldsymbol{\omega}(k) \in \boldsymbol{\Omega}$ with a region of attraction $\boldsymbol{X}_N$, which means there exists a $c > 0$ and a $\gamma \in (0, 1)$ such that for the initial state $\boldsymbol{x}(0) \in \boldsymbol{X}_N$ and admissible disturbance sequence $\boldsymbol{\omega}(k) \in \boldsymbol{\Omega}$, $k \ge 0$, any solution $\boldsymbol{x}(k)$ of above system satisfies $d(\boldsymbol{x}(k), \boldsymbol{Z}) \le c\gamma^k d(\boldsymbol{x}(0), \boldsymbol{Z})$ for all $k \ge 0$, i.e. the real system state $\boldsymbol{x}(k)$ will exponentially approach the disturbance invariant set $\boldsymbol{Z}$, where $d(\boldsymbol{x}, \boldsymbol{Z})$ is the minimal distance of a point $\boldsymbol{x}$ from the set $\boldsymbol{Z}$.

The RMPC approach presented above makes use of the linear and additive character of the original system to divide it into two subsystems. A modified optimization problem for the nominal subsystem is solved online where the initial nominal state is taken as an additional optimization variable. A suitable Lyapunov function is accordingly selected that has value zero in the set $\boldsymbol{Z}$, with which the robust exponential stability of $\boldsymbol{Z}$ can be proved. It is shown that the real state of the controlled system is always along a tunnel with the nominal system state $\bar{\boldsymbol{x}}(k)$ as the center and the disturbance invariant set $\boldsymbol{Z}$ as the cross-section. This idea led to the tube technique, which was later widely used in RMPC synthesis for systems with uncertainty and disturbance.

7.2.2 Synthesis with Mixed H_2/H_∞ Performances

In this section, we introduce another kind of RMPC for systems with disturbance, which adopts H_2 and H_∞ performance in optimization. Consider the following disturbed system:

$$\begin{aligned} \boldsymbol{x}(k+1) &= \boldsymbol{A}(k)\boldsymbol{x}(k) + \boldsymbol{B}(k)\boldsymbol{u}(k) + \boldsymbol{B}_\omega\boldsymbol{\omega}(k) \\ \boldsymbol{z}(k) &= \begin{bmatrix} \boldsymbol{C}\boldsymbol{x}(k) \\ \boldsymbol{R}\boldsymbol{u}(k) \end{bmatrix} \end{aligned} \tag{7.84}$$

where $\boldsymbol{x}(k) \in \mathrm{R}^n$, $\boldsymbol{u}(k) \in \mathrm{R}^m$, and $\boldsymbol{z}(k) \in \mathrm{R}^r$ are system state, control input, and controlled output, respectively. $\boldsymbol{A}(k) \in \mathrm{R}^{n\times n}$, $\boldsymbol{B}(k) \in \mathrm{R}^{n\times m}$, and $[\boldsymbol{A}(k) \quad \boldsymbol{B}(k)] \in \boldsymbol{\Pi}$ with

$$\boldsymbol{\Pi} = \left\{ [\boldsymbol{A}(k)\ \boldsymbol{B}(k)] \,|\, [\boldsymbol{A}(k)\ \boldsymbol{B}(k)] = \sum_{l=1}^{L} \lambda_l [\boldsymbol{A}_l\ \boldsymbol{B}_l], \lambda_l \ge 0, \sum_{l=1}^{L} \lambda_l = 1 \right\} \tag{7.85}$$

The disturbance $\boldsymbol{\omega}(k) \in \mathrm{R}^s$ is bounded as

$$\sum_{k=0}^{\infty} \boldsymbol{\omega}^T(k)\boldsymbol{\omega}(k) \le \bar{\omega}, \quad \boldsymbol{\omega}(k) \in \boldsymbol{\Pi}_q := \{\boldsymbol{\omega} \,|\, \boldsymbol{\omega}^T q^{-1} \boldsymbol{\omega} \le 1\} \tag{7.86}$$

where $\bar{\omega} > 0$ and $q > 0$ are known.

The constraints on the system inputs and maesurable states are given by

$$|u_i(k)| \le \bar{u}_i, \quad i = 1, \ldots, m \tag{7.87}$$

$$|\boldsymbol{\Psi}_j \boldsymbol{x}(k)| \le \bar{x}_j, \quad j = 1, \ldots, s \tag{7.88}$$

where $\boldsymbol{\Psi}_j$ is the jth row of the state constraint matrix $\boldsymbol{\Psi} \in R^{s\times n}$. Note that (7.87) is just the peak bound constraint on the input components, as discussed in Section 6.1.4, while (7.88) is the bound constraint on the linear combination of the states.

Unlike system (7.71) discussed in the last section, system (7.84) includes not only the external disturbance but also the model uncertainty. Although the disturbance invariant set can also be designed for the system with model uncertainty, the resultant design may be conservative.

It is well known that the LQ performance index corresponds to an H_2 optimal control problem, which is generally useful for optimal controller design and has traditionally been adopted as the objective function in RMPC to achieve a better control performance, while the H_∞ performance index is known as a useful criterion on a control system for its ability to reject disturbance. In order to achieve both the desired disturbance rejection and a good dynamic performance, it is a natural idea to combine these two performances, i.e. using a mixed H_2/H_∞ performance, to synthesize RMPC. This idea actually came from research of robust control and was later adopted in RMPC study. In this section we briefly introduce this kind of RMPC synthesis approach presented in [12].

For system (7.84) to (7.88), consider the following two performance requirements:

1) H_∞ performance requirement: under the zero-initial condition, the controlled output $\boldsymbol{z}(k)$ satisfies

$$\sum_{k=0}^{\infty} \boldsymbol{z}^{\mathrm{T}}(k)\boldsymbol{z}(k) \le \gamma^2 \sum_{k=0}^{\infty} \boldsymbol{\omega}^{\mathrm{T}}(k)\boldsymbol{\omega}(k) \tag{7.89}$$

2) H_2 performance requirement: the controlled output $\boldsymbol{z}(k)$ satisfies

$$\|\boldsymbol{z}\|_2^2 := \sum_{k=0}^{\infty} \boldsymbol{z}^{\mathrm{T}}(k)\boldsymbol{z}(k) \le \alpha \tag{7.90}$$

To optimize both performance indices, it is usual to firstly assign an upper bound on one performance index and then design RMPC by online optimizing the other performance index with the above upper bound condition as an additional constraint. Here the H_2 performance index α is taken as the online optimization objective of RMPC and the H_∞ performance index $\gamma > 0$ is set as a fixed scalar to deal with (7.89) as a constraint.

Following the above idea, choose the RMPC control law as $\boldsymbol{u}(k) = \boldsymbol{F}\boldsymbol{x}(k)$. At time k, based on the initial state $\boldsymbol{x}(k)$, an optimization problem with the performance indices (7.89) and (7.90), rewritten with the initial time k, can be solved online. Define the quadratic function $V(\boldsymbol{x}) = \boldsymbol{x}^{\mathrm{T}}\boldsymbol{P}\boldsymbol{x}, \ \boldsymbol{P} > 0$. Denote $\boldsymbol{\phi}(k) = \boldsymbol{A}(k) + \boldsymbol{B}(k)\boldsymbol{F}$ and $\boldsymbol{\phi}_l = \boldsymbol{A}_l + \boldsymbol{B}_l\boldsymbol{F}$, $l = 1, \ldots, L$. We can get

$$
\begin{aligned}
&V(\boldsymbol{x}(k+1)) - V(\boldsymbol{x}(k)) \\
&= \|\boldsymbol{\phi}(k)\boldsymbol{x}(k) + \boldsymbol{B}_\omega\boldsymbol{\omega}(k)\|_{\boldsymbol{P}}^2 - \|\boldsymbol{x}(k)\|_{\boldsymbol{P}}^2 \\
&= \boldsymbol{x}^{\mathrm{T}}(k)\boldsymbol{\phi}^{\mathrm{T}}(k)\boldsymbol{P}\boldsymbol{\phi}(k)\boldsymbol{x}(k) - \boldsymbol{x}^{\mathrm{T}}(k)\boldsymbol{P}\boldsymbol{x}(k) + \boldsymbol{x}^{\mathrm{T}}(k)\boldsymbol{\phi}^{\mathrm{T}}(k)\boldsymbol{P}\boldsymbol{B}_\omega\boldsymbol{\omega}(k) + \boldsymbol{\omega}^{\mathrm{T}}(k)\boldsymbol{B}_\omega^{\mathrm{T}}\boldsymbol{P}\boldsymbol{\phi}(k)\boldsymbol{x}(k) \\
&\quad + \boldsymbol{\omega}^{\mathrm{T}}(k)\boldsymbol{B}_\omega^{\mathrm{T}}\boldsymbol{P}\boldsymbol{B}_\omega\boldsymbol{\omega}(k) + \boldsymbol{z}^{\mathrm{T}}(k)\boldsymbol{z}(k) - \boldsymbol{z}^{\mathrm{T}}(k)\boldsymbol{z}(k) + \gamma^2\boldsymbol{\omega}^{\mathrm{T}}(k)\boldsymbol{\omega}(k) - \gamma^2\boldsymbol{\omega}^{\mathrm{T}}(k)\boldsymbol{\omega}(k) \\
&= [\boldsymbol{x}^{\mathrm{T}}(k)\ \boldsymbol{\omega}^{\mathrm{T}}(k)]\boldsymbol{M}(k)\left[\boldsymbol{x}^{\mathrm{T}}(k)\ \ \boldsymbol{\omega}^{\mathrm{T}}(k)\right]^{\mathrm{T}} + \gamma^2\boldsymbol{\omega}^{\mathrm{T}}(k)\boldsymbol{\omega}(k) - \boldsymbol{z}^{\mathrm{T}}(k)\boldsymbol{z}(k)
\end{aligned}
\tag{7.91}
$$

where

$$
\boldsymbol{M}(k) = \begin{bmatrix} \boldsymbol{\phi}^{\mathrm{T}}(k)\boldsymbol{P}\boldsymbol{\phi}(k) - \boldsymbol{P} + \begin{bmatrix}\boldsymbol{C}\\ \boldsymbol{RF}\end{bmatrix}^{\mathrm{T}}\begin{bmatrix}\boldsymbol{C}\\ \boldsymbol{RF}\end{bmatrix} & \boldsymbol{\phi}^{\mathrm{T}}(k)\boldsymbol{P}\boldsymbol{B}_\omega \\ \boldsymbol{B}_\omega^{\mathrm{T}}\boldsymbol{P}\boldsymbol{\phi}(k) & \boldsymbol{B}_\omega^{\mathrm{T}}\boldsymbol{P}\boldsymbol{B}_\omega - \gamma^2\boldsymbol{I} \end{bmatrix}
$$

Note that from (7.86) the disturbance is energy-bounded. This means that, if the closed-loop system is stable, then $\lim_{k\to\infty}\boldsymbol{x}(k) = \boldsymbol{0}$. Thus, summing (7.91) from k to ∞ can yield

$$
\begin{aligned}
-\boldsymbol{x}^{\mathrm{T}}(k)\boldsymbol{P}\boldsymbol{x}(k) &= \sum_{i=0}^{\infty}\left[-\boldsymbol{z}^{\mathrm{T}}(k+i)\boldsymbol{z}(k+i) + \gamma^2\boldsymbol{\omega}^{\mathrm{T}}(k+i)\boldsymbol{\omega}(k+i)\right] \\
&\quad + \sum_{k=0}^{\infty}\left[\boldsymbol{x}^{\mathrm{T}}(k+i)\ \ \boldsymbol{\omega}^{\mathrm{T}}(k+i)\right]\boldsymbol{M}(k+i)\left[\boldsymbol{x}^{\mathrm{T}}(k+i)\ \ \boldsymbol{\omega}^{\mathrm{T}}(k+i)\right]^{\mathrm{T}}
\end{aligned}
$$

It follows that

$$
\begin{aligned}
\sum_{i=0}^{\infty}\boldsymbol{z}^{\mathrm{T}}(k+i)\boldsymbol{z}(k+i) &= \boldsymbol{x}^{\mathrm{T}}(k)\boldsymbol{P}\boldsymbol{x}(k) + \sum_{i=0}^{\infty}\gamma^2\boldsymbol{\omega}^{\mathrm{T}}(k+i)\boldsymbol{\omega}(k+i) \\
&\quad + \sum_{k=0}^{\infty}\left[\boldsymbol{x}^{\mathrm{T}}(k+i)\ \ \boldsymbol{\omega}^{\mathrm{T}}(k+i)\right]\boldsymbol{M}(k+i)\left[\boldsymbol{x}^{\mathrm{T}}(k+i)\ \ \boldsymbol{\omega}^{\mathrm{T}}(k+i)\right]^{\mathrm{T}}
\end{aligned}
\tag{7.92}
$$

From (7.92), it can be seen that:

1) When $\boldsymbol{x}(k) = \boldsymbol{0}$, the H_∞ performance requirement (7.89), i.e.

$$
\sum_{i=0}^{\infty}\boldsymbol{z}^T(k+i)\boldsymbol{z}(k+i) \le \gamma^2\sum_{i=0}^{\infty}\boldsymbol{\omega}^T(k+i)\boldsymbol{\omega}(k+i)
$$

can be ensured if $\boldsymbol{M}(k+i) \le \boldsymbol{0}$.

2) When $\boldsymbol{M}(k+i) \le 0$, the H_2 performance requirement (7.90) can be guaranteed if the following condition holds:

$$
\sum_{i=0}^{\infty}\left[\boldsymbol{z}^{\mathrm{T}}(k+i)\boldsymbol{z}(k+i)\right] \le \boldsymbol{x}^{\mathrm{T}}(k)\boldsymbol{P}\boldsymbol{x}(k) + \gamma^2\bar{\omega} \le \alpha
\tag{7.93}
$$

where $\boldsymbol{M}(k+i) \leq \mathbf{0}$ can be rewritten as

$$\boldsymbol{M}(k+i) = \begin{bmatrix} \boldsymbol{\phi}^{\mathrm{T}}(k+i)\boldsymbol{P}\boldsymbol{\phi}(k+i) - \boldsymbol{P} + \begin{bmatrix} \boldsymbol{C} \\ \boldsymbol{RF} \end{bmatrix}^{\mathrm{T}} \begin{bmatrix} \boldsymbol{C} \\ \boldsymbol{RF} \end{bmatrix} & \boldsymbol{\phi}^{\mathrm{T}}(k+i)\boldsymbol{P}\boldsymbol{B}_{\omega} \\ \boldsymbol{B}_{\omega}^{\mathrm{T}}\boldsymbol{P}\boldsymbol{\phi}(k+i) & \boldsymbol{B}_{\omega}^{\mathrm{T}}\boldsymbol{P}\boldsymbol{B}_{\omega} - \gamma^2\boldsymbol{I} \end{bmatrix}$$

$$= -\begin{bmatrix} \boldsymbol{P} & \mathbf{0} \\ \mathbf{0} & \gamma^2\boldsymbol{I} \end{bmatrix} + \begin{bmatrix} \boldsymbol{\phi}^{\mathrm{T}}(k+i) \\ \boldsymbol{B}_{\omega}^{\mathrm{T}} \end{bmatrix} \boldsymbol{P} \begin{bmatrix} \boldsymbol{\phi}^{\mathrm{T}}(k+i) \\ \boldsymbol{B}_{\omega}^{\mathrm{T}} \end{bmatrix}^{\mathrm{T}} + \begin{bmatrix} \boldsymbol{C}^{\mathrm{T}} \\ \mathbf{0} \end{bmatrix} \begin{bmatrix} \boldsymbol{C}^{\mathrm{T}} \\ \mathbf{0} \end{bmatrix}^{\mathrm{T}} + \begin{bmatrix} (\boldsymbol{RF})^{\mathrm{T}} \\ \mathbf{0} \end{bmatrix} \begin{bmatrix} (\boldsymbol{RF})^{\mathrm{T}} \\ \mathbf{0} \end{bmatrix}^{\mathrm{T}} \leq \mathbf{0} \tag{7.94}$$

which can be reformulated as the LMIs

$$\begin{bmatrix} \boldsymbol{P} & \mathbf{0} & * & * & * \\ \mathbf{0} & \gamma^2\boldsymbol{I} & * & * & * \\ \boldsymbol{\phi}(k+i) & \boldsymbol{B}_{\omega} & \boldsymbol{P}^{-1} & * & * \\ \boldsymbol{C} & \mathbf{0} & \mathbf{0} & \boldsymbol{I} & * \\ \boldsymbol{RF} & \mathbf{0} & \mathbf{0} & \mathbf{0} & \boldsymbol{I} \end{bmatrix} \geq \mathbf{0}$$

Let $\boldsymbol{Q} = \alpha\boldsymbol{P}^{-1}$, $\boldsymbol{F} = \boldsymbol{Y}\boldsymbol{Q}^{-1}$. Note that $\boldsymbol{\phi}(k+i)$ is a convex combination of $\boldsymbol{\phi}_l$. Left- and right-multiply the above inequality by diag $(\alpha^{1/2}\boldsymbol{P}^{-1}, \alpha^{1/2}\boldsymbol{I}, \alpha^{1/2}\boldsymbol{I}, \alpha^{1/2}\boldsymbol{I}, \alpha^{1/2}\boldsymbol{I})$. Then the following condition can be obtained which can ensure the above inequality:

$$\begin{bmatrix} \boldsymbol{Q} & \mathbf{0} & * & * & * \\ \mathbf{0} & \alpha\gamma^2\boldsymbol{I} & * & * & * \\ \boldsymbol{A}_l\boldsymbol{Q} + \boldsymbol{B}_l\boldsymbol{Y} & \alpha\boldsymbol{B}_{\omega} & \boldsymbol{Q} & * & * \\ \boldsymbol{CQ} & \mathbf{0} & \mathbf{0} & \alpha\boldsymbol{I} & * \\ \boldsymbol{RY} & \mathbf{0} & \mathbf{0} & \mathbf{0} & \alpha\boldsymbol{I} \end{bmatrix} \geq \mathbf{0}, \quad l = 1, \ldots, L \tag{7.95}$$

Meanwhile, (7.93) can be rewritten into LMI as

$$\begin{bmatrix} 1 & * & * \\ \gamma^2\bar{\omega} & \alpha\gamma^2\bar{\omega} & * \\ \boldsymbol{x}(k) & 0 & \boldsymbol{Q} \end{bmatrix} \geq \mathbf{0} \tag{7.96}$$

which implies that $\boldsymbol{x}^{\mathrm{T}}(k)\boldsymbol{Q}^{-1}\boldsymbol{x}(k) + \gamma^2\bar{\omega}/\alpha \leq 1$, and thus the current system state $\boldsymbol{x}(k)$ belongs to an ellipsoidal set $\boldsymbol{\Omega} = \{\boldsymbol{x} \mid \boldsymbol{x}^{\mathrm{T}}\boldsymbol{Q}^{-1}\boldsymbol{x} \leq 1\}$. Therefore, as (6.14) in Section 6.1.4, the input constraints (7.87) can be written as

$$\begin{bmatrix} \boldsymbol{X} & \boldsymbol{Y} \\ \boldsymbol{Y}^{\mathrm{T}} & \boldsymbol{Q} \end{bmatrix} \geq \mathbf{0}, \text{ where } \boldsymbol{X}_{ii} \leq \bar{u}_i^2, \quad i = 1, \ldots, m \tag{7.97}$$

The state constraints (7.88) can be written as

$$\begin{bmatrix} \boldsymbol{V} & * \\ (\boldsymbol{\Psi}\boldsymbol{Q})^{\mathrm{T}} & \boldsymbol{Q} \end{bmatrix} > \mathbf{0}, \text{ where } \boldsymbol{V}_{jj} \leq \bar{x}_j^2, \quad j = 1, \ldots, s \tag{7.98}$$

However, due to the existence of disturbance, the condition $\boldsymbol{M}(k+i) \leq \boldsymbol{0}$, i.e. (7.95), is not enough to ensure $V(\boldsymbol{x}(k+1)) - V(\boldsymbol{x}(k)) \leq 0$ in (7.91), and $\boldsymbol{\Omega} = \{\boldsymbol{x} \mid \boldsymbol{x}^{\mathrm{T}}\boldsymbol{Q}^{-1}\boldsymbol{x} \leq 1\}$ is not a controlled invariant set, i.e. the system state $\boldsymbol{x}(k+1)$ cannot be ensured to be in $\boldsymbol{\Omega}$ even when the current state $\boldsymbol{x}(k) \in \boldsymbol{\Omega}$. This also leads to a problem with recursive feasibility and will be solved by adding some new conditions as follows.

For the current state $\boldsymbol{x}(k) \in \boldsymbol{\Omega}$, it follows that $\boldsymbol{x}(k+1) = \boldsymbol{\phi}(k)\boldsymbol{x}(k) + \boldsymbol{B}_\omega\boldsymbol{\omega}(k)$. To ensure the state $\boldsymbol{x}(k+1) \in \boldsymbol{\Omega}$, divide it into two parts: the nominal substate $\bar{\boldsymbol{x}}(k+1) = \boldsymbol{\phi}(k)\boldsymbol{x}(k)$ and the substate $\boldsymbol{v}(k) = \boldsymbol{B}_\omega\boldsymbol{\omega}(k)$ affected by disturbance. The condition of $\boldsymbol{x}(k+1)$ in the set $\boldsymbol{\Omega} = \{\boldsymbol{x} \mid \boldsymbol{x}^{\mathrm{T}}\boldsymbol{Q}^{-1}\boldsymbol{x} \leq 1\}$ can then be represented by

$$\begin{bmatrix} 1 & (\bar{\boldsymbol{x}}(k+1) + \boldsymbol{v}(k))^{\mathrm{T}} \\ \bar{\boldsymbol{x}}(k+1) + \boldsymbol{v}(k) & \boldsymbol{Q} \end{bmatrix} \geq \boldsymbol{0}$$

Let $0 < b < 1$ be a parameter chosen in advance. The above matrix inequality can be ensured if both of the following two conditions hold:

$$\begin{bmatrix} 1-b & * \\ \bar{\boldsymbol{x}}(k+1) & \boldsymbol{Q}-\boldsymbol{Q}_\omega \end{bmatrix} \geq \boldsymbol{0} \tag{7.99}$$

$$\begin{bmatrix} b & * \\ \boldsymbol{B}_\omega\boldsymbol{\omega}(k) & \boldsymbol{Q}_\omega \end{bmatrix} \geq \boldsymbol{0} \tag{7.100}$$

where (7.100) indicates that $\boldsymbol{\omega}(k)$ should be in an ellipsoidal set $\{\boldsymbol{\omega} \mid b \geq \boldsymbol{\omega}^{\mathrm{T}} B_\omega^{\mathrm{T}} \boldsymbol{Q}_\omega^{-1} B_\omega \boldsymbol{\omega}\}$. According to (7.86), all the disturbances $\boldsymbol{\omega}(k) \in \boldsymbol{\Pi}_q := \{\boldsymbol{\omega} \mid \boldsymbol{\omega}^{\mathrm{T}} q^{-1} \boldsymbol{\omega} \leq 1\}$. Therefore (7.100) can be ensured if $B_\omega^{\mathrm{T}}\boldsymbol{Q}_\omega^{-1}B_\omega \leq bq^{-1}$, which can be rewritten as

$$\begin{bmatrix} bq^{-1} & * \\ \boldsymbol{B}_\omega & \boldsymbol{Q}_\omega \end{bmatrix} \geq \boldsymbol{0} \tag{7.101}$$

Meanwhile, (7.99) means $\bar{\boldsymbol{x}}^{\mathrm{T}}(k+1)\boldsymbol{Q}_b^{-1}\bar{\boldsymbol{x}}(k+1) \leq 1$, where $\boldsymbol{Q}_b = (1-b)(\boldsymbol{Q} - \boldsymbol{Q}_\omega)$. Since $\boldsymbol{x}(k) \in \boldsymbol{\Omega} = \{\boldsymbol{x} \mid \boldsymbol{x}^{\mathrm{T}}\boldsymbol{Q}^{-1}\boldsymbol{x} \leq 1\}$, this can be ensured if

$$\bar{\boldsymbol{x}}^{\mathrm{T}}(k+1)\boldsymbol{Q}_b^{-1}\bar{\boldsymbol{x}}(k+1) \leq \boldsymbol{x}^{\mathrm{T}}(k)\boldsymbol{Q}^{-1}\boldsymbol{x}(k) \tag{7.102}$$

Let $\boldsymbol{Q}_b = \alpha\boldsymbol{P}_b^{-1}$. Referring to (7.91), we get

$$\begin{aligned} &\|\boldsymbol{x}(k+1)\|_{\boldsymbol{P}_b}^2 - \|\boldsymbol{x}(k)\|_{\boldsymbol{P}}^2 \\ &= \left[\boldsymbol{x}^{\mathrm{T}}(k) \;\; \boldsymbol{\omega}^{\mathrm{T}}(k)\right]\boldsymbol{M}_b(k)\left[\boldsymbol{x}^{\mathrm{T}}(k) \;\; \boldsymbol{\omega}^{\mathrm{T}}(k)\right]^{\mathrm{T}} + \gamma^2\boldsymbol{\omega}^{\mathrm{T}}(k)\boldsymbol{\omega}(k) - \boldsymbol{z}^{\mathrm{T}}(k)\boldsymbol{z}(k) \end{aligned} \tag{7.103}$$

where $\boldsymbol{M}_b(k+i) \leq \boldsymbol{0}$ can be ensured by

$$\begin{bmatrix} \boldsymbol{Q} & \boldsymbol{0} & * & * & * \\ \boldsymbol{0} & \alpha\gamma^2\boldsymbol{I} & * & * & * \\ A_l\boldsymbol{Q} + \boldsymbol{B}_l\boldsymbol{Y} & \alpha\boldsymbol{B}_\omega & \boldsymbol{Q}_b & * & * \\ \boldsymbol{C}\boldsymbol{Q} & \boldsymbol{0} & \boldsymbol{0} & \alpha\boldsymbol{I} & * \\ \boldsymbol{R}\boldsymbol{Y} & \boldsymbol{0} & \boldsymbol{0} & \boldsymbol{0} & \alpha\boldsymbol{I} \end{bmatrix} \geq \boldsymbol{0}, \quad l = 1, \ldots, L \tag{7.104}$$

Note that $\bar{\boldsymbol{x}}(k+1)$ is the nominal substate with $\boldsymbol{\omega}(k) = \boldsymbol{0}$; from (7.103) it is known that (7.104) can guarantee (7.102) and then (7.99). It is also tighter than condition (7.95) due to $\boldsymbol{Q}_b < \boldsymbol{Q}$ and thus can cover (7.95). With conditions (7.101) and (7.104), the recursive feasibility, i.e. state $\boldsymbol{x}(k+1) \in \boldsymbol{\Omega}$ if $\boldsymbol{x}(k) \in \boldsymbol{\Omega}$, can be ensured even if a disturbance exists, and the decreasing of $V(\boldsymbol{x}(k))$ can be guaranteed.

As mentioned above, here the H_∞ performance γ is used as a constraint and the H_2 performance α would be optimized. By fixing γ and choosing the parameter b in advance, the online optimization problem of RMPC can be formulated as follows:

$$\begin{aligned} &\min_{Q, Y, X, V, Q_\omega, \alpha} \quad \alpha \\ &\text{s.t.} \quad (7.96) \text{ to } (7.98),\ (7.101), (7.104) \end{aligned} \tag{7.105}$$

In the above, the basic idea of the RMPC synthesis with mixed H_2/H_∞ performance is introduced. It adopts the control startegy with the unique feedback control law, which may be conservative. New design approaches were also developed to improve the performance, which can be found in the literature.

7.3 Strategies for Improving Robust Predictive Controller Design

7.3.1 Difficulties for Robust Predictive Controller Synthesis

The RMPC synthesis approach presented in Section 7.1.1 can also be applied to systems with structural uncertainties. In [1], a kind of system with structural feedback uncertainty is converted into an uncertain LTV system and RMPC for it is synthesized in the same way but with different LMI forms. The basic idea of synthesizing RMPC and the specific technique of handling constraints in [1] have an important impact on subsequent researches of robust predictive control, which have been shown, for example, in the other parts of Section 7.1. However, the approach presented in [1] still has some limitations:

1) To make the design simple, a unique feedback control law is adopted in [1], which strictly restricts the structure of the future control sequence. As the result, the control performance could not be further improved because of the lack of more degrees of freedom for control.
2) In practical applications, the feasible region of the predictive controller, i.e. the region of system initial states with guaranteed stability, is preferred to be as large as possible such that the closed-loop stability can be guaranteed for a larger region of system states. However, with the unique feedback control law as in Section 7.1.1, the closed-loop system is robustly stable only when $\boldsymbol{x}(0) \in \boldsymbol{\Omega}$, i.e. the feasible region is obviously restricted by $\boldsymbol{\Omega}$.
3) The optimization problem in [1] should be solved online according to the rolling style of predictive control. The problem scale increases rapidly with the system dimension and the number of vertices of the uncertain model, which will lead to a heavy computational burden and make it difficult to be implemented in real time. Furthermore, the optimization problem with the LMI formulation cannot be solved as efficiently as that with the QP formulation.

In fact, the above limitations reflect the main difficulty in RMPC synthesis, i.e. the contradiction among the requirements on feasible region (reflecting the region of the system states with guaranteed stability), online computational burden (reflecting whether the

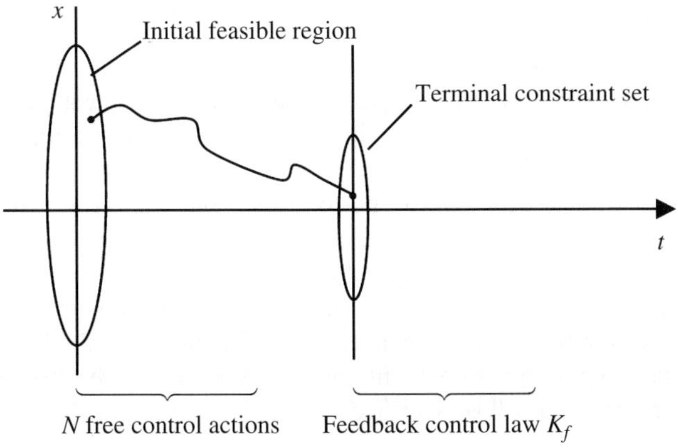

Figure 7.2 Free control actions, initial feasible region, and terminal constraint set in predictive control synthesis.

algorithm can be applied in real time), and control performance (reflecting the optimization effect of the designed predictive controller). In order to illustrate this, consider the general method adopted in the stable and robust predictive controller synthesis shown in Figure 7.2.

At each step of rolling optimization, take the following predictive control strategy: start from the initial system state, adopt N free control inputs in the optimization horizon to steer the system state into the terminal constraint set, and then use the fixed feedback control law in the terminal constraint set to drive the system state to zero.

According to Section 6.2.3, the terminal set constraint is a relaxation of the zero terminal constraint and is easier to implement than it. However, the zero terminal constraint means that the system state will remain at the origin when it arrives at the terminal, while the terminal set constraint means that the system state is not zero when it enters the terminal set and a fixed state feedback control law is still needed to steer the state asymptotically to the origin. In this sense, the performance of the control system corresponding to the terminal set constraint is in general inferior to that of the zero terminal constraint. The larger the terminal set, the worse the control performance will be. From the point of view of improving the control performance, a smaller terminal set is often preferable.

The feasible region of the predictive controller characterizes the applicable region of the system states starting from which the stability or robust stability of the closed-loop system can be theoretically guaranteed. In practice, the initial state outside the feasible region may also lead to stable results, but without a theoretical guarantee. From the point of view of the applicable range of the designed predictive controller, a larger feasible region is often preferable.

With the general control strategy mentioned above, N free control inputs before entering the terminal set and the feedback control law after entering the terminal set need to be solved online. The larger the optimization horizon N, the heavier the online computational burden will be. To reduce the online computational burden, a smaller N is more acceptable.

However, the above three requirements are often contradictory in a specific synthesis algorithm of predictive control. Referring to Figure 7.2, for example,

- A larger feasible region can be achieved by enlarging the terminal set or increasing the optimization horizon, but the control performance will be degenerated or the computational burden will be increased, respectively.
- Reducing the terminal set may improve the control performance, but will reduce the feasible region. If at the same time keeping the feasible region unchanged, the optimization horizon must be increased and thus the computational burden will be increased.
- Decreasing the optimization horizon can reduce the online computational burden, but the feasible region will be reduced if the terminal set is unchanged to maintain the control performance. If the feasible region must be maintained, the terminal set has to be enlarged and the control performance will be sacrificed.

The above-mentioned contradiction between control performance, feasible region, and online computational burden is the critical issue faced by predictive control synthesis, especially in the synthesis of robust predictive controllers. Some typical approaches to improve the design in [1] have been introduced in Section 7.1. In the following, we further introduce some other approaches that also show the variety of the RMPC research.

7.3.2 Efficient Robust Predictive Controller

In order to enlarge the feasible region and reduce the online computational burden, Kouvaritakis et al. proposed an RMPC algorithm named efficient robust predictive control (ERPC) [13]. The basic idea of ERPC is as follows. Firstly, an off-line design using a fixed feedback control law is proposed that satisfies the input constraints and can robustly stabilize the controlled system. Then, additional degrees of freedom for control are introduced and an invariant set for the augmented system is solved off-line, which has the maximum projection set in the original state space. When online, the additional variables are optimized in order to achieve a better control performance.

For the system (7.1) to (7.3) with polytopic uncertainties, consider the input constraints

$$\left|\boldsymbol{u}_j(k)\right| \le u_{j,\max}, \qquad j = 1, \ldots, m \tag{7.106}$$

where $\boldsymbol{u}_j(k)$ is the jth component of control input $\boldsymbol{u}(k) \in \mathrm{R}^m$. The RMPC for this system is formulated as solving an infinite horizon optimization problem at each time k with the cost function

$$J(k) = \sum_{i=0}^{\infty} \left[\|\boldsymbol{x}(k+i+1)\|_{\mathcal{Q}}^2 + \|\boldsymbol{u}(k+i)\|_{\mathcal{R}}^2\right] \tag{7.107}$$

Due to the polytopic uncertainty of the model, the optimization for (7.107) is indeed a "min-max" problem as in (7.6). In the following, the main idea of ERPC is briefly introduced.

(1) Design the feedback control law and handle the input constraints
Firstly, a fixed feedback control law $\boldsymbol{u}(k) = \boldsymbol{K}\boldsymbol{x}(k)$ is designed that stabilizes all the models of system (7.1) to (7.3) without considering the input constraints (7.106). Define the controlled invariant set of system (7.1) to (7.3) with this feedback control law as

$$\boldsymbol{\Omega}_x = \left\{\boldsymbol{x} \in \mathrm{R}^n \mid \boldsymbol{x}^{\mathrm{T}}\boldsymbol{Q}_x^{-1}\boldsymbol{x} \le 1\right\}, \quad \boldsymbol{Q}_x^{-1} > \boldsymbol{0} \tag{7.108}$$

Let $\boldsymbol{\Phi}(k) = \boldsymbol{A}(k) + \boldsymbol{B}(k)\boldsymbol{K}$; then we have $\boldsymbol{\Phi}^{\mathrm{T}}(k)\boldsymbol{Q}_x^{-1}\boldsymbol{\Phi}(k) - \boldsymbol{Q}_x^{-1} < 0$. Due to the polytopic property of the model uncertainty, it follows that

$$\boldsymbol{\Phi}_l^{\mathrm{T}}\boldsymbol{Q}_x^{-1}\boldsymbol{\Phi}_l - \boldsymbol{Q}_x^{-1} < \boldsymbol{0}, \quad l = 1, \ldots, L \tag{7.109}$$

where $\boldsymbol{\Phi}_l = \boldsymbol{A}_l + \boldsymbol{B}_l\boldsymbol{K}$, $l = 1, \ldots, L$. With the control law $\boldsymbol{u}(k) = \boldsymbol{K}\boldsymbol{x}(k)$, the input constraints (7.106) can be written as

$$\left|\boldsymbol{k}_j^{\mathrm{T}}\boldsymbol{x}(k)\right| \le u_{j,\max}, \quad j = 1, \ldots, m$$

where $\boldsymbol{k}_j^{\mathrm{T}}$ is the jth row of $\boldsymbol{K}$. If the system state $\boldsymbol{x}(k)$ belongs to the invariant set $\boldsymbol{\Omega}_x$, it follows that

$$\left|\boldsymbol{k}_j^{\mathrm{T}}\boldsymbol{x}(k)\right| = \left|\boldsymbol{k}_j^{\mathrm{T}}\boldsymbol{Q}_x^{1/2}\boldsymbol{Q}_x^{-1/2}\boldsymbol{x}(k)\right| \le \left\|\boldsymbol{k}_j^{\mathrm{T}}\boldsymbol{Q}_x^{1/2}\right\|\left\|\boldsymbol{Q}_x^{-1/2}\boldsymbol{x}(k)\right\| \le \left\|\boldsymbol{k}_j^{\mathrm{T}}\boldsymbol{Q}_x^{1/2}\right\|, \quad j = 1, \ldots, m$$

Thus the input constraints can be satisfied if

$$\left\|\boldsymbol{k}_j^{\mathrm{T}}\boldsymbol{Q}_x^{1/2}\right\| \le u_{j,\max} \Leftrightarrow u_{j,\max}^2 - \boldsymbol{k}_j^{\mathrm{T}}\boldsymbol{Q}_x\boldsymbol{k}_j \ge 0, \quad j = 1, \ldots, m \tag{7.110}$$

Then, based on (7.109) and (7.110), adopt the method in Section 7.1.1 to design RMPC and obtain $\boldsymbol{K}$, $\boldsymbol{Q}_x$, such that system (7.1) to (7.3) can be robustly stabilized and the input constraints (7.106) are satisfied. This is simply the RMPC strategy presented in [1].

(2) Introduce perturbations to form the augmented system
In order to enlarge the feasible region and improve the control performance, it was suggested in [13] that extra degrees of freedom should be introduced by using perturbations $\boldsymbol{c}(k+i)$ on the fixed state feedback law. The control law was then modified as

$$\boldsymbol{u}(k+i) = \boldsymbol{K}\boldsymbol{x}(k+i) + \boldsymbol{c}(k+i) \tag{7.111}$$

where $\boldsymbol{c}(k+i) = 0$, $i \ge n_c$. This means that $\boldsymbol{c}(k+i)$, $i = 0, \ldots, n_c - 1$ as additional input variables can be used to increase the degree of freedom for RMPC synthesis. Then the closed-loop system can be expressed by

$$\boldsymbol{x}(k+1) = \boldsymbol{\Phi}(k)\boldsymbol{x}(k) + \boldsymbol{B}(k)\boldsymbol{c}(k)$$

By introducing the augmented state,

$$\boldsymbol{z}(k) = \begin{bmatrix} \boldsymbol{x}(k) \\ \boldsymbol{f}(k) \end{bmatrix} \in \mathrm{R}^{n+mn_c}, \quad \text{with } \boldsymbol{f}(k) = \begin{bmatrix} \boldsymbol{c}(k) \\ \boldsymbol{c}(k+1) \\ \vdots \\ \boldsymbol{c}(k+n_c-1) \end{bmatrix} \tag{7.112}$$

it can be formulated as an autonomous system

$$z(k+1)=\Psi(k)z(k) \tag{7.113}$$

where

$$\Psi(k)=\begin{bmatrix} \Phi(k) & B(k) & 0 & 0 & \cdots & 0 \\ 0 & 0 & I & 0 & \cdots & 0 \\ 0 & 0 & 0 & I & \cdots & 0 \\ \vdots & \vdots & \vdots & 0 & \ddots & \vdots \\ & & & \vdots & \ddots & I \\ 0 & 0 & 0 & 0 & \cdots & 0 \end{bmatrix}$$

For the autonomous system (7.113), get the controlled invariant set like (7.108):

$$\Omega_z=\{z\in \mathrm{R}^{n+mn_c} \mid z^{\mathrm{T}}Q_z^{-1}z\le 1\},\quad Q_z^{-1}>0 \tag{7.114}$$

where Q_z satisfies

$$\Psi_l^{\mathrm{T}}Q_z^{-1}\Psi_l-Q_z^{-1}\le 0 \quad\Leftrightarrow\quad \begin{bmatrix} Q_z & Q_z\Psi_l^{\mathrm{T}} \\ \Psi_l Q_z & Q_z \end{bmatrix}\ge 0,\quad l=1,\ldots,L \tag{7.115}$$

Meanwhile, note that the control law is now (7.111), i.e.

$$u(k)=Kx(k)+c(k)=[K\ \ I\ \ 0\ \ \cdots\ \ 0]z(k)$$

Referring to (7.110), the input constraints (7.106) can be ensured if

$$\left\| \left[k_j^{\mathrm{T}}\ \ e_j^{\mathrm{T}}\ \ 0\ \cdots\ 0\right] Q_z^{1/2}\right\| \le u_{j,\max}$$

$$\Leftrightarrow\ u_{j,\max}^2-\left[k_j^{\mathrm{T}}\ \ e_j^{\mathrm{T}}\ \ 0\ \cdots\ 0\right]Q_z\left[k_j^{\mathrm{T}}\ \ e_j^{\mathrm{T}}\ \ 0\cdots 0\right]^{\mathrm{T}}\ge 0,\quad j=1,\ldots,m \tag{7.116}$$

where e_j^T is the jth row of the identity matrix with dimension m.

(3) Design the invariant set to enlarge the feasible region

Now consider the influence of introducing additional input variables to the invariant set of augmented system, which is also the feasible region of the RMPC. The matrix Q_z^{-1} in (7.114) can be divided into four blocks corresponding to x and f: $\hat{Q}_{11}$, $\hat{Q}_{12}$, $\hat{Q}_{21}$, and $\hat{Q}_{22}$, where $\hat{Q}_{12}^{\mathrm{T}}=\hat{Q}_{21}$. Then, from (7.114) it follows that

$$x^{\mathrm{T}}\hat{Q}_{11}x\le 1-2f^{\mathrm{T}}\hat{Q}_{21}x-f^{\mathrm{T}}\hat{Q}_{22}f \tag{7.117}$$

Therefore, if we let $\hat{Q}_{11}=\hat{Q}_x^{-1}$, then for all $z=\left[x^{\mathrm{T}}\ \ 0\right]^{\mathrm{T}}$, the above inequality is degenerated to the condition in (7.108). However, if $f\neq 0$, from (7.117), it is possible to make the projection set of Ω_z on x space larger than Ω_x by properly choosing f. For example, if we let $f=-\hat{Q}_{22}^{-1}\hat{Q}_{21}x$, the projection of Ω_z on x space can be given by

$$\Omega_{xz}=\{x\in \mathrm{R}^n \mid x^{\mathrm{T}}Q_{xz}^{-1}x\le 1\},\quad Q_{xz}^{-1}=\hat{Q}_{11}-\hat{Q}_{12}\hat{Q}_{22}^{-1}\hat{Q}_{21}$$

Since $Q_{xz}^{-1}\le \hat{Q}_{11}$, if $\hat{Q}_{11}=Q_x^{-1}$, it follows that $\Omega_x\subseteq\Omega_{xz}$. That is, the invariant set on x space can be enlarged by the additional input variables. Figure 7.3 shows the enlarged invariant set in the x space after the state space has been augmented.

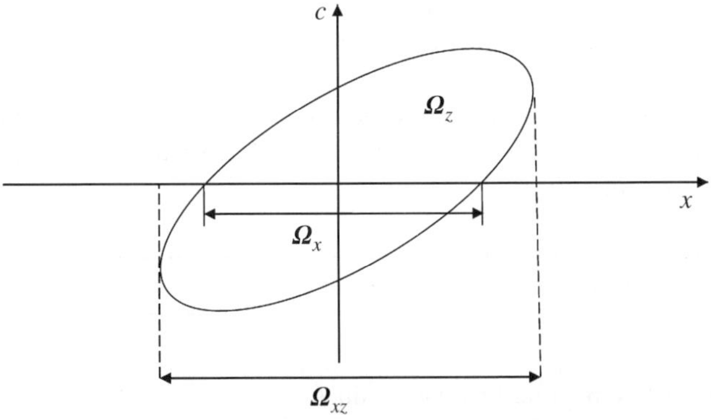

Figure 7.3 Invariant set $\boldsymbol{\Omega}_x$, $\boldsymbol{\Omega}_{xz}$, and $\boldsymbol{\Omega}_z$ [13]. *Source:* Reproduced with permission from Kouvaritakis, Rossiter and Schuurmans of IEEE.

In order to achieve a general formulation, denote the projection set of the invariant set $\boldsymbol{\Omega}_z$, i.e. (7.114) on $\boldsymbol{x}$ space as

$$\boldsymbol{\Omega}_{xz} = \left\{ \boldsymbol{x} \in \mathrm{R}^n \mid \boldsymbol{x}^{\mathrm{T}} \boldsymbol{Q}_{xz}^{-1} \boldsymbol{x} \le 1 \right\}, \quad \boldsymbol{Q}_{xz}^{-1} > \boldsymbol{0} \tag{7.118}$$

Let $\boldsymbol{x} = \boldsymbol{T}\boldsymbol{z}$ with $\boldsymbol{T} = [\boldsymbol{I} \quad \boldsymbol{0}]$ of dimension $n \times (n + mn_c)$ and make a transformation for the LMIs implied in (7.114):

$$\begin{bmatrix} \boldsymbol{T} & \boldsymbol{0} \\ \boldsymbol{0} & 1 \end{bmatrix} \begin{bmatrix} \boldsymbol{Q}_z & \boldsymbol{z} \\ \boldsymbol{z}^{\mathrm{T}} & 1 \end{bmatrix} \begin{bmatrix} \boldsymbol{T}^{\mathrm{T}} & \boldsymbol{0} \\ \boldsymbol{0} & 1 \end{bmatrix} \ge \boldsymbol{0} \iff \begin{bmatrix} \boldsymbol{T}\boldsymbol{Q}_z\boldsymbol{T}^{\mathrm{T}} & \boldsymbol{x} \\ \boldsymbol{x}^{\mathrm{T}} & 1 \end{bmatrix} \ge \boldsymbol{0}$$

Thus it is known that $\boldsymbol{Q}_{xz} = \boldsymbol{T}\boldsymbol{Q}_z\boldsymbol{T}^{\mathrm{T}}$ in (7.118). If $\boldsymbol{Q}_z^{-1}$ is expressed by four blocks defined as above, we can calculate $\boldsymbol{Q}_{xz} = \boldsymbol{T}\boldsymbol{Q}_z\boldsymbol{T}^{\mathrm{T}} = \left(\hat{\boldsymbol{Q}}_{11} - \hat{\boldsymbol{Q}}_{12}\hat{\boldsymbol{Q}}_{22}^{-1}\hat{\boldsymbol{Q}}_{21} \right)^{-1}$ exactly. Therefore, a larger feasible region than $\boldsymbol{\Omega}_x$ can be achieved by making use of the additional input variables to design $\boldsymbol{\Omega}_z$ such that $\boldsymbol{Q}_{xz} \ge \boldsymbol{Q}_x$.

(4) Implementation and algorithm

Then, according to the above analysis, the ERPC synthesis can be summarized as:

1) Design the optimal feedback control law $\boldsymbol{u}(k) = \boldsymbol{K}\boldsymbol{x}(k)$ for the unconstrained system (7.1) to (7.3) by using the usual robust control method with a specific optimization requirement, such as minimizing a worst-case cost or optimizing a nominal performance.
2) Introduce $\boldsymbol{f}$ and construct the augmented autonomous system (7.113) with $\boldsymbol{K}$ and design the invariant set $\boldsymbol{\Omega}_z$ of the autonomous system such that its projection set $\boldsymbol{\Omega}_{xz}$ on $\boldsymbol{x}$ space is as large as possible and $\boldsymbol{Q}_z$ satisfies the invariant set condition (7.115) as well as the input constraint condition (7.116).
3) The current augmented state $\boldsymbol{z}(k)$ belongs to set $\boldsymbol{\Omega}_z$, i.e. condition (7.114) holds.
4) The degree of freedom of $\boldsymbol{f}$ is used to optimize the system performance.

Remark 7.6 Since $\boldsymbol{u}(k) = \boldsymbol{K}\boldsymbol{x}(k)$ is optimal for the system without constraints, $\boldsymbol{f}$ can be regarded as a perturbation on the optimal control law whose role is merely to ensure the feasibility of the predicted trajectories over all model uncertainty $\boldsymbol{\Pi}$ in (7.2). Therefore $\boldsymbol{K}$

is not required to be online tuned by optimizing (7.107) each time. Instead a simple optimization $J_f = \|\boldsymbol{f}\|^2$ subject to feasibility constraints can be online solved for $\boldsymbol{f}$.

Note that (1) and (2) are independent from the system state $\boldsymbol{x}(k)$ and thus can be solved offline to reduce the online computational burden. Therefore the control algorithm can be divided into two parts: offline and online.

Algorithm 7.3 Efficient robust predictive control [13]

Offline part: calculate $\boldsymbol{K}$, $\boldsymbol{Q}_z$, and n_c.

Step 1. Design the optimal robust stabilizing feedback gain $\boldsymbol{K}$.

Step 2. Design $\boldsymbol{Q}_z$ such that the ellipsoid set $\boldsymbol{\Omega}_{xz}$ in (7.118) has the maximum volume, which can be charaterized by maximizing $\det(\boldsymbol{T}\boldsymbol{Q}_z\boldsymbol{T}^{\mathrm{T}})$. This problem can be transferred as the following convex problem to be solved by SDP tools.

$$\begin{aligned} &\min \log \det\left(\boldsymbol{T}\boldsymbol{Q}_z\boldsymbol{T}^{\mathrm{T}}\right)^{-1} \\ &\text{s.t.} \quad (7.115), (7.116) \end{aligned} \tag{7.119}$$

Step 3. If the feasible region $\boldsymbol{\Omega}_{xz}$ achieved in Step 2 is less than the allowable set X_0 for the system states, which is often known in real applications, increase n_c and return to Step 2 to design $\boldsymbol{Q}_z$ again until $X_0 \in \boldsymbol{\Omega}_{xz}$ is satisfied.

Online part: calculate $\boldsymbol{f}(k)$ and $\boldsymbol{u}(k)$ at time k.

Step 1. Solve the following optimization problem to get $\boldsymbol{f}(k)$:

$$\begin{aligned} &\min_{\boldsymbol{f}(k)} \boldsymbol{f}^{\mathrm{T}}(k)\boldsymbol{f}(k) \\ &\text{s.t.} \quad \boldsymbol{z}^{\mathrm{T}}(k)\boldsymbol{Q}_z^{-1}\boldsymbol{z}(k) \le 1 \end{aligned} \tag{7.120}$$

Step 2. Select the first component $\boldsymbol{c}(k)$ from $\boldsymbol{f}(k)$ and construct $\boldsymbol{u}(k) = \boldsymbol{K}\boldsymbol{x}(k) + \boldsymbol{c}(k)$ according to (7.111). Implement $\boldsymbol{u}(k)$ to the real system.

It was proved in [13] that if $\boldsymbol{K}$, $\boldsymbol{Q}_z$, and n_c exist such that $\boldsymbol{x}(0) \in \boldsymbol{\Omega}_{xz}$, ERPC with Algorithm 7.3 will robustly stabilize system (7.1) to (7.3) with satisfaction of the input constraints. Kouvaritakis et al. also pointed out in [14] that the optimization problem (7.120) can be efficiently solved by well-known techniques, such as the Newton–Raphson (NR) method. Compared with the conventional RMPC method in [1], the control performance can be improved and the feasible region is enlarged by using the extra degrees of freedom introduced by $\boldsymbol{f}$, while the online computational burden is greatly reduced because the main part of the computation has been moved to off-line. Thus, ERPC makes a good trade-off among the control performance, feasible region, and online computational burden.

7.3.3 Off-Line Design and Online Synthesis

The EPRC presented in the last section greatly reduces the online computational burden by an off-line design, which shows the possibility to make a trade-off among the control performance, feasible region, and online computational burden by combining off-line and online computations. In view of the heavy online computational burden of the RMPC synthesis approach in [1], Wan and Kothare [15] proposed an offline design/online synthesis method for an RMPC design based on the ellipsoidal invariant set.

Consider the RMPC synthesis problem in Section 7.1.1. For the uncertain system (7.1) to (7.3) with input and state constraints (7.4) and (7.5), given an initial state $\boldsymbol{x}$, if it is feasible for Problem (7.17), then we can solve Problem (7.17) to get the optimal solution γ, $\boldsymbol{Q}$, $\boldsymbol{Y}$, with which an ellipsoidal invariant set $\boldsymbol{\Omega} = \{\boldsymbol{x} \in \mathrm{R}^n \mid \boldsymbol{x}^{\mathrm{T}}\boldsymbol{Q}^{-1}\boldsymbol{x} \le 1\}$ and its corresponding feedback control law $\boldsymbol{u}(k) = \boldsymbol{F}\boldsymbol{x}(k)$ with $\boldsymbol{F} = \boldsymbol{Y}\boldsymbol{Q}^{-1}$ can be obtained. Since all the system states in the set $\boldsymbol{\Omega}$ can be steered to the origin by the feedback control law, the design of an ellipsoidal invariant set is independent from the current system state $\boldsymbol{x}(k)$. Thus, the core idea of [15] is to use off-line to construct a series of ellipsoidal invariant sets and calculate their corresponding feedback control laws and online to determine the most suitable ellipsoidal invariant set and the corresponding feedback control law in terms of an actual system state and then to calculate the required control action. This can be described by the following algorithm in detail.

Algorithm 7.4 Off-line design and online synthesis for Algorithm 7.1 [15]

Offline part: generate a sequence of optimal γ_i, $\boldsymbol{Q}_i$, $\boldsymbol{Y}_i$, $i = 1, \ldots, N$.

Given a feasible initial state $\boldsymbol{x}_1$, let $i = 1$.

Step 1. Calculate the optimal γ_i, $\boldsymbol{Q}_i$, $\boldsymbol{Y}_i$ with respect to $\boldsymbol{x}_i$ by solving Problem (7.17) with an additional constraint $\boldsymbol{Q}_{i-1} > \boldsymbol{Q}_i$ (ignored at $i = 1$), which ensures that the set $\boldsymbol{\Omega}_i = \{\boldsymbol{x} \in \mathrm{R}^n \mid \boldsymbol{x}^{\mathrm{T}}\boldsymbol{Q}_i^{-1}\boldsymbol{x} \le 1\}$ is inside the set $\boldsymbol{\Omega}_{i-1} = \{\boldsymbol{x} \in \mathrm{R}^n \mid \boldsymbol{x}^{\mathrm{T}}\boldsymbol{Q}_{i-1}^{-1}\boldsymbol{x} \le 1\}$, i.e. $\boldsymbol{\Omega}_i \subset \boldsymbol{\Omega}_{i-1}$, and store $\boldsymbol{Q}_i^{-1}$, $\boldsymbol{Y}_i$, $\boldsymbol{F}_i$ in a look-up table.

Step 2. If $i < N$, choose a state $\boldsymbol{x}_{i+1}$ satisfying $\boldsymbol{x}_{i+1}^{\mathrm{T}}\boldsymbol{Q}_i^{-1}\boldsymbol{x}_{i+1} \le 1$. Let $i = i + 1$ and go to Step 1.

Online part: for the current system state $\boldsymbol{x}(k)$, in the look-up table search for the largest index i (i.e. the smallest set $\boldsymbol{\Omega}_i$) such that $\boldsymbol{x}^{\mathrm{T}}(k)\boldsymbol{Q}_i^{-1}\boldsymbol{x}(k) \le 1$. Implement the feedback control law $\boldsymbol{u}(k) = \boldsymbol{F}_i\boldsymbol{x}(k)$.

Using Algorithm 7.4, the online computational burden can be greatly reduced. However, from the online Algorithm 7.1 it is known that the optimal control law and the corresponding invariant set depend on the system state. While in Algorithm 7.4, with a limited number of preserved invariant sets, the same state feedback control law may be used for different states that belong to the same region between two adjacent invariant sets. This means that such a control law is not necessarily the optimal one and the control performance might be degenerated. It is obvious that increasing the number of presented invariant sets may achieve a better control performance. However, due to the limitation of storage as well as the off-line computational burden, the number of chosen initial feasible states is restricted. In the following, two ways to further improve the control performance for Algorithm 7.4 are briefly introduced.

In [15], an additional constraint is imposed on the invariant set obtained by the off-line part of Algorithm 7.4 as follows:

$$\boldsymbol{Q}_i^{-1} - (\boldsymbol{A}_l + \boldsymbol{B}_l\boldsymbol{F}_{i+1})^{\mathrm{T}}\boldsymbol{Q}_i^{-1}(\boldsymbol{A}_l + \boldsymbol{B}_l\boldsymbol{F}_{i+1}) > 0, \quad i = 1, 2, \ldots, N, \quad l = 1, 2, \ldots, L \tag{7.121}$$

If the sets in the look-up table can satisfy conditions (7.121), then the online part of Algorithm 7.4 can be modified as follows.

Online part (modified version 1).
For the current system state $\boldsymbol{x}(k)$, perform a search over $\boldsymbol{Q}_i^{-1}$ in the look-up table to find the largest index i (i.e. the smallest set $\boldsymbol{\Omega}_i$) such that $\boldsymbol{x}^{\mathrm{T}}(k)\boldsymbol{Q}_i^{-1}\boldsymbol{x}(k) \le 1$. If $i < N$, solve $\boldsymbol{x}^T(k)\left(\alpha\boldsymbol{Q}_i^{-1} + (1-\alpha)\boldsymbol{Q}_{i+1}^{-1}\right)\boldsymbol{x}(k) = 1$ for α and apply $\boldsymbol{u}(k) = (\alpha\boldsymbol{K}_i + (1-\alpha)\boldsymbol{K}_{i+1})\boldsymbol{x}(k)$ as the feedback control law. If $i = N$, $\boldsymbol{u}(k) = \boldsymbol{F}_N\boldsymbol{x}(k)$.

Robust stability of the closed-loop system controlled by Algorithm 7.4 or its modified version was proved in [15]. In both cases, the obtained feedback control law between sets $\boldsymbol{\Omega}_i$ and $\boldsymbol{\Omega}_{i+1}$ is guaranteed to keep $\boldsymbol{x}(k)$ within $\boldsymbol{\Omega}_i$ and converge it into $\boldsymbol{\Omega}_{i+1}$ (see [15] for details). Since the feedback control law is online adjusted according to the current system state, a better control performance can be achieved.

Another way to improve Algorithm 7.1 that is different from the method in [15] is based on the fact that the convex combination of multiple ellipsoidal invariant sets is also an invariant set. For n ellipsoidal invariant sets $\boldsymbol{S}_h = \left\{\boldsymbol{x} \mid \boldsymbol{x}^{\mathrm{T}}\boldsymbol{Q}_h^{-1}\boldsymbol{x} \le 1\right\}$, $h = 1, \ldots, n$ with $\boldsymbol{Q}_h, \boldsymbol{Y}_h, \gamma_h$ satisfying (7.14) to (7.16), their convex combination $\hat{\boldsymbol{S}} = \left\{\boldsymbol{x} \mid \boldsymbol{x}^{\mathrm{T}}\hat{\boldsymbol{Q}}^{-1}\boldsymbol{x} \le 1\right\}$ is also an ellipsoidal invariant set with $\hat{\boldsymbol{Q}} \triangleq \sum_{h=1}^{n}\lambda_h\boldsymbol{Q}_h$, $\hat{\boldsymbol{Y}} \triangleq \sum_{h=1}^{n}\lambda_h\boldsymbol{Y}_h$, $\hat{\gamma} \triangleq \sum_{h=1}^{n}\lambda_h\gamma_h$ satisfying (7.14) to (7.16), where $\lambda_h \ge 0$, $h = 1, \ldots, n$, $\sum_{h=1}^{n}\lambda_h \le 1$. This can be easily shown by multiplying a parameter λ_h to the inequalities in (7.14) to (7.16):

$$\begin{bmatrix} \boldsymbol{Q}_h & * & * & * \\ \boldsymbol{A}_l\boldsymbol{Q}_h + \boldsymbol{B}_l\boldsymbol{Y}_h & \boldsymbol{Q}_h & * & * \\ \mathcal{Q}^{1/2}\boldsymbol{Q}_h & \boldsymbol{0} & \gamma_h\boldsymbol{I} & * \\ \mathcal{R}^{1/2}\boldsymbol{Y}_h & \boldsymbol{0} & \boldsymbol{0} & \gamma_h\boldsymbol{I} \end{bmatrix} \ge \boldsymbol{0}, \quad l = 1, \ldots, L$$

$$\begin{bmatrix} u_{\max}^2\boldsymbol{I} & \boldsymbol{Y}_h \\ \boldsymbol{Y}_h^{\mathrm{T}} & \boldsymbol{Q}_h \end{bmatrix} \ge \boldsymbol{0}, \quad \begin{bmatrix} \boldsymbol{Q}_h & (\boldsymbol{A}_l\boldsymbol{Q}_h + \boldsymbol{B}_l\boldsymbol{Y}_h)^{\mathrm{T}}\boldsymbol{C}^{\mathrm{T}} \\ \boldsymbol{C}(\boldsymbol{A}_l\boldsymbol{Q}_h + \boldsymbol{B}_l\boldsymbol{Y}_h) & y_{\max}^2\boldsymbol{I} \end{bmatrix} \ge \boldsymbol{0}, \quad l = 1, \ldots, L$$

and then sum them for $h = 1, \ldots, n$. Note that the condition (7.14) also implies $V(i, k)$ descending and satisfying (7.7) and is tighter than needed by the invariant set.

Since conditions (7.14) to (7.16) are independent from the real-time system state $\boldsymbol{x}(k)$, it is possible to off-line design a series of invariant sets with different performance indices like that in Algorithm 7.4 and then to online optimize the convex combination coefficients to achieve a better control performance. The off-line design is often formulated as several optimization problems under constraints (7.14) to (7.16), each with a specific performance index such as maximizing the covered region of the ellipsoidal invariant set $\boldsymbol{S}_h$ with a fixed γ_h or pursuing a better control performance by setting a much smaller γ_h [9]. With respect to Algorithm 7.1, the online synthesis should take the constraint (7.12) into account with off-line solved $\boldsymbol{Q}_h, \boldsymbol{Y}_h, \gamma_h$, $h = 1, \ldots, n$, and the real-time state $\boldsymbol{x}(k)$, formulated as follows.

Online part (modified version 2).
Solve the following optimization problem for parameter λ_h, $h = 1, \ldots, n$:

$$\min_{\lambda_h} \sum_{h=1}^{n} \lambda_h \gamma_h$$
$$\text{s.t.} \begin{bmatrix} 1 & \boldsymbol{x}^{\mathrm{T}}(k) \\ \boldsymbol{x}(k) & \sum_{h=1}^{n} \lambda_h \boldsymbol{Q}_h \end{bmatrix} > \mathbf{0}, \quad \lambda_h \geq 0, \; h = 1, \ldots, n, \; \sum_{h=1}^{n} \lambda_h = 1 \tag{7.122}$$

Calculate the feedback control gain $\boldsymbol{K} = \left(\sum_{h=1}^{n} \lambda_h \boldsymbol{Y}_h\right)\left(\sum_{h=1}^{n} \lambda_h \boldsymbol{Q}_h\right)^{-1}$ and implement the control law $\boldsymbol{u}(k) = \boldsymbol{K}\boldsymbol{x}(k)$ to the system.

With the above modified version 2, although an optimization problem with LMI formulation still needs to be solved online, the number of online optimization variables and the rows of LMIs decrease greatly because the optimization variables are just the combination coefficients. Therefore, with this algorithm the online computational burden can be greatly reduced. The basic idea of using the convex combination of multiple invariant sets as an invariant set and the synthesis approach of off-line designing multiple invariant sets and online optimizing the combination coefficients can also be applied to other similar cases such as the multistep control sets presented in Section 7.1.4.

7.3.4 Synthesis of the Robust Predictive Controller by QP

In the above sections, LMI is clearly a main tool used to formulate the online optimization problem for RMPC synthesis. This means that the semi-definite programming (SDP) solver should be adopted online, which may lead to a heavy online computational burden. Furthermore, when dealing with asymmetrical constraints, there is often a need to reformulate them into symmetrical ones such that LMIs can be derived through ellipsoidal invariant sets, which obviously makes the result conservative. To overcome these shortcomings, it was also suggested to formulate the RMPC as a QP problem with which a good trade-off among feasible region, control performance, and online computational burden could be achieved. In the following, we briefly introduce the synthesis method used by Li and Xi [16].

Consider the uncertain system (7.1) to (7.3). In order to simplify the presentation, the considered constraints are given in a very general form:

$$-\underline{\boldsymbol{\xi}} \leq \boldsymbol{F}\boldsymbol{x}(k) + \boldsymbol{G}\boldsymbol{u}(k) \leq \bar{\boldsymbol{\xi}} \tag{7.123}$$

where $\bar{\boldsymbol{\xi}}$ and $\underline{\boldsymbol{\xi}}$ are the column vectors with appropriate dimensions and the number of constraints is p.

Consider the objective function of RMPC with an infinite horizon as

$$J_{\infty}(k) = \sum_{i=0}^{\infty} \left[\|\boldsymbol{x}(k+i)\|_{\mathcal{Q}}^2 + \|\boldsymbol{u}(k+i)\|_{\mathcal{R}}^2\right] \tag{7.124}$$

where $\mathcal{Q} \geq \mathbf{0}, \mathcal{R} \geq \mathbf{0}$. To guarantee the closed-loop stability, the control strategy is chosen as

$$\boldsymbol{u}(k+i)=\begin{cases}\boldsymbol{Kx}(k+i)+\boldsymbol{c}(k+i), & 0\le i\le N-1\\ \boldsymbol{Kx}(k+i), & i\ge N\end{cases} \tag{7.125}$$

where $\boldsymbol{K}$ is a feedback gain and $\boldsymbol{c}(k+i)$ is the additional control input. The control strategy in (7.125) is the same as (7.111) in ERPC (see Section 7.3.2). However, in order to enlarge the feasible region, here a dual-mode control strategy is adopted, i.e. to use free control inputs to steer the system state to a terminal invariant set $\boldsymbol{\Omega}_f$ in N steps and then in $\boldsymbol{\Omega}_f$ to use the feedback control law $\boldsymbol{u}=\boldsymbol{Kx}$ to stabilize the system. Using a similar method to that in Section 7.1.1, a terminal cost function can be used as an upper bound to replace the sum of infinite terms after N steps in the cost function. Thus, the online optimization problem can be formulated as

$$\begin{aligned}&\min_{\boldsymbol{c}(i)}\ \max_{[\boldsymbol{A}(k+i)|\boldsymbol{B}(k+i)]\in\boldsymbol{\Pi}} J(k)\\
&J(k)=\sum_{i=0}^{N-1}\left[\|\boldsymbol{x}(k+i\mid k)\|_{\mathcal{Q}}^2+\|\boldsymbol{u}(k+i\mid k)\|_{\mathcal{R}}^2\right]+\|\boldsymbol{x}(k+N\mid k)\|_{\boldsymbol{P}_f}^2\\
&\text{s.t.}\ \ \boldsymbol{x}(k+i+1\mid k)=\boldsymbol{A}(k+i)\boldsymbol{x}(k+i\mid k)+\boldsymbol{B}(k+i)\boldsymbol{u}(k+i\mid k)\\
&\qquad -\underline{\boldsymbol{\xi}}\le \boldsymbol{Fx}(k+i\mid k)+\boldsymbol{Gu}(k+i\mid k)\le\bar{\boldsymbol{\xi}},\ \ i=0,\dots,N-1\\
&\qquad \boldsymbol{x}(k\mid k)=\boldsymbol{x}(k),\ \ \boldsymbol{x}(k+N\mid k)\in\boldsymbol{\Omega}_f\end{aligned} \tag{7.126}$$

where the feedback control gain $\boldsymbol{K}$ and the weighting matrix of the terminal cost function $\boldsymbol{P}_f$ should satisfy the following conditions:

$$\boldsymbol{P}_f-(\boldsymbol{A}_l+\boldsymbol{B}_l\boldsymbol{K})^{\mathrm{T}}\boldsymbol{P}_f(\boldsymbol{A}_l+\boldsymbol{B}_l\boldsymbol{K})>\mathcal{Q}+\boldsymbol{K}^{\mathrm{T}}\mathcal{R}\boldsymbol{K},\quad l=1,\dots,L \tag{7.127}$$

which implies that $V(i,k)=\boldsymbol{x}(k+i\mid k)^{\mathrm{T}}\boldsymbol{P}_f\boldsymbol{x}(k+i\mid k)$ is imposed to descend and to satisfy conditions (7.7) for all possible $[\boldsymbol{A}(k+i)\quad \boldsymbol{B}(k+i)]\in\boldsymbol{\Pi}$ and $J(k)$ after $i=N$ has an upper bound similar to (7.8), $V(\boldsymbol{x}(N,k))=\|\boldsymbol{x}(k+N\mid k)\|_{\boldsymbol{P}_f}^2\le\gamma$.

The terminal invariant set $\boldsymbol{\Omega}_f$ and its corresponding feedback control gain $\boldsymbol{K}$ are necessary for RMPC synthesis. They can be calculated off-line by the following algorithm.

Algorithm 7.5 Off-line algorithm for feedback gain and terminal invariant set [16]

Step 1. Choose $\gamma>0$ and solve the following optimization problem to get the feedback control gain $\boldsymbol{K}=\boldsymbol{Y}_0\boldsymbol{Q}_0^{-1}$ and its corresponding weighting matrix $\boldsymbol{P}_f=\gamma\boldsymbol{Q}_0^{-1}$:

$$\begin{aligned}&\min_{\boldsymbol{Y}_0,\boldsymbol{Q}_0}\ -\det(\boldsymbol{Q}_0)\\
&\text{s.t.}\ \begin{bmatrix}\boldsymbol{Q}_0 & * & * & *\\ \boldsymbol{A}_l\boldsymbol{Q}_0+\boldsymbol{B}_l\boldsymbol{Y}_0 & \boldsymbol{Q}_0 & * & *\\ \mathcal{Q}^{1/2}\boldsymbol{Q}_0 & \boldsymbol{0} & \gamma\boldsymbol{I} & *\\ \mathcal{R}^{1/2}\boldsymbol{Y}_0 & \boldsymbol{0} & \boldsymbol{0} & \gamma\boldsymbol{I}\end{bmatrix}\ge\boldsymbol{0},\ \ l=1,\dots,L\end{aligned} \tag{7.128}$$

(*Continued*)

Algorithm 7.5 (Continued)

$$\begin{bmatrix} W & * & * \\ Q_0 F^{\mathrm{T}} & Q_0 & * \\ Y_0 G^{\mathrm{T}} & \mathbf{0} & Q_0 \end{bmatrix} \geq \mathbf{0}, \qquad W_{ii} \leq \max\left(\bar{\xi}_i^2, \underline{\xi}_i^2\right), \quad i = 1, \ldots, p$$

Step 2. Use the algorithm in Pluymers et al. [17] to calculate the polyhedral invariant set $\boldsymbol{\Omega}_f = \{\boldsymbol{x} \mid \boldsymbol{M}_{as}\boldsymbol{x} \leq \boldsymbol{1}\}$ with the feedback control gain $\boldsymbol{K}$ obtained in Step 1.

Remark 7.7 In the above algorithm, the second condition in (7.128) can be easily deduced from the similar procedure as in above sections, which is used to ensure constraint (7.123). Step 1 of Algorithm 7.5 indeed gives the feedback control gain $\boldsymbol{K}$ and its ellipsoidal invariant set $\boldsymbol{\Omega}_0 = \left\{\boldsymbol{x} \in \mathrm{R}^n \mid \boldsymbol{x}^{\mathrm{T}} \boldsymbol{Q}_0^{-1} \boldsymbol{x} \leq 1\right\}$ of the constrained system. The goal of the optimization (7.128) is also to maximize this set. However, since QP optimization is considered here, this step is only for getting the feedback gain $\boldsymbol{K}$ while the ellipsoidal set $\boldsymbol{\Omega}_0$ will not be used in the following design. Instead a polyhedral invariant set $\boldsymbol{\Omega}_f = \{\boldsymbol{x} \mid \boldsymbol{M}_{as}\boldsymbol{x} \leq \mathbf{1}\}$ is designed in Step 2, which is used as the terminal set of RMPC. The polyhedral invariant set is calculated by the algorithm in [17], aiming at finding the maximal admissible set for which the linear system with polytopic uncertainty (7.1) to (7.3), controlled by the state feedback controller $\boldsymbol{u}(k) = \boldsymbol{K}\boldsymbol{x}(k)$, satisfies the linear state and input constraints (7.123). It is obvious that $\boldsymbol{\Omega}_f \supseteq \boldsymbol{\Omega}_0$ because $\boldsymbol{\Omega}_f$ is the maximal invariant set for above case while $\boldsymbol{\Omega}_0$ is only the maximal invariant set with ellipsoidal form, which shows the advantage of using the polyhedral invariant set for reducing conservativeness.

After $\boldsymbol{P}_f$, $\boldsymbol{\Omega}_f$, $\boldsymbol{K}$ are solved off-line by using Algorithm 7.5, the online optimization problem (7.126) can be formulated as an QP problem, where the constraints of future control inputs should be linear and the objective function should be quadratic.

(1) Linear constraints

The constraints in the optimization problem (7.126) are all linear if the terminal set $\boldsymbol{\Omega}_f$ is selected as a polyhedral one. All these constraints are with respect to the future system states and control inputs. Due to the polytopic uncertainty of the system parameters, if the control strategy (7.125) is adopted, similar to the discussion in Section 7.1.3, the future system state trajectory will belong to a convex hull, whose vertices are the state trajectories along with the vertex trajectory of polytopic models (see (7.45)). For example, at time $k+1$,

$$\boldsymbol{x}_{l_0}(k+1 \mid k) = \boldsymbol{\Phi}_{l_0}\boldsymbol{x}(k) + \boldsymbol{B}_{l_0}\boldsymbol{c}(k) = [\boldsymbol{\Phi}_{l_0} \;\; \boldsymbol{B}_{l_0}]\boldsymbol{\zeta}_0(k), \quad l_0 = 1, \ldots, L$$
$$\boldsymbol{x}(k+1 \mid k) \in \mathrm{Co}\{\boldsymbol{x}_{l_0}(k+1 \mid k),\; l_0 = 1, \ldots, L\}$$

where $\boldsymbol{\Phi}_l = \boldsymbol{A}_l + \boldsymbol{B}_l\boldsymbol{K}$ and $\boldsymbol{\zeta}_i(k) = [\boldsymbol{x}^{\mathrm{T}}(k), \boldsymbol{c}^{\mathrm{T}}(k), \ldots, \boldsymbol{c}^{\mathrm{T}}(k+i)]^{\mathrm{T}}$. Similarly,

$$\begin{aligned} \boldsymbol{x}_{l_i \cdots l_0}(k+i+1 \mid k) &= = \boldsymbol{\Phi}_{l_i}\boldsymbol{x}_{l_{i-1} \cdots l_0}(k+i \mid k) + \boldsymbol{B}_{l_i}\boldsymbol{c}(k+i) \\ &= \boldsymbol{\Phi}_{l_i} \cdots \boldsymbol{\Phi}_{l_0}\boldsymbol{x}(k) + \boldsymbol{\Phi}_{l_i} \cdots \boldsymbol{\Phi}_{l_1}\boldsymbol{B}_{l_0}\boldsymbol{c}(k) + \cdots + \boldsymbol{B}_{l_i}\boldsymbol{c}(k+i) \\ &= [\boldsymbol{\Phi}_{l_i} \cdots \boldsymbol{\Phi}_{l_0} \;\; \boldsymbol{\Phi}_{l_i} \cdots \boldsymbol{\Phi}_{l_1}\boldsymbol{B}_{l_0} \;\; \cdots \;\; \boldsymbol{B}_{l_i}]\boldsymbol{\zeta}_i(k) \triangleq \boldsymbol{\Psi}_{l_i \cdots l_0}\boldsymbol{\zeta}_i(k), \\ & \qquad l_0, \ldots, l_i = 1, \ldots, L, \quad i = 1, \ldots, N-1 \end{aligned} \tag{7.129}$$

$$\boldsymbol{x}(k+i+1\mid k)\in \mathrm{Co}\{\boldsymbol{x}_{l_i\cdots l_0}(k+i+1\mid k),\ l_0,\ldots,l_i=1,\ldots,L\},\quad i=1,\ldots,N-1$$

Due to the polytopic property of the system uncertainty, it is obvious that the future states $\boldsymbol{x}(k+i+1\mid k)$ in the state/input constraints and the terminal set constraint in (7.126) can be replaced by all corresponding state vertices $\boldsymbol{x}_{l_i\cdots l_0}(k+i+1\mid k),\ l_0,\ldots,l_i=1,\ldots,L,\ i=1,\ldots,N$. These constraints can be rewritten as

$$-\underline{\boldsymbol{\xi}}\le \hat{\boldsymbol{M}}_{si}\boldsymbol{\zeta}_i(k)\le \bar{\boldsymbol{\xi}},\quad i=0,\ldots,N-1,\quad l_0,\ldots,l_{i-1}=1,\ldots,L \tag{7.130}$$

$$\hat{\boldsymbol{M}}_{as}\boldsymbol{\zeta}_{N-1}(k)\le \boldsymbol{1},\quad l_0,\ldots,l_{N-1}=1,\ldots,L \tag{7.131}$$

where

$$\hat{\boldsymbol{M}}_{s0}=[\boldsymbol{F}+\boldsymbol{G}\boldsymbol{K}\quad \boldsymbol{G}]$$
$$\hat{\boldsymbol{M}}_{si}=[(\boldsymbol{F}+\boldsymbol{G}\boldsymbol{K})\boldsymbol{\Psi}_{l_{i-1}\cdots l_0}\quad \boldsymbol{G}],\quad i=1,\ldots,N-1$$
$$\hat{\boldsymbol{M}}_{as}=\boldsymbol{M}_{as}\boldsymbol{\Psi}_{l_{N-1}\cdots l_0}$$

(2) Quadratic cost function

Since the optimization problem (7.126) is a min-max problem, the upper bound of $J(k)$ is used as the objective function when converting it to a QP problem. In [16] it has been shown that if a vector $\boldsymbol{w}\in \mathrm{Co}\{\boldsymbol{w}_i,\ i=1,\ldots,L\}$, then $\|\boldsymbol{w}\|_Q^2\le \max\left\{\|\boldsymbol{w}_i\|_Q^2,\ i=1,\ldots,L\right\}$, where $\boldsymbol{Q}>\boldsymbol{0}$. From time k to $k+N$, the future state trajectory belongs to a convex hull and thus the objective function $J(k)$ of the predicted state trajectory must satisfy $J(k)\le \max\{J_i(k)\}$, where $J_i(k)=\boldsymbol{\zeta}_{N-1}^{\mathrm{T}}(k)\ \boldsymbol{P}_i\boldsymbol{\zeta}_{N-1}(k),\ i=1,\ldots,L^N$ and $\boldsymbol{P}_i$ can be calculated according to the vertex sequences of different future state trajectories. Therefore, to get the upper bound of $J(k)$, the following optimization problem should be solved to obtain the matrix $\boldsymbol{P}$:

$$\begin{aligned}&\min_{\boldsymbol{P}}\quad \mathrm{tr}(\boldsymbol{P})\\ &\text{s.t.}\quad \boldsymbol{P}\ge \boldsymbol{P}_i\quad i=1,\ldots,L^N\end{aligned} \tag{7.132}$$

where $\boldsymbol{P}_i$ is determined by the quadratic terms in $J(k)$ and corresponding $\boldsymbol{\Psi}_{l_i\cdots l_0},\ l_0,\ldots,l_i=1,\ldots,L,\ i=1,\ldots,N-1$. If $\boldsymbol{P}$ has been solved, $\boldsymbol{\zeta}_{N-1}^{\mathrm{T}}(k)\boldsymbol{P}\boldsymbol{\zeta}_{N-1}(k)$ can be used as the quadratic cost function for the minimization problem in (7.126).

(3) QP formulation

With the off-line solved feedback gain $\boldsymbol{K}$, terminal set $\boldsymbol{\Omega}_f$, terminal weighting matrix $\boldsymbol{P}_f$, and matrix $\boldsymbol{P}$ in the upper bound of the cost function, the optimization problem (7.126) can be formulated as the following QP problem:

$$\begin{aligned}&\min_{\boldsymbol{c}(k),\ldots,\boldsymbol{c}(k+N-1)}\quad \boldsymbol{\zeta}_{N-1}^{\mathrm{T}}\boldsymbol{P}\boldsymbol{\zeta}_{N-1}\\ &\text{s.t.}\quad (7.130),\ (7.131)\end{aligned} \tag{7.133}$$

From the above analysis it can be seen that since the constraints are commonly linear, the critical issue to transfer the RMPC optimization problem to a QP problem is to adopt the polyhedral invariant set as the terminal set and to design a quadratic function as the upper bound of the objective function.

The condition that ensures the closed-loop system robustly stable is also given in [16], where a Lyapunov function $V(k)=\boldsymbol{\zeta}_{N-1}^{\mathrm{T}}(k)\boldsymbol{P}\boldsymbol{\zeta}_{N-1}(k)$ is adopted and the recursive feasibility is easily proved using the feasible solution at time $k+1$ constructed by shifting forward the last $N-1$ terms in $\boldsymbol{\zeta}_{N-1}^{*}(k)$ and setting its last term as zero.

For the QP problem (7.133), in addition to the usual QP solving methods such as the active set method, the multiparameter programming method is also helpful in reducing the online computational burden. Note that the optimization problem (7.133) is given with respect to the optimization variables $\boldsymbol{c}(k+i)$ and the current system state $\boldsymbol{x}(k)$, i.e. it has the form

$$\min_{\boldsymbol{c}(k+i)} \ f(\boldsymbol{c}(k+i),\ \boldsymbol{x}(k))$$
$$\text{s.t.}\ \ \boldsymbol{g}_i(\boldsymbol{c}(k+i),\ \boldsymbol{x}(k)) \le \boldsymbol{0}$$

Therein, the state $\boldsymbol{x}(k)$ is time-varying and determines the constraints. According to Bemporad et al. [18], the current system state $\boldsymbol{x}(k)$ can be handled as parameters and the explicit solution $\boldsymbol{c}(k) = \boldsymbol{F}_i\boldsymbol{x}(k) + \boldsymbol{g}_i$ can be obtained for the parameter $\boldsymbol{x}(k)$ within the ith subset. Thus, by the multiparameter programing method, the $\boldsymbol{x}$ space is divided into several subsets and then the corresponding explicit solutions can be obtained and stored off-line. When online, instead of solving the QP problem, it is only required to search the subset to which the state $\boldsymbol{x}(k)$ belongs and then to look up the corresponding feedback gain ($\boldsymbol{F}_i$, $\boldsymbol{g}_i$) to construct the current control input.

7.4 Summary

The research of RMPC is an important part of the qualitative synthesis theory of predictive control. In particular, it has become the mainstream of the theoretical research of predictive control since the beginning of this century. The RMPC makes full use of the rich theoretical results on stable predictive controller synthesis and robust controller design, and has been investigated for a wide range of different types of systems. In this chapter, we introduced the fundamental concepts and typical methods for synthesizing robust predictive controllers, focusing on the RMPC for polytopic uncertain systems and the systems with external disturbance.

The RMPC synthesis using LMI in [1] is a classic work in the RMPC field, which triggered much subsequent research and has achieved fruitful theoretical results. In this chapter, the basic version of this method was firstly introduced and how to formulate the problem and how to handle the main difficulties were explained in detail. Based on this, some of its extensions were further introduced in subsequent subsections, i.e. using parameter-dependent Lyapunov functions, adopting dual-mode control schemes with either additional free control sequence or multiple feedback control laws. These improved approaches can reduce conservativeness of the basic RMPC version and improve the control performance, but often at the cost of increasing the computational complexity.

For disturbed systems, how to efficiently handle the external disturbance is the critical issue in RMPC deign. Two catalogs of synthesis approaches for such kinds of RMPC are introduced. The first one adopts the disturbance invariant set and divides the disturbed system into two subsystems. With a proper design for the nominal subsystem and the disturbed subsystem, the real state of the controlled system can be guaranteed to move along a tunnel with the nominal system state as the center and the disturbance invariant set as the cross-section. The second one borrows the idea in robust control and adopts a mixed H_2/H_∞ performance to synthesize RMPC. In this design, an upper bound on the

H_∞ performance index is firstly assigned and RMPC is then designed by online optimizing the H_2 performance index with the upper bound as an additional constraint.

Based on analyzing the limitations of the basic RMPC method in [1], we point out the main difficulty in RMPC synthesis, i.e. the contradiction among the requirements of a feasible region, the online computational burden, and control performance. In the remainder of this chapter, some representative approaches were introduced to illustrate how to handle this contradiction, showing the variety of the RMPC research. It can be seen that when using some strategies with high computational complexity to improve the control performance and enlarge the feasible region, the off-line design and online synthesis is possibly a realistic way to reduce the online computational burden.

References

1 Kothare, M.V., Balakrishnan, V., and Morari, M. (1996). Robust constrained model predictive control using linear matrix inequalities. *Automatica* 32 (10): 1361–1379.
2 Cuzzola, F.A., Geromel, J.C., and Morari, M. (2002). An improved approach for constrained robust model predictive control. *Automatica* 38 (7): 1183–1189.
3 Mao, W.J. (2003). Robust stabilization of uncertain time-varying discrete systems and comments on "an improved approach for constrained robust model predictive control". *Automatica* 39 (6): 1109–1112.
4 Michalska, H. and Mayne, D.Q. (1993). Robust receding horizon control of constrained nonlinear systems. *IEEE Transactions on Automatic Control* 38 (11): 1623–1633.
5 Wan, Z. and Kothare, M.V. (2003). Efficient robust constrained model predictive control with a time varying terminal constraint set. *Systems and Control Letters* 48: 375–383.
6 Ding, B., Xi, Y., and Li, S. (2004). A synthesis approach of on-line constrained robust model predictive control. *Automatica* 40 (1): 163–167.
7 Pluymers, B., Suykens, J.A.K., and De Moor, B. (2005). Min–max feedback MPC using a time-varying terminal constraint set and comments on "Efficient robust constrained model predictive control with a time-varying terminal constraint set". *Systems and Control Letters* 54: 1143–1148.
8 Ding, B.C. (2010). *Modern Predictive Control.* Boca Raton: CRC Press.
9 Li, D. and Xi, Y. (2009). Design of robust model predictive control based on multi-step control set. *Acta Automatica Sinica* 35 (4): 433–437.
10 Lu, Y. and Arkun, Y. (2000). Quasi-min-max MPC algorithms for LPV systems. *Automatica* 36 (4): 527–540.
11 Mayne, D.Q., Seron, M.M., and Raković, S.V. (2005). Robust model predictive control of constrained linear systems with bounded disturbances. *Automatica* 41 (2): 219–224.
12 Huang, H., Li, D., and Xi, Y. (2012). Design and ISpS analysis of the mixed H_2/H_∞ feedback robust model predictive control. *IET Control Theory and Applications* 6 (4): 498–505.
13 Kouvaritakis, B., Rossiter, J.A., and Schuurmans, J. (2000). Efficient robust predictive control. *IEEE Transactions on Automatic Control* 45 (8): 1545–1549.
14 Kouvaritakis, B., Cannon, M., and Rossiter, J.A. (2002). Who needs QP for linear MPC anyway? *Automatica* 38 (5): 879–884.

15 Wan, Z. and Kothare, M.V. (2003). An efficient off-line formulation of robust model predictive control using linear matrix inequalities. *Automatica* 39 (5): 837–846.

16 Li, D. and Xi, Y. (2011). The synthesis of robust model predictive control with QP formulation. *International Journal of Modeling, Identification and Control* 13 (1–2): 1–8.

17 Pluymers, B., Rossiter, J.A., Suykens, J.A.K., and Moor, B.D. The efficient computation of polyhedral invariant sets for linear systems with polytopic uncertainty, *Proceedings of 2005 American Control Conference*, Portland, OR (8–10 June 2005), 804–809.

18 Bemporad, A., Morari, M., Dua, V., and Pistikopoulos, E.N. (2002). The explicit linear quadratic regulator for constrained systems. *Automatica* 38 (1): 3–20.

8

Predictive Control for Nonlinear Systems

Predictive control algorithms were firstly proposed for linear systems. The predictive control algorithms that appeared early on, such as DMC, MAC, GPC, etc. are all based on linear models, where the future outputs are predicted according to the proportion and superposition properties of linear systems. If the plant is weakly nonlinear, it can be approximated by a linear model and the linear predictive control algorithm can be used. In this case, the degeneration of the system performance caused by model mismatch due to nonlinearity is not critical and can be compensated to some extent by introducing appropriate feedback correction. However, if the plant is strongly nonlinear, the output prediction by using a linear model may lead to a large deviation from the actual one and thus the control result may be far from that predicted with the linear model. This means that it cannot be simply handled by using the linear predictive control algorithm. For predictive control applications, how to develop efficient predictive control strategies and algorithms with respect to the characteristics of nonlinear systems has always been the focus of attention. In this chapter, we firstly give the general description of predictive control for nonlinear systems and then introduce some representative strategies and methods. The predictive control algorithms for nonlinear systems are not unified and exhibit a trend of diversification according to different types of plants and different methods or tools. Therefore the main focus will be put on illustrating the specific difficulties faced by predictive control when the plant is nonlinear and the corresponding ideas to solve them.

8.1 General Description of Predictive Control for Nonlinear Systems

The basic principles of predictive control presented in Section 1.2 are also suitable for nonlinear systems. In the following, the predictive control problem for nonlinear systems is explored, starting from these principles.

The model for a nonlinear plant is usually given by one of the following two forms.

1) **State space model**

$$\begin{aligned} \boldsymbol{x}(k+1) &= \boldsymbol{f}(\boldsymbol{x}(k),\ \boldsymbol{u}(k)) \\ \boldsymbol{y}(k) &= \boldsymbol{g}(\boldsymbol{x}(k)) \end{aligned} \tag{8.1}$$

Predictive Control: Fundamentals and Developments, First Edition. Yugeng Xi and Dewei Li.

where $\boldsymbol{x} \in \mathrm{R}^n$, $\boldsymbol{u} \in \mathrm{R}^m$, $\boldsymbol{y} \in \mathrm{R}^p$. With this model, at the sampling time k, the future state $\boldsymbol{x}$ and output $\boldsymbol{y}$ can be predicted if the initial state $\boldsymbol{x}(k)$ and the future control input $\boldsymbol{u}(k)$, $\boldsymbol{u}(k+1)$, ... are known

$$\begin{aligned} &\tilde{\boldsymbol{x}}(k+i \mid k) = \boldsymbol{f}(\tilde{\boldsymbol{x}}(k+i-1 \mid k),\ \boldsymbol{u}(k+i-1)) \\ &\tilde{\boldsymbol{y}}_m(k+i \mid k) = \boldsymbol{g}(\tilde{\boldsymbol{x}}(k+i \mid k)), \quad i = 1,\ 2, \ldots,\ \ \tilde{\boldsymbol{x}}(k \mid k) = \boldsymbol{x}(k) \end{aligned} \tag{8.2}$$

Starting from $i = 1$, by using (8.2) recursively, it follows that

$$\tilde{\boldsymbol{y}}_m(k+i \mid k) = \boldsymbol{F}_i(\boldsymbol{x}(k),\ \boldsymbol{u}(k),\ \boldsymbol{u}(k+1), \ldots, \boldsymbol{u}(k+i-1)), \quad i = 1,\ 2, \ldots \tag{8.3}$$

where $\boldsymbol{F}_i(\cdot)$ is a nonlinear function compounded by $\boldsymbol{f}(\cdot)$ and $\boldsymbol{g}(\cdot)$.

2) **Input–output model**

$$\boldsymbol{y}(k) = \boldsymbol{f}(\boldsymbol{y}(k-1), \ldots, \boldsymbol{y}(k-r),\ \boldsymbol{u}(k-1), \ldots, \boldsymbol{u}(k-r)) \tag{8.4}$$

where $\boldsymbol{u} \in \mathrm{R}^m$, $\boldsymbol{y} \in \mathrm{R}^p$ and r is the model horizon. With this model, at the sampling time k, the future output $\boldsymbol{y}$ can be predicted if $\boldsymbol{y}(k)$, $\boldsymbol{y}(k-1)$, ... , $\boldsymbol{u}(k-1)$, $\boldsymbol{u}(k-2)$, ... and the future control input $\boldsymbol{u}(k)$, $\boldsymbol{u}(k+1)$, ... are known:

$$\begin{aligned} \tilde{\boldsymbol{y}}_m(k+i \mid k) = \boldsymbol{f}[&\tilde{\boldsymbol{y}}_m(k+i-1 \mid k), \ldots, \tilde{\boldsymbol{y}}_m(k+i-r \mid k), \\ &\boldsymbol{u}(k+i-1), \ldots, \boldsymbol{u}(k+i-r)], \quad i = 1,\ 2, \ldots \\ \tilde{\boldsymbol{y}}_m(k+j \mid k) = \boldsymbol{y}(&k+j), \ \ j \le 0 \end{aligned} \tag{8.5}$$

Starting from $i = 1$ and by using (8.5) recursively, it follows that

$$\begin{aligned} \tilde{\boldsymbol{y}}_m(k+i \mid k) = \boldsymbol{G}_i[&\boldsymbol{y}(k), \ldots, \boldsymbol{y}(k-r+1),\ \boldsymbol{u}(k+i-1), \ldots, \boldsymbol{u}(k), \\ &\boldsymbol{u}(k-1), \ldots, \boldsymbol{u}(k-r+1)], \quad i = 1,\ 2, \ldots \end{aligned} \tag{8.6}$$

where $\boldsymbol{G}_i(\cdot)$ is a nonlinear function compounded by $\boldsymbol{f}(\cdot)$.

Through step-by-step substitution with the model prediction formulae (8.2) or (8.5), (8.3) or (8.6) can be obtained, respectively. They both explicitly show how the system future outputs are related to the known information and future control inputs, and seem to be mostly suitable as the prediction model of the nonlinear system. However, note that the nonlinear function $\boldsymbol{F}_i(\cdot)$ or $\boldsymbol{G}_i(\cdot)$ needs to be multiple compounded, if $\boldsymbol{f}(\cdot)$ or $\boldsymbol{g}(\cdot)$ is nonlinear, and the expression of $\boldsymbol{F}_i(\cdot)$ or $\boldsymbol{G}_i(\cdot)$ may become very complicated when i increases. Furthermore, when solving the optimization problem, multiple compound derivatives are involved, which may be difficult to be deduced and used even if the analytic expressions of $\boldsymbol{F}_i(\cdot)$ or $\boldsymbol{G}_i(\cdot)$ have been obtained. Therefore, (8.3) or (8.6) can only express the prediction model in principle, and it is practically not necessary to derive $\boldsymbol{F}_i(\cdot)$ or $\boldsymbol{G}_i(\cdot)$. The prediction model should be directly expressed by (8.2) or (8.5) over the whole prediction horizon.

According to the principles of predictive control, at each time of rolling optimization, the actual system output should be measured and then used for feedback correction. Denote the actual output measured at time k as $\boldsymbol{y}(k)$, construct the prediction error

$$\boldsymbol{e}(k) = \boldsymbol{y}(k) - \tilde{\boldsymbol{y}}_m(k \mid k-1) \tag{8.7a}$$

and then predict the future errors according to historical prediction errors $\boldsymbol{e}(k)$, ..., $\boldsymbol{e}(k-q)$,

$$\tilde{\boldsymbol{e}}(k+i\mid k)=\boldsymbol{E}_i(\boldsymbol{e}(k),\ldots,\boldsymbol{e}(k-q)) \tag{8.7b}$$

which can be used to correct the future system outputs predicted based on the model. In (8.7b) $\boldsymbol{E}_i(\cdot)$ is a linear or nonlinear function, depending on the adopted noncausal prediction method and q is the time length of used historical information. Then the closed-loop output prediction can be given by

$$\tilde{\boldsymbol{y}}_p(k+i\mid k)=\tilde{\boldsymbol{y}}_m(k+i\mid k)+\tilde{\boldsymbol{e}}(k+i\mid k),\quad i=1,\ 2,\ \ldots \tag{8.8}$$

The rolling optimization at the sampling time k is formulated as solving M control inputs starting from the current time $\boldsymbol{u}(k), \ldots, \boldsymbol{u}(k+M-1)$ (similarly assume that $\boldsymbol{u}$ remains unchanged after time $k+M-1$), which minimize the following performance index:

$$\min J(k)=J(\tilde{\boldsymbol{y}}_{PM}(k),\ \boldsymbol{u}_M(k),\ \boldsymbol{w}_P(k))$$

with

$$\tilde{\boldsymbol{y}}_{PM}(k)=\begin{bmatrix}\tilde{\boldsymbol{y}}_p(k+1\mid k)\\ \vdots\\ \tilde{\boldsymbol{y}}_p(k+P\mid k)\end{bmatrix},\quad \boldsymbol{u}_M(k)=\begin{bmatrix}\boldsymbol{u}(k)\\ \vdots\\ \boldsymbol{u}(k+M-1)\end{bmatrix},\quad \boldsymbol{w}_P(k)=\begin{bmatrix}\boldsymbol{w}(k+1)\\ \vdots\\ \boldsymbol{w}(k+P)\end{bmatrix}$$

where $\boldsymbol{w}(k+i)$ is the desired output at time $k+i$ and M and P are the control horizon and the optimization horizon, respectively. Note that $\tilde{\boldsymbol{y}}_p(k+i\mid k)$ in $\tilde{\boldsymbol{y}}_{PM}(k)$ is obtained from the model output $\tilde{\boldsymbol{y}}_m(k+i\mid k)$ given by (8.2) or (8.5) under $\boldsymbol{u}(k+i)=\boldsymbol{u}(k+M-1)$, $i\geq M$, and then through feedback correction (8.8). Then the online optimization problem can be completely expressed as

$$\begin{aligned}&\min_{\boldsymbol{u}_M(k)} J(k)=J(\tilde{\boldsymbol{y}}_{PM}(k),\ \boldsymbol{u}_M(k),\ \boldsymbol{w}_P(k))\\ &\text{s.t.}\ \ (8.2)\ \text{or}\ (8.5),\ (8.7),\ (8.8)\\ &\qquad \boldsymbol{u}(k+i)=\boldsymbol{u}(k+M-1),\ \ i\geq M\end{aligned} \tag{8.9}$$

If constraints on input, output, or state are concerned, they should be put into the optimization problem as a whole. After the optimal $\boldsymbol{u}^*(k), \ldots, \boldsymbol{u}^*(k+M-1)$ have been solved, implement $\boldsymbol{u}^*(k)$ at time k. At the next sampling time, after measuring the actual output and making a feedback correction, the optimization is repeatedly solved. This is the general description of the predictive control problem for nonlinear systems.

It can be seen that the predictive control for nonlinear systems is in principle not different from that for linear systems. However, with respect to specific algorithms, the following two aspects should be emphasized as compared with that for linear systems.

1) With the prediction model of nonlinear systems, (8.2) or (8.5), it is often difficult to explicitly express the output prediction at time k in a simple analytical form of known information and assumed future control inputs. Although it is in principle possible to obtain (8.3) or (8.6) through substitution step by step, as mentioned above, such an analytical form is hard to obtain due to the nonlinear properties of $\boldsymbol{f}(\cdot)$ and $\boldsymbol{g}(\cdot)$. Even if it can be obtained, its multiple compounded complex form is not suitable for solving the optimization problem. Therefore, for model prediction, it is not necessary to

derive a model of the form (8.3) or (8.6); instead the prediction model (8.2) or (8.5) will be established for all sampling times over the optimization horizon and taken as the constraints into the optimization problem (8.9).

2) The difficulty of solving the online optimization problem in predictive control is greatly increased for nonlinear systems. Even when the performance index in (8.9) is quadratic and no constraints on input, output, and state exist, the online optimization is still a general nonlinear optimization problem because the prediction model is included in it as general nonlinear constraints. Since the optimization variables $\boldsymbol{u}(k), \ldots, \boldsymbol{u}(k+M-1)$ appearing in the prediction models are compounded from each other, it is difficult to solve them in an analytical way, and the solution can only be obtained through numerical optimization. Due to the essential difficulty caused by the compounded nonlinearity, the solving methods, either based on parameter optimization or on function optimization, are always complicated and the computational burden is very large, which makes it difficult to meet the requirement on real-time control. Therefore, how to efficiently solve the nonlinear rolling horizon optimization problem in real time is a difficult problem to be solved in predictive control for nonlinear systems.

In summary, although the predictive control problem for nonlinear systems can be described using a clear mathematical formulation, to solve it there still exist essential difficulties caused by nonlinearity. For decades, while predictive control for linear systems has been continually developed, much effort was also put into the research on predictive control for nonlinear systems. In addition to the theoretical research on nonlinear model predictive control (NMPC), many effective strategies and methods were proposed to meet the requirements in applications and successfully used to many real plants such as reactor, industrial robot, etc. The critical issue of these methods is how to overcome the difficulty of solving nonlinear rolling horizon optimization problems. Some of the main strategies and methods are listed as follows.

1) Linearization. After linearizing the nonlinear model, design the controller using the rolling optimization method of predictive control for linear systems, while retaining the nonlinear model for prediction. To overcome the error caused by model linearization, the linearized model should be adjusted through online estimation.
2) Combination of numerical computation and analytics. Use the nonlinear model for simulation to give a corresponding performance index, search the gradient direction for improving the solution through perturbing the control input, and solve the nonlinear optimization problem through repeated iteration using the gradient method.
3) Layered optimization. Transform the predictive control problem for nonlinear systems into that for linear systems through feedback linearization or construct a two-layer hierarchical algorithm; for example, decompose the global optimization problem into several small ones and at the lower layer solve small-scale optimization problems for linearized or nonlinear models while at the high layer coordinate the interactive nonlinear items between sub-problems.
4) Multiple model. Linearize the nonlinear model at different working points and obtain multiple linear models and then use linear predictive control methods. For model selection, both switching mode and fuzzy mode are investigated. With the switching mode, determine online the adopted prediction model according to the actual state. With the fuzzy mode, synthesize online a linear model by determining the

membership of the latest input/output data belonging to each linear model. This method also allows a direct start from the input/output data without the nonlinear model (8.1).

5) Neural networks. Use a neural network to replace the nonlinear system model for prediction and/or design a neural network to solve the online optimization problem.
6) Approximation. Approximate the nonlinear model by using a generalized convolution model or a generalized orthogonal function, etc. and solve the linear or simple nonlinear predictive control problem after truncation.
7) Strategies for specific nonlinear systems, such as predictive control algorithms for the Hammerstein model, bilinear model, etc.

In the following sections, some of above predictive control strategies and methods for nonlinear systems will be introduced, aiming at meeting the requirements of predictive control applications. It should be pointed out that since the 1990s the rapid development of qualitative synthesis theory of predictive control has opened a new perspective for NMPC. Various design methods with guaranteed stability appeared in the literature. These methods focus on ensuring that the controlled system is stable and the online optimization problem involved is quite different from the one presented above, both in formulation and in the solving method. This has been partly shown in Chapters 6 and 7 and therefore will not be discussed further in this chapter.

8.2 Predictive Control for Nonlinear Systems Based on Input–Output Linearization

The control theory for nonlinear systems has achieved great progress, especially the results on feedback linearization, which have been widely used in controlling nonlinear systems. Feedback linearization refers to the introduction of nonlinear state feedback to make the closed-loop system linear or approximately linear. With this concept it is possible to control a nonlinear system using a layered strategy, i.e. firstly to design a nonlinear state feedback control law to get a linearized closed-loop system according to the feedback linearization theory and then to design the required controller with the help of linear control theory for the resultant system. With this layered structure predictive control for nonlinear systems can be transformed into predictive control for linear systems. In the following this layered predictive control strategy is illustrated using input–output linearization for affine nonlinear systems as an example [1].

Consider a multi-input multi-output affine nonlinear system:

$$\begin{aligned}\frac{\mathrm{d}\boldsymbol{x}}{\mathrm{d}t} &= \boldsymbol{f}(\boldsymbol{x}) + \sum_{j=1}^{m} \boldsymbol{g}_j(\boldsymbol{x})u_j \\ y_i &= h_i(\boldsymbol{x}), \qquad i = 1, \ldots, m\end{aligned} \tag{8.10}$$

where $\boldsymbol{f} = (f_1 \ \cdots \ f_n)^{\mathrm{T}}, \boldsymbol{f}, \boldsymbol{g}_j : \mathrm{R}^n \to \mathrm{R}^n$ is a smooth vector field in R^n and $h_i : \mathrm{R}^n \to \mathrm{R}$ is a smooth scalar function in R^n, $\boldsymbol{u} = (u_1 \ \cdots \ u_m)^{\mathrm{T}} \in \mathrm{R}^m$, $\boldsymbol{y} = (y_1 \ \cdots \ y_m)^{\mathrm{T}} \in \mathrm{R}^m$, i.e. the input and the output have the same dimension.

The basic idea of input–output linearization is to differentiate the output function y_i successively until the input u_j appears and then to design the state feedback law to cancel the nonlinear terms. To do that, the Lie derivative of h_i to $\boldsymbol{f}$ is first defined:

$$L_f(h_i) = \langle \mathrm{d}h_i, \boldsymbol{f} \rangle = \frac{\partial h_i}{\partial x_1} f_1 + \cdots + \frac{\partial h_i}{\partial x_n} f_n \tag{8.11}$$

It can be seen that $L_f(h_i)$ is just the directional derivative of h_i along the vector $\boldsymbol{f}$. It is also a scalar function in R^n. High-order Lie derivatives can be similarly defined and expressed using the following recursive form:

$$\begin{aligned} &L_f^0(h_i) = h_i \\ &L_f^k(h_i) = L_f\left(L_f^{k-1}(h_i)\right) = \left\langle \mathrm{d}L_f^{k-1}(h_i), \boldsymbol{f} \right\rangle, \quad k = 1, 2, \ldots \end{aligned} \tag{8.12}$$

For each output component y_i of the multivariable system (8.10), its relative degree r_i can be defined, satisfying

$$\begin{aligned} &L_{g_j}\left(L_f^l(h_i)\right) = \left\langle \mathrm{d}L_f^l(h_i), \boldsymbol{g}_j \right\rangle = 0, \quad l = 0, 1, \ldots, r_i - 2, \; j = 1, \ldots, m \\ &L_{g_j}\left(L_f^{r_i-1}(h_i)\right) = \left\langle \mathrm{d}L_f^{r_i-1}(h_i), \boldsymbol{g}_j \right\rangle \neq 0, \quad \text{where there exists at least one } j \in (1, \ldots, m) \end{aligned} \tag{8.13}$$

According to (8.10), the first-order derivative of y_i can be written as

$$\frac{\mathrm{d}y_i}{\mathrm{d}t} = \frac{\partial h_i}{\partial \boldsymbol{x}}\left(\boldsymbol{f}(\boldsymbol{x}) + \sum_{j=1}^{m} \boldsymbol{g}_j(\boldsymbol{x}) u_j\right) = L_f(h_i) + \sum_{j=1}^{m} L_{g_j}(h_i)\, u_j$$

If $L_{g_j}(h_i) = 0$, $j = 1, \ldots, m$, then $\mathrm{d}y_i/\mathrm{d}t = L_f(h_i)$ has nothing with u_j. Continue to find the second derivative of y_i:

$$\frac{\mathrm{d}^2 y_i}{\mathrm{d}t^2} = \frac{\partial (L_f(h_i))}{\partial \boldsymbol{x}}\left(\boldsymbol{f}(\boldsymbol{x}) + \sum_{j=1}^{m} \boldsymbol{g}_j(\boldsymbol{x}) u_j\right) = L_f^2(h_i) + \sum_{j=1}^{m} L_{g_j}(L_f(h_i))\, u_j$$

If $L_{g_j}(L_f(h_i)) = 0$, $j = 1, \ldots, m$, then $\mathrm{d}^2 y_i/\mathrm{d}t^2 = L_f^2(h_i)$ has nothing with u_j. Continue this procedure until

$$\frac{\mathrm{d}^{r_i} y_i}{\mathrm{d}t^{r_i}} = L_f^{r_i}(h_i) + \sum_{j=1}^{m} L_{g_j}\left(L_f^{r_i-1}(h_i)\right) u_j$$

According to the definition of relative degree r_i, the second term on the right-hand side is generally nonzero. Summarizing the results of the above steps, we get

$$\begin{aligned} &\frac{\mathrm{d}^l y_i}{\mathrm{d}t^l} = L_f^l(h_i), \quad l = 0, 1, \ldots, r_i - 1 \\ &\frac{\mathrm{d}^{r_i} y_i}{\mathrm{d}t^{r_i}} = L_f^{r_i}(h_i) + \sum_{j=1}^{m} L_{g_j}\left(L_f^{r_i-1}(h_i)\right) u_j \end{aligned} \tag{8.14}$$

Denote $\boldsymbol{v} = (v_1 \ \cdots \ v_m)^{\mathrm{T}}$, where

$$v_i = \beta_{i0}^{-1}\sum_{l=0}^{r_i}\beta_{il}L_f^l(h_i) + \beta_{i0}^{-1}\beta_{ir_i}\sum_{j=1}^{m}L_{g_j}\left(L_f^{r_i-1}(h_i)\right)u_j, \quad i=1,\ldots,m \tag{8.15}$$

and β_{il} is a positive number and can be used to assign the desired system dynamics of the linearized system. Rewrite (8.15) into the vector form

$$\boldsymbol{v} = \boldsymbol{p}(\boldsymbol{x}) + \boldsymbol{F}(\boldsymbol{x})\boldsymbol{u}$$

where

$$\boldsymbol{p}(\boldsymbol{x}) = \begin{bmatrix} \beta_{10}^{-1}\sum_{l=0}^{r_1}\beta_{1l}L_f^l(h_1) \\ \vdots \\ \beta_{m0}^{-1}\sum_{l=0}^{r_m}\beta_{ml}L_f^l(h_m) \end{bmatrix} \tag{8.16}$$

$$\boldsymbol{F}(\boldsymbol{x}) = \begin{bmatrix} \beta_{10}^{-1}\beta_{1r_1}L_{g_1}\left(L_f^{r_1-1}(h_1)\right) & \cdots & \beta_{10}^{-1}\beta_{1r_1}L_{g_m}\left(L_f^{r_1-1}(h_1)\right) \\ \vdots & & \vdots \\ \beta_{m0}^{-1}\beta_{mr_m}L_{g_1}\left(L_f^{r_m-1}(h_m)\right) & \cdots & \beta_{m0}^{-1}\beta_{mr_m}L_{g_m}\left(L_f^{r_m-1}(h_m)\right) \end{bmatrix}$$

If the $n \times n$ dimensional matrix $\boldsymbol{F}(\boldsymbol{x})$ is nonsingular, a state feedback law can be obtained according to the above formulae:

$$\boldsymbol{u} = \boldsymbol{F}^{-1}(\boldsymbol{x})(\boldsymbol{v} - \boldsymbol{p}(\boldsymbol{x})) \tag{8.17}$$

Using this feedback law, combine (8.14) and (8.15). It follows that

$$\sum_{l=0}^{r_i}\beta_{il}\frac{\mathrm{d}^l y_i}{\mathrm{d}t^l} = \beta_{i0}v_i, \qquad i=1,\ldots,m \tag{8.18}$$

In this way, the original nonlinear system (8.10) can be transformed into m decoupled single-input single-output linear systems (8.18) with v_i as the input and y_i as the output.

In the layered control structure, based on input–output linearization realized at the first layer, predictive controllers are designed at the second layer for m decoupled single-input single-output linear systems. For the ith subsystem (8.18), its transfer function has the form

$$G_i(s) = \frac{y_i(s)}{v_i(s)} = \frac{\beta_{i0}}{\beta_{ir_i}s^{r_i} + \beta_{i,r_i-1}s^{r_i-1} + \cdots + \beta_{i0}}$$

Sampling and holding with period T, its discrete transfer function is given by

$$G_i(z) = \frac{y_i(z)}{v_i(z)} = \frac{b_{i,1}z^{-1} + \cdots + b_{i,r_i}z^{-r_i}}{1 + a_{i,1}z^{-1} + \cdots + a_{i,r_i}z^{-r_i}}$$

Then the GPC algorithm presented in Section 2.2 can be used to realize model prediction, rolling optimization, and feedback correction. It is also possible to calculate the

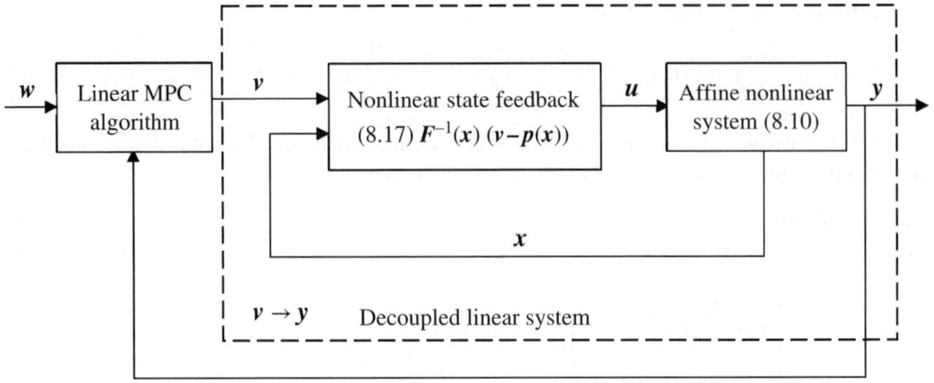

Figure 8.1 Predictive control for nonlinear systems based on input–output linearization.

coefficients of the step response from v_i to y_i according to this discrete transfer function and then use the DMC algorithm in Section 2.1. Specific algorithms can be referred to Chapter 2 and are no longer detailed here.

The input–output linearization-based predictive control strategy for nonlinear systems presented above can be illustrated by the control structure shown in Figure 8.1.

In the following, an example is given to illustrate the application of the presented predictive control strategy.

Example 8.1 [1] Consider the planar two-link manipulator shown in Figure 8.2.

Assume that the mass of each link is concentrated at the end. The length and the quality of the link are both units. The motion equation is given as follows:

$$u_1 = 3\ddot{\theta}_1 + \ddot{\theta}_2 + \left(2\ddot{\theta}_1 + \ddot{\theta}_2\right)\cos\theta_2 - \dot{\theta}_2^2\sin\theta_2 - 2\dot{\theta}_1\dot{\theta}_2\sin\theta_2 + g\cos\left(\theta_1 + \theta_2\right) + 2g\cos\theta_1$$
$$u_2 = \ddot{\theta}_1 + \ddot{\theta}_2 + \ddot{\theta}_1\cos\theta_2 + \dot{\theta}_1^2\sin\theta_2 + g\cos\left(\theta_1 + \theta_2\right)$$

where g is the acceleration of gravity, u_1, u_2 are input torques, and θ_1, θ_2 are output angles. Take $\boldsymbol{x} = \left(\theta_1\ \ \dot{\theta}_1\ \ \theta_2\ \ \dot{\theta}_2\right)^{\mathrm{T}}$ as the state variable; then the affine representation of this nonlinear system can be obtained after a series of calculations:

$$\dot{\boldsymbol{x}} = \boldsymbol{f}(\boldsymbol{x}) + \boldsymbol{g}(\boldsymbol{x})\boldsymbol{u}$$
$$\boldsymbol{y} = \boldsymbol{h}(\boldsymbol{x})$$

where

$$\boldsymbol{f}(\boldsymbol{x}) = \begin{bmatrix} x_2 \\ f_2(\boldsymbol{x}) \\ x_4 \\ f_4(\boldsymbol{x}) \end{bmatrix}, \quad \boldsymbol{g}(\boldsymbol{x}) = \left(\boldsymbol{g}_1(\boldsymbol{x})\ \ \boldsymbol{g}_2(\boldsymbol{x})\right) = \begin{bmatrix} 0 & 0 \\ \dfrac{1}{1+\sin^2 x_3} & -\dfrac{1+\cos x_3}{1+\sin^2 x_3} \\ 0 & 0 \\ -\dfrac{1+\cos x_3}{1+\sin^2 x_3} & \dfrac{3+2\cos x_3}{1+\sin^2 x_3} \end{bmatrix}$$

$$\boldsymbol{h}(\boldsymbol{x}) = \begin{bmatrix} h_1(\boldsymbol{x}) \\ h_2(\boldsymbol{x}) \end{bmatrix} = \begin{bmatrix} x_1 \\ x_3 \end{bmatrix}, \quad \boldsymbol{u} = \begin{bmatrix} u_1 \\ u_2 \end{bmatrix}$$

Figure 8.2 Planar two-link manipulator.

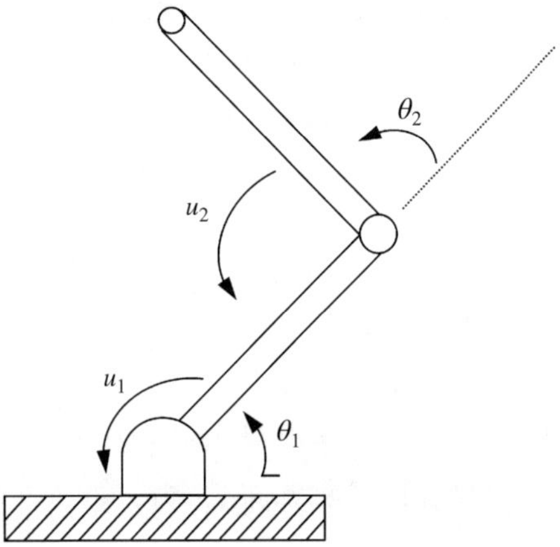

with

$$f_2(\boldsymbol{x}) = \frac{1}{1+\sin^2 x_3}\left\{(x_2+x_4)^2 \sin x_3 + x_2^2 \sin x_3 \cos x_3 - g[\cos x_1 + \sin x_3 \sin(x_1+x_3)]\right\}$$

$$f_4(\boldsymbol{x}) = \frac{1}{1+\sin^2 x_3}\left\{-\left(3x_2^2+2x_2x_4+x_4^2\right)\sin x_3 - \left(2x_2^2+2x_2x_4+x_4^2\right)\sin x_3 \cos x_3 \right.$$
$$\left. + g[2\cos x_1 - \cos x_3 \cos(x_1+x_3) + 2\sin x_1 \sin x_3]\right\}$$

In the following, an input–output linearization is first made for this system.

For $h_1(\boldsymbol{x})$, calculate

$$L_f^0(h_1) = h_1 = x_1$$
$$L_f^1(h_1) = L_f\left(L_f^0(h_1)\right) = \langle \mathrm{d}x_1, \boldsymbol{f} \rangle = x_2$$
$$L_f^2(h_1) = L_f\left(L_f^1(h_1)\right) = \langle \mathrm{d}x_2, \boldsymbol{f} \rangle = f_2(\boldsymbol{x})$$

$$L_{g_1}\left(L_f^0(h_1)\right) = \langle \mathrm{d}x_1, \boldsymbol{g}_1 \rangle = 0, \quad L_{g_2}\left(L_f^0(h_1)\right) = \langle \mathrm{d}x_1, \boldsymbol{g}_2 \rangle = 0$$
$$L_{g_1}\left(L_f^1(h_1)\right) = \langle \mathrm{d}x_2, \boldsymbol{g}_1 \rangle = \frac{1}{1+\sin^2 x_3}, \quad L_{g_2}\left(L_f^1(h_1)\right) = \langle \mathrm{d}x_2, \boldsymbol{g}_2 \rangle = -\frac{1+\cos x_3}{1+\sin^2 x_3}$$

Thus we know that $r_1 = 2$. Similarly calculate

$$L_f^0(h_2) = h_2 = x_3$$
$$L_f^1(h_2) = L_f\left(L_f^0(h_2)\right) = \langle \mathrm{d}x_3, \boldsymbol{f} \rangle = x_4$$
$$L_f^2(h_2) = L_f\left(L_f^1(h_2)\right) = \langle \mathrm{d}x_4, \boldsymbol{f} \rangle = f_4(\boldsymbol{x})$$

$$L_{g_1}\left(L_f^0(h_2)\right) = \langle \mathrm{d}x_3, \boldsymbol{g}_1 \rangle = 0, \quad L_{g_2}\left(L_f^0(h_2)\right) = \langle \mathrm{d}x_3, \boldsymbol{g}_2 \rangle = 0$$
$$L_{g_1}\left(L_f^1(h_2)\right) = \langle \mathrm{d}x_4, \boldsymbol{g}_1 \rangle = -\frac{1+\cos x_3}{1+\sin^2 x_3}, \quad L_{g_2}\left(L_f^1(h_2)\right) = \langle \mathrm{d}x_4, \boldsymbol{g}_2 \rangle = \frac{3+2\cos x_3}{1+\sin^2 x_3}$$

It then follows that $r_2 = 2$. Select the parameters of two decoupled subsystems after input–output linearization:

$$\beta_{i2} = 1,\ \ \beta_{i1} = 8,\ \ \beta_{i0} = 15, \qquad i = 1,\ 2$$

Substitute them into (8.16), which gives

$$\boldsymbol{p}(\boldsymbol{x}) = \frac{1}{15}\begin{bmatrix} 15x_1 + 8x_2 + f_2(\boldsymbol{x}) \\ 15x_3 + 8x_4 + f_4(\boldsymbol{x}) \end{bmatrix}, \qquad \boldsymbol{F}(\boldsymbol{x}) = \frac{1}{15}\begin{bmatrix} \dfrac{1}{1+\sin^2 x_3} & -\dfrac{1+\cos x_3}{1+\sin^2 x_3} \\ -\dfrac{1+\cos x_3}{1+\sin^2 x_3} & \dfrac{3+2\cos x_3}{1+\sin^2 x_3} \end{bmatrix}$$

Thus the state feedback law according to (8.17) can be calculated as

$$\begin{bmatrix} u_1 \\ u_2 \end{bmatrix} = \begin{bmatrix} 3+2\cos x_3 & 1+\cos x_3 \\ 1+\cos x_3 & 1 \end{bmatrix}\left\{ 15\begin{bmatrix} v_1 \\ v_2 \end{bmatrix} - \begin{bmatrix} 15x_1 + 8x_2 + f_2(\boldsymbol{x}) \\ 15x_3 + 8x_4 + f_4(\boldsymbol{x}) \end{bmatrix} \right\}$$

Substitute the expressions of $f_2(\boldsymbol{x})$, $f_4(\boldsymbol{x})$ into the above and the state feedback law is finally given by

$$\begin{aligned} u_1 &= 15(3+2\cos x_3)v_1 + 15(1+\cos x_3)v_2 - (3+2\cos x_3)(15x_1 + 8x_2) \\ &\quad - (1+\cos x_3)(15x_3 + 8x_4) - (2x_2 + x_4)x_4 \sin x_3 + g(2\cos x_1 + \cos(x_1 + x_3)) \\ u_2 &= 15(1+\cos x_3)v_1 + 15v_2 - (1+\cos x_3)(15x_1 + 8x_2) - (15x_3 + 8x_4) \\ &\quad + x_2^2 \sin x_3 + g\cos(x_1 + x_3) \end{aligned}$$

With the above state feedback law, the original nonlinear system is transformed into two input–output decoupled subsystems:

$$\ddot{y}_i + 8\dot{y}_i + 15y_i = 15v_i, \qquad i = 1,\ 2$$

The corresponding discrete input–output model with the sampling period T is given by

$$y_i(k) = -a_1 y_i(k-1) - a_2 y_i(k-2) + b_1 v_i(k-1) + b_2 v_i(k-2), \quad i = 1,\ 2$$

with

$$\begin{aligned} a_1 &= -\mathrm{e}^{-3T} - \mathrm{e}^{-5T}, \qquad a_2 = \mathrm{e}^{-8T} \\ b_1 &= 1 - 2.5\mathrm{e}^{-3T} + 1.5\mathrm{e}^{-5T}, \\ b_2 &= \mathrm{e}^{-8T} + 1.5\mathrm{e}^{-3T} - 2.5\mathrm{e}^{-5T} \end{aligned}$$

After feedback linearization, the problem of predictive control for the original nonlinear system is transformed into designing predictive controllers for two input–output decoupled single variable linear systems. Take $T = 0.1$s, using the GPC design method presented in Section 2.2, and select $N_1 = 1$, $N_2 = 4$, $NU = 1$, $\lambda = 0.05$. The control results are shown in Figure 8.3.

Remark 8.1 In the above the basic idea of predictive control based on feedback linearization is illustrated taking the multivariable affine nonlinear system as an example. The feedback linearization theory of nonlinear systems that has been developed is quite mature and fruitful results have been achieved on input–output linearization, input-state linearization, etc. Even for input–output linearization, the extended linearization method can be applicable to more general nonlinear systems. With the help of feedback linearization the original nonlinear system can be in principle transformed into a linear

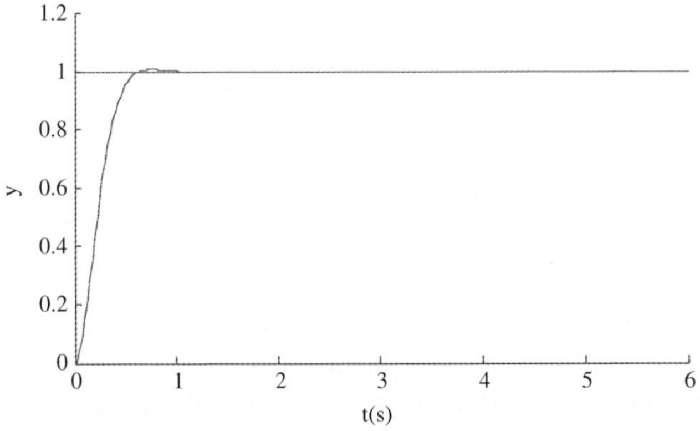

Figure 8.3 GPC control results of Example 8.1 after feedback linearization.

one, and predictive control algorithms for linear systems can then be used. This linearization method is entirely different from the method of approximate linearization by series expansion around the equilibrium point because here the model obtained by state feedback and transformation is strictly linear, rather than through an approximate method.

8.3 Multiple Model Predictive Control Based on Fuzzy Clustering

The predictive control method based on feedback linearization presented above needs to know the nonlinear system model and all the system states should be measurable. Much effort should be paid to calculate the nonlinear state feedback law. In practice, the approximate linearization method is always the simplest one for nonlinear system control, but with a single model it is difficult to accurately approximate the nonlinear system so a multiple model approach is often a better choice for approximating the nonlinear system.

The multiple model approach is based on the divide-and-conquer strategy, suitable to processes with strong nonlinearity and a large operating region. It divides the whole operating region into several separate regions. For each region the corresponding local linear model and controller are established and the global nonlinear system is then approximated and controlled by switching or integration of them. The multiple model strategy with the switching mode selects one local model/controller mostly close to the current system state as the current model/controller, focusing on the switching criterion and on how to ensure that the controlled system is globally stable. Since both model and controller are hard switched, the smoothness of the control will be affected. For the multiple model strategy with the integration mode, control is a weighted sum of local controller outputs, where the weight for each subcontroller is given by the fuzzy membership degree or the probability that the current state belongs to the corresponding local model. This approach can avoid the effect caused by hard switching. In this section, we introduce this kind of multiple model control for nonlinear systems [2, 3].

While the nonlinear system is being controlled by the multiple model approach, local models can be obtained by linearizing the nonlinear model at corresponding operating points. However, how many local models should be selected and how to judge the correlation of the current system state with each local model are always critical and difficult problems for the multiple model approach to deal with in real applications. For such problems, the development of fuzzy control provides a powerful tool. The multiple model in fuzzy control always works with the integration mode. It uses the divide-and-conquer strategy, with which the generalized input space of the system is firstly divided into several subspaces, then the dynamic behaviors of the global system in different subspaces are weighted and combined such that the system can run in a smooth way. Furthermore, this approach does not need to identify the nonlinear system model and can directly start from the process data with fuzziness, which is more important for real applications.

(1) Establishing the multiple model based on fuzzy satisfactory clustering (FSC)

The Takagi-Sugeno (T-S) model is the most widely used multiple model in fuzzy control. It directly uses the data samples of the process rather than its mathematical model. The basic form of the T-S fuzzy multiple model with c rules is as follows:

$$\begin{aligned} &R^i: \text{ if } \boldsymbol{x} \text{ is } A_i \\ &\qquad \text{then } M_i: \; y_i = p_{i0} + p_{i1}x_1 + \cdots + p_{id}x_d \quad i = 1, \ldots, c \end{aligned} \tag{8.19}$$

where $\boldsymbol{x} = (x_1 \; \cdots \; x_d)^{\mathrm{T}}$ is the input variable and A_i is its fuzzy set described either qualitatively or quantitatively. M_i represents the submodel of the ith rule, where y_i is the output of the submodel and p_{ij} is the post-parameter of the ith rule. Then the system output can be given by the weighted submodel outputs

$$y = \sum_{i=1}^{c} \mu_i y_i \Big/ \sum_{i=1}^{c} \mu_i \tag{8.20}$$

where μ_i is the membership degree of $\boldsymbol{x}$ to the fuzzy set A_i.

To identify the structure of the T-S multiple model, it is usually necessary to divide the input space at first and then to identify the parameters in the conclusion part. The main steps include clustering of the data sample set (i.e. to partition the data set $\boldsymbol{Z}$ into several fuzzy subsets by using some clustering algorithm), determination of the cluster number c (i.e. to determine the appropriate number of clusters, which is also the rule number of the T-S fuzzy model), and identification of the fuzzy rules (including the membership function of the premise variable, the parameters of the conclusion part, etc.). Because of different clustering methods for data samples and different representations for submodels, a variety of T-S model identification methods based on fuzzy clustering appeared in the literature. In the following the multiple modeling method based on FSC [3] is introduced.

Consider a multi-input single-output (MISO) system. Its sample set is composed of system input/output data. The jth sample is represented by $\boldsymbol{z}_j = (\boldsymbol{\varphi}_j \; y_j)^{\mathrm{T}}, j = 1, \ldots, N$, where $\boldsymbol{\varphi}_j$ is the regression row vector, called a generalized input vector, which affects the system output. In general, it is composed of a previous system input/output and its dimension is d, y_j is the system output and N is the total number of samples. Then the sample set can be represented by $\boldsymbol{Z} = [\boldsymbol{z}_1, \ldots, \boldsymbol{z}_N]$, where $\boldsymbol{z}_j \in \mathrm{R}^{d+1}$. Then the sample set $\boldsymbol{Z}$ is divided into c clusters $\{\boldsymbol{Z}_1, \ldots, \boldsymbol{Z}_c\}$. For each cluster $\boldsymbol{Z}_i$, denote its cluster center as

$v_i = [v_{i,1} \ \cdots \ v_{i,d+1}]^T \in R^{d+1}$, $i = 1, \ldots, c$. The result of fuzzy clustering can be represented by the membership matrix $U = [\mu_{i,j}]_{c \times N}$, where $\mu_{i,j} \in [0, 1]$ represents the degree to which the sample z_j belongs to the ith cluster with center v_i and satisfies

$$\begin{cases} \sum_{i=1}^{c} \mu_{i,j} = 1, & j = 1, \ldots, N \\ 0 < \sum_{j=1}^{N} \mu_{i,j} < N, & i = 1, \ldots, c \end{cases} \tag{8.21}$$

Gustafson and Kessel [4] proposed an efficient fuzzy clustering approach by using an adaptive distance measure for the clustering covariance matrix, called the GK algorithm; for more details refer to [4] or corresponding textbooks. This algorithm is insensitive to the initial values of the membership matrix and the clustering centers, i.e. different initial conditions do not greatly affect the clustering result. It is not necessary to normalize the sample data due to the use of an adaptive distance measure. Furthermore, it can detect different shapes of clusters because it is not limited to linear clustering. All the above advantages make it an effective clustering algorithm which is widely used in control, image processing, and other fields. However, similar to most other clustering algorithms, in the GK algorithm the cluster number c should be given in advance. Since the cluster number essentially depends on the degree of system nonlinearity, if not enough prior knowledge is available, in general it should be determined by using the trial-and-error method step by step, which will undoubtedly increase the computational burden.

To solve this problem, an FSC approach based on GK fuzzy clustering was proposed in [3]. Its basic idea is as follows. Firstly, start from a lower number of initial clusters, e.g. let $c = 2$. After initial clustering of the system, if the clustering result is not satisfactory, find a sample from the sample set as a new sample center v_{c+1}, which is mostly dissimilar to all the clustering centers $v_1 \sim v_c$. Calculate the new membership matrix U using $v_1 \sim v_{c+1}$ as the initial clustering centers and repeat the GK algorithm to divide the system into $c + 1$ parts. This procedure is repeated according to the requirement on the performance index until a satisfactory performance index is reached. For the above discussed MISO system, the specific T-S modeling procedure based on FSC is given as follows.

Algorithm 8.1 T-S modeling based on FSC [3]

Step 1. Set the initial cluster number $c = 2$.

Step 2. With the initial membership matrix U_0, use the GK algorithm to divide the data sample set Z into c subsets $\{Z_1, \ldots, Z_c\}$ and obtain the corresponding membership matrix $U = [\mu_{i,j}]_{c \times N}$.

Step 3. For each cluster, identify the consequence parameters in the submodel fuzzy rule (8.19) using the stable-state Kalman filter algorithm [5] and then obtain the submodels

(Continued)

Algorithm 8.1 (Continued)

$$
\begin{aligned}
&R_1: \text{ if } \boldsymbol{z} \in \boldsymbol{Z}_1 \text{ then } M_1: \; y_1(\boldsymbol{\varphi}) = p_{10} + p_{11}\varphi(1) + \cdots + p_{1d}\varphi(d) \\
&R_2: \text{ if } \boldsymbol{z} \in \boldsymbol{Z}_2 \text{ then } M_2: \; y_2(\boldsymbol{\varphi}) = p_{20} + p_{21}\varphi(1) + \cdots + p_{2d}\varphi(d) \\
&\quad\vdots \\
&R_c: \text{ if } \boldsymbol{z} \in \boldsymbol{Z}_c \text{ then } M_c: \; y_c(\boldsymbol{\varphi}) = p_{c0} + p_{c1}\varphi(1) + \cdots + p_{cd}\varphi(d)
\end{aligned}
\tag{8.22}
$$

Step 4. Calculate the system output corresponding to input $\boldsymbol{z}_j$ according to the membership matrix $\boldsymbol{U} = [\mu_{i,j}]_{c \times N}$ and (8.20). Note that the parameters in the membership matrix deduced by the GK algorithm satisfy $\sum_{i=1}^{c} \mu_{ij} = 1$; then the system output corresponding to input $\boldsymbol{z}_j$ is given by

$$
\hat{y}\left(\boldsymbol{\varphi}_j\right) = \sum_{i=1}^{c} \mu_{i,j} y_i\left(\boldsymbol{\varphi}_j\right)
\tag{8.23}
$$

Step 5. Calculate the cluster validity index S_c after clustering; use, for example, the root mean square error (RMSE) index

$$
\text{RMSE} = \sqrt{\frac{1}{N}\sum_{j=1}^{N}\left(\hat{y}\left(\boldsymbol{\varphi}_j\right) - y\left(\boldsymbol{\varphi}_j\right)\right)^2}
\tag{8.24}
$$

to evaluate the fitting degree of the model. If $S_c \le S_{TH}$ then end the modeling process; otherwise further clustering is needed so go to Step 6, where S_{TH} is the threshold of the performance index that the user thinks is satisfactory.

Step 6. According to the membership matrix $\boldsymbol{U}$, find a sample $\boldsymbol{z}_n$ from the data set that is mainly different from all cluster centers. The dissimilarity degree can be defined and calculated by the usual formulae in the clustering algorithms, for example

$$
n = \arg\min_{n} \sum_{\substack{1 \le i,j \le c \\ i \ne j}} \left|\mu_{i,n} - \mu_{j,n}\right|
\tag{8.25}
$$

Step 7. With new cluster centers $\boldsymbol{v}_1, \ldots, \boldsymbol{v}_c, \boldsymbol{z}_n$, calculate the corresponding new initial membership matrix $\boldsymbol{U}_0$ according to the usual membership function, rather than randomly selecting $\boldsymbol{U}_0$ through re-initialization.

Step 8. Let $c = c + 1$, $\boldsymbol{U} = \boldsymbol{U}_0$; then go to Step 2.

The above T-S modeling algorithm based on FSC avoids determining $c_{\max}$ in terms of the system nonlinear characteristics, as used in usual clustering integration methods. It directly starts from $c = 2$ and makes the algorithm have a definite number for initial clustering. Furthermore, except for the first clustering, during the clustering process, the initialization parameters such as the membership matrix can be determined mainly according to the result of the last clustering and do not need to be calculated starting from random values. The computational efficiency is obviously increased, particularly for a data set with a great number of samples.

Using the FSC algorithm, the multiple model described by (8.22) can be obtained through off-line determination of the clusters according to data samples, which realizes the multiple modeling for MISO systems. Based on that the multimodel predictive control algorithm for nonlinear systems can be introduced.

(2) Multiple model predictive control algorithm based on fuzzy clustering

A nonlinear multivariable system with r outputs can be decomposed into r MISO subsystems according to its output. After FSC modeling, the lth MISO subsystem is divided into c_l clusters and described by c_l submodels described in (8.22). In order to make the model suitable for predictive control algorithms such as DMC, GPC where control increments are adopted, the submodel in (8.22) is rewritten into increment form with the corresponding fuzzy rules

$$\begin{aligned} &R_i^l: \text{ if } \boldsymbol{\varphi}^l \in \boldsymbol{v}_i^{lx} \\ &\quad \text{then } M_i^l: \ \Delta y_i^l = p_{i1}^l \Delta\varphi^l(1) + \cdots + p_{id_l}^l \Delta\varphi^l(d_l) \\ &\qquad l = 1, \ldots, r, \quad i = 1, \ldots, c_l \end{aligned} \tag{8.26}$$

where $\boldsymbol{\varphi}^l$ is the generalized input vector of the lth MISO system, d_l represents its dimension, y^l is its output, $\boldsymbol{v}_i^{lx} \in \mathrm{R}^{d_l}$ is the projection of its ith clustering center $\boldsymbol{v}_i^l$ on the generalized input space, i.e. the remaining part of the vector $\boldsymbol{v}_i^l$ after removing the output element, p_{ih}^l represents the hth parameter of the ith submodel, y_i^l represents the output of the ith submodel with respect to $\boldsymbol{\varphi}^l$, and c_l, p_{ih}^l have been determined after off-line modeling. The advantage of this increment form is that the constants p_{i0}^l in all linear models can be canceled.

During online control, at each sampling time when the lth MISO system obtains a new data sample $\boldsymbol{\varphi}^l$, its membership to the ith cluster can be calculated by

$$\mu_i^l(\boldsymbol{\varphi}^l) = \frac{\left(d(\boldsymbol{\varphi}^l, \boldsymbol{v}_i^{lx})\right)^{-2/(m-1)}}{\sum_{j=1}^{c}\left(d\left(\boldsymbol{\varphi}^l, \boldsymbol{v}_j^{lx}\right)\right)^{-2/(m-1)}} \tag{8.27}$$

where μ_i^l represents the membership of $\boldsymbol{\varphi}^l$ to the ith submodel M_i^l of y^l, $d\left(\boldsymbol{\varphi}^l, \boldsymbol{v}_j^{lx}\right)$ is the measure of the distance function of the new input vector $\boldsymbol{\varphi}^l$ to $\boldsymbol{v}_i^{lx}$, and $m > 1$ is a tunable parameter characterizing the degree of fuzziness of clustering $\sum_{i=1}^{c_l} \mu_i^l = 1$.

For the generalized input vector $\boldsymbol{\varphi}^l$, after calculating its membership to each submodel by (8.27), the global model of the MISO system can be integrated by (8.23) and (8.27) as follows:

$$\text{if } \boldsymbol{\varphi}^l \text{ then } \Delta y^l = \left(\sum_{i=1}^{c_l} \mu_i^l p_{i1}^l\right) \Delta\varphi^l(1) + \cdots + \left(\sum_{i=1}^{c_l} \mu_i^l p_{id_l}^l\right) \Delta\varphi^l(d_l) \tag{8.28}$$

After obtaining the models represented by (8.28) for all MISO subsystems, the model of the MIMO system can be obtained by integrating them together:

$$\boldsymbol{A}(z^{-1}) \Delta \boldsymbol{y}(t) = \boldsymbol{B}(z^{-1}) \Delta \boldsymbol{u}(t-1) \tag{8.29}$$

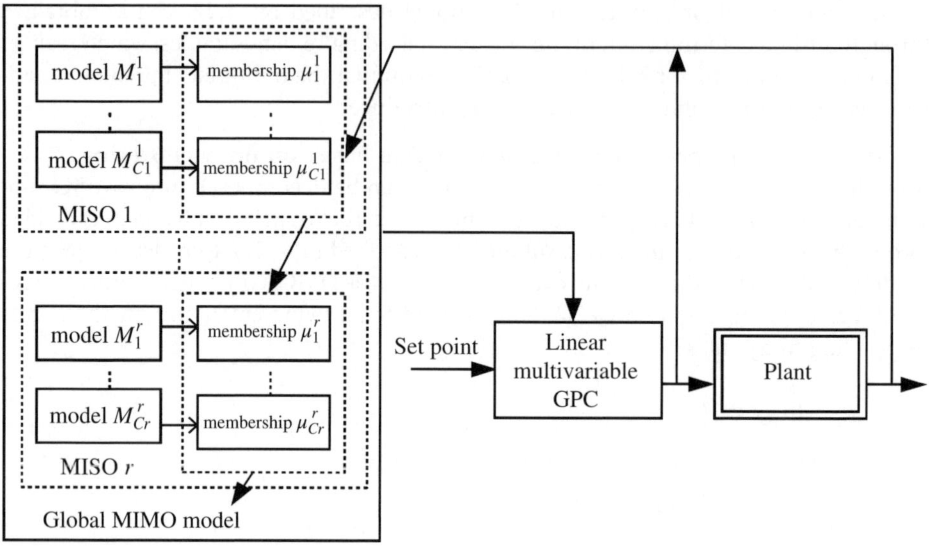

Figure 8.4 Structure of multiple model predictive control based on fuzzy clustering [2].

Then the multivariable GPC algorithm can be adopted to design the multiple model predictive controller. The principle of the multiple model predictive control based on the above fuzzy clustering method is shown in Figure 8.4.

The multiple model predictive control algorithm based on fuzzy clustering for nonlinear systems can be summarized as follows. Firstly, use Algorithm 8.1 (FSC) off-line with data samples to obtain the multiple model description (8.26) for each MISO subsystem, where the number of clusters c_l, the cluster centers $\mathbf{v}_i^{lx}$, and the submodel parameters p_{ih}^l are determined. Then turn to online control. At each step of control, measure the generalized input $\boldsymbol{\varphi}^l$ of each subsystem, calculate the membership μ_i^l of it to the corresponding submodel by (8.27). Then obtain the parametric model of the whole multivariable system according to (8.28) and (8.29), and take it as the prediction model to perform predictive control using the conventional GPC algorithm.

Remark 8.2 For nonlinear systems and time-varying systems, at each step the membership μ_i^l of $\boldsymbol{\varphi}^l$ to the ith submodel always changes and thus the obtained parameters of the linear model (8.29) are also time varying. Therefore, the output prediction formula should be updated at each step, which is similar to the adaption part in the GPC algorithm caused by the nonlinear characteristics of the system.

Example 8.2 [3] Consider the pH neutralization process in the CSTR (Continuously Stirred Tank Reactor) with three reaction streams HNO_3, NaOH, and $NaHCO_3$ as inputs shown in Figure 8.5. The control purpose is to make the pH value of the liquid in the stirred tank reach the desired one while keeping the liquid level at the set height. Take the flow rates F_a, F_b corresponding to HNO_3, NaOH as the control inputs, the flow rate F_{bf} of $NaHCO_3$ as the disturbance, and the liquid level h, pH value as the outputs; then we have a two inputs (F_a, F_b) and two outputs (h, pH) system with disturbance F_{bf}.

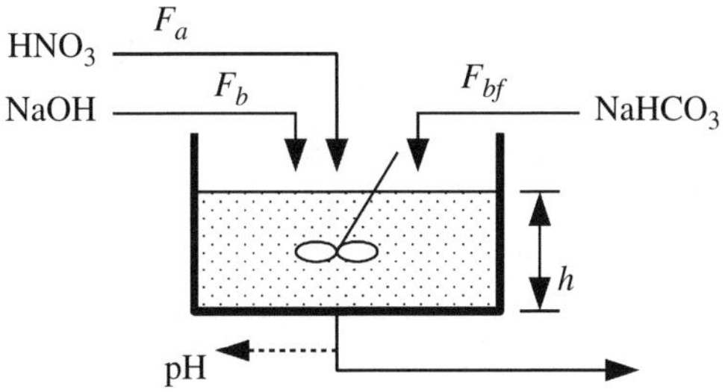

Figure 8.5 pH neutralization process [3]. *Source:* Reproduced with permission from Ning Li, Shao-Yuan Li and Yu-Geng Xi of Elsevier.

The static titration curve of the system shows that there exists serious nonlinearity between the pH variation and the system input, and the disturbance F_{bf} has a large impact on the variation of the nonlinearity.

For easy identification, define the relationship of system inputs and outputs as follows:

$$\hat{y}_h(t) = \Psi_h\left(F_a(t-1), F_b(t-1), F_{bf}(t-1), y_h(t-2), y_h(t-1)\right)$$
$$\hat{y}_{\text{pH}}(t) = \Psi_{\text{pH}}\left(F_a(t-1), F_b(t-1), F_{bf}(t-1), y_{\text{pH}}(t-2), y_{\text{pH}}(t-1)\right)$$

Firstly, let the flow rate variation of the streams be

$$F_a(t) = 16 + 4\sin(2\pi t/15)$$
$$F_b(t) = 16 + 4\cos(2\pi t/25)$$
$$F_{bf}(t) = 0.55 + 0.055\sin(2\pi t/10)$$

The data sample set is produced, from which the multiple model can be obtained using Algorithm 8.1 based on FSC. Figure 8.6a and b show the modeling results for the pH channel and the h channel, respectively, when the cluster number is six. It can be seen that for both channels the model has a good fitting degree. The RMSE for the pH channel and the h channel is 0.209 and 0.0318, respectively, which indicates that the model can well approximate the nonlinear characteristics of the system.

After obtaining the local models of two MISO systems off-line (each has six submodels; refer to [3] for details), the global model of this multivariable system can be obtained online according to (8.27), (8.28) and (8.29). For GPC control, select the design parameters $N_1 = 1, N_2 = 5, NU = 2, \boldsymbol{Q} = 0.1\boldsymbol{I}, \boldsymbol{R} = 0.01\boldsymbol{I}$. In Figure 8.7, the variations of the control inputs and the system outputs by using the unconstrained multivariable GPC algorithm are shown for tracking pH ($9 \rightarrow 8 \rightarrow 9 \rightarrow 8$) and h ($15 \rightarrow 17 \rightarrow 16 \rightarrow 18$) with a constant disturbance $F_{bf} = 0.55$ ml/s. With other simulations it is also found that the system outputs exhibit good tracking performance for different disturbance amplitudes, and there is no steady-state error if the disturbance remains constant.

The above-introduced multiple model predictive control algorithm based on fuzzy clustering does not need the nonlinear model of the system and can directly start from the input and output data. Through fuzzy clustering, the problem of identifying a complex nonlinear system is solved by firstly finding a set of simple linear models for corresponding subregions defined by fuzzy boundaries and then establishing the system global model by integrating them through fuzzy weighting. In this way the predictive control problem for nonlinear systems can be transformed into that for linear systems, where the predictive control algorithm is not based on one of multiple models selected

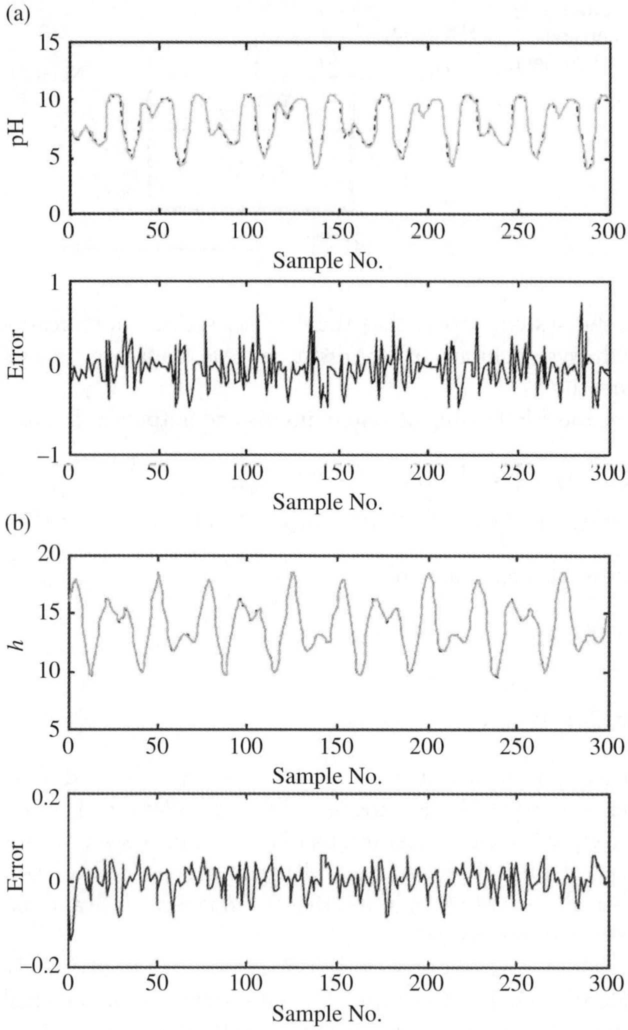

Figure 8.6 Fuzzy modeling of pH neutralization process [3]: (a) pH channel, (b) *h* channel. *Source:* Reproduced with permission from Ning Li, Shao-Yuan Li and Yu-Geng Xi of Elsevier.

according to the system input/output but on the system global model integrated according to the degree of input/output belonging to each model. Thus the online switching problem can be avoided. This multiple model predictive control method has a wide selection space both in specific clustering strategies and in model forms. More algorithms and effective applications can be referred to the relevant literature.

8.4 Neural Network Predictive Control

Artificial neural network provides a powerful tool for modeling and control of nonlinear systems and has been widely used in many fields. Since the 1980s, the application of neural networks in the control field has rapidly developed and a number of survey papers

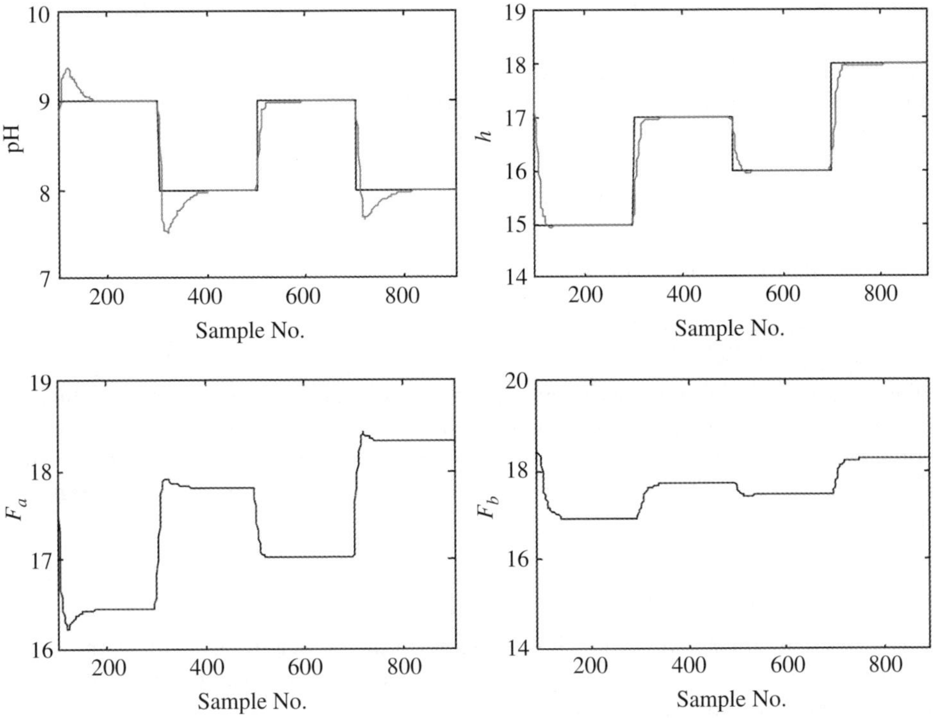

Figure 8.7 Result of multiple model predictive control for pH neutralization process with constant disturbance $F_{bf} = 0.55$ ml/s [3]. *Source:* Reproduced with permission from Ning Li, Shao-Yuan Li and Yu-Geng Xi of Elsevier.

appeared in which the neural network predictive control is also a typical example for neural network applications [6]. The applications of neural networks in predictive control include neural network modeling for nonlinear systems and optimization solving using neural networks. In relevant literature, it is usual to adopt a neural network for modeling and then to realize the predictive control in two different ways. One is directly using a neural network to solve the rolling optimization problem, while the other is firstly to identify the system dynamic response by a neural network model and then to solve the online rolling optimization using the parametric optimization method. In this section, we briefly introduce the predictive control algorithm directly using neural networks to modeling and on-line optimization for single-input single-output nonlinear systems [7].

(1) Neural network for modeling and prediction

Assume that the input–output model of a nonlinear system can be represented by

$$y(k) = f(u(k-1), \ldots, u(k-n_b), y(k-1), \ldots, y(k-n_a)) \tag{8.30}$$

For neural network modeling, this specific representation (8.30) is unknown and only the input and output sample data are available. Consider the commonly used backpropagation (BP) network with one layer of hidden nodes. It is composed of three layers of nodes, i.e. the input node layer, the hidden node layer, and the output node layer. Nonlinear transformation is considered only in the hidden node layer. Denote the output of the input node as o_j, $j = 1, \ldots, n_a + n_b \triangleq n_t$. They are indeed the variables in the bracket on the right side of (8.30). Denote the input of the hidden node i as x_i and its output as z_i, $i = 1, \ldots, m$, where m is the number of hidden nodes. Denote the output of the output node as $\hat{y}(k)$. According to the working principle of the BP network, it follows that

$$\begin{aligned} x_i &= w_{i0} + \sum_{j=1}^{n_t} w_{ij} o_j, \quad i = 1, \ldots, m \\ z_i &= \varphi(x_i), \qquad i = 1, \ldots, m \\ \hat{y}(k) &= w_0 + \sum_{i=1}^{m} w_i z_i \end{aligned} \tag{8.31}$$

where w_{ij} is the weighting coefficient of connection from the input node j to the hidden node i, w_{i0} is the input bias of the hidden node i, w_i is the weighting coefficient of connection from the hidden node i to output node, w_0 is the input bias of the output node, and $\varphi(\cdot)$ is the action function of the neuron, usually taken as the Sigmoid function

$$\varphi(x) = \frac{1}{1 + e^{-x}} \tag{8.32}$$

The task of neural network modeling is to determine the weighting coefficients and the input biases best matched with the given sample data set, which can be described as the following optimization problem:

$$\min_{w} E = \frac{1}{2} \sum_{l=1}^{N} (\hat{y}_l(\boldsymbol{w}) - y_l)^2 \tag{8.33}$$

where N is the number of samples, y_l and $\hat{y}_l$ represent, respectively, the system output in the lth group of samples and the neural network output calculated by (8.31) of this sample, and $\boldsymbol{w}$ is a parameter vector containing all the weighting coefficients and input biases.

The weighting coefficients of the BP network can be obtained as follows. Let

$$\begin{aligned} E_l &= \frac{1}{2}(\hat{y}_l(\boldsymbol{w}) - y_l)^2, \quad e_l = \hat{y}_l(\boldsymbol{w}) - y_l, \quad l = 1, \ldots, N \\ \delta_i &= \frac{\partial E_l}{\partial z_i} = \frac{\partial E_l}{\partial \hat{y}_l}\frac{\partial \hat{y}_l}{\partial z_i} = e_l w_i, \quad i = 1, \ldots, m, \quad l = 1, \ldots, N \\ \xi_i &= \frac{\partial E_l}{\partial x_i} = \frac{\partial E_l}{\partial z_i}\frac{dz_i}{dx_i} = \delta_i \dot{\varphi}(x_i) = \delta_i z_i (1 - z_i), \quad i = 1, \ldots, m \end{aligned} \tag{8.34}$$

then

$$\begin{aligned} &\frac{\partial E_l}{\partial w_0} = \frac{\partial E_l}{\partial \hat{y}_l}\frac{\partial \hat{y}_l}{\partial w_0} = e_l, \qquad \frac{\partial E_l}{\partial w_i} = \frac{\partial E_l}{\partial \hat{y}}\frac{\partial \hat{y}}{\partial w_i} = e_l z_i \\ &\frac{\partial E_l}{\partial w_{i0}} = \frac{\partial E_l}{\partial x_i}\frac{\partial x_i}{\partial w_{i0}} = \xi_i, \qquad \frac{\partial E_l}{\partial w_{ij}} = \frac{\partial E_l}{\partial x_i}\frac{\partial x_i}{\partial w_{ij}} = \xi_i o_j \\ &l = 1, \ldots, N, \quad i = 1, \ldots, m, \quad j = 1, \ldots, n_t \end{aligned} \tag{8.35}$$

After setting the initial parameter vector $\boldsymbol{w}$ for each sample, the network output $\hat{y}_l$ can be calculated by (8.31) for input o_j and compared with the actual output y_l to construct the error e_l. Starting from the output side, use the back propagation of error e_l to calculate δ_i, ξ_i using (8.34) and the partial derivative of E_l to w (w refers to any network parameter in $\boldsymbol{w}$) using (8.35). Then construct

$$\frac{\partial E}{\partial w} = \sum_{l=1}^{N} \frac{\partial E_l}{\partial w} \tag{8.36}$$

based on which the network parameters are improved by the gradient method

$$w^{\text{new}} = w^{\text{old}} - \eta \frac{\partial E}{\partial w} \tag{8.37}$$

This process is repeated until the performance index (8.33) reaches the minimum. Then the obtained BP network is the best match to the sample data set. It implicitly establishes the nonlinear mapping (8.30) between historical data and the current output, which can be directly used to one-step output prediction when the current control input is given.

However, in predictive control, the above neural network needs to be improved for a multistep prediction. The simplest way is to establish P simple BP networks, as above, when the prediction horizon is P, where the output of the sth BP network is the system output $\hat{y}(k+s)$ predicted at k, $s = 1, \dots, P$. For the sake of symbol brevity, assume $n_a = n_b \triangleq n$ in the nonlinear model (8.30). According to (8.4) and (8.6), from (8.30) it follows that

$$\hat{y}(k+s) = G_s(y(k), \dots, y(k-n+1),\ u(k+s-1), \dots, u(k), \dots, u(k-n+1)) \tag{8.38}$$

Thus the sth BP network should be an approximation of the nonlinear mapping $G_s(\cdot)$ from the input/output information available at time k and the future control inputs to $\hat{y}(k+s)$. Refer to (8.31) and consider the control horizon $M \le P$; the sth BP network can then be expressed by

$$\begin{aligned}
\text{BP}_s: \quad & x_i^s = w_{i0}^s + \sum_{j=1}^{n_1} w_{ij}^s u(k+j-n) + \sum_{l=n_1+1}^{n_2} w_{il}^s y(k+l-n-n_1) \\
& \qquad n_1 = n + \min(s, M) - 1, \quad n_2 = n_1 + n \\
& z_i^s = \varphi\left(x_i^s\right) \\
& \hat{y}(k+s) = w_0^s + \sum_{i=1}^{m} w_i^s z_i^s, \qquad s = 1, \dots, P
\end{aligned} \tag{8.39}$$

These BP networks work using the same principle and the whole network outputs can reflect the predicted future outputs at different times over the prediction horizon. Since both the learning process and the real-time prediction of these BP networks can be done in parallel, it is a practical and efficient way for multistep prediction for nonlinear systems.

(2) Neural network for online optimization

After establishing the neural network model for the nonlinear system, we now discuss the predictive control method based on that. Predictive control runs in a rolling style, i.e. at each sampling time the control action is obtained by online solving a nonlinear optimization problem. In addition to output prediction, the neural network model established above can also be used for online optimization, which can be solved by the same gradient optimization process as that in model parameter identification.

At the sampling time k, let the optimization performance index $J(k)$ have the form

$$\min J(k) = \frac{1}{2}\sum_{s=1}^{P}(\hat{y}(k+s) - y_r(k+s))^2 \tag{8.40}$$

where $\hat{y}(k+s)(s = 1, \ldots, P)$ are the outputs of the basic BP prediction models when future inputs are $u(k+h-1)(h = 1, \ldots, M)$, $y_r(k+s)(s = 1, \ldots, P)$ are the desired outputs. Note that

$$\frac{\partial J(k)}{\partial u(k+h-1)} = \sum_{s=1}^{P}\left\{\frac{\partial J(k)}{\partial \hat{y}(k+s)}\frac{\partial \hat{y}(k+s)}{\partial u(k+h-1)}\right\}$$

and it is known from (8.38) that $\hat{y}(k+s)$ is not related to $u(k+h-1)$ when $h > s$, so the above equation can be rewritten into

$$\frac{\partial J(k)}{\partial u(k+h-1)} = \sum_{s=h}^{P}\frac{\partial J(k)}{\partial \hat{y}(k+s)}\frac{\partial \hat{y}(k+s)}{\partial u(k+h-1)} \tag{8.41}$$

where

$$\frac{\partial \hat{y}(k+s)}{\partial u(k+h-1)} = \sum_{i=1}^{m}\frac{\partial \hat{y}(k+s)}{\partial z_i^s}\frac{\mathrm{d}z_i^s}{\mathrm{d}x_i^s}\frac{\partial x_i^s}{\partial u(k+h-1)} = \sum_{i=1}^{m} w_i^s z_i^s(1-z_i^s)w_{i,h+n-1}^s$$

Furthermore, according to the performance index (8.40), it follows that

$$\frac{\partial J(k)}{\partial \hat{y}(k+s)} = \hat{y}(k+s) - y_r(k+s), \quad s = 1, \ldots, P$$

Then the gradient can be given by

$$\frac{\partial J(k)}{\partial u(k+h-1)} = \sum_{s=h}^{P}\left\{(\hat{y}(k+s) - y_r(k+s))\sum_{i=1}^{m} w_i^s z_i^s(1-z_i^s)w_{i,h+n-1}^s\right\} \tag{8.42}$$

Therefore, one can initially set a group of controls $\boldsymbol{u}_M(k)$, calculate $\tilde{\boldsymbol{y}}_{PM}(k)$ using the model (8.39), and then substitute it into the performance index (8.40) to calculate $\hat{y} - y_r$ in $J(k)$. Based on that premise, the control can be improved by the gradient method

$$u^{\text{new}}(k+h-1) = u^{\text{old}}(k+h-1) - \alpha\frac{\partial J(k)}{\partial u(k+h-1)}, \quad h = 1, \ldots, M \tag{8.43}$$

where α is the step length and the gradient can be calculated by (8.42). This iteration process should be repeated until $J(k)$ reaches a minimum. Then $u(k)$ as the current optimal control can act for the system for control.

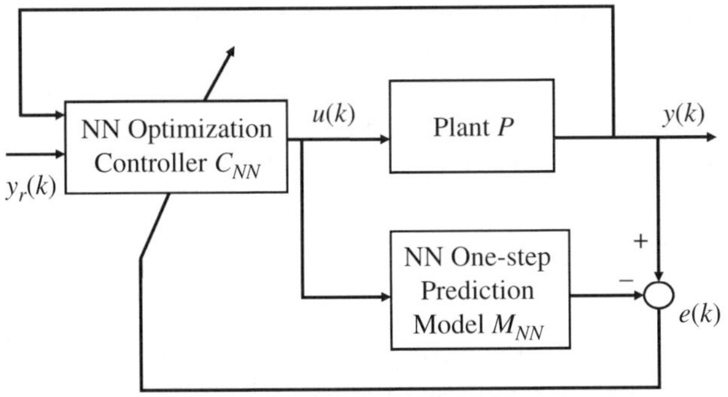

Figure 8.8 Structure of the neural network based predictive control for nonlinear systems.

The above predictive control algorithm with neural network modeling and online optimization can be described by the internal model control (IMC) structure shown in Figure 8.8, where M_{NN} is the one-step neural network prediction model. It only provides the predicted model output at the next sampling time in terms of current control and (8.39), and then constructs the output error using the measured real output to make a feedback correction. The core part in the figure is the neural network online optimization controller C_{NN}. It uses the neural network prediction model (8.39) and the optimization algorithm (8.42), (8.43) to iteratively calculate the optimal control and then implements the current control, where the weighting coefficients of the model (8.39) have been obtained through off-line learning.

The presented algorithm is only a fundamental one of the neural network based on predictive control algorithms. In fact, there are a large variety of methods both for neural network modeling and for predictive control based on the neural network model: for example, using the Hopfield network model, considering a general nonlinear performance index instead of a quadratic one, putting constraints on the system input and output, identifying step response coefficients after establishing the neural network model and then using traditional predictive control algorithm, and so on. This can be referred to much of the literature on neural network predictive control. Later in Chapter 10, we will also introduce an application example of using the simplified dual neural network (SDNN) to solve the quadratic programming (QP) problem in predictive control and implementing the algorithm in DSP.

8.5 Predictive Control for Hammerstein Systems

The nonlinearity of many systems is caused by the input component, such as input saturation, dead zone, hysteresis, etc. Among those the Hammerstein model is typical to describe the system with input nonlinearity. It is composed of a static nonlinear component and a dynamic linear component. Since a large variety of nonlinear processes such as pH neutralization, high purity separation, etc. can be described by the Hammerstein model, identification and control of such systems have been widely investigated. The control strategies for Hammerstein systems can be roughly classified into two catalogs. One is the global solving strategy, i.e. including the input nonlinear part in the optimization problem and solving the control action directly, which is of course very difficult. Another is the nonlinear separation strategy, that is, firstly using some control algorithm for the dynamic linear model to calculate its input, which is an intermediate variable but also the output of the static nonlinear component, and then conversely calculating the actual control according to the static nonlinearity. This strategy fully uses the specific structure of the Hammerstein model and transforms the controller design problem into a linear control one, and is thus much simpler than the global solving strategy.

An unconstrained single variable predictive control problem for the Hammerstein model was investigated early in the 1980s. In recent years, the research on predictive control for this kind of system is more in-depth, including adopting different linear models and different predictive control laws, considering input saturation constraints and other input nonlinearities, and studying the stability of the closed-loop system, etc. Here we only introduce a two-step predictive control strategy for the Hammerstein model while considering the input saturation constraints [8, 9].

Consider the Hammerstein model composed of a memoryless static nonlinear component

$$\boldsymbol{v}(k) = \boldsymbol{\Phi}(\boldsymbol{u}(k)), \quad \boldsymbol{\Phi}(\mathbf{0}) = \mathbf{0} \tag{8.44}$$

and a dynamic linear component

$$\boldsymbol{x}(k+1) = \boldsymbol{A}\boldsymbol{x}(k) + \boldsymbol{B}\boldsymbol{v}(k) \tag{8.45}$$

where $\boldsymbol{x} \in \mathrm{R}^n$, $\boldsymbol{v} \in \mathrm{R}^m$, $\boldsymbol{u} \in \mathrm{R}^m$ are the state, intermediate variable, and input, respectively, and $\boldsymbol{\Phi}$ represents the nonlinearity between the input and intermediate variable. Assume that state $\boldsymbol{x}$ is measurable and the pair $(\boldsymbol{A} \quad \boldsymbol{B})$ is controllable. Also assume the nonlinearity $\boldsymbol{\Phi} = \boldsymbol{f} \circ \mathrm{sat}$, i.e. compounded by an invertible static nonlinearity $\boldsymbol{f}(\cdot)$ and a physical input saturation constraint sat represented by

$$-\underline{\boldsymbol{u}} \le \boldsymbol{u}(k) \le \bar{\boldsymbol{u}}$$
$$\text{with} \quad \underline{\boldsymbol{u}} = (\underline{u}_1, \ldots, \underline{u}_m)^{\mathrm{T}}, \quad \bar{\boldsymbol{u}} = (\bar{u}_1, \ldots, \bar{u}_m)^{\mathrm{T}}, \quad \underline{u}_i > 0, \quad \bar{u}_i > 0, \quad i = 1, \ldots, m \tag{8.46}$$

According to the system model given by (8.44) and (8.45), it can be found that the control input $\boldsymbol{u}$ affects system state $\boldsymbol{x}$ through intermediate variables $\boldsymbol{v}$. Although the relationship between $\boldsymbol{u}$ and $\boldsymbol{x}$ is nonlinear, the relationship between $\boldsymbol{v}$ and $\boldsymbol{x}$ is linear. Therefore, the control law can be calculated in two steps. Firstly, the desired intermediate variable $\boldsymbol{v}(k)$ is solved by using the unconstrained predictive control algorithm for the linear model (8.45). Then the control action $\boldsymbol{u}(k)$ can be obtained by solving the nonlinear algebraic equation group (NAEG) (8.44) and the saturation constraints can be satisfied by using the desaturation method. This two-step predictive control strategy is indeed a nonlinear separation strategy. In the following the solving procedure of this strategy is given in some detail.

Firstly, consider the predictive control problem for the linear model (8.45) with input $\boldsymbol{v}$ and state $\boldsymbol{x}$. It is a conventional unconstrained predictive control problem. Let $\boldsymbol{x}(k+i \mid k)$, $\boldsymbol{v}(k+i \mid k)$ be the future state and intermediate variable at time $k+i$, both predicted at time k, and define the optimization performance index at time k as

$$J_N(k) = \sum_{i=0}^{N-1} \left[\|\boldsymbol{x}(k+i \mid k)\|_{\boldsymbol{Q}}^2 + \|\boldsymbol{v}(k+i \mid k)\|_{\boldsymbol{R}}^2 \right] + \|\boldsymbol{x}(k+N \mid k)\|_{\boldsymbol{Q}_N}^2 \tag{8.47}$$

where N is the optimization horizon, $\boldsymbol{Q} \geq \boldsymbol{0}$, $\boldsymbol{R} > \boldsymbol{0}$ are weighting matrices of state and intermediate variables, and the terminal state weighting matrix $\boldsymbol{Q}_N > \boldsymbol{0}$. At each time k, with measured state $\boldsymbol{x}(k)$, the following optimization problem should be solved:

$$\begin{aligned} &\min_{\boldsymbol{v}(k|k),\ \cdots,\ \boldsymbol{v}(k+N-1|k)} J_N(k) \\ &\text{s.t.}\ \ \boldsymbol{x}(k+i+1 \mid k) = \boldsymbol{A}\boldsymbol{x}(k+i \mid k) + \boldsymbol{B}\boldsymbol{v}(k+i \mid k), \qquad i = 0, \ldots, N-1 \\ &\qquad \boldsymbol{x}(k \mid k) = \boldsymbol{x}(k) \end{aligned} \tag{8.48}$$

This is a finite horizon standard LQ problem and the optimal LQ control law can be given by

$$\boldsymbol{v}(k+i \mid k) = -\left(\boldsymbol{R} + \boldsymbol{B}^{\mathrm{T}}\boldsymbol{P}_{i+1}\boldsymbol{B}\right)^{-1}\boldsymbol{B}^{\mathrm{T}}\boldsymbol{P}_{i+1}\boldsymbol{A}\boldsymbol{x}(k+i \mid k), \ \ i = 0, \ldots, N-1$$

where $\boldsymbol{P}_i$ can be calculated using the following Riccati iteration:

$$\begin{aligned} &\boldsymbol{P}_N = \boldsymbol{Q}_N \\ &\boldsymbol{P}_i = \boldsymbol{Q} + \boldsymbol{A}^{\mathrm{T}}\boldsymbol{P}_{i+1}\boldsymbol{A} - \boldsymbol{A}^{\mathrm{T}}\boldsymbol{P}_{i+1}\boldsymbol{B}\left(\boldsymbol{R} + \boldsymbol{B}^{\mathrm{T}}\boldsymbol{P}_{i+1}\boldsymbol{B}\right)^{-1}\boldsymbol{B}^{\mathrm{T}}\boldsymbol{P}_{i+1}\boldsymbol{A}, \ \ i = N-1, \ldots, 1 \end{aligned} \tag{8.49}$$

Thus, the optimal predictive control law $\boldsymbol{v}(k) = \boldsymbol{v}(k \mid k)$ can be given by

$$\boldsymbol{v}(k) = -\left(\boldsymbol{R} + \boldsymbol{B}^{\mathrm{T}}\boldsymbol{P}_1\boldsymbol{B}\right)^{-1}\boldsymbol{B}^{\mathrm{T}}\boldsymbol{P}_1\boldsymbol{A}\boldsymbol{x}(k) \tag{8.50}$$

After $\boldsymbol{v}(k)$ has been solved, the task of the second step is to back-calculate the control $\boldsymbol{u}(k)$ according to (8.44) and make it satisfy the input saturation constraint (8.46). Note that the control law (8.50) may be unable to be implemented via $\boldsymbol{u}(k)$, so the $\boldsymbol{v}(k)$ given by (8.50) is only a desired intermediate variable and is denoted as $\boldsymbol{v}^L(k)$ in the following. The NAEG $\boldsymbol{v}^L(k) - \boldsymbol{\Phi}(\boldsymbol{u}(k)) = \boldsymbol{0}$ can be solved for $\boldsymbol{u}(k)$ by iteration and there exist various methods depending on the specific form of $\boldsymbol{f}(\cdot)$. In order to reduce the computational burden, it is suggested to use the Newton iteration method or its simple modified forms. Sometimes the solution of NAEG is not unique. In this case, it can be determined by additional conditions, such as choosing $\boldsymbol{u}(k)$ mostly close to $\boldsymbol{u}(k-1)$, choosing $\boldsymbol{u}(k)$ with the smallest amplitude, etc. Thus we can get the solution of the above NAEG and formally denote it as $\hat{\boldsymbol{u}}(k) = \boldsymbol{g}(\boldsymbol{v}^L(k))$. Then the control input $\boldsymbol{u}(k)$ can be obtained by desaturation of $\hat{\boldsymbol{u}}(k)$ with $\boldsymbol{u}(k) = \mathrm{sat}(\hat{\boldsymbol{u}}(k))$ such that (8.46) is satisfied.

The control structure of the two-step predictive control algorithm for the Hammerstein model (8.44), (8.45) is shown in Figure 8.9.

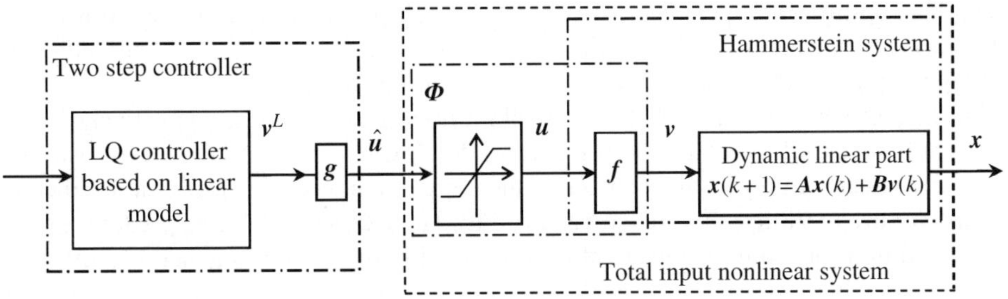

Figure 8.9 Structure of the two step predictive control system [9].

It is obvious that when implementing $\boldsymbol{u}(k) = \text{sat}(\boldsymbol{g}(\boldsymbol{v}^L(k))) \triangleq (\text{sat} \circ \boldsymbol{g})(\boldsymbol{v}^L(k))$ on the input of the Hammerstein system, the resultant $\boldsymbol{v}(k) = \boldsymbol{f}(\boldsymbol{u}(k)) = (\boldsymbol{f} \circ \text{sat} \circ \boldsymbol{g})(\boldsymbol{v}^L(k))$ is in general not the same as $\boldsymbol{v}^L(k)$, i.e. the control law in terms of the intermediate variable $\boldsymbol{v}(k)$ is formally given by

$$\boldsymbol{v}(k) \triangleq \boldsymbol{h}(\boldsymbol{v}^L(k)) = \boldsymbol{h}\left(-\left(\boldsymbol{R} + \boldsymbol{B}^{\mathrm{T}}\boldsymbol{P}_1\boldsymbol{B}\right)^{-1}\boldsymbol{B}^{\mathrm{T}}\boldsymbol{P}_1\boldsymbol{A}\boldsymbol{x}(k)\right) \tag{8.51}$$

which is not exactly the optimal LQ control law (8.50). Under certain conditions on the LQ design parameters and on $\boldsymbol{h}(\boldsymbol{v}^L(k))$ [8], some theoretical results were achieved on the stability of the closed-loop system and on the domain of attraction.

It can be seen that for the Hammerstein system described by (8.44), (8.45), using the idea of dynamic-static separation can divide its predictive control problem into two parts: an unconstrained predictive control problem for the linear dynamic model and an NAEG solving problem for the nonlinear static model. With the specific form of this model, the rolling optimization is actually implemented for a linear model rather than for a nonlinear one. The input nonlinearity is only handled as an additional static component. Here only some simple cases have been discussed, aiming at illustrating how the two-step control strategy is adopted by taking advantage of such a model structure into account. In fact, for predictive control of the Hammerstein model and even for systems with more general input nonlinearity, there exist abundant research results on both strategies and theory.

8.6 Summary

Predictive control for nonlinear systems has the same principles as that for linear systems, i.e. predicting the future dynamic behavior of the system based on the prediction model, implementing real-time control through online solving nonlinear optimization problems using model prediction, and correcting the model prediction by using real-time measured system information. However, compared with predictive control for linear systems, deriving an explicit causal prediction model without intermediate variables becomes complicated because the proportion and superposition properties no longer hold for nonlinear systems and the nonlinearity is often complex and diverse. In this case, the fundamental nonlinear model should be established for all the time instants over the optimization horizon and put as equality constraints into the optimization problem. With respect to online optimization, even without input, output, and state constraints, the analytical solution is almost impossible to derive, and the nonlinear programming method is often needed to find the solution iteratively. Therefore, how to effectively solve the online optimization problem is the most difficult problem in predictive control for nonlinear systems. In accord with this, this chapter has been to focus on the efficient strategies and methods of predictive control for nonlinear systems.

The layered structure of predictive control for nonlinear systems can effectively utilize the mature technology of predictive control for linear systems. There exist various layered strategies in predictive control applications, such as the hierarchical predictive control for urban traffic networks to be introduced in Section 10.2.4. The predictive control based on input–output linearization presented here provides a general framework where the feedback linearization at the lower layer aims at transforming the nonlinear system into a linear one for which mature predictive control algorithms can be adopted.

The fuzzy multiple model and neural network are common tools to deal with nonlinear systems and provide a new way for its predictive control. The use of them in predictive control indicates that in addition to the predictive control based on usual mathematical models, it is also possible to make prediction and optimization directly starting from the process data through establishing a neural network model or fuzzy model. This is the concrete embodiment of the predictive control principles and provides an open space for developing new kinds of prediction model, rolling optimization, and feedback correction. It is particularly advantageous for practical use when the model identification is difficult or expensive but the process data are rich and available.

The description of nonlinear systems is much more complex than that of linear systems, which prevents the general formulation and solution of predictive control for nonlinear systems. However, for nonlinear systems with special nonlinear forms or with specific characters, it is still possible to develop effective predictive control strategies or algorithms. The Hammerstein model is a typical nonlinear system with a linear dynamic model and static input nonlinearity. The two-step predictive control presented here makes dynamic-static separation in terms of the specific character of the model, and thus greatly simplifies the controller design.

Predictive control for nonlinear systems has been always the focus of attention since the appearance of predictive control. In addition to fruitful results in theoretical research of NMPC, many effective algorithms and strategies were developed to meet the needs from practical applications. In this chapter, only a few of these research works were introduced. Compared with the maturity of predictive control for linear systems, predictive control for nonlinear systems is still the most challenging topic in the field of predictive control.

References

1 Sun, H. Layered predictive control strategies for a class of nonlinear systems. PhD thesis, 1990. Shanghai Jiao Tong University (in Chinese).

2 Li, N. Multi-model-based modeling and control. PhD thesis, 2002. Shanghai Jiao Tong University (in Chinese).

3 Li, N., Li, S., and Xi, Y. (2004). Multiple model predictive control based on the Takagi-Sugeno fuzzy models: a case study. *Information Sciences* 165 (3–4): 247–263.

4 Gustafsson, D. and Kessel, W.C. Fuzzy clustering with a fuzzy covariance matrix, *Proceedings of 1978 IEEE Conference on Decision and Control*, San Diego (10–12 January 1979), 761–766.

5 Takagi, T. and Sugeno, M. (1985). Fuzzy identification of systems and its applications to modeling and control. *IEEE Trans. SMC* 15 (1): 116–132.

6 Hunt, K.J., Sbarbaro, D., Zbikowski, R. et al. (1992). Neural networks for control systems – a survey. *Automatica* 28 (6): 1083–1112.

7 Li, J., Xu, X., and Xi, Y. Artificial neural network-based predictive control, *Proceedings of IECON'91*, Kobe, Japan (28 October–1 November 1991), 2: 1405–1410.

8 Ding, B.C., and Xi, Y.G. (2006). A two-step predictive control design for input saturated Hammerstein systems. *International Journal of Robust and Nonlinear Control* 16 (7): 353–367.

9 Ding, B.C. Methods for stability analysis and synthesis of predictive control. PhD thesis, 2003. Shanghai Jiao Tong University (in Chinese).

9

Comprehensive Development of Predictive Control Algorithms and Strategies

Predictive control brings new ideas and powerful tools to optimization-based control in industrial processes. To meet various requirements of complex industrial processes, not only new algorithms but also efficient control structures, as well as application-oriented new optimization concepts and strategies, have been developed, which construct an important part of predictive control research and show the diversity of predictive control in applications. In this chapter, we will give some examples of such developments, which not only helps to better understand the basic principles of predictive control but also exhibits the flexibility of predictive control in practical use, and thus can provide a useful reference for predictive control applications.

9.1 Predictive Control Combined with Advanced Structures

9.1.1 Predictive Control with a Feedforward–Feedback Structure

It is well known that in addition to control inputs, there are also uncontrollable system inputs, including measurable or predicable external actions that cannot be manipulated, and unknown disturbance. The purpose of control is to adjust the control inputs continuously to suppress the influence of these uncontrollable inputs such that the system outputs have a desired dynamic behavior. For an unknown disturbance with no rule to follow, the adjustment is only available when the disturbed effect appears in measurable variables. Thus this kind of disturbance must be suppressed by feedback control. For uncontrollable inputs with known variation rules, because of the model prediction function of predictive control, the influence of them on system outputs can be predicted to some extent and this part of the information can be introduced to compensate for their influence in advance. This motivated the feedforward–feedback control structure, which is suitable for the above-mentioned case and has been widely used in industrial processes.

In Chapter 3, feedback correction as an important part of predictive control algorithms such as DMC has been discussed in details. It uses the error between the measured information and the predicted one to construct the future prediction errors

$$\tilde{e}(k+i \mid k+1) = h_i e(k+1), \quad i = 1, \ldots, N \tag{9.1}$$

and then uses them to correct the predicted future outputs

$$\tilde{y}_{\mathrm{cor}}(k+i \mid k+1) = \tilde{y}_{N1}(k+i \mid k) + \tilde{e}(k+i \mid k+1)$$

Predictive Control: Fundamentals and Developments, First Edition. Yugeng Xi and Dewei Li.

Note that feedback correction is not necessarily to take the weighting form (9.1). It can be realized in a more general form, for example with the time series forecasting formula

$$\tilde{e}(k+i\mid k+1)=E_i(e(k+1),\ e(k),\ \ldots,\ e(k-l+1)) \tag{9.2}$$

where $E_i(\cdot)$ is a linear or nonlinear function and l is the time length of the historical data used by prediction. No matter which form is used, feedback correction can only be done with a noncausal prediction because nothing is known about the cause of the error.

However, for the uncontrollable inputs with some known variation rules, it is unreasonable to regard them also as an unknown disturbance and to suppress their influence on system outputs only by feedback correction. In terms of the known information it is possible to establish an additional causal model to put the influence of these inputs on system outputs into the prediction model. In this way, except for the already existing feedback path, a feedforward path also appears in the predictive control structure, with which the known causal information can be fully used and the influence of this kind of uncontrollable inputs on system outputs can be compensated timely and effectively.

Now take the single variable DMC algorithm as an example to illustrate the feedforward compensation, where the key issue is how to include the influence of known but uncontrollable input on system output as causal information in model prediction. Denote y, u, v as the system output, controllable input, and uncontrollable input but with known variation rules, respectively. To do that, in addition to the step response sequence $\{a_i\}$ from u to y, the step response sequence $\{b_i\}$ from v to y should also be measured. At each sampling time, the plant output variation is caused by both the control input u and the uncontrollable input v. Thus the influence of the uncontrollable input can be superposed in the prediction model (2.4) such that

$$\tilde{\boldsymbol{y}}_{PM}(k)=\tilde{\boldsymbol{y}}_{P0}(k)+\boldsymbol{A}\Delta\boldsymbol{u}_M(k)+\boldsymbol{B}\Delta\boldsymbol{v}_P(k) \tag{9.3}$$

where

$$\boldsymbol{B}=\begin{bmatrix} b_1 & \cdots & 0 \\ \vdots & \ddots & \vdots \\ b_P & \cdots & b_1 \end{bmatrix},\qquad \Delta\boldsymbol{v}_P(k)=\begin{bmatrix} \Delta v(k) \\ \vdots \\ \Delta v(k+P-1) \end{bmatrix}$$

Here $\Delta v(k) = v(k) - v(k-1)$ is the increment of the uncontrollable input v.

Furthermore, similar to (2.10), the one-step output prediction can be given in vector form

$$\tilde{\boldsymbol{y}}_{N1}(k)=\tilde{\boldsymbol{y}}_{N0}(k)+\boldsymbol{a}\Delta u(k)+\boldsymbol{b}\Delta v(k) \tag{9.4}$$

where $\boldsymbol{b} = (b_1\ \cdots\ b_N)^{\mathrm{T}}$.

Take the same optimization performance index (2.5). If constraints are not considered, the optimal control vector can be given by

$$\Delta\boldsymbol{u}_M(k)=\left(\boldsymbol{A}^{\mathrm{T}}\boldsymbol{Q}\boldsymbol{A}+\boldsymbol{R}\right)^{-1}\boldsymbol{A}^{\mathrm{T}}\boldsymbol{Q}[\boldsymbol{w}_P(k)-\tilde{\boldsymbol{y}}_{P0}(k)-\boldsymbol{B}\Delta\boldsymbol{v}_P(k)]$$

and the current control increment is

$$\Delta u(k)=\boldsymbol{d}^{\mathrm{T}}[\boldsymbol{w}_P(k)-\tilde{\boldsymbol{y}}_{P0}(k)-\boldsymbol{B}\Delta\boldsymbol{v}_P(k)] \tag{9.5}$$

where $\boldsymbol{d}^{\mathrm{T}}$ is also expressed by (2.8) and can be calculated off-line.

Compare (9.5) with the control law given by (2.7) and a new item $\boldsymbol{B}\Delta\boldsymbol{v}_P(k)$ appears in (9.5). It can be physically explained as the item to compensate for the influence of the uncontrollable input on the system output over the optimization horizon, which is equivalent to constructing a new desired value vector

$$\boldsymbol{w}_P^*(k) = \boldsymbol{w}_P(k) - \boldsymbol{B}\Delta\boldsymbol{v}_P(k)$$

and then to consider the rolling optimization problem only with controllable inputs. It is obvious of feedforward characteristics, as shown in Figure 9.1.

The control law (9.5) is realizable and can fully compensate the influence of the uncontrollable input v on system output y if the dynamic response $\{b_i\}$ from v to y is known and the variation of v is fully predictable. However, in practice, this condition is hardly to be satisfied and some adjustments should be made. Two common cases are discussed as follows.

1) The dynamic response $\{b_i\}$ is known and $v(k)$ is measurable, but the future $v(k+i)$ is unpredictable.

An example of this case is the additional flow for liquid control in a tank. In this case, at the sampling time k, $\Delta v(k)$ can be obtained by measuring $v(k)$. However, $\Delta\boldsymbol{v}_P(k)$ in (9.3) is unavailable due to the unpredictable future $v(k+i)$. The prediction model (9.3) should then be revised to give

$$\tilde{\boldsymbol{y}}_{PM}(k) = \tilde{\boldsymbol{y}}_{P0}(k) + \boldsymbol{A}\Delta\boldsymbol{u}_M(k) + \boldsymbol{b}_P\Delta v(k) \tag{9.6}$$

where $\boldsymbol{b}_P = (b_1 \ \cdots \ b_P)^{\mathrm{T}}$. Accordingly, the control law should be written as

$$\Delta u(k) = \boldsymbol{d}^{\mathrm{T}}[\boldsymbol{w}_P(k) - \tilde{\boldsymbol{y}}_{P0}(k) - \boldsymbol{b}_P\Delta v(k)]$$

Obviously, without a priori knowledge on future v values, the prediction model (9.3) has to be replaced by (9.6) even if the prediction horizon $P > 1$. Thus (9.6) implies the assumption that $v(k)$ remains constant in the whole optimization horizon, which may not match the actual situation. However, as long as v varies smoothly and infrequently, the control result with (9.6) may still be superior to that without feedforward compensation. It should be further pointed out that once the current control $u(k)$ has been calculated and implemented, the output prediction at $k+1$ only concerns the

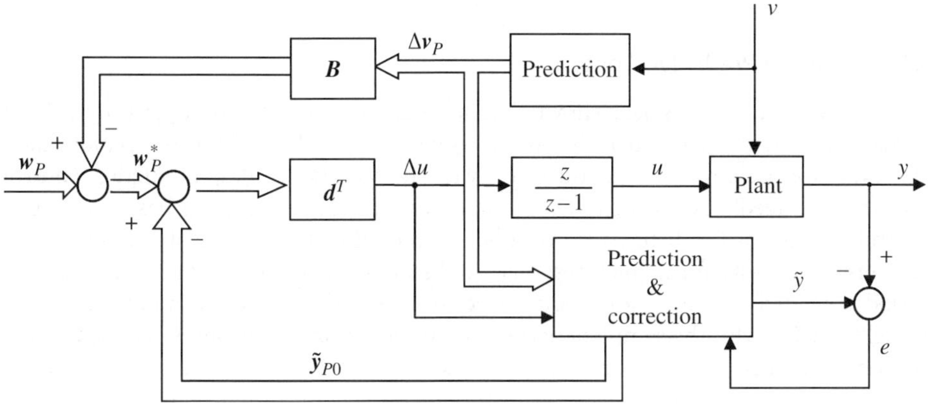

Figure 9.1 Scheme of DMC algorithm with feedforward compensation.

known $v(k)$. Therefore, the influence of v has been fully addressed in $\tilde{y}_1(k+1\mid k)$ calculated by (9.4). It will not be included in the output error $e(k+1)$ and does not need to be considered in feedback correction.

2) The variation of an uncontrollable input v is somewhat predictable but the step response $\{b_i\}$ from v to y lacks clarity.

Examples of this situation widely exist, as in a boiler combustion system where the user load as an uncontrollable input may have some known daily statistic rules, but the accurate model of the load affecting the controlled variables is difficult to obtain. In this case, using data analysis to establish a rule-based or fuzzy model between v and y is more acceptable than identifying an accurate step response model $\{b_i\}$. This kind of model shows that, although rougher than the step response, the main tendency of the dynamic influence of uncontrollable input on system output can be reflected. If it could be included in the prediction model as a feedforward part, the influence of uncontrollable input would be compensated to a certain extent.

The above DMC algorithm with a feedforward–feedback structure is easy extend to multivariable systems and the same principle is also suitable for other predictive control algorithms. It can be seen that the prediction model in predictive control can include a known uncontrollable input and its influence can be compensated through the feedforward part in model prediction, while the feedback correction part mainly compensates the influence of unknown disturbance by noncausal prediction. Therefore, predictive control in principle integrates feedforward and feedback in one algorithm and adopts different prediction strategies for different kinds of uncontrollable input. With this structure the known causal information is fully used to compensate the influence of uncontrollable input on system dynamics, and thus better control results can be achieved.

It should be pointed out that the model prediction with uncontrollable input aims at utilizing causal information of its influence on the process dynamics as much as possible. It is not limited to a mathematical model. With the help of modern data-driven, artificial intelligence techniques, causal prediction of an uncontrollable input on the system dynamics can be implemented in a more general way. For example, in decentralized predictive control, the interactions between subsystems can be established by a data-driven model or a rule-based model. For predictive control in subsystems, it is available to construct the feedforward part based on causal prediction using these models to improve its control performance.

9.1.2 Cascade Predictive Control

In the above section it has been shown that in predictive control feedforward control can be adopted to handle uncontrollable inputs with known models or known variation rules. However, for an unknown disturbance without variation rules, predictive control can only use feedback correction to compensate it. But when using basic predictive control algorithms for disturbance rejection, some difficulties still exist. Firstly, disturbance rejection asks for a quick response to disturbance, and usually a small sampling period is needed. However, for predictive control algorithms based on a nonparametric model such as DMC, the sampling period cannot be too small. Secondly, as discussed in Section 3.3.1, during feedback correction, the selection of the correction coefficients h_i is difficult to take both robustness and disturbance rejection into account because it is unclear whether the error is caused by a model mismatch or disturbance. Thirdly, feedback correction starts to work only when disturbance affects the system output and the output error occurs. If the disturbance has a longer transmission time to the system output, it is difficult for control to have a quick response to the disturbance in time.

In process control, a cascade control structure is often used to enhance the system's ability to reject disturbance if some intermediate variable is measurable. Usually feedback of the measured intermediate variable constructs the inner loop and feedback of the system output constructs the outer loop. Since the secondary disturbance acting on the front part of the process is suppressed in time by the inner loop feedback, the disturbance rejection ability of the controlled system can be greatly improved.

With the help of the idea of cascade control, predictive control can be combined with the cascade control structure to form cascade predictive control, which can effectively solve the above mentioned difficulties. Firstly, the inner loop feedback control with an intermediate variable can respond to the disturbance in a timely way. Secondly, the cascade control has a layered structure, where the inner loop and the outer loop can adopt different sampling periods, such that different goals such as robustness or disturbance rejection, etc. can be accordingly addressed in different layer designs. In this way, contradictions on selecting sampling time and correction strategy encountered by the basic predictive control algorithms can be solved. The design of cascade predictive control is illustrated as follows by using the DMC-PID cascade control as an example.

Firstly, since the densely sampled PID controller can well suppress the influence of disturbance, take the measurable intermediate variable, which is close to where the disturbance is most likely to occur, to construct PID closed-loop control. This inner loop control adopts a frequency much higher than single-layer DMC control, aiming at quickly and efficiently suppressing a suddenly appearing disturbance. Then this closed loop, together with the remaining part of the plant, can be regarded as a generalized plant and controlled by usual DMC in the outer loop. Since the main part of the disturbance has been suppressed in a timely manner in the inner loop, its influence on the generalized plant can be regarded as a slight mismatch of the model, so the outer loop DMC should make good dynamic performance and robustness as the design goals. This kind of layered DMC-PID control has the same structure as cascade control (see Figure 9.2), but in the outer loop the DMC algorithm is adopted to replace the PID controller that is usually used. This cascade predictive control strategy not only maintains the structural advantage of cascade control but also makes full use of the functional advantage of predictive control.

In terms of the goals of layered control described above, in the cascade control structure shown in Figure 9.2, the key issues for designing the inner loop (also called the secondary loop) and the outer loop (also called the major loop) are given, respectively, as follows.

(1) Design of the secondary loop

The design of the secondary loop should make the front part of the plant $G_2(s)$ contain the main secondary disturbance and have a small time delay or small time constant. Digital PID control with denser sampling frequency is often adopted in this loop, mainly for timely suppression of the secondary disturbance entering into the plant.

For a quasi-continuous digital PID controller adopted at the secondary loop, usual engineering setting methods such as the critical proportion method, the response

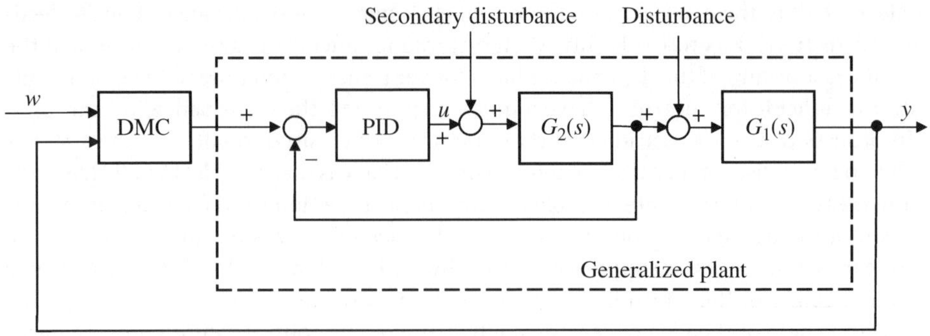

Figure 9.2 DMC-PID cascade control structure.

curve method, etc. can be combined with an empirical formula to determine the PID parameters. Since a great number of industrial processes can be approximated by a first-order inertia process with a time delay, and $G_2(s)$ only contains the part with a small time delay or small time constant, so a P-type controller is in general enough to get a quick response. This is not only conducive to reducing deviation caused by disturbance but is also helpful when designing the outer loop control.

(2) Design of the major loop

The controlled plant in the major loop is a generalized plant (see the dotted line frame in Figure 9.2) including the whole closed inner loop together with the remaining part of the plant $G_1(s)$. The DMC algorithm is adopted in this loop mainly for achieving good tracking performance and strong robustness against model mismatch.

It should be noted that in DMC control of the major loop, the step response model should be established for the generalized plant rather than the original plant. During testing, the unit step signal should be imposed as the desired value of the secondary loop rather than the control input directly acting on the plant. In the whole generalized plant, the secondary plant $G_2(s)$ has a smaller time constant than that of the major plant $G_1(s)$. After inner loop control, the dynamic response of the closed inner loop would be further accelerated, so the dynamics of the generalized plant is dominantly determined by the major plant $G_1(s)$, i.e. the plant part with a larger time delay or a large time constant, for which the DMC algorithm with prediction and optimization functions is obviously more effective than PID control. Since the influence from the secondary disturbance has been greatly reduced after inner loop control, it can be regarded as a slight mismatch of the generalized plant model, so the correction strategy in the DMC algorithm of the major loop should focus on strengthening the robustness. Then the DMC parameters can be selected according to the tuning rules presented in Section 3.4.

Remark 9.1 The DMC-PID cascade control strategy makes full use of the structural advantage of cascade control and can quickly suppress the secondary disturbance entering into the plant by using dense PID control in the inner loop. Meanwhile, with the functional advantage of predictive control, it can effectively deal with time delay in the outer loop and has a good tracking performance and robustness. Therefore, its comprehensive control performance is better than that of only using the DMC algorithm or PID cascade control. Due to the use of a layered structure, different design goals may be assigned to different layers and handled separately, which not only resolves the contradiction that the single-layer DMC is difficult to handle but also improves design redundancy to make the DMC parameters easy to tune. Therefore the DMC-PID cascade structure is attractive and practical in complex industrial processes.

Example 9.1 Cascade predictive control for the conversion section of synthetic ammonia production

The conversion section is an important part of synthetic ammonia production whose process flowchart is shown in Figure 9.3.

The semi-water-gas from the compression section is preheated in the saturation tower and mixed with an appropriate amount of steam at the outlet of the saturation tower to reach an appropriate steam-to-water ratio. It is then separated by a steam-water separator and passes through a series of heat exchangers (not drawn in the figure) to heat up to around 380°C; it then enters the converter. The following transformation reaction occurs in the converter:

$$CO + H_2O \text{ (vapor)} \overset{\text{catalyst}}{\rightleftarrows} CO_2 + H_2 + Q \text{ (heat release)}$$

In this way, not only is the useless CO in the semi-water-gas removed but also the H_2 needed in the synthetic reaction is obtained. The converted gas is cooled in a cooling tower and moved to the next section for carbonization. In order to ensure the efficiency of the conversion process, it is technologically required that the residual CO content in the converted gas is < 3%.

In this conversion process, the added steam flow is the control variable. Both the speed and the balance of the chemical reaction are affected by various factors, such as pressure,

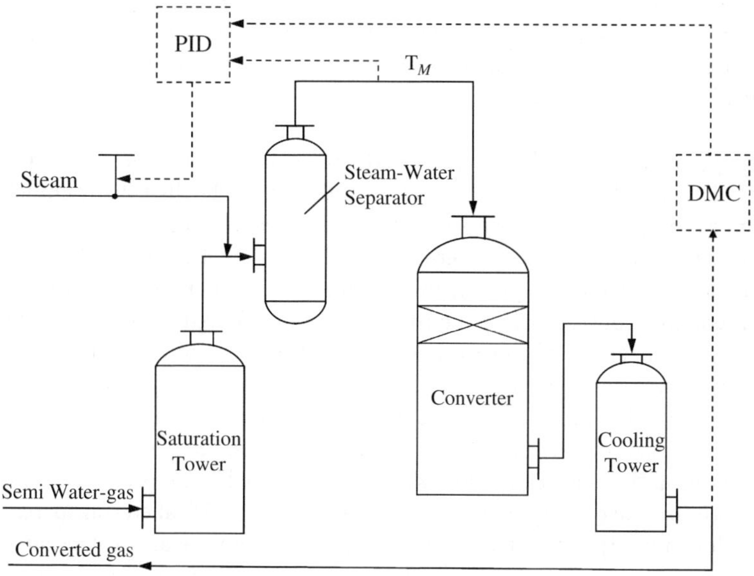

Figure 9.3 Flowchart and control scheme of conversion section.

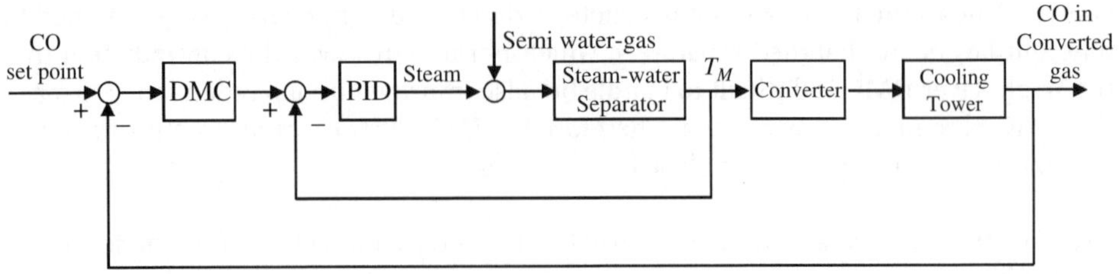

Figure 9.4 DMC-PID cascade control scheme of the conversion section.

flow, catalyst, etc. but are mostly affected by the uncertain variation of the steam pressure caused by mixing of the semi-water-gas. Therefore the DMC-PID cascade control structure is chosen. In the inner loop, the gas temperature T_M at the outlet of the steam-water separator is selected as the controlled variable because it can quickly reflect the influence of the main disturbance of steam pressure. A digital P controller is designed to suppress the disturbance. In the outer loop the DMC algorithm is adopted to control the CO content in the converted gas and to handle the model mismatch caused by operation condition variations that cannot be handled by the inner loop. The block diagram of this control structure is shown in Figure 9.4.

This control scheme was tested at the converter section of a synthetic ammonia plant and achieved good results. The fluctuation around the desired value of the CO content in the converted gas was under 2% and smaller than that under PID cascade control. Meanwhile the steam quantity used per day was also obviously saved. The purpose of improving the control performance and reducing energy consumption was achieved.

This cascade predictive control strategy can also be extended to multivariable systems, in which the inner loop contains multiple single PID control loops with the purpose of suppressing the influence of the main disturbance in each loop. In the outer loop, multivariable DMC control can be adopted for the generalized plant composed of closed inner loops with the purpose of handling couplings between single loops and achieving global optimization control for the whole system.

Remark 9.2 The precondition of the cascade predictive control is the existence of a measurable intermediate variable, which can reflect the influence of the disturbance much earlier than the process output. If the process is composed of a series of dynamic components from input to output, where multiple intermediate variables are measurable, then the cascade control can be successively designed from the inner to the outer part to constitute multiple cascade predictive control, where the disturbance at different locations can be suppressed timely by feedback of the measurable information nearest to it. If there is no measurable intermediate variable in the process, according to the layered structure of cascade control, the inner loop can still be constituted by dense digital PID control with direct feedback from the plant output to suppress the disturbance as soon as possible, and in the outer loop, taking all the closed inner loops as the generalized plant, predictive control is adopted to optimize the desired values of all inner loops. This is called transparent control and is mentioned in [1]. These two cases are different from the above presented cascade predictive control, but reflects its principles, i.e. feed back the measurable intermediate information to suppress the inner disturbance in a timely and effective way and adopt a layered structure to deal with different requirements on control.

9.2 Alternative Optimization Formulation in Predictive Control

9.2.1 Predictive Control with Infinite Norm Optimization

In most predictive control algorithms, the quadratic performance index is adopted in optimization, which corresponds to the two-norm of the prediction error and reflects the requirement on performance of system output tracking the desired value. However, in some application environments, control does not focus on tracking, but on good stabilization performance, i.e. all the outputs should be kept near their set points and excessive deviation is not allowed. In this case, the mathematical measure to characterize the control requirement directly is to minimize the maximal possibly appearing deviation, given by

$$\min_{\Delta \boldsymbol{u}_M(k)} \max_{l=1,\cdots,P} \max_{i=1,\cdots,p} |w_i(k+l)-\tilde{y}_i(k+l)| \tag{9.7}$$

This is equivalent to using the following infinite norm performance index to replace the usual quadratic performance index

$$\min J(k) = \|\boldsymbol{w}(k)-\tilde{\boldsymbol{y}}_{PM}(k)\|_\infty \tag{9.8}$$

where the infinite norm of vector $\boldsymbol{x} = (x_1 \ \cdots \ x_n)^{\mathrm{T}}$ is defined as $\|\boldsymbol{x}\|_\infty = \max_i |x_i|$. When the infinite norm in (9.8) is doubly valued both on space (for all outputs) and on time (for all sampling instants in the optimization horizon), its specific expression is (9.7). Obviously, this kind of optimization aims at minimizing the maximal possible deviation of all outputs in the optimization horizon, and thus directly reflects the physical requirement of good stabilization.

Due to the linear form of the performance index (9.7), if the plant model and the physical constraints are both linear, this optimization problem could be solved by linear programming (LP). Note that as (9.7) involves items of extreme values and absolute values, it is necessary to firstly transform it into a standard LP problem. In the following, referring to the idea of Campo and Morari [2, 3], we introduce the DMC algorithm with infinite norm optimization as well as the design method for a robust predictive controller with parameter uncertainty.

(1) DMC algorithm with infinite norm optimization

In Section 5.2, we give the online optimization formulation of the constrained multivariable DMC algorithm (5.18). Write the input and output inequality constraints separately and substitute the quadratic performance index by the infinite norm performance index (9.8); it follows that

$$\begin{aligned}
&\min_{\Delta \boldsymbol{u}_M(k)} J(k) = \|\boldsymbol{w}(k)-\tilde{\boldsymbol{y}}_{PM}(k)\|_\infty \\
&\text{s.t.} \ \ \tilde{\boldsymbol{y}}_{PM}(k) = \tilde{\boldsymbol{y}}_{P0}(k) + \boldsymbol{A}\Delta\boldsymbol{u}_M(k) \\
&\qquad \boldsymbol{\alpha} \le \Delta\boldsymbol{u}_M(k) \le \boldsymbol{\beta} \\
&\qquad \boldsymbol{y}_{\min} \le \tilde{\boldsymbol{y}}_{P0}(k) + \boldsymbol{A}\Delta\boldsymbol{u}_M(k) \le \boldsymbol{y}_{\max}
\end{aligned} \tag{9.9}$$

Substitute the output prediction formula into the performance index and denote

$$\boldsymbol{f} = \boldsymbol{w}(k) - \tilde{\boldsymbol{y}}_{P0}(k) - \boldsymbol{A}\Delta\boldsymbol{u}_M(k) \tag{9.10}$$

then the above optimization problem can be rewritten as

$$\begin{aligned}&\min_{\Delta \boldsymbol{u}_M(k)} \ \max_i |f_i| \\ &\text{s.t.}\ \ \boldsymbol{\alpha} \le \Delta \boldsymbol{u}_M(k) \le \boldsymbol{\beta} \\ &\qquad \boldsymbol{\gamma} \le \boldsymbol{f} \le \boldsymbol{\delta}\end{aligned}$$

where f_i is the element of $\boldsymbol{f}$, $\boldsymbol{\gamma} = \boldsymbol{w}(k) - \boldsymbol{y}_{\max}$, $\boldsymbol{\delta} = \boldsymbol{w}(k) - \boldsymbol{y}_{\min}$. Let $\mu = \max_i |f_i|$; then (9.9) can be further formulated by

$$\begin{aligned}&\min_{\Delta \boldsymbol{u}_M(k),\,\mu} \ \mu \\ &\text{s.t.}\ \ \boldsymbol{\alpha} \le \Delta \boldsymbol{u}_M(k) \le \boldsymbol{\beta} \\ &\qquad \boldsymbol{\gamma} \le \boldsymbol{f} \le \boldsymbol{\delta} \\ &\qquad -\boldsymbol{l}\mu \le \boldsymbol{f} \le \boldsymbol{l}\mu, \quad \boldsymbol{l} = (1 \ \cdots \ 1)^{\mathrm{T}}\end{aligned} \tag{9.11}$$

It is known from the expression of $\boldsymbol{f}$ in (9.10) that the performance index and all the constraint conditions of the optimization problem (9.11) are linear expressions of variable $\Delta \boldsymbol{u}_M(k)$ and μ. Thus it is a typical LP problem and can be easily solved by standard software. Note that the number of constraints in (9.11) may be much larger than that of the optimization variables. If so, the dual LP problem can be at first solved to reduce the computational burden and the optimal solution of the original problem can then be obtained by the complementary relaxation condition. These methods of solving the LP problem can be found in the LP related literature.

(2) Robust predictive control with parameter uncertainty

Using the thinking of predictive control with infinite norm optimization and the above algorithm, the problem of robust predictive controller design with parameter uncertainty was studied further in [3], which is a very early work on robust predictive control.

Assume that the plant step response model $\boldsymbol{A}$ depends on the parameter $\boldsymbol{\theta}$, denoted by $\boldsymbol{A}(\boldsymbol{\theta})$. For the sake of simplification, let the nominal parameter $\boldsymbol{\theta}_0 = \boldsymbol{0}$ and denote the variation space of the parameter as

$$\boldsymbol{\Omega} = \{\boldsymbol{\theta} \mid |\theta_j| < \varepsilon_j,\ j = 1, \ldots, q\}$$

where q is the number of parameters. Obviously, $\boldsymbol{\Omega}$ is a polyhedron in the $\boldsymbol{\theta}$ space with the center $\boldsymbol{\theta}_0 = \boldsymbol{0}$. For each point $\boldsymbol{\theta}$ in the parameter space $\boldsymbol{\Omega}$, a corresponding step response model $\boldsymbol{A}(\boldsymbol{\theta})$ can be obtained. Therefore, when the parameter $\boldsymbol{\theta}$ is perturbed in $\boldsymbol{\Omega}$, the corresponding models also construct a model set $\boldsymbol{\Pi} = \{\boldsymbol{A}(\boldsymbol{\theta}) \mid \boldsymbol{\theta} \in \boldsymbol{\Omega}\}$.

Since there are an infinite number of points in the parameter space $\boldsymbol{\Omega}$, this model set $\boldsymbol{\Pi}$ is of an infinite dimension and the nominal model $\boldsymbol{A}(\boldsymbol{0})$ is only one in the set. The robust predictive control with infinite norm optimization pursues not only minimization of the maximal output derivation of the system under standard model $\boldsymbol{A}(\boldsymbol{0})$ but also minimization of all possible maximal output derivations when $\boldsymbol{\theta}$ is perturbed in the parameter space $\boldsymbol{\Omega}$. This means that the infinite norm is valued not only for different output variables and different time instants in the optimization horizon, but also for different models caused by variations of parameter $\boldsymbol{\theta}$. Therefore, the online optimization of this kind of robust predictive control can be formulated by

$$\begin{aligned}&\min_{\Delta \boldsymbol{u}_M(k)} J(k) = \|\boldsymbol{w}(k) - \tilde{\boldsymbol{y}}_{PM}(k)\|_\infty \\ &\text{s.t. } \tilde{\boldsymbol{y}}_{PM}(k) = \tilde{\boldsymbol{y}}_{P0}(k) + \boldsymbol{A}(\boldsymbol{\theta})\Delta \boldsymbol{u}_M(k) \\ &\qquad \boldsymbol{\alpha} \le \Delta \boldsymbol{u}_M(k) \le \boldsymbol{\beta} \\ &\qquad \boldsymbol{y}_{\min} \le \tilde{\boldsymbol{y}}_{P0}(k) + \boldsymbol{A}(\boldsymbol{\theta})\Delta \boldsymbol{u}_M(k) \le \boldsymbol{y}_{\max}\end{aligned} \tag{9.12}$$

where it is assumed that the previous parameters $\boldsymbol{\theta}$ are known through identification and $\tilde{\boldsymbol{y}}_{P0}(k)$ is fixed. Since the variation of the parameter $\boldsymbol{\theta}$ at a future time is uncertain, the model prediction formula and corresponding constraint conditions are all related to $\boldsymbol{\theta}$, and accordingly the infinite norm performance index should be changed as

$$\|\boldsymbol{w}(k) - \tilde{\boldsymbol{y}}_{PM}(k)\|_\infty = \max_{\boldsymbol{\theta}\in\boldsymbol{\Omega}} \max_{l=1,\cdots,P} \max_{i=1,\cdots,p} |w_i(k+l) - \tilde{y}_i(k+l)|$$

In order to solve this problem, similar to above, define

$$\boldsymbol{f}(\boldsymbol{\theta}) = \boldsymbol{w}(k) - \tilde{\boldsymbol{y}}_{P0}(k) - \boldsymbol{A}(\boldsymbol{\theta})\Delta \boldsymbol{u}_M(k)$$

Then (9.12) can be rewritten as

$$\begin{aligned}&\min_{\Delta \boldsymbol{u}_M(k)} \max_{\theta\in\Omega} \max_i |f_i(\boldsymbol{\theta})| \\ &\text{s.t. } \boldsymbol{\alpha} \le \Delta \boldsymbol{u}_M(k) \le \boldsymbol{\beta} \\ &\qquad \gamma \le \boldsymbol{f}(\boldsymbol{\theta}) \le \boldsymbol{\delta}\end{aligned}$$

Let

$$\mu = \max_{\theta\in\Omega} \max_i |f_i(\boldsymbol{\theta})|$$

We can then obtain

$$\begin{aligned}&\min_{\Delta \boldsymbol{u}_M(k),\,\mu} \mu \\ &\text{s.t. } \left.\begin{aligned}&\boldsymbol{\alpha} \le \Delta \boldsymbol{u}_M(k) \le \boldsymbol{\beta} \\ &\gamma \le \boldsymbol{f}(\boldsymbol{\theta}) \le \boldsymbol{\delta} \\ &-\boldsymbol{l}\mu \le \boldsymbol{f}(\boldsymbol{\theta}) \le \boldsymbol{l}\mu\end{aligned}\right\} \boldsymbol{\theta} \in \boldsymbol{\Omega}\end{aligned} \tag{9.13}$$

Unlike (9.11), the constraint conditions in (9.13) may have an infinite number because the relationship of $\boldsymbol{A}$ and the parameter $\boldsymbol{\theta}$ is given in general form and $\boldsymbol{\theta}$ may take an arbitrary value in the parameter space $\boldsymbol{\Omega}$. In general, it is unable to solve this problem. However, if the matrix $\boldsymbol{A}(\boldsymbol{\theta})$ is an affine function of $\boldsymbol{\theta}$, it can be deduced that $\boldsymbol{f}(\boldsymbol{\theta})$ is also an affine function of $\boldsymbol{\theta}$. In this case, it was shown in [3] that the solution of (9.13) lies on the boundary vertexes of $\boldsymbol{\Omega}$. Denote the set of these boundary vertexes as $\boldsymbol{S}$, which contains the 2^q vertexes $\boldsymbol{\theta}_i$ of the polyhedron $\boldsymbol{\Omega}$. Thus the optimization problem with an infinite number of constraints can be simplified to that with a limited number of constraints.

$$\begin{aligned}&\min_{\Delta \boldsymbol{u}_M(k),\,\mu} \mu \\ &\text{s.t. } \left.\begin{aligned}&\boldsymbol{\alpha} \le \Delta \boldsymbol{u}_M(k) \le \boldsymbol{\beta} \\ &\gamma \le \boldsymbol{f}(\boldsymbol{\theta}) \le \boldsymbol{\delta} \\ &-\boldsymbol{l}\mu \le \boldsymbol{f}(\boldsymbol{\theta}) \le \boldsymbol{l}\mu\end{aligned}\right\} \boldsymbol{\theta} \in \boldsymbol{S} = \{\boldsymbol{\theta}_i \mid i = 1, \ldots, 2^q\}\end{aligned} \tag{9.14}$$

Write the specific constraint form for each of the 2^q elements in $\boldsymbol{S}$; then the above optimization problem can be rewritten into a standard LP problem. Since the number of constraints is very large, it needs to be simplified by solving its dual LP problem.

As an affine function of $\boldsymbol{\theta}$, $\boldsymbol{A}(\boldsymbol{\theta})$ can be formulated as a linear combination of known step responses weighted by $\boldsymbol{\theta}$. For example, it can be taken as

$$\boldsymbol{A}(\boldsymbol{\theta}) = \boldsymbol{A}(\boldsymbol{0}) + \sum_{j=1}^{q} \theta_j \boldsymbol{A}_j$$

where $\boldsymbol{A}_j$ represents nonmodeled dynamics and is known, while θ_j is the element of uncertain $\boldsymbol{\theta}$ and characterizes a perturbation on the nominal model. The above formula means that the actual model is based on the standard response $\boldsymbol{A}(\boldsymbol{0})$ but also perturbed by the nonmodeled dynamics $\boldsymbol{A}_j$. The predictive controller designed by the above method can keep good robustness when this kind of parameter uncertainty exists.

9.2.2 Constrained Multiobjective Multidegree of Freedom Optimization and Satisfactory Control

The predictive control with infinite norm optimization presented above indicates that the rolling optimization and the performance index in predictive control do not need to be of fixed form, as they can be formulated according to the practical requirements of applications. In a complex industrial environment, the constraint types and the optimization goals may be quite diverse. The user also may have different concerns on satisfying specific requirements. Such an optimization problem is quite different from the standard one. In order to characterize this kind of optimization problems, the concept of constrained multiobjective multidegree of freedom optimization (CMMO) was proposed in [4] and based on that the scheme of satisfactory control was given for better understanding of predictive control applications in a complex industrial environment.

The predictive control algorithms presented above all have an optimization proposition with specific constraints and performance index, often expressed as a QP problem with linear constraints and quadratic performance index. However, in industrial practice, the optimization proposition is not ready but comes from the requirements of process operators on optimization control. These requirements are various, including traditional ones such as controlling the output to its set point, as well as others such as controlling some variable in a certain range or keeping it above or below a certain bound. The traditional degree of freedom for control has also a different meaning under this kind of requirements. In order to illustrate the optimization control with different requirements in the industrial environment, we refer to the manual of the IDCOM software package developed by SETPOINT INC. Some examples are given as follows.

Example 9.2 Optimization control with various requirements in an industrial environment.

Case 1. Use one manipulated variable to control two outputs.
Figure 9.5a shows a case of using one manipulated variable u to control two outputs y_1, y_2, where y_1 is required to be controlled below the given bound $y_{1,\max}$, while y_2 is to be controlled to the set point y_{2d}. From the figure it is shown that when $t < k$ this control goal

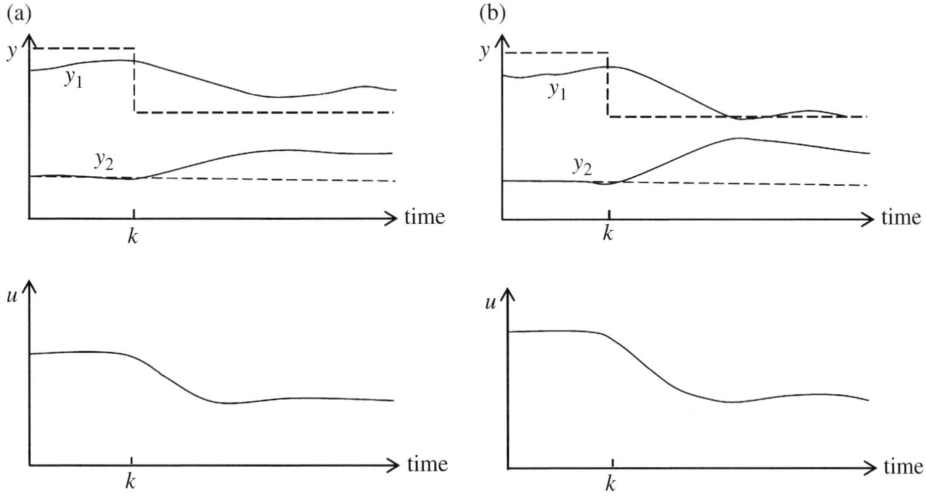

Figure 9.5 One manipulated variable controls two outputs. (a) Control ability cannot satisfy the requirements after the control goal changes. (b) Trade-off of control if the control ability is insufficient.

can be reached, i.e. one control variable can control two output variables meeting given requirements. However, if $y_{1,\max}$ is decreased at time k, using only a single manipulated variable u may make it impossible to control both y_1 and y_2 as required. In this case, the control ability cannot satisfy practical requirements. Without an additional control variable, the operator can only make a trade-off between the requirements on y_1 and y_2. For example, if it is particularly expected to decrease y_1 such that it could be below $y_{1,\max}$, then the deviation of y_2 from y_{2d} would be larger (see Figure 9.5b).

Case 2. Control with input constraint and requirements on output and input.
Figure 9.6 is another example where the manipulated variable u has a low bound constraint and the output y is required to be controlled to its set point. Furthermore, in addition to controlling the output y, the manipulated variable u is also asked to have minimum energy consumption, i.e. min u. During $t < k$, due to the restriction on the degree of freedom, u can only maintain y at its set point and is unable to take min u into account. However, if the control requirement on y is released from the set point to below a certain given bound, u could be gradually decreased for minimizing energy consumption while satisfying the bound constraint on y. If the bound constraint on y can be further relaxed, u can be further decreased until it reaches the lower bound itself.

Case 3. Control with multidegrees of freedom for various requirements.
Figure 9.7 demonstrates a case of using three manipulated variables u_1, u_2, u_3 to control two outputs y_1, y_2, with set point control for y_1 and (upper) bound control for y_2. Meanwhile, for economic consideration, u_1, u_3 are expected to be close to their desired values (see the dotted lines in the figure). It can be seen from the figure that, at the beginning of control, u_1, u_3 maintain their desired values and only u_2 is used to meet the control requirements on y_1, y_2, which is similar to the case in Figure 9.5. Although all the above requirements can be fully satisfied from the beginning, when the upper bound for y_2 is decreased at time k, the control requirements on y_1, y_2 cannot be simultaneously satisfied. In this case, some degree of freedom of u_1 or u_3 should be released for satisfying the requirements on

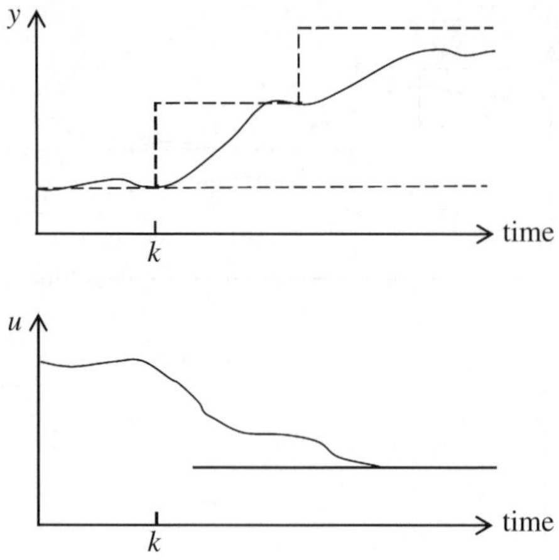

Figure 9.6 Control with input constraint and requirements on output and input.

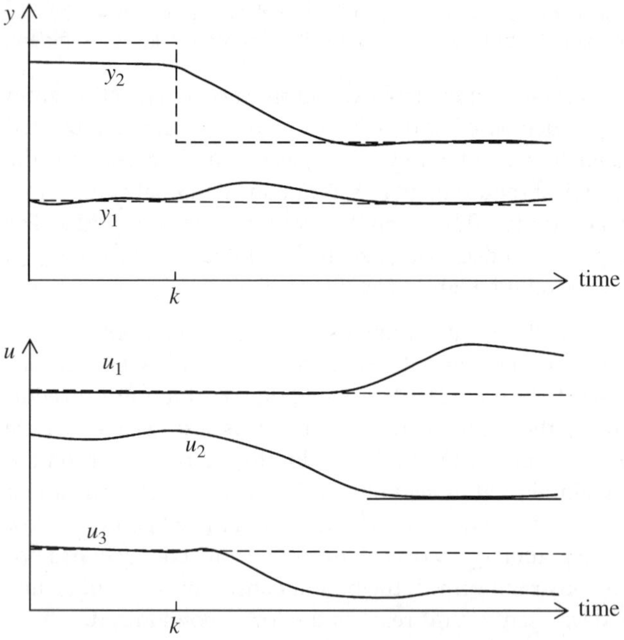

Figure 9.7 Control with multidegrees of freedom and various requirements.

controlling y_1, y_2. If it is expected that u_1 maintains its desired value and remains unchanged, u_3 must deviate from its desired value. Furthermore, with the process going on, u_2 decreases continuously until it reaches its constraint lower bound. Due to the loss of the degree of freedom of u_2, u_1 should no longer maintain its desired value. With the released degree of freedom of u_1 the requirements on y_1, y_2 can still be satisfied.

The above examples indicate that, in the complex industrial environment, due to the existence of various kinds of constraints and multiple control objectives, the optimization control problem has some new characters when compared with traditional optimal control theory.

1) In traditional control theory, the performance index and the constraint conditions are given independently and have different meanings. In industrial applications, the optimization proposition is not given early and should be established according to the user's requirements. From the viewpoint of process operators, the difference between the performance index and the constraints is not so strict. For example, the performance index of minimizing some specific items can be handled as soft constraints by setting all of these items as zero. Various constraints can also be regarded as control objectives that must be satisfied or should be satisfied as far as possible. Therefore, what is important is how to start from the user's requirements to reasonably define the performance and constraints in the optimization problem.
2) The various requirements put on the optimization problem are not of the same importance. They can be prioritized from hard to soft according to the concerns of the process operator and classified into the following three catalogs:
   - Hard requirements. Due to physical properties of facilities and instruments and from safety considerations, some hard requirements need to be put on manipulated variables or controlled variables. These requirements must be satisfied over the whole control process, otherwise the control is not realizable or not allowed.
   - Adjustable hard/soft requirements. The technological requirements ensuring the product quality, such as setting the desired value for system output, should be satisfied as far as possible, but with some flexibility. For example, if it is not possible to control the output to the desired value, sometimes it is allowable to control it to an admissible range centered at the desired value. However, the admissible range is a hard requirement because output beyond this range may lead to a waste product. This kind of soft requirement is adjustable but limited by some hard requirements.
   - Soft requirements. The additional expectations on control quality or manipulated variables due to economic requirements could be further considered only when the above two kinds of requirements have been satisfied. This is a kind of more soft requirements.
3) The optimization problem is formulated and solved by utilizing the degrees of freedom of all the manipulated variables to meet all or part of the requirements according to their priority from hard to soft. The hard requirements due to physical, safety, and technological conditions are set as constraints that are the necessary conditions of optimization and must be satisfied, while the soft ones, due to economic considerations and other expectations, should be satisfied as far as possible by using the remaining degrees of freedom and will be addressed in the performance index.
4) The adjustable hard/soft requirements play an important role and can be adjusted by the process operator according to real status. Sometimes they need to be relaxed from hard to soft to a certain extent to make the optimization problem feasible (such as relaxing the set point condition for some output variable to a certain range) or to achieve a trade-off between conflict requirements (such as trade-off when the output variables cannot all reach their set points). In other cases where the control degrees of freedom still exist, they can also be set from soft to hard to make the solution closer to the user's expectations. The obtained solution is a satisfactory one that reflects the process operator paying different degrees of attention to various requirements.

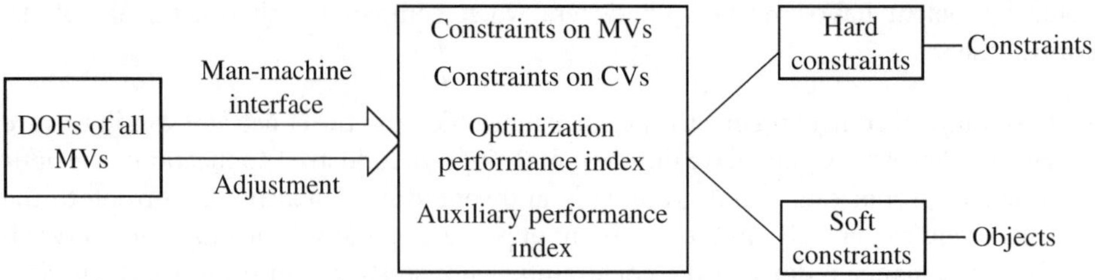

Figure 9.8 Constrained multiobjective multidegrees of freedom optimization (CMMO).

In the complex industrial environment, the optimization problem with the above characteristics is called the CMMO. The principle of CMMO is shown in Figure 9.8.

Note that the optimization proposition of CMMO with a performance index and constraints established by the requirement handling of the process operator is not guaranteed to be solvable. In the following, we discuss the feasibility problem of CMMO and show how to adjust hard/soft constraints to make the CMMO feasible [5].

For a stable linear time-invariant multivariable system, the constraints in its CMMO problem contain the input–output process model and constraints on process inputs and outputs. Although these constraints are closely related to the real-time input and output variables, it is shown that the optimization problem is solvable if the initial and the steady-state inputs and outputs satisfy these constraints. Let $\boldsymbol{u}_s \in \mathrm{R}^m$, $\boldsymbol{y}_s \in \mathrm{R}^r$ be the steady-state values of process input and output, respectively, and $\boldsymbol{H} \in \mathrm{R}^{r\times m}$ is the known steady-state gain matrix. Then the CMMO is called feasible if there exist $\boldsymbol{u}_s$, $\boldsymbol{y}_s$ satisfying

$$\begin{aligned} &\boldsymbol{y}_s = \boldsymbol{H}\boldsymbol{u}_s \\ &\underline{\boldsymbol{u}} \le \boldsymbol{u}_s \le \bar{\boldsymbol{u}} \\ &\underline{\boldsymbol{y}} \le \boldsymbol{y}_s \le \bar{\boldsymbol{y}} \end{aligned} \tag{9.15}$$

where $\bar{\boldsymbol{u}}$, $\underline{\boldsymbol{u}}$ and $\bar{\boldsymbol{y}}$, $\underline{\boldsymbol{y}}$ are the upper and lower bounds for input and output, respectively. and can be specified by the operator as illustrated in Example 9.2. Since (9.15) is composed of linear equality and inequalities with respect to $\boldsymbol{u}_s$, $\boldsymbol{y}_s$, its feasibility can be easily judged by LP. However, in order to generally judge or adjust the feasibility of (9.15), we define the following relaxing amount of the constraints to characterize the adjustability of the hard/soft constraints

$$\boldsymbol{\Delta} = \left[\Delta\bar{\boldsymbol{u}}^{\mathrm{T}},\ \Delta\underline{\boldsymbol{u}}^{\mathrm{T}},\ \Delta\bar{\boldsymbol{y}}^{\mathrm{T}},\ \Delta\underline{\boldsymbol{y}}^{\mathrm{T}}\right]^{\mathrm{T}} \ge \boldsymbol{0}$$

where for a hard constraint the corresponding $\Delta_i = 0$ and for a hard/soft constraint $\Delta_i \ge 0$ is allowed. Then (9.15) can be reformulated by

$$\begin{aligned} &\boldsymbol{y}_s = \boldsymbol{H}\boldsymbol{u}_s \\ &\underline{\boldsymbol{u}} - \Delta\underline{\boldsymbol{u}} \le \boldsymbol{u}_s \le \bar{\boldsymbol{u}} + \Delta\bar{\boldsymbol{u}} \\ &\underline{\boldsymbol{y}} - \Delta\underline{\boldsymbol{y}} \le \boldsymbol{y}_s \le \bar{\boldsymbol{y}} + \Delta\bar{\boldsymbol{y}} \\ &\boldsymbol{\Delta} = \left[\Delta\bar{\boldsymbol{u}}^{\mathrm{T}},\ \Delta\underline{\boldsymbol{u}}^{\mathrm{T}},\ \Delta\bar{\boldsymbol{y}}^{\mathrm{T}},\ \Delta\underline{\boldsymbol{y}}^{\mathrm{T}}\right]^{\mathrm{T}} \ge \boldsymbol{0} \end{aligned} \tag{9.16}$$

To get a standard form, denote $\boldsymbol{x}_1 = \boldsymbol{u}_s - \underline{\boldsymbol{u}} + \Delta\underline{\boldsymbol{u}}$, $\boldsymbol{x}_2 = \bar{\boldsymbol{u}} + \Delta\bar{\boldsymbol{u}} - \boldsymbol{u}_s$, $\boldsymbol{x}_3 = \boldsymbol{H}\boldsymbol{u}_s - \underline{\boldsymbol{y}} + \Delta\underline{\boldsymbol{y}}$, $\boldsymbol{x}_4 = \bar{\boldsymbol{y}} + \Delta\bar{\boldsymbol{y}} - \boldsymbol{H}\boldsymbol{u}_s$ and let $\boldsymbol{z} = \left[\boldsymbol{x}_1^{\mathrm{T}}, \boldsymbol{x}_2^{\mathrm{T}}, \boldsymbol{x}_3^{\mathrm{T}}, \boldsymbol{x}_4^{\mathrm{T}}, \boldsymbol{\Delta}^{\mathrm{T}}\right]^{\mathrm{T}}$. Cancel $\boldsymbol{u}_s$ by linear combination of $\boldsymbol{x}_i$; then it follows that

$$\boldsymbol{R}\boldsymbol{z} = \boldsymbol{b}: \quad \begin{cases} \boldsymbol{x}_1 + \boldsymbol{x}_2 = \bar{\boldsymbol{u}} + \Delta\bar{\boldsymbol{u}} - \underline{\boldsymbol{u}} + \Delta\underline{\boldsymbol{u}} \\ \boldsymbol{H}\boldsymbol{x}_1 + \boldsymbol{x}_4 = \bar{\boldsymbol{y}} + \Delta\bar{\boldsymbol{y}} - \boldsymbol{H}\underline{\boldsymbol{u}} + \boldsymbol{H}\Delta\underline{\boldsymbol{u}} \\ \boldsymbol{H}\boldsymbol{x}_2 + \boldsymbol{x}_3 = \boldsymbol{H}\bar{\boldsymbol{u}} + \boldsymbol{H}\Delta\bar{\boldsymbol{u}} - \underline{\boldsymbol{y}} + \Delta\underline{\boldsymbol{y}} \end{cases}$$

Now the following LP problem can be defined:

$$\begin{aligned} &\min \ w = \boldsymbol{c}^{\mathrm{T}}\boldsymbol{\Delta} \qquad \text{with } \boldsymbol{c}^{\mathrm{T}} = [1, \ldots, 1] \\ &\text{s.t. } \boldsymbol{R}\boldsymbol{z} = \boldsymbol{b} \\ &\qquad \boldsymbol{z} \geq \boldsymbol{0} \end{aligned} \tag{9.17}$$

It is obvious that the original CMMO (9.15) is feasible if the LP problem (9.17) has an optimal solution $w = 0$, while $w \neq 0$ means that (9.15) is infeasible and some hard/soft constraints should be relaxed. For the hard/soft constraint adjustment, consider following two cases.

If the admissible ranges of all the constraints are known as $[\boldsymbol{u}_{\min}\ \boldsymbol{u}_{\max}]$ and $[\boldsymbol{y}_{\min}\ \boldsymbol{y}_{\max}]$, then put constraints $\boldsymbol{u}_{\min} \leq \underline{\boldsymbol{u}} - \Delta\underline{\boldsymbol{u}}$, $\bar{\boldsymbol{u}} + \Delta\bar{\boldsymbol{u}} \leq \boldsymbol{u}_{\max}$, $\boldsymbol{y}_{\min} \leq \underline{\boldsymbol{y}} - \Delta\underline{\boldsymbol{y}}$, $\bar{\boldsymbol{y}} + \Delta\bar{\boldsymbol{y}} \leq \boldsymbol{y}_{\max}$ into (9.16) and establish corresponding standard LP problems like (9.17). If the LP problem has an optimal solution $\boldsymbol{\Delta}^*$, then the CMMO is feasible after relaxing the hard/soft constraints in (9.15) by

$$\begin{aligned} &\underline{\boldsymbol{u}} - \Delta\underline{\boldsymbol{u}}^* \leq \boldsymbol{u}_s \leq \bar{\boldsymbol{u}} + \Delta\bar{\boldsymbol{u}}^* \\ &\underline{\boldsymbol{y}} - \Delta\underline{\boldsymbol{y}}^* \leq \boldsymbol{y}_s \leq \bar{\boldsymbol{y}} + \Delta\bar{\boldsymbol{y}}^* \end{aligned}$$

If the LP problem is unsolvable, then the CMMO is infeasible even if all the adjustable hard/soft constraints have been relaxed to their admissible bounds and become hard constraints.

Another case is where the process operator has no exact knowledge of the admissible ranges of the adjustable hard/soft constraints and needs to adjust the CMMO through human–computer interaction. In this case, the process operator firstly determines the priority of relaxing the constraint margin Δ_i and then adopts a weighting vector $\boldsymbol{c}^{\mathrm{T}} = [c_1, \ldots, c_N]$ in the performance index (9.17), where the relaxing amount for the constraint to be firstly adjusted is weighted by a smaller c_i and the hard constraint that is not allowed to be adjusted is assigned with $\Delta_i = 0$ or weighted by a very large number. With the weighted performance index, solving the LP problem (9.17) will give a new $\boldsymbol{\Delta}$ with some specified $\Delta_i > 0$. If the process operator is satisfactory with the result, complete the adjustment. Otherwise set an acceptable bound for the nonsatisfactory Δ_i, put it as an additional constraint in the LP problem, and then solve it again. This procedure continues until the process operator is satisfied.

The feasibility problem for online optimization discussed above is actually addressed in industrial predictive control technologies, where a QP problem of steering the output to its set point under input and output constraints should be solved online at the dynamic control level:

$$\min_{\Delta u(k+i)} \sum_{i=0}^{N-1} \|y_T - y(k+i+1)\|_Q^2 + \|\Delta u(k+i)\|_R^2$$
$$\text{s.t.}\ \ \Delta u(k+i) \in \Omega_{\Delta u},\quad y(k+i+1) \in \Omega_y$$

where the output target y_T is often assigned by the local economic optimization level, which provides the desired steady state with optimal economic benefits under given technological conditions. Since this level works over a long period (often several hours), it is possible that the given y_T is unreachable because of the variation of technological conditions or disturbance. Therefore, in the dynamic control level, the feasibility of the given steady state should first be checked and, if not suitable, a new feasible steady state should be determined for predictive control algorithms. Thus the dynamic control level is indeed a two-stage structure composed of steady-state optimization (SSO) and model predictive control (MPC).

At time k, if the given target steady state (y_T, u_T) is infeasible, a new feasible steady state ($y_s(k)$, $u_s(k)$) can be found by solving the following optimization problem (it is a QP problem if the constraints are linear):

$$\begin{aligned} &\min_{u_s(k),\ y_s(k)} \ \|y_s(k) - y_T\|_{Q_S}^2 + \|u_s(k) - u_T\|_{R_S}^2 \\ &\text{s.t.}\ \ y_s(k) = Hu_s(k) \\ &\qquad u_s(k) \in \Omega_u,\ \ y_s(k) \in \Omega_y \end{aligned} \tag{9.18}$$

which means driving the steady-state inputs and outputs as closely as possible to targets determined by the local economic optimization level without violating input and output constraints. Another way to do that is by solving the following LP problem [6]:

$$\begin{aligned} &\min_{u_s(k),\ y_s(k)} \ c_y^{\mathrm{T}}(y_s(k) - y_T) + c_u^{\mathrm{T}}(u_s(k) - u_T) + c_\Delta^{\mathrm{T}}\Delta_y \\ &\text{s.t.}\ \ y_s(k) = Hu_s(k) + b(k) \\ &\qquad b(k) = b(k-1) + (y(k) - y(k \mid k-1)) \\ &\qquad u_{\min} \le u_s(k) \le u_{\max} \\ &\qquad y_{\min} - \Delta_y \le y_s(k) \le y_{\max} + \Delta_y \\ &\qquad \Delta_y \ge 0 \end{aligned} \tag{9.19}$$

where $b(k)$ is the model bias due to disturbance, etc., Δ_y is the admissible relaxing amount of the output used to guarantee a feasible solution to the LP problem, c_Δ^{T} is a tuning parameter for adjustment of Δ_y. Note that both the weighting matrices Q_S, R_S in the QP problem (9.18) and the weighting vectors c_y, c_u in the LP problem (9.19) are selected according to the economic significance of the corresponding steady-state output or input.

It has been shown that the optimization concept implied in CMMO is indeed embedded into the rolling optimization in predictive control. Based on this concept, [4] gave the satisfactory control scheme to characterize predictive control in industrial applications, which is composed of the following parts, not only including the basic elements of predictive control algorithms but also taking the implementation techniques into account:

- The process model used for prediction, optimization, and fault diagnosis, etc.
- The online performed CMMO for optimization of the dynamic process
- The learning mechanism for online correcting the state and adjusting the model
- The CAD system assisting the process operator for decision making
- The man–machine interface for the process operator to express and adjust requirements

Satisfactory control is a kind of optimization oriented control but is so named because we emphasize the satisfactory rather than the optimal during control. Firstly, the optimization problem solved during control is not fixed but is adjusted online by the process operator according to actual operation status. The optimization control is not for a given performance index but for subjective satisfaction of the process operator. Secondly, during optimization at each sampling time, the CMMO is essentially a multiobjective optimization. Its solution is a satisfactory one obtained by a trade-off between different requirements. Thirdly, throughout the whole control procedure, the system behavior is not optimized over an infinite horizon but over a limited future time horizon in a rolling style, which means that this kind of control does not pursue optimality in the theoretical sense, but focuses on satisfaction in an uncertain environment from an application viewpoint.

9.3 Input Parametrization of Predictive Control

Predictive control performs online optimization in a rolling style. At each sampling time, a nonlinear constrained optimization problem should be solved, which often needs iteration many times. The online computational burden greatly depends on the number of optimization variables as well as constraints. In order to reduce the number of online optimization variables to improve the real-time control ability, some specific optimization algorithms and strategies were proposed in predictive control applications, where the problem of optimizing the input variables at different sampling times is transferred into that of optimizing less input variables or other new variables, called input parametrization [7]. In this section, we introduce two typical input parametrization methods that have been widely adopted in predictive control applications.

9.3.1 Blocking Strategy of Optimization Variables

In predictive control, blocking technology was firstly proposed by Ricker [8]. Blocking refers to dividing the variation of the control variables in the optimization horizon into several blocks. In the interior of each block the control variable remains unchanged and the changes only take place between the blocks. This can generally be described by

$$\boldsymbol{u}(k+i)=\begin{cases}\boldsymbol{u}(k), & i=0,\ 1,\ \dots,\ l_1-1\\ \boldsymbol{u}(k+l_1), & i=l_1,\ l_1+1,\dots,\ l_1+l_2-1\\ \vdots & \vdots\\ \boldsymbol{u}(k+l_{s-1}), & i=l_1+\cdots+l_{s-1},\ \dots,\ l_1+\cdots+l_s-1\end{cases} \tag{9.20}$$

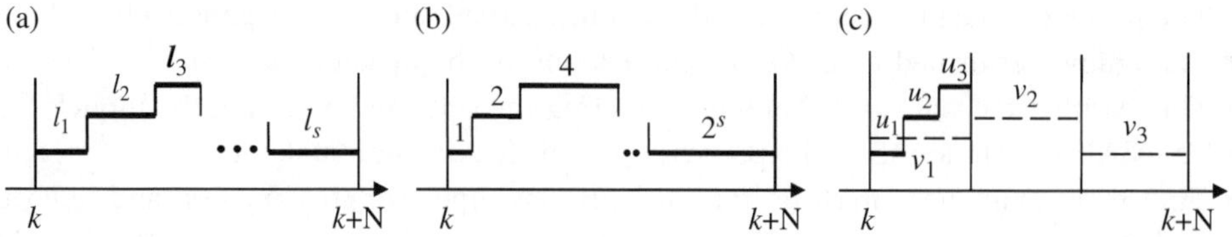

Figure 9.9 Blocking strategy for optimization variables in predictive control. (a) General blocking strategy; (b) horizontal blocking strategy with "accurate for near, rough for far"; (c) layered blocking strategy with "rough for long horizon and fine for short horizon."

where s is the number of blocks and l_i is the number of optimization time instants in the ith block. The summation of all the time instants in all blocks is the length of the optimization horizon, i.e. $l_1 + \cdots + l_s = P$. With this strategy, the number of variations of the optimization variables along time can be reduced from the original optimization horizon length P to the block number s. This blocking strategy is shown in Figure 9.9a.

The blocking strategy for optimization variables given by (9.20) covers some strategies that appeared in early predictive control literature, such as those given below.

1) **Introducing the control horizon**
The concept of the control horizon M was introduced in original predictive control algorithms such as DMC, MAC to reduce the online computational burden. With $M \le P$ the number of optimization variables can be reduced from P to M (for the SISO case) while the optimization is still over a long range. This is equivalent to the case of $s = M$ and $l_1 = \cdots = l_{s-1} = 1$ in the blocking strategy (9.20).
2) **Average blocking strategy with "fine optimization, rough control"**
Select $l_1 = \cdots = l_s = l > 1$ in (9.20), which is simply the case of the "fine optimization, rough control" strategy presented in [9], where the control period is enlarged to l times the sampling period, while optimization is still for predicted outputs at all sampling instants over the optimization horizon.
3) **Horizontal blocking strategy with "accurate for near, rough for far"** [10]
With rolling optimization of predictive control, only the solved current control action is actually implemented while others will be recalculated at a later time. Thus the control variables close to the current time should be calculated more accurately while those far from the current time can be roughly considered. This idea can be realized by setting the lengths of the blocks in (9.20) in the form of power series, i.e. $l_i = q^{i-1}$, $i = 1, \ldots, s$, where q can be selected as an integer such as 2, 3, etc. (see Figure 9.9b). For example, for $q = 2$, the lengths of the blocks are 1, 2, 4, ... In this way, the number of control variations at $2^s - 1$ time instants can be reduced to s. If q is larger, the number of optimization variables can be greatly reduced.
4) **Layered blocking strategy with "rough for long term and fine for short term"** [10]
In predictive control, decreasing the optimization horizon can also reduce the number of optimization variables, but at the cost of performance degeneration because of short-term optimization. To solve this problem, [10] adopts a coordination strategy combining rough optimization for long-term and fine optimization for the short term (see Figure 9.9c). Firstly, as a usual blocking strategy, the whole optimization horizon is divided into s equal blocks according to (9.20), with each block containing l sampling time instants, i.e. $P = sl$. Solve the optimization problem and obtain

the control variable $\boldsymbol{v}_1$ at the first block. This rough optimization can reduce the number of optimization variables while keeping a long-range optimization. However, the solved $\boldsymbol{v}_1$ is not optimal because setting a unique control action for the first block implies a restriction on the degrees of freedom of controls at different times in the block. Therefore a fine optimization for the control variables in the first block is needed, i.e. to solve the following optimization problem:

$$\min_{\boldsymbol{u}(k)\ \cdots\ \boldsymbol{u}(k+l-1)} \sum_{i=1}^{l} \left\| \boldsymbol{w}(k+i) - \tilde{\boldsymbol{y}}_p(k+i) \right\|^2 + \lambda \sum_{i=1}^{l} \left\| \boldsymbol{u}(k+i-1) - \boldsymbol{v}_1 \right\|^2$$

The optimization horizon for fine optimization is compressed into the first block, where the optimization variable changes only l times. The first item in the performance index reflects the accuracy requirement of the fine optimization on the control variables, while the second item indicates that the control variables should be as close as possible to the optimal one given by the rough optimization. Through adjusting the weighting coefficient λ it is possible to coordinate global performance and local accuracy. In this layered strategy, two optimization problems should be solved, with the control variable changing s times and l times, respectively. Compared with the original optimization problem with control variable changing $P = sl$ times, the online computational burden can be greatly reduced.

9.3.2 Predictive Functional Control

Predictive functional control (PFC) is a kind of predictive control algorithm suitable for fast dynamic systems proposed by Richalet et al. in [11] with application to an industrial robot. Different from directly solving the control input variables in traditional predictive control algorithms, the online optimization problem in PFC is to solve some combination coefficients with which the control variables can be calculated. Since the number of these combination coefficients is small and independent from the length of an optimization horizon, PFC can reduce the online computational burden while still keeping long-range optimization.

PFC is based on the structural analysis for control inputs. It is assumed that with a limited input spectrum, the control input must belong to a set of specific functions $\{f_n, n = 1, \dots, s\}$. Taking the single variable system as an example, the future control input can be represented by a linear combination of these functions:

$$u(k+i) = \sum_{n=1}^{s} \mu_n(k) f_n(i), \quad i = 0, \dots, P-1 \tag{9.21}$$

where $f_n(n = 1, \dots, s)$ is called the base function, $f_n(i)$ represents the value of the base function f_n at $t = iT$, and P is the length of the optimization horizon in predictive control. The selection of base functions depends on the plant dynamics and the desired trajectory. In general they can be selected as step, ramp, exponent function, etc. For each selected base function f_n, when acting on the plant, the output response can be calculated off-line and denoted as $g_n(i)$.

At sampling time k, with the future control input $u(k + i)$ acting on the plant, the output prediction can be given by

$$y_M(k+i) = y_0(k+i) + y_f(k+i), \quad i = 1, \dots, P \tag{9.22}$$

where

$$y_0(k+i) = F(y(k), \ldots, y(k-n_a), u(k-1), \ldots, u(k-n_b)) \tag{9.23}$$

is called the free model output and $F(\cdot)$ can be taken as various different forms such as a linear difference equation, a convolution model, etc. Here "free" refers to this part of the output prediction only depending on the historical information of plant input/output and is free from the newly added control inputs. Also take

$$y_f(k+i) = \sum_{n=1}^{s} \mu_n(k) g_n(i), \quad i = 1, \ldots, P \tag{9.24}$$

as the model function output, representing the output response caused by the future control input $u(k+i)$ with the form (9.21) starting from the sampling time k. Thus (9.21) and (9.24) show the difference between PFC and other predictive control algorithms. The newly added future inputs are linear combinations of the base functions rather than independent variables along time. Therefore the output variation caused by them is the superposition of responses from various base functions, rather than from control inputs at different times. Note that the sampled values of the base functions $f_n(i)$ and their responses $g_n(i)$ are already calculated off-line, so only the linear combination coefficients $\mu_n(k)$ are unknown and taken as the optimization variables.

Based on (9.22), the closed-loop output prediction can be given through feedback correction

$$y_p(k+i) = y_M(k+i) + e(k+i), \quad i = 1, \ldots, P \tag{9.25}$$

where

$$e(k+i) = h_i e(k), \quad i = 1, \ldots, P$$
$$e(k) = y(k) - y_M(k)$$

Given the desired reference trajectory for system output

$$y_r(k+i) = \alpha^i y(t) + \left(1-\alpha^i\right) c(k+i), \quad i = 1, \ldots, P \tag{9.26}$$

where $c(k+i)$ is the desired output at future time $k+i$, then the online optimization problem at sampling time k can be formulated as

$$\min_{\mu_1, \cdots \mu_s} J\left(y_r(k), y_p(k)\right) \tag{9.27}$$

After obtaining $\mu_1, \ldots, \mu_s$ through optimization, the current control input can be calculated according to (9.21):

$$u(k) = \sum_{n=1}^{s} \mu_n(k) f_n(0) \tag{9.28}$$

and this control input is then put to implementation. The principle of PFC to solve the new control input is shown in Figure 9.10.

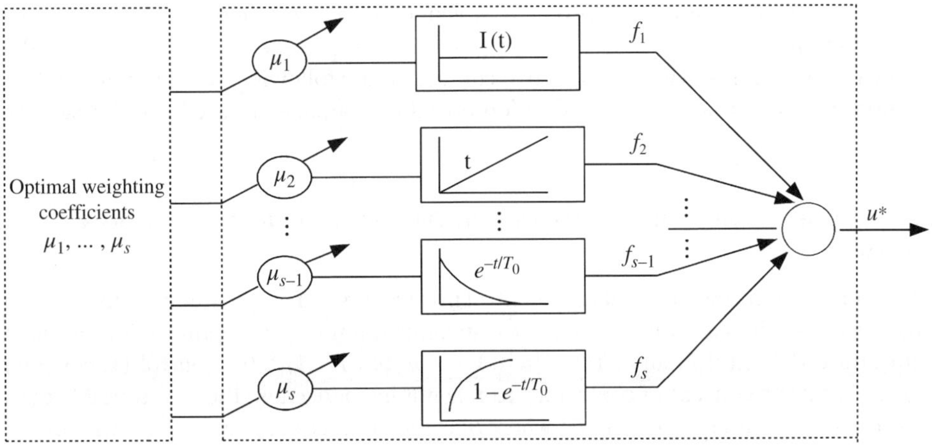

Figure 9.10 Optimization using weighted base functions in PFC.

Remark 9.3 It can be seen that PFC has the same principles as those in other predictive control algorithms, i.e. prediction model, rolling optimization, and feedback correction. The only difference is that its control input can be represented by the linear combination of known base functions, and thus the optimization variables are the combination coefficients $\mu_1(k), \dots, \mu_s(k)$ instead of the control inputs $u(k), \dots, u(k+P-1)$, which means that the control degree of freedom is transferred from time into space. Since the number of μ_i is independent of the optimization horizon length P while the number of $u(\cdot)$ increases with P, the online computational burden of PFC would be much lower than that of other predictive control algorithms, particularly for long-range optimization.

The design parameters of the PFC algorithm include the optimization horizon P, the parameter α related to the time constant of the reference trajectory, and the base functions $f_1, \dots, f_s$, all affect the control accuracy, stability (robustness), and dynamic response. Fortunately [11] indicates that these design parameters affect the control performance with different emphasis. Control accuracy mainly depends on the selection of the base functions. Dynamic response is mainly affected by the reference trajectory and the optimization horizon P plays a key role in control stability and robustness. Therefore, during control system design, these parameters can be quickly tuned according to the performance requirements. This algorithm has been successfully used in many real systems such as high speed and high accuracy tracking control of an industrial robot with six degrees of freedom [11].

9.4 Aggregation of the Online Optimization Variables in Predictive Control

The two-input parameterization methods presented in the last section reduce the number of online optimization variables by introducing specific strategy and algorithm, respectively. It seems that there is no connection between them. For achieving a more general input parameterization to unify these two methods, [12, 13] introduced the aggregation concept developed in large-scale system theory into predictive control for solving the online optimization problem. A general framework of optimization variable aggregation in predictive control was proposed, which provides a new insight into the online optimization solving in predictive control. In the following, we give a brief introduction, where the abbreviation OV (optimization variable) is used for the sake of brevity.

9.4.1 General Framework of Optimization Variable Aggregation in Predictive Control

The online optimization of predictive control involves system outputs, control inputs, and possible constraints at future P steps. It is in general a nonlinear optimization problem with future P control inputs $\boldsymbol{U}(k) = [\boldsymbol{u}^{\mathrm{T}}(k) \ \cdots \ \boldsymbol{u}^{\mathrm{T}}(k+P-1)]^{\mathrm{T}}$ (or control increments) as the OVs if the concept of control horizon is not introduced. In the multivariable case, $\boldsymbol{U}(k)$ is a vector with dimension mP, where m is the number of system inputs. In order to get a better control result, the optimization horizon P should not be selected too small. However, a large P may result in a large number of OVs and a heavy online computational burden.

In order to reduce the number of online OVs, consider the following linear aggregation transformation:

$$\boldsymbol{U}(k) = \boldsymbol{H}\boldsymbol{V}(k) \tag{9.29}$$

where $\boldsymbol{V}(k) = [v_1(k) \ \cdots \ v_s(k)]^{\mathrm{T}}$ is an s-dimensional vector, called as aggregation variable. Take $s \ll mP$, which means that $\boldsymbol{V}(k)$ has a much lower dimension than $\boldsymbol{U}(k)$. $\boldsymbol{H}$ is an $mP \times s$ constant matrix, called as aggregation matrix, and can be represented by

$$\boldsymbol{H} = \begin{bmatrix} h_{11} & \cdots & h_{1s} \\ \vdots & \ddots & \vdots \\ h_{mP,1} & \cdots & h_{mP,s} \end{bmatrix} \tag{9.30}$$

If (9.29) is substituted into the original optimization problem and $\boldsymbol{V}(k)$ is taken as a new OV instead of $\boldsymbol{U}(k)$, the scale of the online optimization problem can be greatly reduced. For example, in the constrained multivariable DMC algorithm discussed in Section 5.2, although the control horizon $M(M \le P)$ is introduced to reduce the number of the OVs (from mP to mM), the control increment $\Delta\boldsymbol{u}_M(k)$ as the OV can be further aggregated based on the above principle. Let

$$\Delta\boldsymbol{u}_M(k) = \boldsymbol{H}\boldsymbol{V}(k)$$

where

$$\boldsymbol{H} = \begin{bmatrix} h_{11} & \cdots & h_{1s} \\ \vdots & \ddots & \vdots \\ h_{mM,1} & \cdots & h_{mM,s} \end{bmatrix}, \quad \boldsymbol{V}(k) = \begin{bmatrix} v_1(k) \\ \vdots \\ v_s(k) \end{bmatrix}$$

Substituting this into (5.18), the optimization problem with $\boldsymbol{V}(k)$ as the OV can be obtained as

$$\min_{\boldsymbol{V}(k)} J(k) = \|\boldsymbol{w}(k) - \tilde{\boldsymbol{y}}_{P0}(k) - \boldsymbol{A}\boldsymbol{H}\boldsymbol{V}(k)\|_{\boldsymbol{Q}}^2 + \|\boldsymbol{H}\boldsymbol{V}(k)\|_{\boldsymbol{R}}^2$$

$$\text{s.t.} \ \ \boldsymbol{C}\boldsymbol{H}\boldsymbol{V}(k) \le \boldsymbol{l}$$

Since the dimension of the OV is reduced from mM to s, the solving of the optimization problem can be greatly simplified.

Equation (9.29) is essentially an input parameterization transformation with the property of decreasing dimension. It gives a general framework for OV aggregation, which not only covers various strategies and algorithms developed for reducing the number of OVs in predictive control but also provides a novel way to develop new optimization strategies. In the following some cases are discussed.

1) **Blocking strategy**

The blocking strategy presented in Section 9.3.1 can be obtained by setting the following aggregation matrix and aggregation variable in (9.29):

$$\boldsymbol{H} = \underbrace{\begin{bmatrix} \boldsymbol{I}_m & \mathbf{0}_m & \cdots & \mathbf{0}_m \\ \vdots & \vdots & \ddots & \vdots \\ \boldsymbol{I}_m & \mathbf{0}_m & \cdots & \mathbf{0}_m \\ \vdots & \vdots & \cdots & \vdots \\ \mathbf{0}_m & \cdots & \mathbf{0}_m & \boldsymbol{I}_m \\ \vdots & \ddots & \vdots & \vdots \\ \mathbf{0}_m & \cdots & \mathbf{0}_m & \boldsymbol{I}_m \end{bmatrix}}_{ms} \begin{matrix} \left.\vphantom{\begin{matrix} a \\ b \\ c \end{matrix}}\right\} l_1 \\ \\ \left.\vphantom{\begin{matrix} a \\ b \\ c \end{matrix}}\right\} l_s \end{matrix}, \quad \boldsymbol{V}(k) = \begin{bmatrix} \boldsymbol{u}(k) \\ \boldsymbol{u}(k+l_1) \\ \vdots \\ \boldsymbol{u}(k+l_{s-1}) \end{bmatrix}$$

where $\boldsymbol{I}_m$ is an $m \times m$ identity matrix and $\mathbf{0}_m$ is an $m \times m$ zero matrix.

2) **PFC**

The PFC algorithm presented in Section 9.3.2 can be obtained by setting the following aggregation matrix and aggregation variable in (9.29):

$$\boldsymbol{H} = \begin{bmatrix} f_1(0) & \cdots & f_s(0) \\ \vdots & \ddots & \vdots \\ f_1(P-1) & \cdots & f_s(P-1) \end{bmatrix}, \quad \boldsymbol{V}(k) = \begin{bmatrix} \mu_1(k) \\ \vdots \\ \mu_s(k) \end{bmatrix}$$

This is obvious when compared with (9.21).

3) **Input decaying strategy**

In many control problems it is expected that the control action smoothly and steadily approaches its steady-state value. Particularly in a regulation problem the control increment should asymptotically decay to zero. Under the framework of OV aggregation, [12] proposed an input decaying strategy for OVs, which defines the future controls $\boldsymbol{u}(k+1), \ldots, \boldsymbol{u}(k+P-1)$ as the amplitude decaying series of the current control $\boldsymbol{u}(k)$:

$$\boldsymbol{u}(k+i) = \rho^i(k)\boldsymbol{u}(k), \quad i = 0, \ldots, P-1 \tag{9.31}$$

where $\rho(k)$ is the decaying coefficient, which can be determined in advance. Then the aggregation matrix $\boldsymbol{H}$ has the form

$$\boldsymbol{H} = \left[\boldsymbol{I}_m\ \rho(k)\boldsymbol{I}_m\ \cdots\ \rho^{P-1}(k)\boldsymbol{I}_m\right]^{\mathrm{T}}, \quad \boldsymbol{V}(k) = \boldsymbol{u}(k)$$

After aggregation in this way, the online optimization is only solved for $\boldsymbol{u}(k)$, such that the dimension of the online OVs can be reduced from mP to m.

The above analysis shows that the OV aggregation strategy of predictive control (9.29) gives a feasible input parameterization strategy from a more general perspective, which not only covers the commonly used blocking technology and PFC algorithm but also provides a way to derive new input parameterization strategies such as input decay. More important is the fact that (9.29) focuses on the mathematical form of the input parameterization rather than on its physical meaning and selection of the aggregation variable $\boldsymbol{V}(k)$ is not restricted to physical variables. Thus more degrees of freedom exist for selecting s to improve the control performance. In the following, we discuss how to guarantee the performance of the predictive control system when using the OV aggregation strategy.

9.4.2 Online Optimization Variable Aggregation with Guaranteed Performances

The OV aggregation strategy of predictive control transforms the online optimization problem with a high-dimensional variable into that with a lower one and thus reduces the computational complexity for its online solving. It is advantageous, particularly when handling constrained predictive control problems. However, it is also obvious that after using the aggregation strategy, the whole optimal control sequence $\boldsymbol{U}(k)$ is different from before and the optimization performance will in general be degenerated due to the decrease in the control degrees of freedom. Note that the online optimization in predictive control takes a rolling style and what is actually significant is the fact that only the current control is implemented among all the derived optimal control actions. It is naturally worth exploring how to select a proper aggregation matrix such that the current control $\boldsymbol{u}(k)$ solved by using aggregation strategy can be equivalent to the original one. If so, the performance of the controlled system can be maintained and is unaffected by aggregation. We call aggregation with this feature equivalent aggregation [14] and discuss it for the following cases.

(1) Optimization based on equivalent aggregation in unconstrained predictive control

Consider a controllable linear time invariant system

$$\boldsymbol{x}(k+1) = \boldsymbol{A}\boldsymbol{x}(k) + \boldsymbol{B}\boldsymbol{u}(k) \tag{9.32}$$

where $\boldsymbol{A} \in \mathrm{R}^{n\times n}$, $\boldsymbol{B} \in \mathrm{R}^{n\times m}$. Assume that the current system state $\boldsymbol{x}(k)$ is known. Taking the multivariable predictive control algorithm in state space form and referring to Section 2.3, with the optimization horizon and the control horizon uniformly denoted as N, then the online optimization problem in the unconstrained case can be described by

$$\begin{aligned} \text{OP 1}: \quad & \min_{\boldsymbol{U}(k)} J(k) = \boldsymbol{X}^{\mathrm{T}}(k)\boldsymbol{Q}\boldsymbol{X}(k) + \boldsymbol{U}^{\mathrm{T}}(k)\boldsymbol{R}\boldsymbol{U}(k) \\ & \text{s.t.} \quad \boldsymbol{X}(k) = \boldsymbol{S}\boldsymbol{x}(k) + \boldsymbol{G}\boldsymbol{U}(k) \end{aligned} \tag{9.33}$$

where

$$X(k)=\begin{bmatrix} x(k+1\mid k)\\ x(k+2\mid k)\\ \vdots\\ x(k+N\mid k)\end{bmatrix},\qquad U(k)=\begin{bmatrix} u(k)\\ u(k+1)\\ \vdots\\ u(k+N-1)\end{bmatrix}$$

$$S=\begin{bmatrix} A\\ A^2\\ \vdots\\ A^N\end{bmatrix},\qquad G=\begin{bmatrix} B & 0 & \cdots & 0\\ AB & B & \ddots & \vdots\\ \vdots & \ddots & \ddots & 0\\ A^{N-1}B & \cdots & AB & B\end{bmatrix}$$

Q and R are positive definite weighting matrices with appropriate dimensions for state and control, respectively. Using the OV aggregation strategy (9.29), the online optimization problem of predictive control becomes

$$\begin{aligned} \text{OPA 1:}\quad & \min_{V(k)} J(k)=X^{\mathrm{T}}(k)QX(k)+U^{\mathrm{T}}(k)RU(k)\\ & \text{s.t.}\quad X(k)=Sx(k)+GU(k)\\ & \qquad\;\; U(k)=HV(k)\end{aligned}\tag{9.34}$$

where $H\in \mathrm{R}^{Nm\times s}$ is the aggregation matrix to be determined and s is the number of the OVs after aggregation, $s<Nm$.

In order to investigate the properties of the solutions of the optimization problems OP 1 and OPA 1, the following lemmas [14] are first given without proof.

Lemma 9.1 For the homogeneous linear equations $\boldsymbol{\Gamma} x=0$, $\boldsymbol{\Gamma}=[a_1\ a_2\cdots a_i\cdots a_d]\in \mathrm{R}^{k\times d}$ and $k<d$, if a_i is linearly independent of other column vectors, then the ith element x_i of the solution of the equation group is identical to zero.

Lemma 9.2 Matrix $W\in \mathrm{R}^{Nm\times Nm}$, $H\in \mathrm{R}^{Nm\times s}$ and rank(W) = Nm, rank(H) = s. Divide W into two blocks by column $[W_1 \vdots W_2]$, $W_1\in \mathrm{R}^{Nm\times m}$, $W_2\in \mathrm{R}^{Nm\times(Nm-m)}$. If rank $(H^{\mathrm{T}}W_2)=s-m$, then the column vector in $H^{\mathrm{T}}W_1$ is linearly independent of other columns in $H^{\mathrm{T}}W$.

Now we discuss how to realize equivalent aggregation for unconstrained predictive control. Since both OP 1 and OPA 1 are unconstrained optimization problems, the optimal solutions can be directly obtained by letting $\partial J(k)/\partial U(k)=\mathbf{0}$ and $\partial J(k)/\partial V(k)=\mathbf{0}$. It follows that

$$\left(G^{\mathrm{T}}QG+R\right)U(k)=-G^{\mathrm{T}}QSx(k)$$
$$\left[(GH)^{\mathrm{T}}Q(GH)+H^{\mathrm{T}}RH\right]V(k)=-(GH)^{\mathrm{T}}QSx(k)$$

Left-multiply H^{T} to both sides of the first equation and then compare it with the second one. It follows that

$$H^{\mathrm{T}}\left(G^{\mathrm{T}}QG+R\right)[HV(k)-U(k)]=0 \tag{9.35}$$

Then the following theorem can be obtained.

Theorem 9.1 Consider unconstrained predictive control with m-dimensional input and control horizon N; for its online optimization problem OP 1, there always exists an equivalent aggregation matrix with the number of aggregation variables $s \geq m$.

Proof: First rewrite (9.35) into

$$H^{\mathrm{T}}\left(G^{\mathrm{T}}QG+R\right)\xi=0, \qquad \xi=HV(k)-U(k)$$

Equivalent aggregation refers to that the current control $u(k)$ given by $U(k)$ before aggregation is equal to the current control $\tilde{u}(k)$ in $\tilde{U}(k)=HV(k)$ after aggregation, i.e. the first m elements in ξ are identical to zero. Since both Q and R are positive definite matrices, $(G^{\mathrm{T}}QG+R)$ is full rank. Divide $(G^{\mathrm{T}}QG+R)$ into two blocks in column sequence $[W_1 \vdots W_2]$, $W_1 \in \mathrm{R}^{Nm\times m}$, $W_2 \in \mathrm{R}^{Nm\times(Nm-m)}$. It is known from Lemmas 9.1 and 9.2 that it should be $\mathrm{rank}(H^{\mathrm{T}}W_2)=s-m$ if equivalent aggregation is asked for.

Since $W_2=[\nu_1\ \nu_2 \cdots \nu_{Nm}]^{\mathrm{T}}$ is an $Nm\times(Nm-m)$ matrix with full column rank where ν_i^{T} is the ith row of W_2, there must exist $Nm-m$ row vectors in W_2 that are linearly independent. For convenience in description, assume them to be $\nu_1^{\mathrm{T}},\nu_2^{\mathrm{T}},\ldots,\nu_{Nm-m}^{\mathrm{T}}$; then the other rows ν_i^{T} can be expressed as their linear combinations

$$\nu_i^{\mathrm{T}}=\lambda_{i,1}\nu_1^{\mathrm{T}}+\cdots+\lambda_{i,Nm-m}\nu_{Nm-m}^{\mathrm{T}}, \qquad i=Nm-m+1,\ldots,Nm \tag{9.36}$$

If $s \geq m$, take the first m rows of H^{T} as $[\lambda_{i,1}\ \ \lambda_{i,2}\cdots\lambda_{i,Nm-m}\ \ 0\ \cdots\ 0\ \ -1\ \ 0\ \cdots\ 0]$, where -1 appears at the ith column, $i=Nm-m+1,\ldots,Nm$, the corresponding rows in $H^{\mathrm{T}}W_2$ are $\lambda_{i,1}\nu_1^{\mathrm{T}}+\cdots+\lambda_{i,Nm-m}\nu_{Nm-m}^{\mathrm{T}}-\nu_i^{\mathrm{T}}=0$. Furthermore, take the remaining part of H^{T} after removing these m rows as $[I \vdots 0]$. Since $H^{\mathrm{T}}W_2 \in \mathrm{R}^{s\times(Nm-m)}$ and $Nm-m \gg s$, it is obvious that $\mathrm{rank}(H^{\mathrm{T}}W_2)=s-m$, and thus the equivalent aggregation matrix exists.

Remark 9.4 Theorem 9.1 indicates that for the online optimization problem OP 1 of unconstrained predictive control, if the OV aggregation strategy (9.29) is adopted, as long as the number of aggregation variables is not less than the input dimension, i.e. $s \geq m$, it is always possible to find an appropriate aggregation matrix H such that the current control solved by aggregation-based optimization is identical to that solved without aggregation, i.e. equivalent aggregation can be achieved. In the case $s<m$, since $\mathrm{rank}(H^{\mathrm{T}}(G^{\mathrm{T}}QG+R))<m$, the dimensionality of the solution space of equation $H^{\mathrm{T}}(G^{\mathrm{T}}QG+R)\xi=0$ is greater than $Nm-m$, it cannot be guaranteed that the first m elements in ξ are fixed as zero, and thus equivalent aggregation cannot be guaranteed.

The proof of Theorem 9.1 actually shows the way to construct the equivalent aggregation matrix H, which can be described by the following algorithm.

Algorithm 9.1 Calculation of the equivalent aggregation matrix for unconstrained predictive control

Step 1. Divide $\boldsymbol{W}=\boldsymbol{G}^{\mathrm{T}}\boldsymbol{Q}\boldsymbol{G}+\boldsymbol{R}$ into two blocks in column sequence $[\boldsymbol{W}_1 \vdots \boldsymbol{W}_2]$, $\boldsymbol{W}_1 \in \mathrm{R}^{Nm\times m}$, $\boldsymbol{W}_2 \in \mathrm{R}^{Nm\times(Nm-m)}$.

Step 2. Denote

$$\boldsymbol{W}_2=\begin{bmatrix}\boldsymbol{\nu}_1^{\mathrm{T}}\\ \boldsymbol{\nu}_2^{\mathrm{T}}\\ \vdots\\ \boldsymbol{\nu}_{Nm}^{\mathrm{T}}\end{bmatrix}_{(Nm\times(Nm-m))}$$

where $\boldsymbol{\nu}_i^{\mathrm{T}}$ is an $Nm-m$-dimensional row vector, $i=1,\ldots,Nm$. Remove m rows $\boldsymbol{\nu}_i^{\mathrm{T}}$, $i=i_1,\ldots,i_m$, from $\boldsymbol{W}_2$ to make the $(Nm-m)\times(Nm-m)$ matrix composed of the remaining part of $\boldsymbol{W}_2$ full rank.

Step 3. Similar to (9.36), represent each $\boldsymbol{\nu}_i^{\mathrm{T}}$, $i\in\{i_1,\ldots,i_m\}$, as the linear combination of other $\boldsymbol{\nu}_j^{\mathrm{T}}$, i.e.

$$\boldsymbol{\nu}_i^{\mathrm{T}}=\sum_{\substack{j=1\\ j\neq i_1,\ldots,i_m}}^{Nm}\lambda_{i,j}\boldsymbol{\nu}_j^{\mathrm{T}},\quad i\in\{i_1,\ldots,i_m\}$$

Calculate the coefficients of linear combinations.

Step 4. Construct the transpose of the aggregation matrix, the $s\times Nm$-dimensional matrix $\boldsymbol{H}^{\mathrm{T}}$. Its $i\in\{i_1,\ldots,i_m\}$th row has altogether Nm elements, where the ith is -1, the $i_1,\ldots,i_m$th (except the ith) are 0, and other $Nm-m$ elements are sequentially taken as the coefficients $\lambda_{i,j}$, $j=1,\ldots,Nm$, $j\notin\{i_1,\ldots,i_m\}$, of the linear combination corresponding to $\boldsymbol{\nu}_i^{\mathrm{T}}$.

Step 5. If $s>m$, the remaining $s-m$ rows in $\boldsymbol{H}^{\mathrm{T}}$ can be set arbitrarily only under the condition of full row rank. Then the aggregation matrix $\boldsymbol{H}$ can be obtained by transposing $\boldsymbol{H}^{\mathrm{T}}$.

(2) Optimization based on equivalent aggregation of predictive control with zero terminal constraints

It is known from predictive control theory that in order to ensure the stability of the controlled system, some artificial constraints need to be imposed even if there is no physical constraint. A zero terminal constraint is one of the typical selections. The online optimization problem of predictive control with an additional zero terminal constraint has the form

$$\begin{aligned}\text{OP 2:}\quad & \min_{\boldsymbol{U}(k)} J(k)=\boldsymbol{X}^{\mathrm{T}}(k)\boldsymbol{Q}\boldsymbol{X}(k)+\boldsymbol{U}^{\mathrm{T}}(k)\boldsymbol{R}\boldsymbol{U}(k)\\ & \text{s.t.}\quad \boldsymbol{X}(k)=\boldsymbol{S}\boldsymbol{x}(k)+\boldsymbol{G}\boldsymbol{U}(k)\\ & \qquad\ \ \boldsymbol{x}(k+N)=0\end{aligned}\tag{9.37}$$

Compared with the unconstrained problem OP 1 expressed by (9.33), an artificial constraint of the zero terminal state is added in (9.37).

In order to obtain an analytical solution for the OP 2 problem as easy as that in solving the OP 1 problem, the added zero terminal constraint condition should be firstly resolved. Note that according to the linear system theory, this condition means that

$$\boldsymbol{x}(k+N)=\boldsymbol{A}^N\boldsymbol{x}(k)+\boldsymbol{A}^{N-1}\boldsymbol{B}\boldsymbol{u}(k)+\cdots+\boldsymbol{B}\boldsymbol{u}(k+N-1)=\boldsymbol{0}$$

It follows that

$$\boldsymbol{W}_C\boldsymbol{U}(k)=-\boldsymbol{A}^N\boldsymbol{x}(k) \tag{9.38}$$

where $\boldsymbol{W}_C = [\boldsymbol{A}^{N-1}\boldsymbol{B}\ \ \boldsymbol{A}^{N-2}\boldsymbol{B}\cdots\boldsymbol{B}]$ is the controllability matrix of the system. Since for the controllable system rank $\boldsymbol{W}_C = n$, there are n linearly independent column vectors in $\boldsymbol{W}_C$ that form the base of $\boldsymbol{W}_C$. With no loss of generality, assume that they are the last n columns of $\boldsymbol{W}_C$, denoted by $\boldsymbol{h}_1, \dots, \boldsymbol{h}_n$. Denote $\boldsymbol{\Theta}_0 = [\boldsymbol{h}_1\ \cdots\ \boldsymbol{h}_n]$, $\boldsymbol{W}_C = [\boldsymbol{\Theta}_1\ \ \boldsymbol{\Theta}_0]$; then $\boldsymbol{\Theta}_0 \in \mathrm{R}^{n\times n}$, $\boldsymbol{\Theta}_1 \in \mathrm{R}^{n\times(Nm-n)}$. According to the theory of linear equations, the solution of the linear equations (9.38) can be composed of the null solution space and the particular solution of $\boldsymbol{W}_C$, i.e.

$$\boldsymbol{U}(k)=\boldsymbol{U}_0\boldsymbol{K}+\boldsymbol{U}^*(k) \tag{9.39}$$

where the Nm-dimensional vector $\boldsymbol{U}^*(k)$ is a particular solution of (9.38), satisfying $\boldsymbol{W}_C\boldsymbol{U}^*(k) = -\boldsymbol{A}^N\boldsymbol{x}(k)$. Since $\boldsymbol{\Theta}_0$ is reversible, $\boldsymbol{U}^*(k)$ can be taken as

$$\boldsymbol{U}^*(k)=\begin{bmatrix}\boldsymbol{0}\\ -\boldsymbol{\Theta}_0^{-1}\boldsymbol{A}^N\boldsymbol{x}(k)\end{bmatrix} \tag{9.40}$$

As $\boldsymbol{U}_0 \in \mathrm{R}^{Nm\times(Nm-n)}$ satisfies $\boldsymbol{W}_C\boldsymbol{U}_0 = [\boldsymbol{\Theta}_1\ \ \boldsymbol{\Theta}_0]\boldsymbol{U}_0 = \boldsymbol{0}$ and is the base of the null solution space of $\boldsymbol{W}_C$, it can therefore be selected as

$$\boldsymbol{U}_0=\begin{bmatrix}\boldsymbol{I}\\ -\boldsymbol{\Theta}_0^{-1}\boldsymbol{\Theta}_1\end{bmatrix} \tag{9.41}$$

Note that, for a given system, $\boldsymbol{W}_C = [\boldsymbol{A}^{N-1}\boldsymbol{B}\ \boldsymbol{A}^{N-2}\boldsymbol{B}\ \cdots\ \boldsymbol{B}]$ is known and thus $\boldsymbol{U}_0$ is known. The $Nm-n$-dimensional vector $\boldsymbol{K}$ in (9.39) can be freely chosen, and $\boldsymbol{U}^*(k)$ depends on the system state at time k.

Replace the zero terminal condition in the OP 2 by (9.39). The OP 2 problem can then be rewritten as

$$\begin{aligned}\text{OP 2:}\quad & \min_{\boldsymbol{K}} J(k)=\boldsymbol{X}^{\mathrm{T}}(k)\boldsymbol{Q}\boldsymbol{X}(k)+\boldsymbol{U}^{\mathrm{T}}(k)\boldsymbol{R}\boldsymbol{U}(k)\\ & \text{s.t.}\ \ \boldsymbol{U}(k)=\boldsymbol{U}_0\boldsymbol{K}+\boldsymbol{U}^*(k)\\ & \qquad \boldsymbol{X}(k)=\boldsymbol{S}\boldsymbol{x}(k)+\boldsymbol{G}\boldsymbol{U}(k)=\boldsymbol{S}\boldsymbol{x}(k)+\boldsymbol{G}_1\boldsymbol{K}+\boldsymbol{G}\boldsymbol{U}^*(k)\end{aligned}$$

where $\boldsymbol{G}_1 = \boldsymbol{G}\boldsymbol{U}_0$. Solving this problem, it follows that

$$-\left(\boldsymbol{G}_1^{\mathrm{T}}\boldsymbol{Q}\boldsymbol{G}_1+\boldsymbol{U}_0^{\mathrm{T}}\boldsymbol{R}\boldsymbol{U}_0\right)\boldsymbol{K}=\boldsymbol{G}_1^{\mathrm{T}}\boldsymbol{Q}\boldsymbol{S}\boldsymbol{x}(k)+\boldsymbol{G}_1^{\mathrm{T}}\boldsymbol{Q}\boldsymbol{G}\boldsymbol{U}^*(k)+\boldsymbol{U}_0^{\mathrm{T}}\boldsymbol{R}\boldsymbol{U}^*(k) \tag{9.42}$$

Similarly, the online optimization problem of aggregation-based predictive control with an added zero terminal constraint has the form

$$\text{OPA 2:}\quad \min_{V(k)} J(k) = X^{\mathrm{T}}(k)QX(k) + U^{\mathrm{T}}(k)RU(k) \tag{9.43}$$
$$\text{s.t.}\quad X(k) = Sx(k) + GU(k)$$
$$U(k) = HV(k)$$
$$x(k+N) = 0$$

After adopting the aggregation strategy (9.29), the system controllability matrix becomes the $n \times s$-dimensional matrix $W_C H$. In order to guarantee that the system is still controllable and with degrees of freedom for optimization, let $s > n$ and let the $Nm \times s$-dimensional matrix H be

$$H = \begin{bmatrix} H_1 & \mathbf{0} \\ \mathbf{0} & I \end{bmatrix} \tag{9.44}$$

where $H_1 \in \mathrm{R}^{(Nm-n)\times(s-n)}$ and $I \in \mathrm{R}^{n\times n}$. Then $W_C H = [\Theta_1 H_1 \quad \Theta_0]$, rank$(W_C H) = n$, so the system is still controllable and the last n columns of $W_C H$ are still $\Theta_0 = [h_1 \cdots h_n]$.

The solution of the optimization problem OPA 2 should satisfy the zero terminal constraint

$$W_C HV(k) = -A^N x(k)$$

where the matrix $W_C H$ is of dimension $n \times s$. Since $s > n$, $V(k)$ can be similarly given by

$$V(k) = V_0 \bar{K} + V^*(k) \tag{9.45}$$

where the s-dimensional $V^*(k)$ is a particular solution, satisfying $W_C HV^*(k) = -A^N x(k)$. Therefore $HV^*(k) = U^*(k)$. When $U^*(k)$ is taken as the form of (9.40), in terms of the structure of H in (9.44), it follows that

$$V^*(k) = \begin{bmatrix} \mathbf{0} \\ -\Theta_0^{-1} A^N x(k) \end{bmatrix} \tag{9.46}$$

Since the base of the null solution space $V_0 \in \mathrm{R}^{s\times(s-n)}$ satisfies $W_C HV_0 = [\Theta_1 H_1 \quad \Theta_0] V_0 = \mathbf{0}$, we can select

$$V_0 = \begin{bmatrix} I \\ -\Theta_0^{-1}\Theta_1 H_1 \end{bmatrix} \tag{9.47}$$

where the $s-n$-dimensional vector $\bar{K}$ can be freely selected.

According to (9.41), (9.44), and (9.47), it follows that $HV_0 = U_0 H_1$ and then

$$HV(k) = U_0 H_1 \bar{K} + U^*(k)$$

Substitute the zero terminal constraint condition by (9.45) in the optimization problem OPA 2 and combine it with the above equation. The OPA 2 problem can then be rewritten as

$$\text{OPA 2:}\quad \min_{\bar{K}} J(k) = X^{\mathrm{T}}(k)QX(k) + U^{\mathrm{T}}(k)RU(k)$$
$$\text{s.t.}\quad U(k) = U_0 H_1 \bar{K} + U^*(k)$$
$$X(k) = Sx(k) + GU(k) = Sx(k) + G_1 H_1 \bar{K} + GU^*(k)$$

By solving it we can obtain

$$-\boldsymbol{H}_1^{\mathrm{T}}(\boldsymbol{G}_1^{\mathrm{T}}\boldsymbol{Q}\boldsymbol{G}_1+\boldsymbol{U}_0^{\mathrm{T}}\boldsymbol{R}\boldsymbol{U}_0)\boldsymbol{H}_1\bar{\boldsymbol{K}}=\boldsymbol{H}_1^{\mathrm{T}}(\boldsymbol{G}_1^{\mathrm{T}}\boldsymbol{Q}\boldsymbol{S}\boldsymbol{x}(k)+\boldsymbol{G}_1^{\mathrm{T}}\boldsymbol{Q}\boldsymbol{G}\boldsymbol{U}^*(k)+\boldsymbol{U}_0^{\mathrm{T}}\boldsymbol{R}\boldsymbol{U}^*(k)) \tag{9.48}$$

Left-multiply $\boldsymbol{H}_1^{\mathrm{T}}$ on both sides of (9.42) and then compare it with (9.48); it follows that

$$\boldsymbol{H}_1^{\mathrm{T}}(\boldsymbol{G}_1^{\mathrm{T}}\boldsymbol{Q}\boldsymbol{G}_1+\boldsymbol{U}_0^{\mathrm{T}}\boldsymbol{R}\boldsymbol{U}_0)(\boldsymbol{H}_1\bar{\boldsymbol{K}}-\boldsymbol{K})=\boldsymbol{0} \tag{9.49}$$

Then the following theorem can be obtained.

Theorem 9.2 Consider the zero terminal constrained predictive control with input dimension m and control horizon N, for its online optimization problem OP 2, an equivalent aggregation matrix always exists with the number of aggregation variables $s \geq n + m$.

Proof: Note that $\boldsymbol{H}_1 \in \mathrm{R}^{(Nm-n)\times(s-n)}$ and the similarity of (9.49) and (9.35). It is known from Theorem 9.1 that there must exist a matrix $\boldsymbol{H}_1$ such that the first m elements of $\boldsymbol{H}_1\bar{\boldsymbol{K}}-\boldsymbol{K}$ are identical to zero if $s-n \geq m$. Since

$$\begin{aligned}\boldsymbol{H}\boldsymbol{V}(k)-\boldsymbol{U}(k)&=\boldsymbol{U}_0\boldsymbol{H}_1\bar{\boldsymbol{K}}+\boldsymbol{U}^*(k)-\boldsymbol{U}_0\boldsymbol{K}-\boldsymbol{U}^*(k)=\boldsymbol{U}_0(\boldsymbol{H}_1\bar{\boldsymbol{K}}-\boldsymbol{K})\\&=\begin{bmatrix}\boldsymbol{I}\\-\boldsymbol{\Theta}_0^{-1}\boldsymbol{\Theta}_1\end{bmatrix}(\boldsymbol{H}_1\bar{\boldsymbol{K}}-\boldsymbol{K})=\begin{bmatrix}\boldsymbol{H}_1\bar{\boldsymbol{K}}-\boldsymbol{K}\\-\boldsymbol{\Theta}_0^{-1}\boldsymbol{\Theta}_1(\boldsymbol{H}_1\bar{\boldsymbol{K}}-\boldsymbol{K})\end{bmatrix}\end{aligned}$$

there must exist a corresponding $\boldsymbol{H}_1$ to realize an equivalent aggregation if the aggregation matrix $\boldsymbol{H}$ takes the form of (9.44) and $s \geq n + m$.

In the following an algorithm is given to calculate the equivalent aggregation matrix $\boldsymbol{H}$ for zero terminal constrained predictive control.

Algorithm 9.2 Calculation of the equivalent aggregation matrix for predictive control with zero terminal constraint

---

Step 1. Find the $\{i_1, \ldots, i_n\}$th columns of $\boldsymbol{W}_C = [\boldsymbol{A}^{N-1}\boldsymbol{B}\ \boldsymbol{A}^{N-2}\boldsymbol{B}\ \cdots\ \boldsymbol{B}]$ (the first m columns are not included), which are linearly independent and form the base of $\boldsymbol{W}_C$. Move them to the last n columns of $\boldsymbol{W}_C$. Calculate $\boldsymbol{U}_0$ according to (9.41).

Step 2. Use Algorithm 9.1 to form the matrix $(\boldsymbol{G}_1^{\mathrm{T}}\boldsymbol{Q}\boldsymbol{G}_1+\boldsymbol{U}_0^{\mathrm{T}}\boldsymbol{R}\boldsymbol{U}_0)$, and obtain the $(Nm-n)\times(s-n)$ dimensional matrix $\boldsymbol{H}_1$.

Step 3. Set the $Nm \times s$ dimensional matrix

$$\hat{\boldsymbol{H}} \triangleq \begin{bmatrix}\boldsymbol{H}_1 & \boldsymbol{0}\\ \boldsymbol{0} & \boldsymbol{I}\end{bmatrix}$$

Step 4. Adjust the last n rows of $\hat{\boldsymbol{H}}$ to its $\{i_1, \ldots, i_n\}$th row; the resultant matrix is $\boldsymbol{H}$.

---

(3) Quasi-equivalent aggregation of predictive control with input and state constraints

The above discussions on equivalent aggregation for unconstrained and zero terminal constrained predictive control are all based on an available analytical solution of the online optimization problem, which indicates that aggregation of OVs does not necessarily result in degeneration of the control performance because of the rolling optimization character of predictive control. However, the use of aggregation strategy is mainly for constrained cases where the online optimization has no analytical solution and the computational complexity becomes critical. In practical applications, in addition to commonly used constraints on control inputs and system states, sometimes artificial constraints such as terminal sets are needed to guarantee the stability of the predictive control system. The online optimization problem with these constraints is unable to be solved in an analytical way and it is impossible to analyze exactly whether the control inputs solved by optimization problems with and without OV aggregation are identical. In this case, the aggregation strategy can still be utilized, but focuses on making the current control solved by aggregation-based optimization close to the original one as far as possible. Thus this kind of aggregation is called quasi-equivalent aggregation.

Firstly consider the predictive control with terminal set constraints; the online optimization problem can be described by

$$\begin{aligned} \text{OP 3:} \quad & \min_{\boldsymbol{U}(k)} J(k) = \boldsymbol{X}^{\mathrm{T}}(k)\boldsymbol{Q}\boldsymbol{X}(k) + \boldsymbol{U}^{\mathrm{T}}(k)\boldsymbol{R}\boldsymbol{U}(k) \\ & \text{s.t.} \quad \boldsymbol{X}(k) = \boldsymbol{S}\boldsymbol{x}(k) + \boldsymbol{G}\boldsymbol{U}(k) \\ & \qquad \boldsymbol{x}(k+N) \in \boldsymbol{X}_f \end{aligned} \tag{9.50}$$

where $\boldsymbol{X}_f$ is a known set containing the origin, such as an ellipsoid or a polyhedron. For this predictive control problem, assume that the terminal states caused by optimization without and with OV aggregation are $\boldsymbol{x}_1$ and $\boldsymbol{x}_2$, respectively, $\boldsymbol{x}_1, \ \boldsymbol{x}_2 \in \boldsymbol{X}_f$ and $\boldsymbol{x}_2 = \boldsymbol{x}_1 + \Delta \boldsymbol{x}$. With these terminal states, similar to the zero terminal constraint case discussed above, when selecting the aggregation matrix $\boldsymbol{H}$ as (9.44), the same $\boldsymbol{U}_0$ and $\boldsymbol{V}_0$ as (9.41) and (9.47) can be obtained, respectively. However, different from (9.40) and (9.46), the corresponding particular solution becomes

$$\boldsymbol{U}^*(k) = \begin{bmatrix} \boldsymbol{0} \\ -\boldsymbol{\Theta}_0^{-1}\left(\boldsymbol{x}_1 - \boldsymbol{A}^N\boldsymbol{x}(k)\right) \end{bmatrix}; \quad \boldsymbol{V}^*(k) = \begin{bmatrix} \boldsymbol{0} \\ -\boldsymbol{\Theta}_0^{-1}\left(\boldsymbol{x}_2 - \boldsymbol{A}^N\boldsymbol{x}(k)\right) \end{bmatrix}$$

Denote

$$\boldsymbol{F} = \begin{bmatrix} \boldsymbol{0} \\ -\boldsymbol{\Theta}_0^{-1} \end{bmatrix}$$

Then $\boldsymbol{U}^*(k) - \boldsymbol{H}\boldsymbol{V}^*(k) = -\boldsymbol{F}\Delta\boldsymbol{x} \triangleq \Delta\boldsymbol{U}^*(k)$. After a similar deduction as in the zero terminal constraint case, it follows that

$$\boldsymbol{H}_1^{\mathrm{T}}\boldsymbol{U}_0^{\mathrm{T}}\left(\boldsymbol{G}^{\mathrm{T}}\boldsymbol{Q}\boldsymbol{G} + \boldsymbol{R}\right)\left(\boldsymbol{U}_0\boldsymbol{H}_1\bar{\boldsymbol{K}} - \boldsymbol{U}_0\boldsymbol{K} - \Delta\boldsymbol{U}^*(k)\right) = \boldsymbol{0}$$

For the above equation, Lemma 9.2 is unavailable to get the corresponding conclusion as in above case because the matrix $\boldsymbol{U}_0^{\mathrm{T}}\left(\boldsymbol{G}^{\mathrm{T}}\boldsymbol{Q}\boldsymbol{G} + \boldsymbol{R}\right)$ is not a square one. However, note that if $\Delta\boldsymbol{U}^*(k) = \boldsymbol{0}$, the above equation is exactly (9.49), where an equivalent aggregation is reachable as in the zero terminal constraint case. Therefore, if $\Delta \boldsymbol{U}^*(k)$ can be as small as possible, its influence on the solution of the above equation could be reduced, and the solution might be very close to that with an equivalent aggregation, i.e. achieving quasi-equivalent aggregation. Therefore, when selecting the base vectors in $\boldsymbol{W}_C$ to form $\boldsymbol{\Theta}_0$, $\|\boldsymbol{\Theta}_0^{-1}\|$ should be as small as possible, and thus $\Delta \boldsymbol{U}^*(k)$ will be as small as possible. In this way the influence of $\Delta \boldsymbol{U}^*(k)$ on $\boldsymbol{H}_1\bar{\boldsymbol{K}} - \boldsymbol{K}$ can be reduced.

With physical constraints on control inputs and system states, due to the appearance of inequality constraints, online optimization problem should be solved by QP or other nonlinear programming methods, for which the above analysis based on analytical solutions is no longer available. Note that in constrained predictive control, if the solution of unconstrained optimization does not violate the constraints, it is identical to the solution solved by programming algorithms. Therefore, the difference between these two only exists when the solution of unconstrained optimization violates the constraints, while this constraint violation case mostly only appears at the first several steps in the control process. Based on this consideration, the control variables at the first several steps can remain without aggregation and only the later control variables are handled by the equivalent aggregation method. According to the specific requirements of the optimization problem, first solve the equivalent or quasi-equivalent $Nm \times s$-dimensional aggregation matrix $\boldsymbol{H}_0$ by using the aggregation strategy for unconstrained, zero terminal constrained, or terminal set constrained cases, respectively, and then divide them into

$$\boldsymbol{H}_0 = \begin{bmatrix} \boldsymbol{H}_{01} \\ \boldsymbol{H}_{02} \end{bmatrix}$$

where the row number q of $\boldsymbol{H}_{01}$ corresponds to the number of control variables to be retained. Then construct the $Nm \times (s+q)$-dimensional quasi-equivalent aggregation matrix

$$\boldsymbol{H} = \begin{bmatrix} \boldsymbol{I}_{q\times q} & \boldsymbol{0} \\ \boldsymbol{0} & \boldsymbol{H}_{02} \end{bmatrix}$$

Note that now the number of aggregation variables has been increased from s to $s + q$.

For the quasi-equivalent aggregation mentioned above, the number of free variables q should be selected as small as possible to reduce the online computational burden if the control quality can be ensured. A design method was proposed in [15] to determine q based on simulation analysis.

Example 9.3 [15] Consider a continuous stirred tank reactor (CSTR). The nonlinear system is linearized at the steady state $T_s = 394$ K, $C_{As} = 0.265$ mol/l, where T_s is the reactor temperature and C_{As} is the liquid concentration in the reactor. Take the sampling period as 0.15 min; we then have the following discrete time state equation:

$$\boldsymbol{x}(k+1) = \begin{bmatrix} -0.2164 & -0.0123 \\ 98.1479 & 1.3210 \end{bmatrix} \boldsymbol{x}(k) + \begin{bmatrix} 0.0055 \\ -1.1434 \end{bmatrix} u(k)$$

where the control input $u(k)$ is the coolant flow and is constrained by $|u(k)| \le 1$ m^3/min.

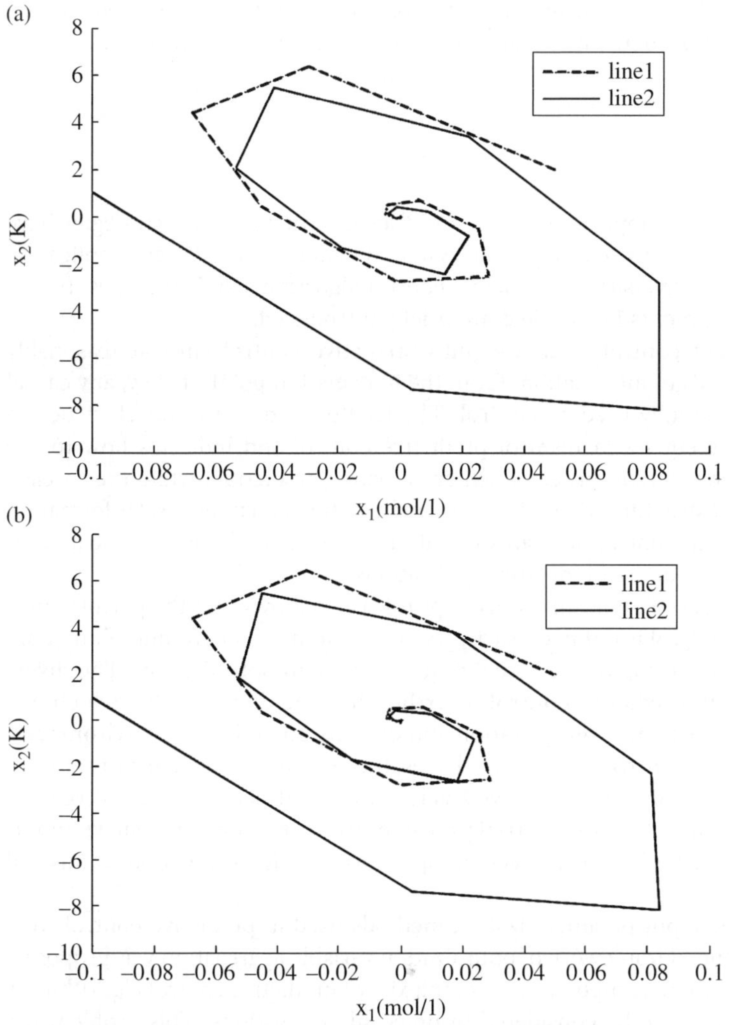

Figure 9.11 Comparison of predictive control without/with quasi-equivalent aggregation [15]. (a) Original predictive control without aggregation; (b) predictive control with quasi-equivalent aggregation.

In the optimization problem (9.50), set the optimization horizon $N = 20$, $\boldsymbol{Q} = \boldsymbol{I}, \boldsymbol{R} = \boldsymbol{I}$. In order to guarantee the stability of the controlled system, impose the terminal constraint $\boldsymbol{x}^{\mathrm{T}}(k+20)\boldsymbol{x}(k+20) \leq 1$. Using the algorithm in [15], the number of free variables in the quasi-equivalent aggregation can be determined as $q = 4$. The dimension of the OVs after aggregation is reduced from $Nm = 20$ to $n + m + q = 7$. In Figure 9.11, the equivalence of the designed controller and the original one is verified, where the dotted line and the solid line represent the state trajectories starting from different initial states $\boldsymbol{x}(0) = [0.05 \quad 2]^{\mathrm{T}}$ and $[-0.1 \quad 1]^{\mathrm{T}}$, respectively. It is shown that the control performance of the predictive controller with a quasi-equivalent aggregation strategy is almost the same as that of the original predictive controller. It obviously reduces the online computational burden while still guarantees the control performance.

9.5 Summary

During predictive control applications, various efficient structures and strategies have been developed that are oriented to specific system dynamics and different application scenarios, showing the diversity of predictive control algorithms and strategies. In this chapter, some developments in this field are briefly introduced.

Introducing efficient control structures into predictive control aims at reasonably handling different kinds of information. From the information point of view, any causal information should be fully used by control. The feedforward and feedback structure comes from the inherent mechanism of predictive control and indicates how to use the causal information fully in order to achieve a better prediction, while the cascade control uses a layered structure where the causal information and noncausal information are handled by different control mechanisms with respective emphasis. Both structures have a strong practical value in industrial applications.

Predictive control based on infinite norm optimization shows that the performance index in predictive control is not necessarily fixed as in some standard ones, but could be flexibly defined according to the actual requirements on optimization. The diversity of optimization in predictive control is further illustrated by CMMO, which was proposed to characterize the optimization pursued in an industrial environment. Referring to predictive control software packages, the concept and scheme of satisfactory control were proposed, taking CMMO as its core and describing predictive control implementation in complex industrial processes from an application viewpoint. It is also of reference value in many practical application fields other than industrial process control.

In this chapter, the input parameterization methods used in predictive control were discussed and the general framework of optimization variable aggregation was proposed. This is because the on-line computational complexity of predictive control algorithms is often an important issue to be considered in practical applications. This problem has been concerned from the appearance of predictive control, and some heuristic input parameterization methods have been proposed. The unified framework of optimization variable aggregation provides a more general way, not only by covering existing input parameterization methods but also for developing more new algorithms and strategies. In terms of the specific rolling style of predictive control, it is even possible to pursue equivalent or quasi-equivalent aggregation such that the control performance can be guaranteed while the complexity of online optimization is reduced.

References

1 Richalet, J., Rault, A., Testud, J.L. et al. (1978). Model predictive heuristic control: applications to industrial processes. *Automatica* 14 (5): 413–428.

2 Campo, P.J. and Morari, M. Infinite-norm formulation of model predictive control problems, *Proceedings of the 1986 American Control Conference*. Seattle, WA (18–20 June 1986), 339–343.

3 Campo, P.J. and Morari, M. Robust model predictive control, *Proceedings of the 1987 American Control Conference*. Minneapolis (10–12 June 1987), 2: 1021–1026.

4 Xi, Y. (1995). Satisfactory control of complex industrial process. *Information and Control* 24 (1): 14–20. (in Chinese).

5 Xi, Y. and Gu, H. (1998). Feasibility analysis and soft constraints adjustment of CMMO. *Acta Automatica Sinica* 24 (6): 727–732. (in Chinese).

6 Ying, C.M. and Joseph, B. (1999). Performance and stability analysis of LP-MPC and QPMPC cascade control systems. *AIChE Journal* 45 (7): 1521–1534.

7 Qin, S.J. and Badgwell, T.A. (2003). A survey of industrial model predictive control technology. *Control Engineering Practice* 11 (7): 733–764.

8 Ricker, N.L. (1985). Use of quadratic programming for constrained internal model control. *Industrial and Engineering Chemistry Process Design and Development* 24 (4): 925–936.

9 Reid, J.R., Chaffin, D.E., and Silverthorn, J.T. (1981). Output predictive algorithmic control: precision tracking with application to terrain following. *Journal of Guidance and Control* 4 (5): 502–509.

10 Yang, J., Xi, Y., and Zhang, Z. Segmentized optimization strategy for predictive control, *Proceedings of the 1st Asian Control Conference*. Tokyo (27–30 July 1994), 2: 205–208.

11 Richalet, J., Abu El Ata-Doss, S., Arber, C. et al. Predictive functional control: application to fast and accurate robots. *Automatic Control Tenth Triennial World Congress of IFAC* (ed. R. Isermann). Oxford: Pergamon Press, 1988, 4: 251–258.

12 Du, X. and Xi, Y. (2002). Aggregation optimization strategy in model predictive control. *Control and Decision* 17 (5): 563–566. (in Chinese).

13 Li, D. and Xi, Y. The general framework of aggregation strategy in model predictive control and stability analysis, *Proceedings of the 11th IFAC Symposium on Large Scale Complex Systems: Theory and Applications*. Gdansk, Poland (23–25 July 2007).

14 Li, D., Xi, Y., and Qin, H. (2007). An equivalent aggregation optimization strategy in model predictive control. *Acta Automatica Sinica* 33 (3): 302–308. (in Chinese).

15 Jiang, S., Xi, Y., and Li, D. (2009). Research of quasi-equivalent aggregation strategy in model predictive control based on simulation. *Journal of System Simulation* 21 (20): 6547–6551. (in Chinese).

10

Applications of Predictive Control

It is well recognized that predictive control has not resulted from the development of control theory, but was promoted by the requirements of complex industrial processes on optimization control. It appeared in the industrial field as a kind of new control technology from the date of birth. Early in the 1970s predictive control was successfully applied in process industries such as oil refinery, chemical, and power industries, and exhibited an excellent capability of handling complex constrained optimization problems. Since then, with the improvement and generalization of commercial predictive control software, predictive control has been successfully applied to thousands of large industrial facilities. It is widely accepted in the industrial process control field as the core of advanced process control (APC), and is even regarded as the only control technology that can handle the multivariable constrained optimization in a systematical and intuitive way. Since entering this century, with the increasing requirements on the constrained optimization control for complex systems, the application of predictive control rapidly expanded to advanced manufacturing, power electronics, energy, environment, medical, and other fields. In this chapter, we will firstly give an overview of predictive control industrial applications and introduce the key technologies of its implementation in industrial processes, before giving a brief overall description of its expansion to other application fields. Some application examples are also given to show the great potential of predictive control in solving constrained optimization problems of various fields.

10.1 Applications of Predictive Control in Industrial Processes

10.1.1 Industrial Application and Software Development of Predictive Control

Predictive control or Model Predictive Control (MPC) firstly appeared in industrial processes as a new kind of computer control algorithm. It is now commonly regarded that the successful application of predictive control in industrial processes in the 1970s is the symbol of its birth. The representative pioneer works include Model Predictive Heuristic Control (MPHC) proposed by Richalet et al. and Dynamic Matrix Control (DMC) proposed by Cutler and Ramaker. Richalet et al. first reported the principles of MPHC and the industrial applications of the predictive control software package IDCOM

Predictive Control: Fundamentals and Developments, First Edition. Yugeng Xi and Dewei Li.

(IDentification and COMmand) in the Fourth IFAC Symposium on Identification and System Parameter Estimation in 1976, and then published the first journal paper on predictive control in Automatica [1]. In the meantime, the Shell Company independently started to develop its own predictive control technology DMC in the early 1970s and got a preliminary application in 1973. Culter and Ramaker reported the principles of DMC and its application in the catalytic cracking unit of the oil refinery industry in the AIChE Meeting in 1979 and the JACC Conference in 1980 [2]. Based on these advanced technologies with successful industrial applications, predictive control began to appear in the field of industrial process control as a kind of new optimization control technology.

Over more than 30 years, predictive control has been widely used in industrial processes, and the commercial software package has also been continuously developed and improved. In a large number of the literature involving predictive control industrial applications, there were some survey papers that comprehensively and deeply introduced the application environment and status of predictive control, such as [3–6], etc. In [5], Qin and Badgwell reviewed the history of industrial MPC technology, surveyed the development and the technical details of main industrial MPC products, summarized industrial MPC application areas, and pointed out the limitations of current technology and development direction of industrial MPC. It is a classical literature for understanding the MPC technology in industrial applications.

The statistics analysis in [5] shows that the application of linear MPC technology covers a wide range of industrial fields, particularly in the process industry, such as oil refinery, petrochemical, chemical, and others. The listed number of application examples is continuously increasing. The scale of the problem to be controlled can reach hundreds of variables. However, compared with linear MPC technology, the maturity of nonlinear MPC technology seems low, and the variety of commercial software is very limited.

In industrial MPC applications, according to the needs of industry for optimization control, some professional software vendors investigated and developed commercial software with MPC as its core. They also proposed systematic solution for APC of large-scale industrial processes to promote the generalization of predictive control technology in industrial processes, which constitutes the mainstream of predictive control industrial applications, and the corresponding commercial software also represents the development level of industrial MPC technology. The evolution of industrial MPC software vendors and products is described in detail in [5], and the development of industrial MPC commercial software was divided into four stages as follows [5].

The first generation of the industrial MPC software is represented by IDCOM and DMC accompanied by the appearance of initial MPHC and DMC algorithms in the 1970s. These two algorithms are based on plant impulse and step response, respectively, and are quite different in implementation details. For example, IDCOM uses a reference trajectory while DMC does not. In IDCOM the input and output constraints are included in the optimization problem, which should be solved by heuristic iteration, while the early DMC was for the unconstrained case and the optimal control can be solved as a least squares problem. However, both of them embody the principles of predictive control and solve a finite horizon optimization problem online with a quadratic performance index. Their appearance marked the birth of predictive control technology and had a great impact on industrial process control.

The second generation of the industrial MPC software is represented by QDMC (Quadratic Dynamic Matrix Control) developed in the middle of the 1980s. It addressed the shortcomings of constraint handling in early algorithms by formulating the DMC algorithm as a quadratic programming (QP) problem in which input and output constraints appear explicitly. It provides a systematic way to implement input and output constraints. Successful applications in many industrial processes were reported.

The third generation of the industrial MPC software appeared around 1990, represented by IDCOM-M (a multivariable version of IDCOM) and SMCA (Setpoint Multivariable Control Architecture, a software integrating identification, simulation, configuration, and control all in one) developed by Setpoint Inc., HIECON (Hierarchical Constrained Control) by Adersa Inc., and SMOC (Shell Multivariable Optimizing Controller) by Shell Research in France, etc. This generation of MPC software made many improvements to MPC technology according to the needs of industrial application practice, such as distinguishing different priorities of constraints from hard to soft, providing some mechanism to recover to feasible when the optimization problem is infeasible, addressing the issues resulting from control structure variation in real time, providing a richer set for selecting the feedback, and allowing for a wider range of process models and controllers, and so on, thus making up for the shortcomings of the original technology.

The fourth generation of the industrial MPC software was launched in the middle of the 1990s after a series of mergers and acquisitions of MPC software vendors, represented by DMC-plus by Aspen Technology Inc. and RMPCT (Robust Model Predictive Control Technology) by Honeywell Inc., with the following features:

- Windows-based graphical user interfaces.
- Multiple optimization levels to address prioritized control objectives.
- Additional flexibility in the steady-state target optimization, including QP and economic objectives.
- Direct consideration of model uncertainty (robust control design).
- Improved identification technology based on prediction error method and sub-space ID methods.

In addition to the above products, there are also PFC (Predictive Functional Control) and GLIDE (an identification package) developed by Adersa Inc., SMOC-II by Shell Global Solutions, Connoisseur (control and identification package) by Invensys Inc., etc. They constitute the main market of current industrial MPC technology. In [5] the application areas of these products are also analyzed with statistics.

In order to provide a comprehensive solution for APC, these software vendors offer not only MPC control algorithms but also systematic software with all required functions including system configuration, model identification, system simulation, multivariable constrained control, steady-state optimization (SSO), process monitoring, human–computer interaction, etc. Furthermore, they also provide experienced professional teams to assist the users to complete the on-site implementation, make technical training, and provide customer service. Details on the features of these software products as well as their specific models, modeling and identification methods, and control techniques, etc., can be found in [5].

10.1.2 The Role of Predictive Control in Industrial Process Optimization

In the early literature of predictive control [1], Richalet et al. clearly pointed out the role of predictive control technology in large-scale industrial processes, which was also illustrated in later literature such as [5]. According to such literature, the optimization-based control of modern large-scale industrial processes is often based on the hierarchical structure shown in Figure 10.1, where the functions from top to bottom can be described as follows [5]:

Level 3. Global economic optimization. The production tasks are assigned to each production unit through a plant-wide optimizer, which determines optimal steady-state settings for each unit in the plant. The decision cycle is usually "day."

Level 2. Local economic optimization. According to the decision from the upper level, the optimal steady state with respect to the unit economic index is calculated with a more refined unit model and higher frequency and is sent to the next level in detail. It is usually performed by the unit computer with the cycle in the level of "hour."

Level 1. Dynamic constraint control. The multivariable system with constraints and uncertain disturbances is controlled dynamically, such that the system runs at the

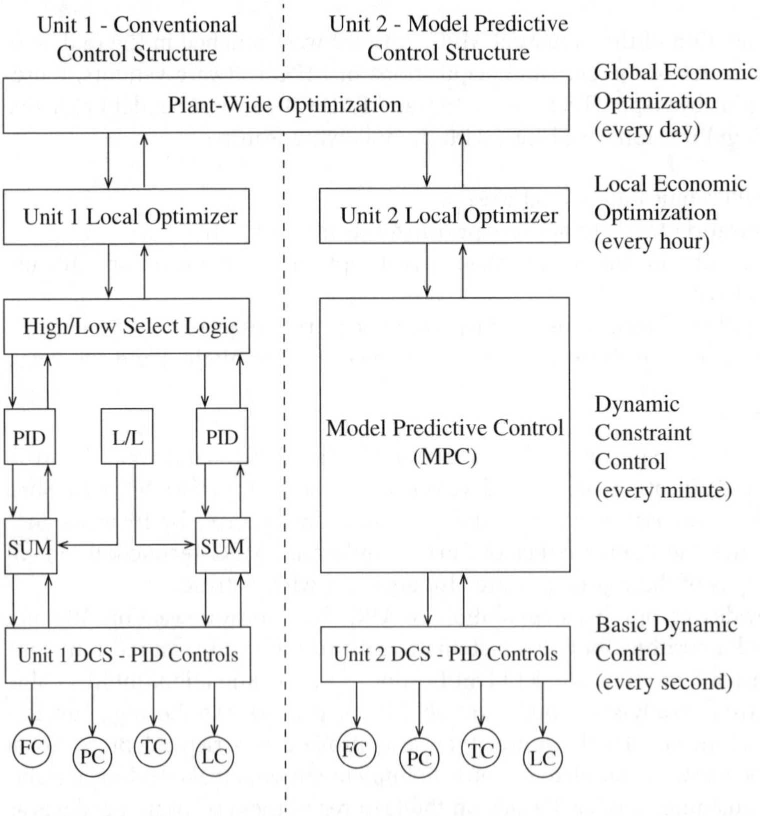

Figure 10.1 Hierarchy of control system functions in a typical processing plant [5]. (A conventional structure is shown on the left; an MPC structure is shown on the right.). *Source:* Reproduced with permission from S. Joe Qin and Thomas A, Badgwell of Elsevier.

desired steady state assigned by the optimization level while the probability for constraint violation is minimized. In a conventional control structure this is accomplished by using a combination of PID algorithms, lead–lag (L/L) compensations, and high/low select logic. In the APC systems, however, this combination is replaced by the MPC technology. It has a cycle of several "minutes."

Level 0. Basic dynamic control. This is mainly to ensure a quick tracking to the control command given by the upper level through closed-loop control of the supplementary system (such as the servo valve). Generally the PID control algorithm is adopted at a low level of the Distributed Control System (DCS). Its cycle is often taken at several "seconds."

As [1] pointed out, in this multilayer structure, the economic benefits induced by levels 0 and 1 are in practice usually negligible. In contrast, optimization at level 2 can bring valuable improvements in the economics of the systems. However. a necessary but not sufficient condition for optimization at level 2 to be satisfactory is firstly to have levels 0 and 1 optimized. This can be illustrated by Figure 10.2.

In Figure 10.2, the curve $J(q)$ characterizes the relationship between the economic performance index J (transformed to a minimization index) with the steady-state operating point q determined by the optimization level. The smaller the J, the higher are the economic profits. Due to the constraint on the operating point q, the ideal case is to set the steady-state operating point close to the constraint boundary q_0 such that minimal J or maximal economic profits can be achieved without violating constraints. However, this ideal "boundary control" is not realistic in the industrial environment where uncertain disturbances on states and structures exist. In this case, dynamic control cannot ensure the operating point strictly at its set-point. The actual system state will be distributed at a certain range around the set-point because of the uncertain disturbance. To ensure safety and product quality, the probability of the constraint violation must be lower than a certain threshold. If the control quality at the dynamic control level is bad, the variance of the actual state deviating from the set-point will have a wide distribution, as shown by curve 1 in the figure. Thus at the optimization level the steady-state operating point q_1 has to be set far from the constraint boundary, which will lead to a worse economic

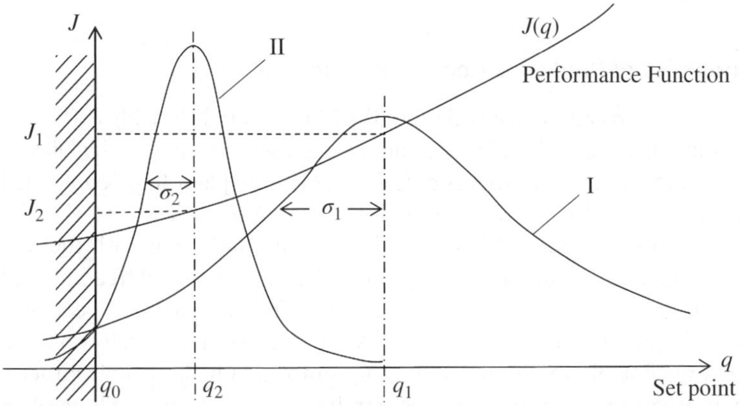

Figure 10.2 Effect of dynamic control quality to achieving "boundary control" [1]. *Source:* Reproduced with permission from J. Richalet, A. Rault, J.L Testud and J. Papon of Elsevier.

performance index J_1. In contrast, if strict dynamic control can be achieved at the dynamic control level, the control variance is small, as shown by curve II in the figure. Then at the optimization level, the steady-state operating point q_2 can be set more closely to the constraint boundary such that the "boundary control" could be approximately achieved. Under the precondition of safety and product quality, higher economic profits with performance index J_2 can be obtained.

It has been shown that the control quality at the dynamic control level plays an indirect but important role for improving the economic benefit of the industrial production process. Considering that predictive control usually uses the performance index of minimizing the output variance in optimization, its model-based prediction is helpful to including various practical constraints into optimization, and the mechanism of rolling optimization combined with feedback correction has a strong ability to reject disturbance, so it is suitable to be taken as the dynamic control algorithm at this level.

The application of predictive control at the dynamic control level can take two schemes [1]:

1) Direct Digit Control (DDC), i.e. the outputs of the predictive controller are the set points of cascaded PID controllers at level 0. The prediction model is in this case the process model.
2) Transparent Control (TC), i.e. at level 1, firstly using conventional analog PID controllers to improve the plant dynamics, and then using predictive control for the closed PID control loop to achieve high-quality dynamic control. In this case, the outputs of the predictive controller are the set-points of the PID controller at level 1. An overall model including the PID closed loops will be derived as the prediction model. This control scheme can also be regarded as using predictive control to dynamically adjust the set-points of the closed PID control systems. This operation scheme is much in favor because there are less risks and more tolerance.

No matter which kind of control scheme is adopted, the main task of the predictive controller at the dynamic control level is always to pursue strict dynamic control to reduce the variance of the controlled variables deviating from their set-points. It should be closer to the ideal "boundary control" and thus should provide a better basis for the economic optimization at level 2 to achieve more economic benefits.

10.1.3 Key Technologies of Predictive Control Implementation

In industrial practice, we often call the control input the Manipulated Variable (MV), the system output the Controlled Variable (CV) and the measurable disturbance the Disturbance Variable (DV). In terms of the numbers of CV and MV, the plant to be controlled can be divided into three cases: the "thin" plant with the MV number less than CV, the "square" plant with the equal number of MV and CV, and the "fat" plant with the MV number more than CV. In a traditional sense, they have different degrees of freedom and correspond to the overdetermined, well-posed and underdetermined cases, respectively. For a "thin" plant, it is impossible to control all the CVs to their set-points. However, as described by satisfactory control in Section 9.2.2, in the complex industrial environment, due to the existence of various soft and hard constraints as well as multiple control objectives, the optimization control problem has different features as compared with the traditional optimal control, and should be solved by CMMO (Constrained

Multiobjective Multidegree of freedom Optimization). In general, the task of a predictive controller is to receive the optimal steady-state operating point from the upper optimization level and to drive the process from the current steady-state operating point to the desired one. The requirements on the predictive controller are given as follows in order of importance [5]:

- Prevent violation of input and output constraints.
- Drive the CVs to their steady-state optimal values (dynamic output optimization).
- Drive the MVs to their steady-state optimal values using the remaining degrees of freedom (dynamic input optimization).
- Prevent excessive movement of MVs.
- When signals and actuators fail, control as much of the plant as possible.

The implementation of predictive control in industrial processes is not only to execute the algorithm but also to have a comprehensive procedure including system analysis, model identification, optimization problem formulation, constraints handling, and parameter tuning, etc. The common properties of industrial application software of predictive control can be summarized by:

- Face the multivariable control problem with constraints and multiple goals.
- Adopt at the dynamic constraint control level with the set-point of the lower level control loop as its MV.
- Mostly use an impulse or step response as the prediction model and further modeling for a parametric model is unnecessary,
- The obtained impulse or step response model should be tested, validated, and modified over and over, which is sometimes costly.
- Mathematical programming methods such as LP, QP, NLP are usually adopted in online optimization, also with additional know-hows.
- The software has three functions: model identification, off-line simulation and tuning, and online implementation.
- The user interface is friendly and easy to use for revising the constraints and adjusting their priority.
- It embodies the idea of satisfactory control and can flexibly handle the constraints and objectives with different importance.

In the following, referring to [5], we further illustrate some key techniques in industrial applications of predictive control.

(1) Identification techniques for the prediction model

Predictive control is a model based control and firstly needs to establish a suitable prediction model. With the DDC or the TC scheme, the MV of the prediction model is the input of the plant or the generalized plant, respectively, i.e. the set-point of the PID control loop, while CV is the plant output.

The prediction model can be either a first-principle model derived from theoretical analysis or an empirical model derived from process data. However, in industrial practice, the most widely used model for linear or nearly linear plants is the impulse response or step response model obtained using experimental data. According to [5], in order to get the process input and output data for identifying the process model, a test signal should be imposed on the MV in steady-state conditions to excite the variation of CV.

The usually used test signals include the Pseudo Random Binary Signal (PRBS) or PRBS-like stepping signal. In most industrial identification software packages, through one by one imposing the test signal on a single MV while keeping other MVs unchanged, the response data of all CVs to MVs can be obtained. In order to ensure sufficient excitation and to reduce the noise effect, it generally needs to make 8 to 15 step jumps for each MV in the test and the signal-to-noise ratio of CV to noise should be higher than 6. The data obtained from different test ranges need to be further integrated, and the data destroyed due to serious disturbance should be eliminated. The industry community regards this testing procedure as an important prerequisite for successful implementation of predictive control. It needs to be continued under the supervision of engineers for 5 to –15 days. During the test, in order to maintain the model fixed, the structure of the PID controller in the loop is not allowed to change or its parameters adjusted. The operator's intervention is only for an emergency case.

With the input and output data obtained from the process test, if PRBS is used as the test signal, the system impulse response and then the step response can be obtained using the correlation analysis method, while if the step test signal is used, the finite impulse response (FIR) model represented in increment form can be identified by the least squares method, and then the finite step response (FSR) model can be obtained. The dimensionality of the model is often selected from 30 to 120. Since the increment form is sensitive to high-frequency noise, the data should be smoothed before identification.

(2) Real-time implementation techniques of predictive control

At each control cycle, the typical steps of predictive control include:

1) **Read real-time data**. Real-time data of process inputs (MVs and DVs) and outputs (CVs) are firstly read through process measurement. In addition to their numerical values, the sensor status is also checked to see whether it is functioning properly or not. For each MV, the status of the associated lower level control and the valve will also be checked; if saturated then the MV will be permitted to move in one direction only. If the MV controller is disabled then the MV cannot be used for control but can be considered as a measured disturbance (DV).
2) **Output feedback**. Put the current prediction and optimization on the basis of real-time information of the process. Most industrial MPC techniques do not involve the concept of process state and cannot use the method of separately designing state estimation and state feedback control in state space. Aside from a few products exploiting Kalman filter technology for output feedback, most of them rely upon ad-hoc biasing schemes to incorporate feedback. For stable systems, the feedback is often obtained by comparing the measured outputs with the predicted ones, and the errors between them are used to correct future output prediction values, as mentioned in the feedback correction part of the DMC algorithm in Section 2.1. A feedback correction strategy is heuristic, depending on how to model the prediction errors. For example, when the error is caused by a continuous step disturbance acting on the plant output, the equal correction strategy can achieve good results. However, when the step disturbance actually appears at the plant input side, it may be rejected slowly. The discussion on different disturbance models and the corresponding feedback correction strategies can be referred to Section 3.3.1 or related literature.

3) **Determining the controlled subprocess**. Define the variables and constraints of the current control problem according to the actual operation status. During system analysis, the MVs, CVs, and DVs of the controlled system should be determined. Although some variables do not belong to the CVs such as the low-level control outputs (e.g. valve positions), in practice it is always needed to define them as the additional CVs due to the existence of the constraints. These CVs are then treated by range or zone control instead of set point control, with the purpose of keeping the lower level controllers away from valve constraints; otherwise the control performance might be degenerated.

   The process variables determined by system analysis may be changed during real-time operation. For example, if the lower level controller is saturated at a high or low bound, a temporary hard MV constraint needs to be imposed to prevent the MV moving in the wrong direction. If the lower level control function is disabled, the corresponding MV cannot be used for control and should be treated as a DV. Therefore, at each control execution, the system status should be checked and a subprocess for control needs to be defined.

   Sensor faults considered in most MPC products are limited to complete failure that goes beyond prespecified control limits. If a noncritical CV fails, the control action can be continued but the failed CV measurement should be replaced by the model predicted value. If the failure is maintained for a specified period of time, then this CV will be dropped from the control performance index. However, if a critical CV fails, the predictive controller will turn off immediately.
4) **Removal of ill-conditioning**. Avoid solving a singular control problem. If two controlled outputs respond in an almost identical way to the available inputs, to independently control them will need excessive control actions. This is called an ill-conditioned process. Its gain matrix has a high condition number, which means that small changes in controller error will lead to large MV moves.

   Although the ill-conditioning problem will certainly be checked at the design phase, it is still important to check it for the controlled subprocess at each control execution because the subprocess is updated with time. If necessary, ill-conditioning in the model should be removed. Two strategies are often used by MPC controllers, i.e. the singular value thresholding and the input move suppression. The former decomposes the process model based on singular values using URV decomposition. The singular values below a threshold magnitude represent directions along which the process hardly moves. They are discarded and then a process model with a lower condition number is reassembled and used for control. The latter directly reduces the condition number by increasing the magnitude of the diagonal elements of the matrix to be inverted in the least-squares solution.
5) **Local steady-state optimization (SSO)**. Determine the feasible steady-state setpoint that is closest to the desired one according to the actual process status. At each control cycle it is necessary to recalculate the steady-state input, state, or output targets because they may become infeasible due to disturbances entering the loop or an operator redefining the control problem. The optimization problem is typically formulated so as to drive steady-state inputs and outputs as closely as possible to targets determined by the local economic optimization level without violating input and

output constraints. The SSO uses a steady-state model and explicitly includes constant disturbances estimated through output feedback such that they can be removed. As described in Section 9.2.2, the SSO often uses LPs or QPs. For example, in DMC-plus, a sequence of LPs and/or QPs are used to solve the local SSO, where CVs and MVs are ranked by priority. Firstly, CVs with the highest priority are optimized subject to hard and soft output constraints put on them as well as all input hard constraints. Subsequent optimizations will be solved by preserving the future trajectory of high-priority CVs as equality constraints. Likewise inputs with a higher priority can be moved sequentially toward their optimal values when extra degrees of freedom exist.

During SSO, one can always find steady-state targets that are feasible with respect to the input constraints. However, the same is not true for output constraints because it may not be possible to completely remove the disturbance at a steady state by available control if a large disturbance enters the process. Therefore, soft output constraints should be introduced in the SSO that allow for some violation of an output constraint, but the violation is minimized in the performance index.

6) **Dynamic optimization**. Solve the optimal MVs for a determined optimization problem of the subprocess. When the controlled sub-process and the steady-state target are determined, the task of dynamic optimization is to calculate a set of MV increments that will drive the process to the desired steady-state operating point without violating constraints. This is usually attributed to solving a quadratic optimization problem under model equality constraints and other inequality constraints, where the performance index may include four penalized terms: deviations from the desired output trajectory, output constraint violations, future input deviations from the desired steady-state input and rapid input changes. Their relative importance in the performance index is controlled by setting various weighting matrices. Most of the MPC controllers use a quadratic performance index and solve the QP reliably using standard software. However, for very large problems, or very fast processes, fast sub-optimal algorithms are used to generate approximate solutions. For example in the DMC-plus algorithm, when an input is predicted to violate a maximum or minimum limit, then it is set equal to this limit and the optimization is repeated with this MV removed. While in the PFC algorithm, unconstrained optimal solution is firstly solved, and then clipped if it exceeds hard constraints. Both of these techniques can prevent violation of hard input constraints but will lead to a loss of performance.

For dynamic optimization, various industrial MPC algorithms have different strategies. The DMC-plus algorithm penalizes only the last input deviation large enough, which is approximately equivalent to having a terminal state constraint. A predefined output reference trajectory is adopted by some other algorithms to avoid aggressive MV moves. The RMPCT controller defines the optimization problem subject to the funnel constraints. In the case when the output error term in the performance index is vanish due to funnel formulation, the optimization will be only for the input term. In order to resolve conflicting dynamic control objectives, the HIECON algorithm solves CV and MV optimization problems separately. A quadratic optimization problem only with output error term is solved first. For the fat plants this will remain degrees of freedom to optimize the input settings. So a quadratic optimization problem only with input deviation term is again solved with a set of equality constraints that preserve the future output trajectories given by the output optimization. With this method it is not required to set weights to determine the trade-off between output and input errors, but at the cost of additional computation.

7) **Control implementation**. Act the solved MV to the process.

(3) Techniques in the optimization problem

In the optimization problem of predictive control at each sampling time, the following techniques are also involved by industrial predictive control algorithms.

1) **Constraint formulations**. There are three types of constraints commonly used in industrial MPC technology: hard, soft, and set-point approximation. Hard constraints should never be violated and must be imposed as the constraints of the optimization problem. Soft constraints allow some violation and the violation is typically minimized using a quadratic penalty in the performance index. Another way to handle soft constraints is to use a set-point approximation, i.e. define set-points for both limits of each soft constraint and put additional penalties on output deviations from these set-points in the performance index. However, the weight for this kind of output penalty should be adjusted dynamically because it becomes significant only when the CV comes close to the constraint limit. In a normal case when the CV is within the constraint limit, the weight for this additional penalty term is rather small and the steady-state target as the set-point becomes critical in the performance index. However, if a violation is predicted the corresponding weight should be increased to a large value so that the control can bring the CV back to its constraint limit.
2) **Output trajectories**. Industrial MPC technology provides four basic options to specify future CV behavior: set-point, zone, reference trajectory, or funnel. Zone control is designed to keep the CV within a zone defined by upper and lower boundaries. It can be implemented by defining upper and lower soft constraints or using the set-point approximation of soft constraints. The reference trajectory refers to a desired future path for each CV and is often assigned by a first- or second-order curve starting from the current CV value to the set-point with a specified time constant. Future CV deviations from the reference trajectory instead of the set-point are penalized in the optimization performance index. A funnel control summarizes the above two strategies. When the CV goes outside the zone, one can define a CV funnel with gradually contracting upper and lower bounds to bring the CV back within its range. The slope of the funnel is determined by a performance ratio, i.e. the desired time to return to the zone divided by the open-loop response time.
3) **Input–output horizon and input parameterization**. In the online optimization problem of industrial MPC algorithms, the prediction horizon (also called the optimization horizon) P and the control horizon M are generally used. The former refers to a finite set of future time intervals over which the future CV behavior is predicted and optimized, while the latter refers to a finite set of future time intervals in which the MV can be changed. In general, $M \leq P$, which are both the basic tuning parameters of predictive controllers.

   The prediction horizon P is in general set long enough to capture the steady-state effects of all future input variations. In most cases, the quadratic term in the performance index penalizing the deviations of future outputs from the set-point is calculated for all time points in the prediction horizon. However, it is also allowed to calculate this term only for a subset of these points. Particularly when the dynamic responses of various CVs are quite different, a separate set of time points in the prediction horizon can be defined for each output for calculating the corresponding penalty terms in the performance index.

The control horizon M determines the number of optimization variables in the optimization problem, and is thus closely related to the computational burden of online optimization. In order to reduce the computational burden, industrial predictive controllers use three different methods to parameterize the MV profile. Using a blocking technique the user can specify points on the control horizon where the control moves will not be changed. This reduces the dimension of the resulting optimization problem but at the expense of degeneration of the control performance. As its special case, taking a single future input move can greatly simplify the calculation but is only applicable in limited cases. The third one is to parameterize the future input profile using a set of polynomial basis functions as in the PFC algorithm. This allows a relatively complex input profile over a large control horizon to be specified by using a small number of unknown parameters. These methods have been introduced and discussed in Section 9.3.

In the above, the specific implementation techniques of predictive control in industrial applications have been introduced. For more details, readers can refer to [5] and other literature. Because of the function of model-based prediction and optimization, the ability to realize overall optimization for multivariable systems and the characteristics of directly and explicitly handling constraints in optimization, predictive control exhibits obvious advantages as compared with PID loop control in conventional control structures, particularly when applying it to the complex industrial processes with multivariable, constraints, and large time delays. In large industrial productions, predictive control is commonly recognized as the preferred algorithm at the dynamic constraint control level. Furthermore, in a large number of applications for individual unit processes or facilities, predictive control also achieved better results than a conventional PID controller because of its unique advantages in handling constrained multivariable optimization control. In recent years, the application of predictive control in large industrial processes is further extended to the local economic optimization level and combined with real-time optimization (RTO) so that the production process can always be optimized in a complex and changeable industrial environment.

10.1.4 QDMC for a Refinery Hydrocracking Unit

After generally introducing the industrial predictive control technology, in this section a hydrocracking reactor will be taken as an example to illustrate the application of predictive control in large industrial processes [7]. This example was reported in the 1980s using the second generation software QDMC, but is typical in understanding basic considerations and implementation details of industrial MPC technology when applying it to large complex industrial processes.

10.1.4.1 Process Description and Control System Configuration

In the petroleum industry, hydrocracking is a key process, which has an important impact on product generation. In the late 1970s, some companies started to use multivariable predictive control technology to control it. An early literature [7] reported the application of using QDMC to control hydrocracking reactors.

A significant portion of the synthetic crude charge to this refinery is processed by two parallel hydrocracking trains. Each train consists of two multibed reactors: a first-stage hydrotreater and a second-stage hydrocracker. Since the same control strategy is used for all four reactors, [7] focused on the first-stage reactor shown in Figure 10.3 to illustrate the predictive control design and implementation procedures.

The first-stage reaction is primarily a denitrification step, although a minimal amount of cracking does occur. In Figure 10.3, high-purity hydrogen from an on-site hydrogen plant is added to the hydrocarbon feed at high pressure after being heated in a furnace. The combined stream is reacted over a hydrotreating catalyst. Since it is an exothermic reaction, unheated hydrogen is introduced at the top of each bed as an intermediate quench to ensure that the reaction operates under the desired temperature. A quench stream on the first reactor bed provides fine-tuning of the reactor inlet temperature control.

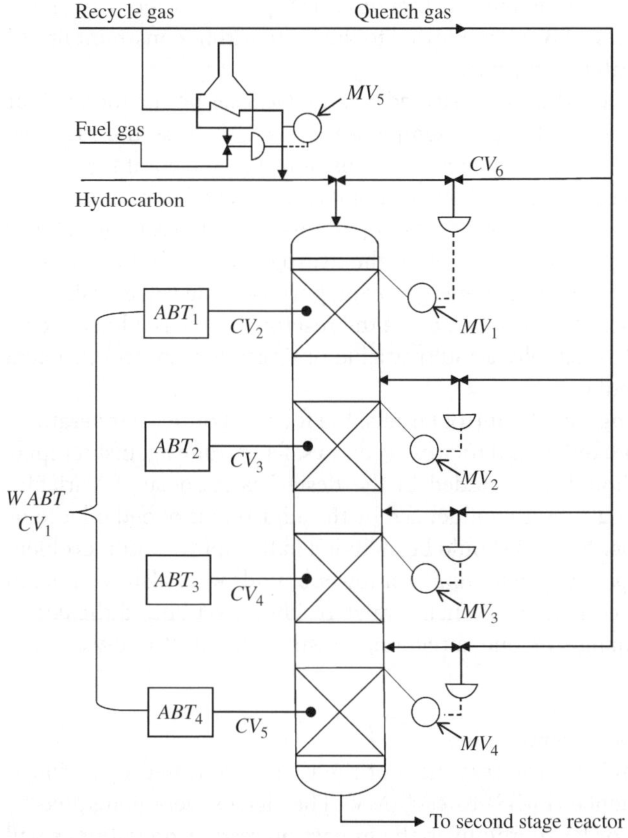

Figure 10.3 Control of the first-stage reactor [7]. *Source:* Reproduced with permission from S.J Kelly, M.D Rogers and D.W. Hoffman of IEEE.

The control system was designed according to the hierarchical structure described in Section 10.1.2. At level 0, regulatory control is performed by a FOXBORO SPECTRUM DCS. The reactor inlet temperature is controlled by fuel gas flow to the preheated furnace. Temperature controllers at the top of the reactor beds adjust the hydrogen quench valves. These valves provide safeguarding against temperature excursions in the reactor. Outlet temperatures and differential temperatures are measured for each reactor bed. Although the DCS system has controlled the temperatures at the reactor inlet as well as at the top of the reactor beds, reaction cannot be guaranteed to meet strict process conditions in the complex production environment. Therefore, at level 1, QDMC control is further implemented, aiming at obtaining high-quality dynamic control performance to ensure the reaction meets normal process conditions, thus improving productivity.

10.1.4.2 Problem Formulation and Variable Selection

The reaction severity of the process is measured by the Weighted Average Bed Temperature (WABT), which is the weighted average of the Average Bed Temperatures (ABTs) of the four beds in the reactor. For QDMC, the primary control objective is the regulation of WABT, while the secondary control objective is online selectable as either a bed temperature profile mode or an energy minimization mode. Under profile control, individual ABTs are controlled to a predetermined profile that optimizes product selectivity. In the energy minimization mode, these ABTs are allowed to fluctuate within constraint limits, while minimizing the feed heater firing rate.

For a QDMC application, five MVs were selected, i.e. the DCS set points for the four bed inlet temperatures and the reactor inlet temperature. Six CVs were the four bed ABTs, the reactor WABT, and the inlet quench valve position. The set-point for WABT is given according to process requirements. The set-points for each bed temperature can be calculated from the WABT set-point and a set of operator-entered bias targets. Since the reactor inlet quench indirectly reflects whether the hydrogen mixed with hydrocarbon feed into the reactor is overheated, for energy saving purposes, the valve position of the reactor inlet quench is also taken as a CV. Since the reactor WABT is a linear combination of the four ABTs, it is actually a multivariable optimization control problem with five inputs and five outputs.

In addition to the CVs having direct control targets, the four bed outlet temperatures should be maintained below a fixed bound to provide extra safeguarding against temperature excursions and thus should be included in the design as Associated Variables (AVs). Since bed outlet temperatures are influenced by the adjustment of bed inlet temperature set-points, these upper bounds should be included in the optimization problem as constraints. Quench valve positions with upper or lower bounds were also considered in the QDMC design as hard constraints. When a valve reached a set bound the corresponding MV could only change in one direction, causing QDMC to re-work the solution.

10.1.4.3 Plant Testing and Model Identification

In order to implement QDMC control, responses of the CVs were tested by adding a Pseudo-Random Binary Sequence (PRBS) to each MV. The signals were generated by the supervisory computer. In order to minimize the impact on reactor operation, small signals were used and tests were conducted over a 24 hour period. A time series analysis package was used to generate step response weights of CVs, as shown in Figure 10.4.

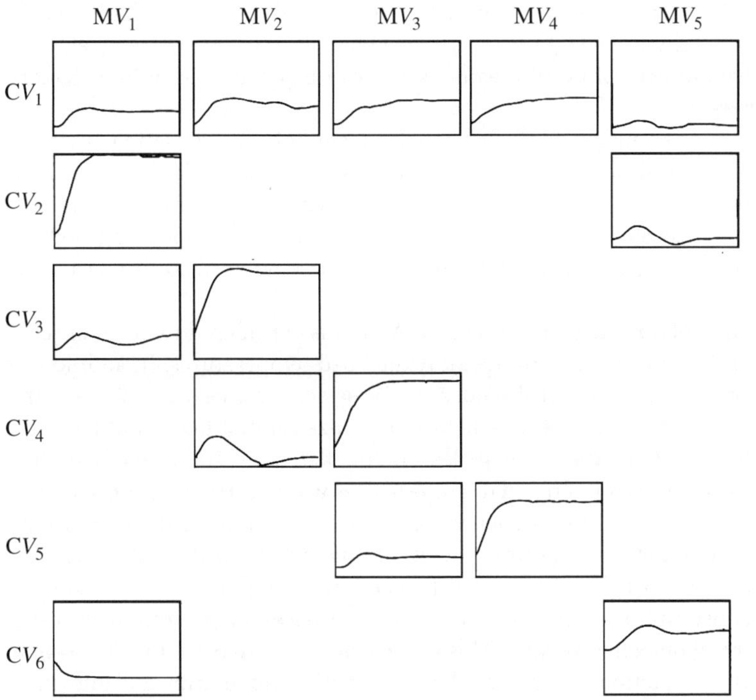

Figure 10.4 Dynamic step responses of CVs [7]. *Source:* Reproduced with permission from S.J Kelly, M.D Rogers and D.W. Hoffman of IEEE.

It can be seen from the figure that the set-point variation of each bed inlet temperature (MV1–MV4) has not only a direct influence on the ABT (CV2–CV5) of the reaction bed itself, but also a disturbance to the ABTs of the lower reaction beds (CV3–CV5), but these disturbances will tend to a zero steady state under the DCS closed-loop control at the bottom level. It is also found that it is difficult to express all of the dynamic responses in the figure as low-order models. However, QDMC can directly take the actual step weights as the elements of the dynamic matrix without identifying the minimum model. Furthermore, in order to ensure safe operation, four bed outlet temperatures have upper bounds and their future variations need to be predicted. Therefore the step responses of AVs to each MV should be obtained through a similar test procedure. The generated dynamic matrix is not shown here.

10.1.4.4 Off-Line Simulation and Design

Based on the above step response models, a PC-based off-line QDMC simulation package was made to test several aspects of the reactor control program. At each sampling time, this constrained multivariable DMC needs to solve an optimization problem similar to (5.18), which can be converted into the following standard QP problem:

$$\begin{aligned} &\min_{\Delta \boldsymbol{U}} J(k) = \frac{1}{2}\Delta \boldsymbol{U}^{\mathrm{T}} \boldsymbol{H} \Delta \boldsymbol{U} - \boldsymbol{g}^{\mathrm{T}} \Delta \boldsymbol{U} \\ &\text{s.t.} \quad \boldsymbol{C} \Delta \boldsymbol{U} \geq \boldsymbol{c} \end{aligned} \tag{10.1}$$

where $\boldsymbol{H} = \boldsymbol{A}^{\mathrm{T}}\boldsymbol{\Gamma}^{\mathrm{T}}\boldsymbol{\Gamma}\boldsymbol{A} + \boldsymbol{\Lambda}^{\mathrm{T}}\boldsymbol{\Lambda}$, $\boldsymbol{g} = \boldsymbol{A}^{\mathrm{T}}\boldsymbol{\Gamma}^{\mathrm{T}}\boldsymbol{\Gamma}\boldsymbol{e}$, $\boldsymbol{e} = \boldsymbol{w}(k) - \tilde{\boldsymbol{y}}_0(k)$, and $\boldsymbol{A}$ is the dynamic matrix. $\boldsymbol{\Gamma}^{\mathrm{T}}\boldsymbol{\Gamma}$ and $\boldsymbol{\Lambda}^{\mathrm{T}}\boldsymbol{\Lambda}$ are actually the weighting matrices $\boldsymbol{Q}$ and $\boldsymbol{R}$ in (5.18), respectively. Elements of them have intuitive physical meanings and can be off-line or online changed as tuning parameters.

This performance index includes all the penalty terms of CVs deviating from the desired values and all the suppression terms on MV increments, while the constraints give admissible variation ranges for MVs, CVs, AVs, and MV increments. The AVs are not controlled to a target but must be maintained within safe operating bounds. In the case when extra degrees of freedom exist, they may be used to drive the AVs toward their bounds.

As the primary control objective, the reactor WABT was considered to be an order of magnitude more important than any of the individual bed temperatures. In addition, as mentioned above, two different control modes can be selected as the secondary control objective. The profile mode would be selected to maximize catalyst life or influence the reactor yield pattern through a specific (e.g. descending, flat, or ascending) temperature profile. In this mode, the control of the bed temperatures was 5 to 10 times more important than control of the reactor feed quench valve. In the case of potentially serious consequences of high bed outlet temperatures, the weighting factors for these AVs were set 10–100 times higher than for WABT control. The energy saving mode was designed to minimize fuel gas consumption in the hydrogen preheat furnaces on the premise of ensuring the primary control objective WABT. This mode would take maximum advantage of the exothermicity of the reaction by reducing the heater outlet temperature and individual bed quenches. In this mode, the weight of each ABT would be set to zero. The reactor inlet temperature set-point was treated as a permanently constrained MV and would be slowly ramped down until a quench valve constraint was reached.

10.1.4.5 Online Implementation and Results

After off-line design and simulation, online control can be implemented. The supervisory computer calculates the individual ABTs and the reactor WABT online according to real-time information and specifies the set-points of ABTs according to the WABT set-point and a set of bed temperature biases entered by the operator. Online tuning of QDMC parameters, modifying the constraints limits, or selecting an optimal control mode are also available through a customized display.

Implementation of the QDMC strategies has resulted in a number of quantifiable benefits in the hydrocracker unit. The servo response is far more precise and rapid than operator control. Figure 10.5 includes first-stage reactor trends from a 30 hour period, during which time two WABT set-point changes were made using the QDMC control strategy. Figure 10.5a shows the smooth response of the reactor during the profile mode for two WABT set-point changes. The temperature profile of the reactor given by the ABTs slowly exhibits the desired state without disturbing the tight WABT control. Fuel gas consumption is substantially increased in this mode. Figure 10.5b shows a switch from the profile mode to the energy minimization mode with two WABT set-point changes. The servo response still seems to be good, while the fuel gas consumption has been consistently reduced by 25%. Additionally, reductions in reactor bed quench flows will permit higher unit throughputs because the unit is generally limited by the hydrogen supply.

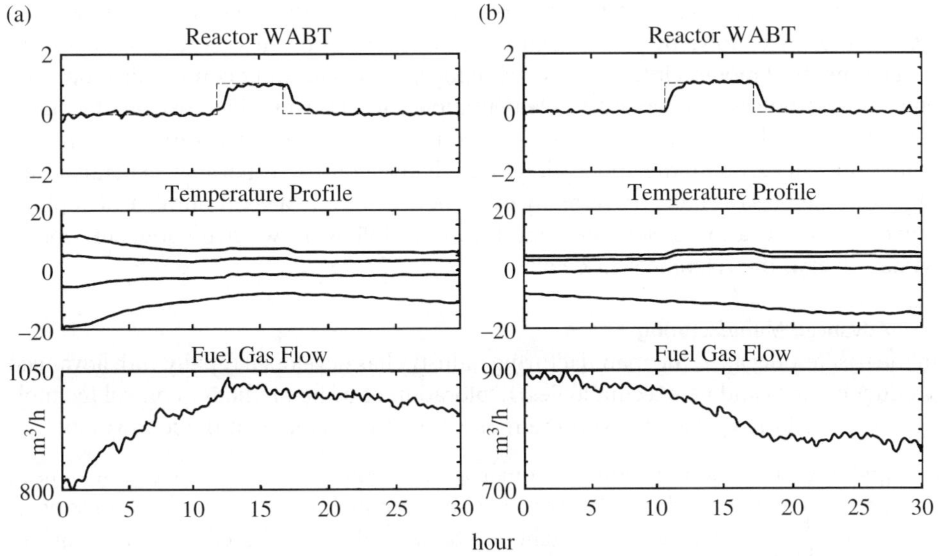

Figure 10.5 Control results by QDMC [7]: (a) profile mode and (b) energy minimization mode. *Source:* Reproduced with permission from S.J Kelly, M.D Rogers and D.W. Hoffman of IEEE.

10.2 Applications of Predictive Control in Other Fields

10.2.1 Brief Description of Extension of Predictive Control Applications

Predictive control deals with control problems of constrained systems with optimization purpose, which promotes the control forward from feedback stabilization to system optimization. A number of survey papers indicate that the most attractive feature of predictive control lies in its ability to explicitly handle constraints. This ability comes from the model-based prediction for the future behavior of the dynamic system. The constraints on future inputs, outputs, or states can be explicitly expressed and incorporated into the online optimization problem, which can be easily solved by mature software products. Successful applications on thousands of worldwide industrial facilities show that predictive control, as a practical constrained control approach, has been widely accepted by the industrial process control community.

Although predictive control emerged from the process industry and has been mainly applied to industrial processes for quite a long time, its ability to handle control problems with constrained optimization in uncertain environments has been fully exhibited and attracts researchers working in many fields other than industrial processes. Particularly since entering this century, with the scientific and technological progress and social development, the requirement for control is getting higher and higher. Rather than traditionally satisfying stabilizing design, control design is more about incorporating optimization so as to achieve a better control performance. In the meantime, optimization is restricted by more and more factors. Besides traditional physical constraints such as actuator saturation, various constraints brought about by technology, safety, economics (quality, energy consumption, etc.), and social factor (environment, urban management, etc.) indices should be incorporated. The contradiction between the higher requirements and the more complicated constraints motivates the exploration of using predictive control technology to solve these constrained optimization problems. Furthermore, recent developments of modern information technology, such as network communication, high efficient computing, embedded implementation, etc., also provide stronger technical support for predictive control implementation than was the case several decades ago. Under this background, the application of predictive control technology expanded rapidly from the process industry to many other new fields and a great number of application examples in various fields were reported. In the following, we give a brief description without listing the references in detail.

(1) Advanced Manufacturing

In the past few decades, the manufacturing industry has developed rapidly with advanced production lines and new technologies. Exploration of using predictive control technology in this field is very active with the motivation, for example, of the following:

- Higher requirement on production quality, such as nonlinear predictive control used in path following of industrial robots, PFC of a parallel robot for complex tracking in pick-and-place applications or machining tasks, PDE model based predictive control used in an autoclave molding process, etc.
- New production technology, such as predictive control in advanced combustion control, predictive control for desired 3D weld pool surface geometry in gas tungsten arc welding (GTAW), predictive control of thin film surface roughness and mean slope in thin film manufacturing etc.
- Mechanical and electrical facilities, such as aircraft gas turbine engine, boiler turbine, injection molding machine, diesel engine, overhead crane, etc.

(2) Power Electronics

There is a growing interest in the use of predictive control in the field of power electronics. The reason is, as indicated by [8], for electrical and mechanical systems in this field, very good mathematical models exist to predict the behavior of the variables under control. In addition, today's powerful microprocessors can perform the large amount of calculations needed in predictive control at a high speed and reduced cost.

The researches of predictive control applications in power electronics cover a wide range and can be referred to some review papers. Four main categories of predictive control applications for PWM power converters are presented in [8]: grid connected converters, inverters with RL output load, inverters with output inductor–capacitor (LC) filters, and high-performance drives. The predictive control applications in power electronics are classified in [9] according to the categories of power quality, machine drive, grid connected converters, and controllable power supply. Specific applications such as finite control set model predictive control (FCS-MPC) in power electronics were surveyed in [10], where the author concluded that MPC offers a new and very attractive alternative for control in power electronics and drives.

The technology readiness of predictive control in medium- and high-voltage power electronics is also assessed and it is strongly believed that model-based optimization techniques will play a pivotal role in the design and reliable operation of next generation power electronics systems. Although there are some factors that influence product development, such as implementation complexity, business value of the proposed solution, ease of maintenance, and skills and competence development, the last step of MPC to becoming mature is currently being done, and hopefully, just one step ahead, the MPC will be extensively applied to control complex electrical systems [8].

(3) Buildings

As [11] indicated, buildings are dynamical systems with several control challenges: large storage capacities, switching aggregates, technical and thermal constraints, and internal and external disturbances (occupancy, ambient temperature, solar radiation). Conflicting optimization goals naturally arise in buildings, where the maximization of user comfort versus the minimization of energy consumption poses the main tradeoff to be balanced. MPC is the ideal control strategy to deal with such problems.

Early in the middle 1990s, predictive control was applied to residential building heating systems. Since this century, with the modernization and intelligence of residential and commercial buildings, the applications of predictive control to buildings, particularly to HVAC (Heating, Ventilation, and Air Conditioning) systems, have been rapidly expanded, from individual facilities, such as boilers in heating systems, to overall buildings, such as energy efficient buildings with thermal storage systems.

Several review papers have given details of various explorations around the use of predictive control in buildings. It has been shown that predictive control may be used for local control loops at the low level, but is more often used for high-level implementations to improve building operation in tandem with an existing building automation system. Some critical issues of implementation are specially discussed, such as developing a model of the building suitable to predictive control, estimating the internal and external disturbances, resolving the conflicting goals, etc. It was also pointed out that predictive control in buildings is at present an individual design procedure for a specific building. Although there have been a number of reports of the control of single rooms, few results have been reported on real buildings of a significant scale.

(4) Critical Infrastructure Systems

The everyday life of modern societies relies heavily on desired operation of Critical Infrastructure Systems (CIS). The CIS are spatially distributed and of a network structure. Their dynamic models are often high dimensional and heterogeneous with nonlinearity, uncertainty, and several timescales. Both inputs and states or outputs are constrained. Hence, the CIS are large-scale complex systems with a variety of different objectives to be met under a wide range of operational conditions [12]. Due to the above characters, predictive control is extensively studied and applied to control and optimization of such systems.

In addition to a few early reports on predictive control application to CIS (for example, see Section 10.2.2), most of these explorations have appeared during this century due to the rapid development of network and communication technology, covering a wide range of different kinds of CIS, for example:

- Water supply systems, such as setting a dynamic target trajectory for each pool in irrigation canals and efficient management of a large-scale river.
- Drainage systems, such as multiple MPC for a drainage canal system, predictive control of urban drainage systems for flood prevention, and CSO (combined sewer overflows) reduction.

- Urban sewer system and sewage treatment, such as hybrid MPC of a sewer network minimizing flooding and maximizing sewage treatment and hierarchical predictive control for integrated wastewater treatment plant–sewer systems (IWWTS).
- Gas transportation network, such as linear MPC for an industrial-scale oxygen pipeline network and economic nonlinear MPC for periodic optimal operation of gas pipeline networks.

Research on integrated monitoring, control and security of CIS was surveyed [12] and a two-level hierarchical predictive control structure with different timescales was proposed. Taking a drinking water distribution system (DWDS) as an application case study, an integrated optimizing control of water quality and quantity was designed, where the upper level controller optimized the integrated quantity and quality, and outputted the operating schedules for the next 24 hours, while the lower level controller improved the quality control by adopting the robustly feasible MPC (RFMPC).

(5) Power Networks

In the last 10 years research reports of applying predictive control strategy to power networks appeared in rapidly increasing numbers. In addition to the large-scale property of the power networks, the new installation of ever growing renewable energy resources also brought new technological problems and strongly motivated the exploration of using predictive control strategies. Some concerned topics are given as follows:

- Large-scale power networks, such as distributed MPC for large-scale power networks and predictive control of voltage profiles in a medium voltage network with constraints on the voltage profiles and/or on the reactive power flows along the network.
- New energy resources and microgrids (MGs), such as predictive control for photovoltaic and wind energy systems, and economic/environmental operation management of MGs.
- Smart grids, such as hierarchical predictive control to balance grid load and consumers demand, event-driven MPC for managing charging operations of electric vehicles (EV), and scheduling for optimal power exchanges among MGs.

(6) Automobiles

The recent development of the automotive industry calls for a new generation of vehicles with intelligence, safety, and reduction of fuel consumption and emission. Particularly since the last decade, the exploration of applying predictive control technologies to unmanned ground vehicles (UGV) and new energy cars has been dramatically increased. For example:

- Active steering, such as predictive control of the vehicle front steering angle, model predictive active steering, and obstacle avoidance for UGVs.
- Adaptive cruise control (ACC), such as explicit MPC for a parameterized ACC of cars, and nonlinear MPC for controlling the throttle and brake to achieve smooth switching of ACC/CC.
- Hybrid electrical vehicle (HEV), such as predictive control used for boost converter, ultracapacitor manager and motor drive of a fuel cell HEV, nonlinear MPC to improve both the fuel economy and safety of Plug-in HEVs, and MPC for the power management strategy (PMS) utilized in HEVs.

(7) Autonomous Vehicles

The exploration of using a predictive control strategy to control vehicles appeared very early, just after its appearance in industrial processes, such as the terrain following tracking of an aircraft with nonminimum phase characteristics by using output predictive algorithmic control. It was firstly motivated by the requirement on a high precise control, but was later also used to solve complex tasks with optimization and constraints for different kinds of vehicles (see, for example, the survey paper [13]):

- Aircraft control, such as longitudinal axis control of a nonlinear F-16 aircraft, nonlinear MPC for a fixed-wing unmanned aerial vehicle (UAV), and active MPC for an unmanned helicopter in full flight envelope.
- Spacecraft control, such as MPC for magnetic attitude control, for all-electric spacecraft precision pointing, decentralized predictive control for spacecraft formation, and MPC for spacecraft rendezvous and docking.
- Underwater vehicle control, such as nonlinear MPC for tracking of autonomous underwater vehicles (AUVs), for dynamic positioning of an underwater vehicle, and depth and speed control of a hover-capable hybrid AUV.

(8) Medical Treatment Systems

There are a number of areas in the medical field where predictive control technology has been taken as an effective way to achieve a more accurate operation, optimal therapy as well as to guarantee the implementation of new therapy plans and devices.

- Clinical control, such as adjusting the infusion rates of nitroprusside and dopamine to simultaneously regulate mean arterial pressure and cardiac output in congestive heart failure subjects and adjusting the ultrafiltration rate (UFR) to regulate the relative blood volume (RBV) and heart rate (HR) during hemodialysis.
- Anesthesia control, such as on-line administration of anesthetic drugs during surgery, closed-loop regulation of hypnosis using isoflurane with bispectral index (BIS) as the CV, simultaneous control of muscle relaxation using the drug atracurium, as well as unconsciousness using the inhalational drug isoflurane.
- Glucose control of Type 1 Diabetes Mellitus (T1DM), such as optimal insulin delivery, artificial pancreas devices considering individualization (intersubject variability), and disturbance rejection (meal compensation).
- Others, such as teleoperated surgical robot for beating-heart surgery with motion prediction, and pharmacological therapy optimization for drug usage reduction while respecting safety and therapeutic constraints imposed by physicians.

Remark 10.1 There are also many other examples and more areas where predictive control application has been explored. It has been found that the applications reported in the literature were at different levels, many shown by simulations or real data-based simulations, some by individual experiment tests, and a few were applied to real plants. Compared with the predictive control applications in industrial processes, the technical maturity of predictive control in most other fields is still low and not at the commercial stage. Even for some fields such as power electronics, it still needs a step forward when becoming a systematic technique with commercial products. Nevertheless, the growing interest in various fields reflects the high expectation of people for this powerful control tool. In the following, we will introduce some application examples to show how predictive control is applied to solving different challenging problems in these fields.

10.2.2 Online Optimization of a Gas Transportation Network

In the last few decades, natural gas has become one of the main energy sources in the world. Compared with traditional solid and liquid fuels, natural gas has the advantages of low cost and little pollution, but is difficult to store because of its large volume. Therefore most of the natural gas is transmitted through the pipelines. In a large natural gas transmission network, in order to meet the needs of the end users and easy to transport, the natural gas needs to be maintained at a high pressure. Because the gas pressure drops during a long-distance transmission process, a number of compressor stations should be set in the transmission process to increase the gas pressure. Therefore, the transmission cost is very high and becomes a considerable part of the cost of natural gas.

In a large natural gas transmission network, the set-points of different compressor stations should be adjusted according to the actual network operation status. The main control objective is to maintain the transmitted natural gas at an appropriate pressure to meet the requirement of normal transmission and the user's demand. However, in order to improve economic efficiency, there is also a need to consider the objective of reducing the transmission cost. In the past with a manual operation, it was often difficult to balance both objectives. Usually the operator could only firstly ensure the user's demand and then use his experience to keep the energy consumption in a certain range as far as possible. The difficulty of the problem is that the natural gas transmission network is a high-dimensional nonlinear system with a large number of physical constraints and thus is very difficult to calculate. Furthermore, due to the small number of measurements and poor knowledge of the system states and external disturbances, off-line optimization for such systems is in practice unavailable. Therefore, investigating an online optimization strategy for such networked systems has become an urgent and challenging problem. In this section we will briefly introduce the predictive control strategy for online optimization of gas transportation networks presented earlier by Marques and Morari in [14].

10.2.2.1 Problem Description for Gas Transportation Network Optimization

For a long time people have made a lot of effort on modeling, simulation, and state estimation of natural gas transmission networks and have obtained considerable achievements. For example, the simulation software package GANESI and the state estimation software package developed by the Technical University of Munich in the 1980s have been successfully used by industry for off-line simulation and online estimation of the natural gas transmission network. These studies provide a solid basis for developing a predictive control strategy for gas transmission networks.

A natural gas transmission network is typically composed of sources, loads, and compressor stations, connected through the transmission pipelines. The source collects gases transmitted from different wells or other parts of the network, supplying the required gas at a constant and known pressure. Loads refer to customers, including civil and industrial users. They have different demands for pressure and the amount of natural gas. Their consumption of gas is quite random. Although it is possible to obtain some statistics on natural gas consumption according to a large amount of operation data, and to make forecasting for future load variations accordingly, this forecasting is often quite inaccurate. A compressor station is used to increase the gas pressure to make it transmitted normally and to meet the requirements of the users. It is an energy consumption unit. The power consumption for compressing the gas from the pressure P_s to the pressure P_d at a flow rate q is given by

$$HP = C_1 q(t)\left\{\left[\frac{P_d(t)}{P_s(t)}\right]^{C_2} - 1\right\} \tag{10.2}$$

where C_1, C_2 depend on the compressor and the gas pumped.

In the whole network, the most important component is the pipeline, which defines the major dynamic characteristics of the transmission process. The pipelines usually have a length from a few tens of km to more than 100 km, with a diameter from 300 to 1200 mm. For long transmission pipelines, let z represent the coordinate extending along the pipeline and t represent the time; the gas status at the pipeline can then be characterized by pressure $p(z, t)$ and mass flow rate $q(z, t)$, which are correlated with time and space. According to the continuity equation, the motion equation, and the gas state equation, the dynamics of the gas at the pipeline can be described by the following partial differential equations:

$$\begin{aligned} &\frac{\partial p}{\partial t} + \frac{b^2}{A}\frac{\partial q}{\partial z} = 0 \\ &\frac{\partial q}{\partial t} + A\frac{\partial p}{\partial z} + f\frac{b^2}{DA}\frac{|q|}{p}q = 0 \end{aligned} \tag{10.3}$$

where D and A are the diameter and cross-sectional area of the pipeline and b and f are parameters depending on the gas type and pipeline characteristics.

For such a networked system, its optimization problem can be described as follows:

1) The pressures of the source points are known, and the demands of the loads can be predicted online.
2) In each pipeline in the network, the pressure and the flow of high-pressure gas flow can be described and calculated by the aforementioned state equation.
3) There are a variety of constraints arising from physical limits or user demands. The physical constraints refer to the working range of the compressor restricted by surge, chock, and maximum and minimum speeds, which can be approximately described by the upper and lower bound constraints on the flow and the compression ratio for each compressor station [14]. Demand constraint refers to certain requirements on the gas pressure from users, and can usually be represented by the lower bound of the gas pressure at each load node.
4) On the premise of meeting the user's demand for gas, make an optimization for energy saving and consumption reduction with the performance index

$$\min J = \sum_{i=1}^{m}\int_{t_p}^{t_f} C_{1i} q_i(t)\left\{\left[\frac{u_i(t)}{P_{si}(t)}\right]^{C_{2i}} - 1\right\}dt \tag{10.4}$$

where m is the number of compressor stations and $[t_p, t_f]$ is the optimization time period, usually at the level of hours to dozens of hours. It can be partitioned into N time intervals with equal or unequal lengths, where the control input u_i (i.e. the discharge pressure P_{di} of each compressor station) remains constant at each time interval. The purpose of optimization is to find out the pressure setting P_{di} for each compressor station to minimize the above performance index under the physical constraints and the demand constraints. In the optimization process, the length of the optimization time period, the partition number N, and the length of each interval can be used as adjustable parameters.

For computer control, the model, the constraints, and the optimization performance index described in continuous form need to be transferred into discrete form through time (and space) discretization. For example, in the dynamic simulation program GANESI, each transmission pipeline is first divided into finite segments through space discretization, and the states of limited points representing these segments are used to characterize the state of the whole pipeline. In this way, the partial differential equations are transformed into ordinary differential equations. Using a numerical computation, the ordinary differential equations are further transformed into nonlinear algebraic equations through time discretization. Finally, the resultant nonlinear algebraic equations are solved iteratively. At each iteration step they are transformed into linear algebraic equations which are easy to solve. With the above model transformations, a GANESI software package can efficiently calculate the evolution process of gas states at the pipelines in the network.

10.2.2.2 Black Box Technique and Online Optimization

The above optimization problem can be attributed to solving the limited parameters of the optimal set-points for compressor stations at all the intervals in the prediction horizon. However, compared with predictive control for a single device or process in industrial processes, this optimization problem has a special complexity. Firstly, its model, constraints, and performance index are strongly nonlinear and cannot be solved analytically. Secondly, the model given by the software package GANESI can effectively calculate and forecast the future system states with a given initial state and boundary conditions. However, when using the usual optimization method, as indicated in Section 8.1, the whole discrete network model (including the nonlinear discrete dynamic models of all the pipelines, the correlation models between the pipelines, the relationship between power consumption and gas pressure, etc.) should be as equality constraints imposed into the optimization problem. Because of the large scale of the network, the huge number of equality and inequality constraints makes the optimization problem almost intractable by usual nonlinear optimization algorithms.

Although it is difficult to solve the optimization problem with the complex gas transmission network model using usual mathematical programming tools, note that the software package GANESI can effectively make computation and forecasting with this model. An optimization method proposed in [14] was based on a "black box" technique that takes the dynamic simulation program GANESI as a "black box" to predict the future performance and constraints, and then uses the gradient method to search for the optimal control actions iteratively. It starts from a given initial state and a group of assumed future control inputs $u_i(1), \ldots, u_i(N)$. The evolution process of the gas states in the network can be calculated using the dynamic simulation program GANESI according to the known boundary conditions (i.e. the gas pressures at the source points and the forecasted user's demands) from which the values of the objective function and the constraint functions can be obtained. Then perturb each control input and use GANESI to calculate the objective function and the constraint functions after perturbation, and then the gradient both of the objective function and of each constraint function to the control input can be estimated. Based on the values of these functions and gradients, the control inputs can be iteratively improved by the gradient optimization method until the optimal solution is obtained. This is a gradient optimization algorithm with $m \times N$ optimization variables. Since every control input needs to be perturbed for calculating the gradient, a large amount of simulation computations by GANESI are required. However, due to the high efficiency of this software package for model computation, the computational burden for this kind of optimization is still far less than that by directly solving it using the nonlinear programming method. This is the "combination of numerical computation and analytics" strategy of predictive control for nonlinear systems presented in Section 8.1.

The above optimization method can be used to calculate off-line the optimal control action. However, due to the inaccuracy of the model and the roughness of the user's demand forecasting, implementing the results from off-line optimization cannot guarantee optimal performance and constraint satisfaction. Therefore, closed-loop optimization should be adopted in predictive control instead of the above open-loop optimization. At each sampling time, feedback correction is first performed using real-time information. On the one hand, in order to reduce the influence of model uncertainty and disturbance, the estimated network states are updated by using observer technology according to current measured output information such that they can be close to the immeasurable real states and provide more exact initial conditions for optimization. On the other hand, the forecasting for future user's demands can be updated according to the errors between the measured actual demands and the forecasted ones, which can make the boundary conditions of optimization closer to actual scenarios. After feedback correction, the optimization problem can be solved by the above black box technique. Among the optimal solutions only the current control action needs to be implemented. This procedure is repeated at the next sampling time. Although the optimization problem at each time is solved as an open loop one, for the whole process closed-loop optimization is achieved because the feedback correction is embedded into the rolling process.

10.2.2.3 Application Example

The applications of the above online optimization strategy for some examples were investigated in [14]. As shown in Figure 10.6, a simple network is composed of one source, one load, two compressor stations, and three pipelines, where the gas transmission pipelines P1, P2, P3 have lengths of 80 km, 100 km, 100 km, respectively, and all with the diameter of 800 mm. Source A supplies the gas at a constant pressure of 5×10^6 Pa. Load B has a user demand for gas flow varying from $320 \sim 470$ km^3/h, with an average flow of 399 km^3/h and exhibits cyclic behavior statistically with a 24 hour period (not shown). The maximum and minimum flow rates for both the compressors K1 and K2 are 550 km^3/h and 150 km^3/h, respectively. The system is required to deliver the gas at a pressure of not less than 4×10^6 Pa.

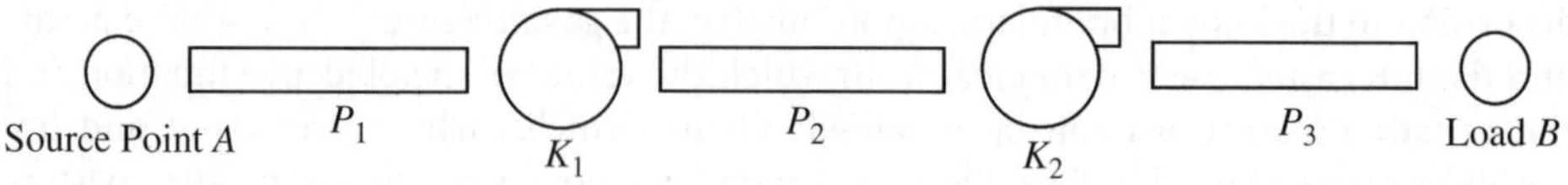

Figure 10.6 Example of a natural gas transmission network [14]. *Source:* Reproduced with permission from Dardo Marques and Manfred Morari of Elsevier.

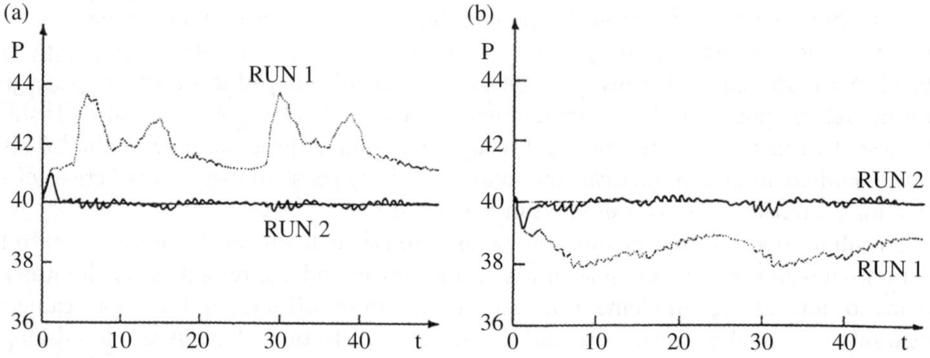

Figure 10.7 Delivery pressure at the load point B [14]: (a) real demand lower than forecasted one and (b) real demand higher than forecasted one. *Source:* Reproduced with permission from Dardo Marques and Manfred Morari of Elsevier.

Use predictive control with a rolling optimization strategy for the above system where the optimization horizon is divided into three intervals with two of 1 hour and the third of 2 hours. For comparison, two optimization schemes are adopted:

Scheme I: open-loop optimization. Only use the model and the off-line demand forecast.
Scheme II: closed-loop optimization. Use online error correction and prediction.

In the case when the demand forecast exceeds the real demand (real demand is 32 ∼ 47 km^3/h lower than the forecasted one, i.e. 10% prediction error), the delivery pressures at load point B with two schemes are shown in Figure 10.7a, where the fine line (RUN 1) and the rough line (RUN 2) represent the results from Schemes I and II, respectively. It is shown that both Schemes I and II can meet the requirements of the users for gas pressures. However, the pressure with Scheme I is greater than necessary for keeping the system in the feasible operating region. On the premise of guaranteeing the user's demand, Scheme II can make the total energy consumption of the two compressors reduce to 87.6% of that with Scheme I.

In the case when the demand forecast falls under the real demand (real demand is 32 ∼ 47 km^3/h higher than the forecasted one), the delivery pressures at the load point B are shown in Figure 10.7b. It is shown that since the off-line optimization in Scheme I does not sufficiently estimate the user's demand, and the real-time feedback mechanism is lacking, the delivery pressure at B is not enough to meet the requirements of the users for gas pressures. However, if scheme II with feedback correction is adopted, the online optimization can timely update the demand forecast through a closed-loop correction, and thus can still provide the natural gas at the required pressure to users.

The above results indicate that if rolling optimization is only based on the model and prior information, it is still an open-loop optimization. Only when the feedback correction is embedded into the rolling optimization procedure can the closed-loop optimization be realized through online correction, which can keep the system states in a feasible operating region and meanwhile reduce the energy consumption.

10.2.2.4 Hierarchical Decomposition for a Large-Scale Network

For a large-scale natural gas transmission network, optimization using the above method will produce a large computational burden. In order to make the online optimization feasible, [14] proposed a hierarchical decomposition method to reduce the computational complexity. At the upper level of the hierarchical structure, the source pressures and the demand forecasts for all the delivery points are used and the optimal discharge pressures (assume the control horizon to be one, i.e. only set once) are computed off-line for all network compressor stations. This optimization has in general a long period corresponding to the demand forecasting update frequency (e.g. 24 hours) and can be performed by the supervisory computer at the upper level. The scale of the problem is large, but does not affect the real-time performance due to the long period. At the lower level, online optimization control is performed only for such compressor stations that are directly linked to a delivery point (i.e. there is only a pipeline but no other compressor station between the compressor station and the delivery point), with a high rolling frequency (e.g. every hour) and a short optimization horizon (e.g. four hours), while keeping the discharge pressures of other compressor stations with constant values given by the upper level optimization. With a short horizon and less optimization variables, the online computational time of this small-scale optimization problem can be reduced to the level of a few minutes, which can satisfy real-time computing requirements as compared with the rolling period of one hour. If a large deviation is found for a demand forecast, to improve the energy savings, except for the compressor stations directly linked to delivery points, the compressor stations adjacent to them could also be put into online optimization. In this way, the energy saving can be improved because more power is involved in the optimization.

10.2.3 Application of Predictive Control in an Automatic Train Operation System

In modern cities, the subway system plays an important role in human daily life, where train control is a core technology for system operation. The traditional manual driving makes an interval driving plan and drives the train according to the traffic conditions, train operation status, as well as driving experience. Thus a modern automatic train control (ATC) system has become an integrated information system to help (or substitute) the driver control the train movements automatically and guarantee the safety and efficiency of train operation.

In general, ATC contains three subsystems [15], i.e. automatic train protection (ATP), automatic train operation (ATO), and automatic train supervision (ATS). These three subsystems work together to ensure the safe and efficient running of trains. ATS gives train routing and schedule adherence instructions to ATO according to the train state and schedule. ATO gathers the relevant information, such as train speed, programmed stop and dwelling time, and then decides the brake or accelerating rates of the train.

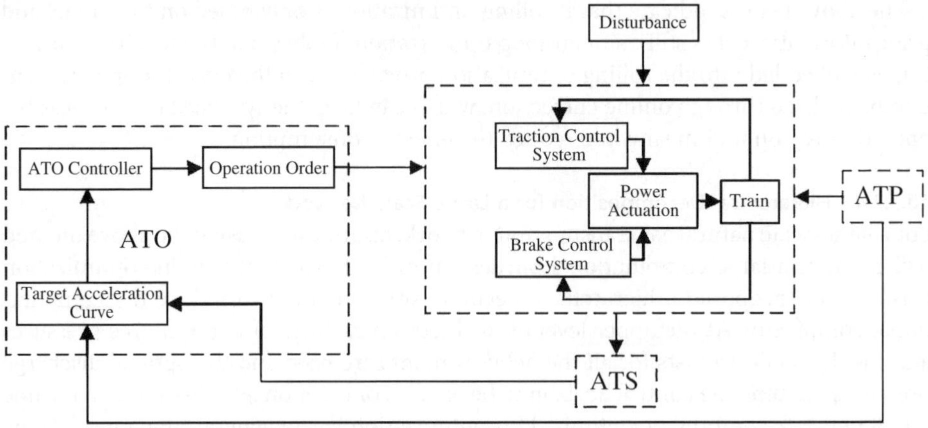

Figure 10.8 The structure of the automatic train control (ATC) system.

It focuses on train operational strategies that are directly related to the train operation efficiency. Meanwhile, ATP keeps monitoring the real-time train running status, and corrects the train operation commands or triggers the emergency brake if necessary. The structure of ATC with these three subsystems is shown in Figure 10.8.

ATO is responsible for all the train traction and braking control commands, and thus is a key to the operational efficiency and profitability of train operation systems. In general, ATO involves tasks at two levels. The task at the first level is to obtain the offline energy-efficient speed profile between successive stations while at the second level it tracks the optimal speed profile given by the first level. In this section, we only briefly introduce the application of predictive control in the ATO system to track a given speed profile of the train [16].

Usually the train runs in three different modes: traction, braking, and coasting. Because of the different dynamics of the train running in driving and braking modes, two control systems, i.e. the traction control system and the brake control system are already set up, respectively, with whatever quick tracking can be achieved. The train together with these two control loops forms a black box that receives the operation command and outputs the train state such as real acceleration $a(t)$ and speed $v(t)$. During manual driving, the driver only needs to give operation commands in the form

$$U_a(t) = F(d(t), b(t), p(t)) \tag{10.5}$$

where $d(t)$ is a (0, 1) variable, representing running traction control when $d(t) = 1$ and not running traction control when $d(t) = 0$; $b(t)$ has the same meaning but for running brake control. If both $d(t) = 0$ and $b(t) = 0$, the train will run in the coasting operation mode. $p(t)$ represents the driving or braking level given as a percentage from 0 to 100% with positive for driving and negative for braking, respectively. Note that if a command acceleration $a_c(t)$ is given and uses symbols with the above meaning, it is easy to convert it into a command $U_a(t)$ of (10.5).

Now consider using ATO to substitute for the driver. The ATO controller will face the above-mentioned black box, which is fixed and unknown, and should produce a command acceleration $a_c(t)$ and convert it into an operation command for the train system. This is very similar to the case in industrial applications where the level 0 is the basic PID control for dynamic tracking and the level 1 is the multivariable constrained predictive control, which provides the dynamic set-points for the low-level PID controllers.

In order to track the given speed profile more accurately and take constraints into account, the ATO controller uses a predictive control strategy. To do that a prediction model should first be established for the black box, which can be used to predict the future train states if the current state is known and the future control commands are given. Note that the black box will have quite different dynamics when the train runs in driving and braking operation modes. Thus two prediction models can be established for driving and braking operation modes separately, each with the command acceleration $a_c(t)$ as the input and the response acceleration $a_r(t)$ as the output, which approximates the real acceleration $a(t)$. In practice it has been found that for both models the input–output dynamic response can be described by a simplified first-order inertia model with a time delay as follows:

$$G(s) = \frac{a_r(s)}{a_c(s)} = e^{-\tau s}\frac{K}{Ts+1} \tag{10.6}$$

where K is the steady-state gain, T is the inertia time constant, and τ is the time delay. These parameters are different for the driving model and the braking model, and can be easily identified according to the running data of the corresponding operation.

According to the measured data from a real train, the simplified driving and braking models are established as follows:

$$\begin{cases} \text{Driving}: G(s) = e^{-0.6s}\dfrac{1.4}{0.4s+1} \\ \text{Braking}: G(s) = e^{-0.8s}\dfrac{1}{0.4s+1} \end{cases} \tag{10.7}$$

Due to the slopes of the train on the way, the real acceleration also contains an extra slope acceleration $\eta(t)$ in addition to the response acceleration caused by the traction or brake. Therefore the model acceleration is more accurately given by

$$a_m(t) = a_r(t) + \eta(t) \tag{10.8}$$

where $\eta(t)$ is known for a given train journey. Furthermore, the relationship between the model speed $v_m(t)$ and the model acceleration $a_m(t)$ is given by

$$\frac{v_m(s)}{a_m(s)} = \frac{1}{s} \tag{10.9}$$

Taking the model speed and acceleration as the state $\boldsymbol{X}_m(t) = [v_m(t) \quad a_m(t)]^{\mathrm{T}}$, according to the above equations, the prediction model can be written in the following state space form:

$$\begin{aligned} \dot{\boldsymbol{X}}_m(t) &= \begin{bmatrix} 0 & 1 \\ 0 & -\dfrac{1}{T} \end{bmatrix}\boldsymbol{X}_m(t) + \begin{bmatrix} 0 \\ \dfrac{K}{T} \end{bmatrix}u(t-\tau) + \begin{bmatrix} 0 \\ \dot{\eta}(t) + \dfrac{\eta(t)}{T} \end{bmatrix} \\ y_m(t) &= v_m(t) = [1 \quad 0]\boldsymbol{X}_m(t) \\ u(t) &= a_c(t) \end{aligned} \tag{10.10}$$

where $y_m(t)$ is the model output and $u(t)$ is the model input. This model can be used both as a driving model and as a braking model, but with different parameters. In the following, when designing the predictive controller, which model will be taken as the prediction model will depend on the operation mode of the train.

For sampling control the above continuous model (10.10) should be changed into a discrete-time model through discretization with the sampling period T_s. Furthermore, to reduce the possibility of a steady tracking error, it is transformed into an input incremental form as follows:

$$\begin{aligned}\tilde{\boldsymbol{x}}(k+1) &= \tilde{\boldsymbol{A}}_i\tilde{\boldsymbol{x}}(k) + \tilde{\boldsymbol{B}}_i\Delta u(k-n) + \tilde{\boldsymbol{J}}_i\tilde{D}_i(k) \\ y_m(k) &= \boldsymbol{C}\tilde{\boldsymbol{x}}(k+1)\end{aligned} \tag{10.11}$$

where $n = \tau/T_s$ is assumed to be an integer and $\tilde{\boldsymbol{x}}(k) = [v_m(k)\ \ a_m(k)\ \ u(k-n-1)]^{\mathrm{T}}$, $\Delta u(k-n) = u(k-n) - u(k-n-1)$, and $\boldsymbol{C} = [1\ \ 0\ \ 0]$. $\tilde{\boldsymbol{A}}_i$, $\tilde{\boldsymbol{B}}_i$, $\tilde{\boldsymbol{J}}_i$ are constants related to the model parameters, $\tilde{D}_i(k)$ is given by a known slope acceleration, where $i = 1, 2$ correspond to the driving operation and braking operation, respectively.

Referring to Section 2.3, based on the above state space model (10.11), a prediction model for future output can be established. At sampling time k, due to the existence of an n step time delay, only the future model outputs starting from $k+n+1$ need to be considered in optimization. They are predicted through two steps. Firstly, the state $\tilde{\boldsymbol{x}}(k+n)$ is calculated using the measured current state $\boldsymbol{x}(k)$, past control increments $\Delta u(k-1), \ldots, \Delta u(k-n)$, and known $\tilde{D}_i(\cdot)$. With $\tilde{\boldsymbol{x}}(k+n)$ as the initial state, the future outputs can be predicted by using (10.11)

$$\boldsymbol{Y}(k+n) = \boldsymbol{F}\tilde{\boldsymbol{x}}(k+n) + \boldsymbol{G}\Delta\boldsymbol{U}(k) + \boldsymbol{H}\boldsymbol{D}(k+n)$$

where

$$\boldsymbol{Y}(k+n) = \begin{bmatrix} y_m(k+n+1) \\ \vdots \\ y_m(k+n+P) \end{bmatrix},\ \Delta\boldsymbol{U}(k) = \begin{bmatrix} \Delta u(k) \\ \vdots \\ \Delta u(k+M-1) \end{bmatrix},\ \boldsymbol{D}(k+n) = \begin{bmatrix} \tilde{D}(k+n) \\ \vdots \\ \tilde{D}(k+n+P-1) \end{bmatrix}$$

and $\boldsymbol{F}$, $\boldsymbol{G}$, $\boldsymbol{H}$ are known matrices with proper dimensions. With known $\tilde{\boldsymbol{x}}(k+n)$ and $\boldsymbol{D}(k+n)$, $\boldsymbol{Y}(k+n)$ can be calculated according to the future control input $\Delta\boldsymbol{U}(k)$.

The predictive controller can be designed for the driving operation and the braking operation, respectively. At each sampling time k an optimization problem is formulated as follows:

$$\begin{aligned}&\min_{\Delta\boldsymbol{U}(k)} J = \|\boldsymbol{W}(k+n) - \boldsymbol{Y}(k+n)\|_{\boldsymbol{Q}}^2 + \|\Delta\boldsymbol{U}(k)\|_{\boldsymbol{R}}^2 \\ &\text{s.t. } \boldsymbol{Y}(k+n) = \boldsymbol{F}\tilde{\boldsymbol{x}}(k+n) + \boldsymbol{G}\Delta\boldsymbol{U}(k) + \boldsymbol{H}\boldsymbol{D}(k+n) \\ &\quad\ \ \boldsymbol{Y}_{\min} \le \boldsymbol{Y}(k+n) \le \boldsymbol{Y}_{\max} \\ &\quad\ \ \boldsymbol{U}_{\min} \le \boldsymbol{U}(k) \le \boldsymbol{U}_{\max} \\ &\quad\ \ \Delta\boldsymbol{U}_{\min} \le \Delta\boldsymbol{U}(k) \le \Delta\boldsymbol{U}_{\max}\end{aligned} \tag{10.12}$$

where $\boldsymbol{W}(k+n)$ is the P-dimensional target speed vector of the given speed profile at the sampling time $k+n+1$ to $k+n+P$. The cost function is composed of two parts, representing the requirements of accurate tracking (minimizing the tracking error) and comfort riding (reducing the jerk, i.e. the increment of acceleration), with $\boldsymbol{Q}$, $\boldsymbol{R}$ as the weighting matrices, respectively. The equality constraint represents the model output prediction, and other inequality constraints are bound restrictions for command accelerations and jerks as well as for model output speeds. As indicated above, $\tilde{\boldsymbol{x}}(k+n)$ appearing in (10.12) should be calculated based on the measured current state $\boldsymbol{x}(k)$ and other known information before optimization. Thus (10.12) is a typical QP problem and can be solved by a mature software package.

The model prediction and optimization discussed above can be used for both the driving mode and the braking mode with corresponding model parameters. In a real application a switching strategy can be designed to select the appropriate model for control. However, since the model prediction and optimization are across several future time instants in which different operation modes may occur, a multimodel predictive control strategy is proposed as follows.

At each sampling time k, solve the optimization problem (10.12) for the driving model and braking model, and obtain the current control input $u_d(k)$ and $u_b(k)$, respectively. Then, according to the target speed profile, find the target accelerations at the sampling instants in the optimization horizon $a_w(k+n+1) \quad \cdots \quad a_w(k+n+P)$ and take the real control input according to the following rules:

1) If the target accelerations are all zero, the train will run in coasting mode, $u(k) = 0$.
2) If the target accelerations are all positive or partly zero, the train should be accelerated. Thus it will run in the driving mode and $u(k) = u_d(k)$.
3) If the target accelerations are all negative or partly zero, the train should be braked. Thus it will run in braking mode and $u(k) = u_b(k)$.
4) If the target accelerations are some positive and some negative, it is the transition time between the driving mode and the braking mode. The real control input $u(k)$ can be calculated using the following equation:

   $$u(k) = \lambda u_d(k) + (1-\lambda)u_b(k), \quad \lambda \in [0,1]$$

   where λ is a factor to make the transition smooth. It varies from 1 to 0 when switching from the driving mode to the braking mode and from 0 to 1 when switching from the braking mode to the driving mode.

   After the current control input $u(k)$, i.e. the command acceleration $a_c(k)$ with a proper symbol is obtained, it can be easily converted into the command $U_a(t)$ to control the train.

This multimodel predictive control strategy has been used for the train system with the approximation models (10.7). Take the sampling period $T_s = 0.1s$. Select the optimization horizon $P = 10$, control horizon $M = 3$, and proper weighting matrices $\boldsymbol{Q}$, $\boldsymbol{R}$. In the multimodel strategy let $\lambda = 0.5$. The control result is shown in Figure 10.9. Compared with the curve by PID control, the resultant curve by multimodel predictive control is closer to the target curve. It demonstrates that the predictive controller can respond quickly and perform well when controlling time-delay systems.

The above application case study introduced predictive control into the ATO system of a train in a subway system. It should be pointed out that both the problem formulation and the control strategy of the predictive control for ATO are very similar to those of the

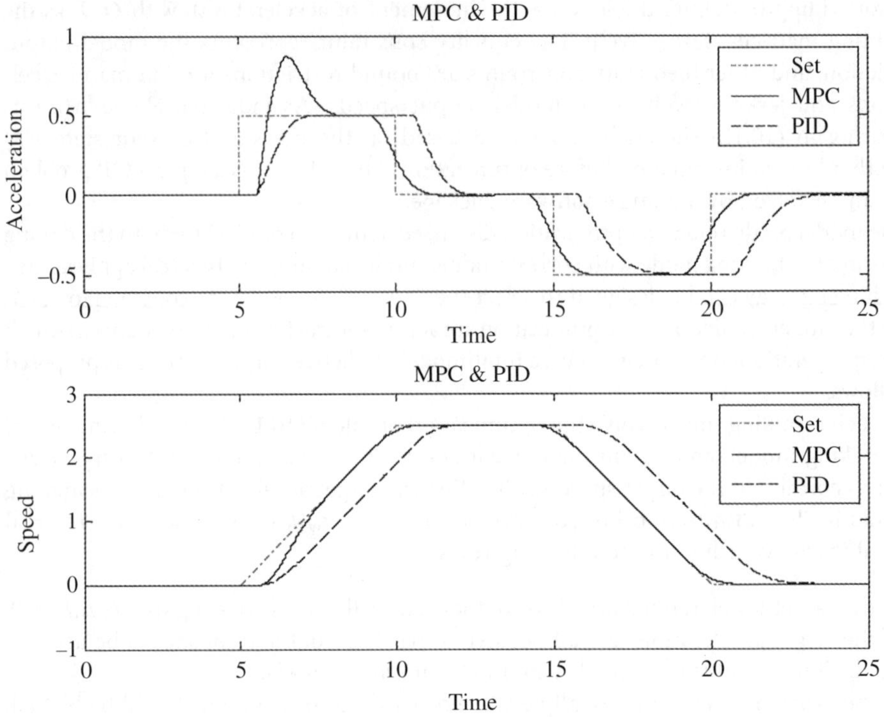

Figure 10.9 Comparison of predictive control with PID control for a train [16]. *Source:* Reproduced with permission from Yan Luo, Dewei LI and Yugeng XI of IEEE.

predictive control for the ACC system of unmanned cars or other vehicles, where the predictive controller in the outer loop determines a desired acceleration/deceleration profile for the vehicle, while the throttle and brake controllers in the inner loop ensure tracking of this profile. The readers can refer to relevant literature.

10.2.4 Hierarchical Predictive Control of Urban Traffic Networks

Due to the rapid development of society and the economy, traffic congestion in large-scale urban traffic networks becomes a growing problem all over the world. Although various control strategies with different traffic flow models have been developed for controlling the urban traffic networks, most of these works use a centralized control with detailed modeling such that there is little space for obtaining a trade-off between the modeling accuracy and the computational complexity. Since the urban traffic network consists of many links and signalized intersections and its scale becomes larger, hierarchical or distributed control is more tractable than centralized control for implementation in practice. In this section, we introduce a hierarchical predictive control approach for large-scale urban traffic networks [17] that can efficiently improve mobility and reduce traffic congestion.

10.2.4.1 Two-Level Hierarchical Control Framework

This hierarchical control architecture consists of two levels with one network controller at the upper level and several subnetwork controllers at the lower level, as shown in Figure 10.10. The optimization problem for a whole traffic network is assigned to

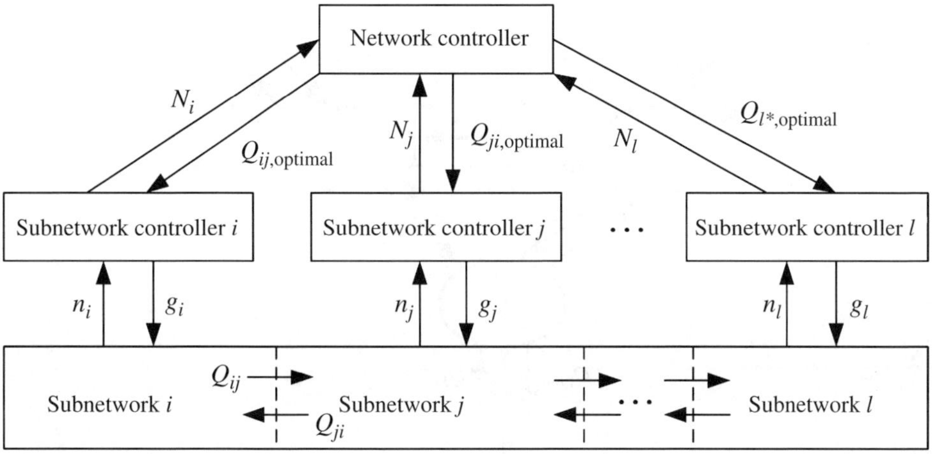

Figure 10.10 Hierarchy of predictive control for urban traffic networks [17]. *Source:* Reproduced with permission from Zhao Zhou, Bart De Schutter, Shu Lin and Yugeng Xi of IEEE.

different levels, in which different optimal control problems with specific tasks are solved using the predictive control approach.

To do that, a heterogeneous large-scale urban traffic network is first decomposed into several homogeneous subnetworks with an appropriate scale using some partition methods. At the upper level, the traffic status of the subnetworks is balanced from the network-wide point of view, where the Macroscopic Fundamental Diagram (MFD) is adopted to characterize the macroscopic characteristics of traffic status of the subnetworks. The traffic flows among the subnetworks are optimized using predictive control, driving the traffic status of each subnetwork to the unimpeded section in its MFD curve. Taking these optimal traffic flows between subnetworks as soft constraints, at the lower level, each subnetwork controller focuses on its own region to coordinate the intersection signals within the area using the predictive control approach. To make the whole system reach a better performance, information communication and coordination are necessary between the two levels. With this hierarchical control structure, the high-dimensional optimization problem for the whole traffic network can be decomposed into low-dimensional ones for small-scale subnetworks and are thus applicable to real traffic networks.

10.2.4.2 Upper Level Design

In order to coordinate the traffic flows among subnetworks (see Figure 10.11a), an aggregate traffic model that can macroscopically describe the dynamic behavior of the whole traffic system is needed. The concept of MFD provides a tractable way for this modeling. MFD was proposed in the traffic engineering area to describe the relationship between the number of vehicles (or densities) and the space-mean traffic flow in the network. All MFD curves have the same shape and a critical point. This point gives the critical number of vehicles and the maximum space-mean traffic flow. The left part of this critical point corresponds to the uncongested state, in which the average traffic flow is in a free flow status and increases with the number of vehicles until the critical point is reached. The right part corresponds to the congested status, in which the network becomes heavily congested with an increasing number of vehicles. For a fixed network and a given signal

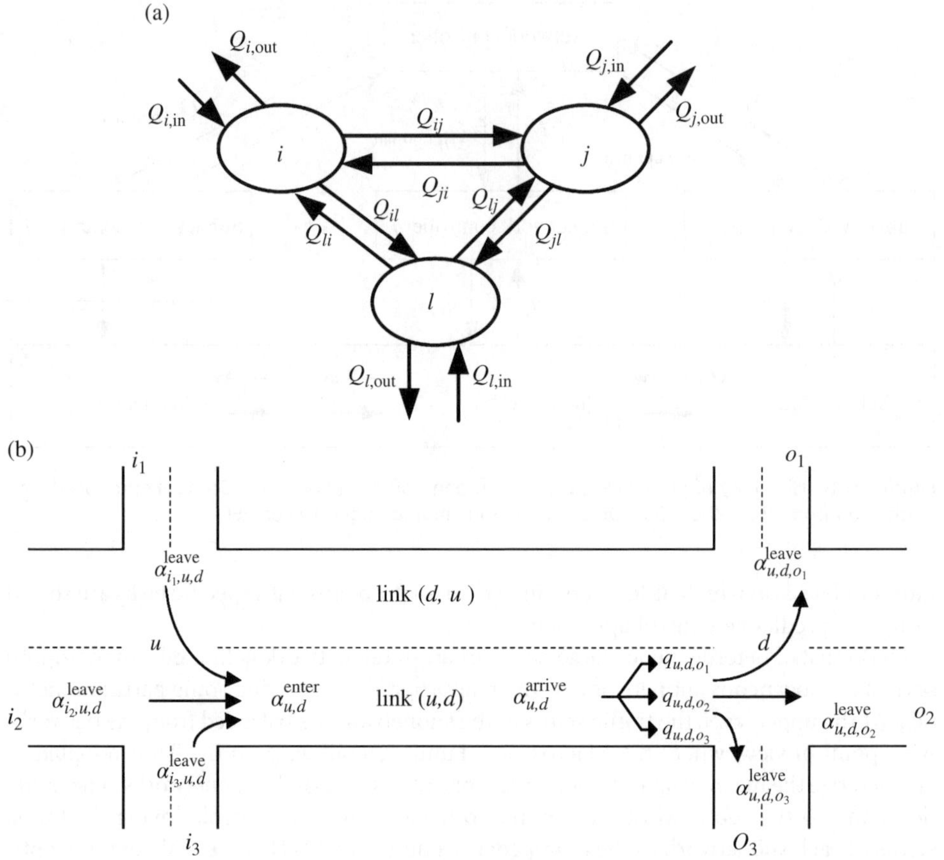

Figure 10.11 Modeling basis for predictive control at two levels [17]: (a) upper level and (b) lower level. *Source:* Reproduced with permission from Zhao Zhou, Bart De Schutter, Shu Lin and Yugeng Xi of IEEE.

timing plan, MFD can be obtained based on the traffic data by using some signal processing techniques.

Since the upper level design aims at improving the mobility and maximizing the throughput of the subnetworks, the objective function of the predictive controller is composed of two parts: minimizing the number of vehicles in the subnetworks to mitigate the traffic congestion and keeping the number of vehicles in all subnetworks below their critical points to reduce the risk of oversaturation. In order to predict the future vehicle number of the subnetwork, a simple conservation equation is established as the prediction model, which describes the dynamic evolution of the vehicle number of the subnetwork with its inflows and outflows, including the traffic flows between it and its neighbor subnetworks.

In this way, the upper level predictive control problem can be formulated as a rolling horizon optimization problem with vehicle numbers of all the subnetworks as the states and the traffic flows between subnetworks as the control variables to be optimized. The performance index at each sampling time is taken for future limited time intervals and for all subnetworks, composed of weighted TTS (Total Time Spent, proportional to the vehicle number) of the subnetwork and a penalty function for vehicle numbers exceeding the critical one, corresponding to the above-mentioned two parts, respectively. The constraints include the prediction model, the static function of the vehicle number with the space-mean traffic flow as described by the MFD curve, as well as various constraints on the traffic flows. The solved optimal traffic flows between subnetworks will be sent to the subnetwork controllers through communication as reference targets.

10.2.4.3 Lower Level Design

At the lower level of the hierarchical structure, each subnetwork controller coordinates the intersection signals within the area to regulate traffic flows and mitigate congestion. For this purpose, a more detailed urban traffic model for links and intersections is needed. Here the macroscopic simplified urban traffic model (S model) is adopted as the prediction model because it can describe the dynamic process of traffic flows in links and intersections in a more accurate way (see Figure 10.11b). For more details about the S model, readers can refer to [18].

Since the controllers in this level aim at optimizing the traffic flows and making the outflow of each subnetwork as close as possible to the optimal one provided by the upper level controller, the performance index over the prediction horizon is accordingly composed of two parts: TTS (proportional to the vehicle number) for all the links in the subnetwork and quadratic penalty terms constructed by the deviations of total outflows to its neighbor subnetworks from the corresponding desired ones given by the upper level controller. Combined with the prediction model and some constraints, this optimization problem can be solved by SQP or, after further decomposition, can be solved by the distributed multiagent predictive control approach, as presented in [17], where the subnetwork is decomposed into several subregions and the interactions between subregions are as constraints included in the optimization problem of each subregion and handled by the multiagent coordination approach.

10.2.4.4 Example and Scenarios Setting

Reference [17] studied the application of this two-level hierarchical predictive control strategy to a middle scale urban traffic network, shown in Figure 10.12. It has 55 nodes including 21 source nodes providing traffic demands and 34 intersections controlled by traffic signals, as well as 154 two-way links with different lengths (from 216 to 366 m). All the links have two lanes.

The entire traffic network is first divided into three subnetworks by using some partition method, as shown in Figure 10.12. At the upper level, for each subnetwork five MFDs under different signal timing plans of undersaturated, saturated, and oversaturated traffic situations are obtained. A fifth-order polynomial function is then derived to describe the averaged nonsymmetric unimodal MFD curve of each subnetwork, which will be used as a nonlinear function in the constraints in upper level optimization. At the lower level, in order to implement distributed predictive control, each subnetwork is partitioned into two subregions of the same size. The parameter setting for cycle times, tuning parameters in control, weights and constraint bounds in optimization, etc., can be found in [17].

Two scenarios with different traffic demands (i.e. the traffic network input flow rates) are considered. Scenario I corresponds to the oversaturated traffic condition with an increasing, high traffic demand from 2000 to 3000 vehicles/h, while Scenario II describes

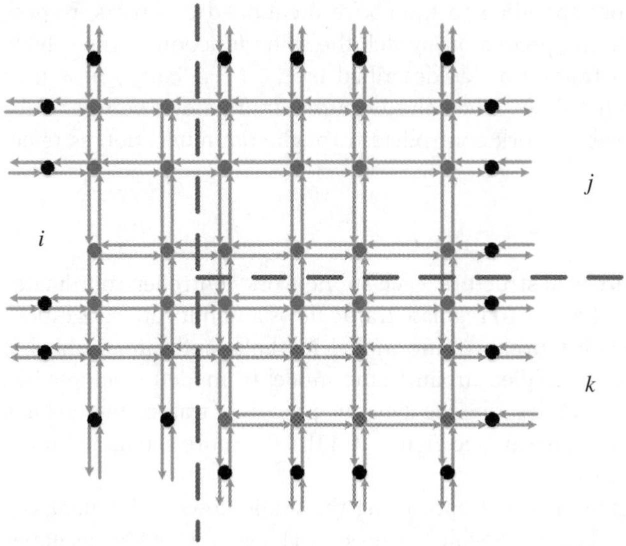

Figure 10.12 Urban traffic network studied in [17]. *Source:* Reproduced with permission from Zhao Zhou, Bart De Schutter, Shu Lin and Yugeng Xi of IEEE.

a peak hour situation with the traffic demand increasing from 800 to 1500 vehicles/h and then decreasing back to 1000 vehicles/h.

For comparison the following four control methods are adopted:

- Fixed-time control. The green time split is predefined with the green time durations 30 and 30 seconds for the two phases of all the intersections.
- Centralized predictive control. The whole urban traffic network is controlled without decomposition. The S model is utilized as the prediction model and the TTS in the network is minimized at each control step.
- Decentralized predictive control. The whole urban traffic network is decomposed into three subnetworks. Each subnetwork adopts a centralized predictive control strategy independently without information (traffic demands) from its neighbors.
- Two-level hierarchical predictive control.

For evaluating the control performance, four estimation criteria are considered:

- TTS (total time spent). The accumulated amount of the TTS by all the vehicles inside the traffic network for total running time (5400 seconds).
- TDT (total delay time). The difference between the total travel time of all vehicles inside the traffic network and the total free-flow travel time.
- Weighted average flow, which represents the mobility of the traffic network.
- Number of vehicles in each subnetwork.

10.2.4.5 Results and Analysis

To evaluate the effectiveness of the hierarchical predictive control strategy, the TTS and TDT results of an entire traffic network for all control approaches in the two scenarios are shown in Figure 10.13 and Table 10.1. It can be seen that compared with fixed-time

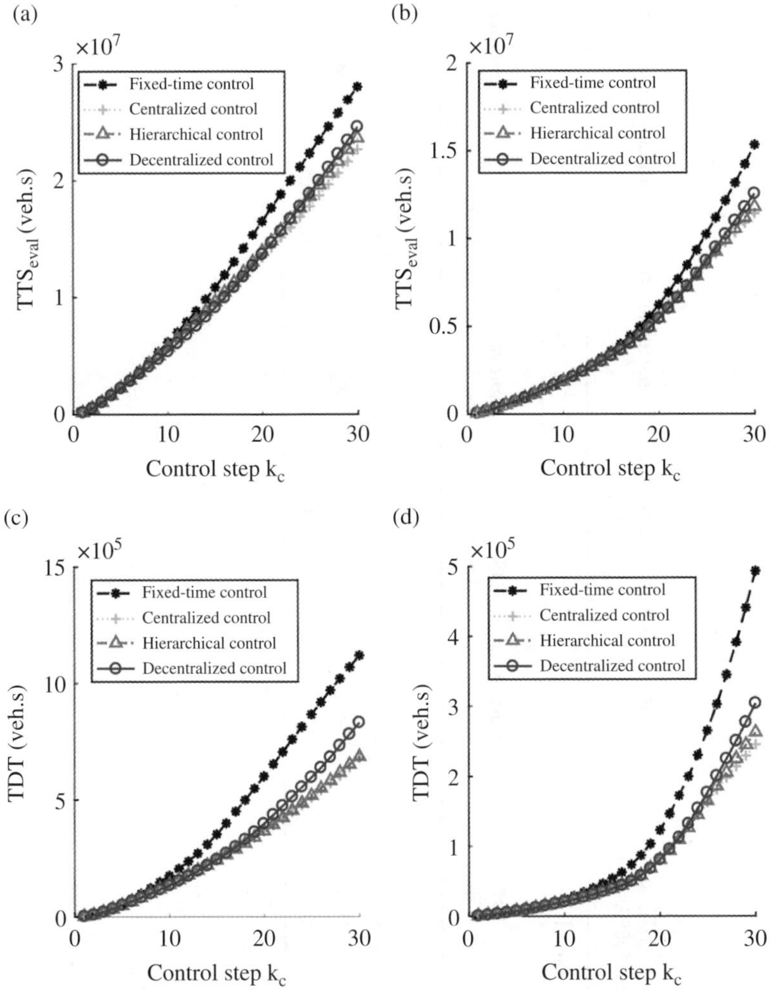

Figure 10.13 Total time spent (TTS) and total delay time (TDT) comparison for all control approaches [17]: (a) and (c) Scenario I and (b) and (d) Scenario II. *Source:* Reproduced with permission from Zhao Zhou, Bart De Schutter, Shu Lin and Yugeng Xi of IEEE.

control, three other control approaches yield a significant decrease in the TTS. The centralized control exhibits a better performance than the other two control approaches when the traffic situation becomes oversaturated after the 20th control step. The hierarchical control is slightly better than the decentralized control with a smaller average difference to centralized control. From Figure 10.13b, the improvement in TDT by using hierarchical control compared with the decentralized control is more obvious, especially in the oversaturated case. This means that the hierarchical control can improve the mobility in the network because the local predictive controllers under the guidance of the upper level coordination are capable of obtaining information from neighbor subnetworks through communication. The results illustrate that the hierarchical control can approximate the performance of centralized control and is able to reduce the TTS and the TDT more than the decentralized control.

Table 10.1 Total time spent (TTS) and total delay time (TDT) for all control approaches in the two scenarios [17]. *Source:* Reproduced with permission from Zhao Zhou, Bart De Schutter, Shu Lin and Yugeng Xi of IEEE.

	Scenario I				Scenario II			
Control approach	**TTS (Veh.s)**	**Improve (%)**	**TDT (Veh.s)**	**Improve (%)**	**TTS (Veh.s)**	**Improve (%)**	**TDT (Veh.s)**	**Improve (%)**
Fixed time	2.80×10^7	–	11.2×10^5	–	1.53×10^7	–	4.93×10^5	–
Centralized	2.26×10^7	19.3	6.81×10^5	39.2	1.14×10^7	25.5	2.45×10^5	50.3
Decentralized	2.46×10^7	12.1	8.31×10^5	25.8	1.26×10^7	17.6	3.04×10^5	38.3
Hierarchical	2.36×10^7	15.7	6.87×10^5	38.7	1.18×10^7	22.9	2.62×10^5	36.9

The hierarchical predictive control is able to balance the traffic demand and to coordinate the traffic flow, which can be inferred from the weighted average flows and the numbers of vehicles in the subnetworks. Figure 10.14a to c and d to f give the weighted average flows of all the subnetworks for the two scenarios, respectively. The results demonstrate that the hierarchical predictive control and the centralized predictive control are both able to keep a relatively higher traffic flow in each subnetwork compared with the other two control schemes in an increasing traffic demand situation (Scenario I). For the peak hour situation with a firstly increasing and then decreasing traffic demand (Scenario II), all three predictive control schemes yield a high average traffic flow in the subnetworks, which is better than fixed-time control. The difference between subnetwork i and l is not obvious, but in subnetwork j, hierarchical control yields a better performance than decentralized control due to the coordination mechanism of the hierarchical control scheme.

Since the scale of the optimization problem in hierarchical control is decreased, the CPU time for obtaining the optimal solutions is much less than that of the centralized and decentralized approaches. The average/maximum CPU time spent for solving the online optimization problem at each control step by the hierarchical control approach is 80.7/105.8 s, respectively, much less than 797.2/1055.3 s of the centralized control and 149.6/193.0 s of the decentralized control, which illustrates that the computational complexity of hierarchical control is significantly reduced.

The results of the case studies show that the hierarchical predictive control approach can increase the weighted average traffic flow by keeping the number of vehicles approaching to the critical point of the MFD, and yield an efficient performance that is comparatively close to the results of centralized predictive control, but the computation times required for solving the optimization problems are much lower than the other two predictive control methods.

10.3 Embedded Implementation of Predictive Controller with Applications

Although predictive control has been widely applied to many complex industrial processes, for quite a long time it was rarely used by field controllers where the equipped hardware and software were unable to support the complex computation for solving the constrained optimization problems in real time. With the rapid development of microelectronic technology, more and more high-performance chips appear in the market. Currently, there are several commercially available embedded hardware platforms for field controllers, including digital signal processor (DSP), application-specific integrated circuit (ASIC), field programmable gate array (FPGA), etc. It is now possible to integrate them with microprocessors and related peripherals to form a complete embedded system with flexibility and computational efficiency for field control. With this background, implementing a predictive control algorithm on the embedded system has attracted wide interest in the last decade. This greatly extends the application areas of predictive control.

Regarding the implementation of the predictive controller, two key points have to be considered, i.e. the QP algorithm and its corresponding hardware platform. In this

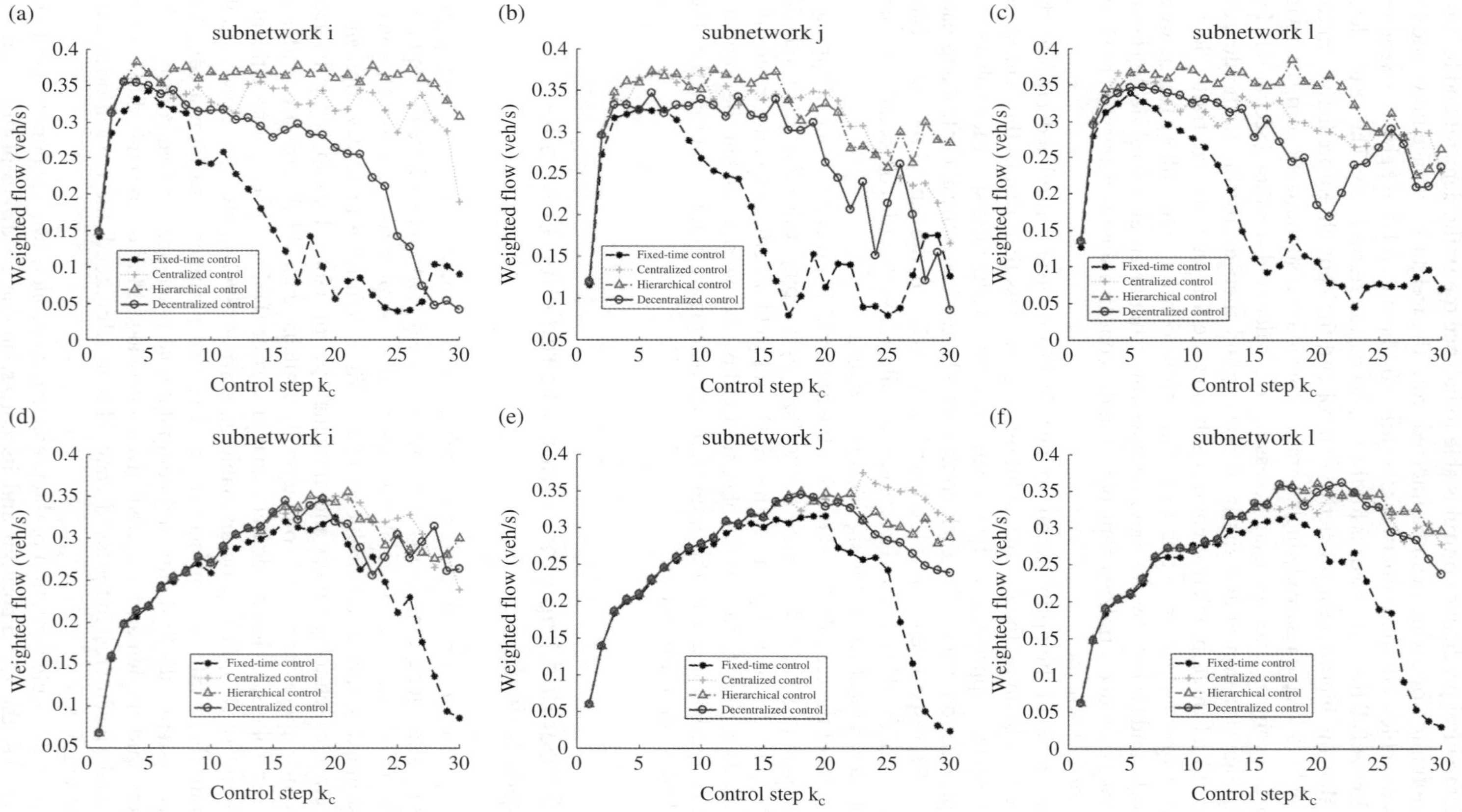

Figure 10.14 Weighted average flows of subnetworks for four control strategies [17]: (a) to (c) Scenario I and (d) to (f) Scenario II. *Source:* Reproduced with permission from Zhao Zhou, Bart De Schutter, Shu Lin and Yugeng Xi of IEEE.

section, we briefly introduce implementation of the predictive control algorithm on two kinds of embedded platforms, focusing on QP problem solving.

10.3.1 QP Implementation in FPGA with Applications

In general, as described in previous chapters, the online optimization problem of predictive control can be commonly formulated as a QP problem in standard form:

$$\begin{aligned}&\min f(\boldsymbol{x})=\frac{1}{2}\boldsymbol{x}^{\mathrm{T}}\boldsymbol{H}\boldsymbol{x}+\boldsymbol{c}^{\mathrm{T}}\boldsymbol{x}\\&\text{s.t. } \boldsymbol{G}\boldsymbol{x}\le\boldsymbol{b}\end{aligned} \tag{10.13}$$

where the matrices and vectors are determined by specific problems. The time to solve this problem is the main online computation time of predictive control. How to efficiently solve the QP problem with limited hardware resources is critical for predictive control implementation on embedded systems.

FPGA has a number of programmable logic resources that can be directly configured to perform complex computations in hardware. In addition, with the adroitly designed architecture including features such as pipelining and parallel computing, FPGA can achieve a much higher processing speed than software implementations. This makes FPGA particularly attractive for computation intensive tasks. The key point when designing a predictive controller on FPGA is how to arrange the control algorithm on the structure of the system hardware and software as much as possible to utilize the parallel processing capabilities of FPGA.

There are many effective methods to solve the QP problem, among which the active set method and interior point method are most commonly used. Some of the literature have compared the suitability of these two methods for FPGA implementation and highlighted the advantages of the active set method for small and medium problems. Since the optimization problem involved in field control is usually of a small scale, the primal active set method is selected for solving the QP problem in the embedded systems.

According to the procedure of the active set method mentioned in Section 5.2.3, it is found that the general QP problem is actually solved by a sequence of QP problems with equality constraints (ECQP). The main computations for solving ECQP are matrix operations including matrix inversion, matrix multiplication, and matrix addition, where the matrix inversion consists of divisions and branch operations and is resource-expensive.

Considering that the matrix inversion is only executed once at each iteration, in [19] it is implemented by software to balance efficiency and resource utilization. Focus is put on designing a hardware matrix multiplier and adder to improve efficiency. Since the matrix multiplication is essentially calculated by a number of vector multiplications, a vector multiplier is designed and shown in Figure 10.15.

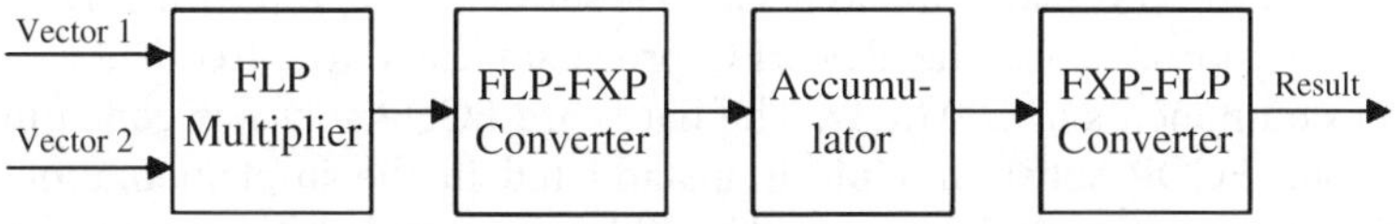

Figure 10.15 Pipeline structure of a vector multiplier [19]. *Source:* Reproduced with permission from Nan Yang, Dewei Li, Jun Zhang and Yugeng Xi of Elsevier.

As shown in the figure, this vector multiplier reads two vectors in an element-by-element fashion and then processes the data with four stages: FPGA built-in floating point (FLP) multiplier, FLP to fixed point (FXP) conversion, FXP accumulation, and FXP to FLP conversion. FXP numbers are used here to increase the throughput of a vector multiplier because FLP accumulation is hard to accomplish within one clock cycle. All of these stages have a throughput of one floating point operation per clock cycle, which means that they can accept new data at every clock cycle, even if the result of previous data is not yet ready. By this pipelined structure, the computing power of each component can be fully exploited.

The matrix adder can be designed in a similar structure as the matrix multiplier to use an FLP adder to perform addition and subtraction. Furthermore, a special Direct Memory Access (DMA) block is designed to guarantee that the correct data is delivered and calculated in every cycle by visiting each row or column of a matrix in a predefined order. It can fetch matrix data in either a normal or transposed form to avoid unnecessary transpositions. This design is suitable for matrix multiplication and addition with arbitrary dimensions. Under the condition of abundant hardware resources, more than one vector multiplier and FLP adder can be operated in parallel to further improve the computing efficiency.

Besides matrix operations, the data transfer between a microprocessor and QP solver via a data bus is also time-consuming. To overcome this problem, the on-chip block RAMs are combined into a large memory region as the local memory to store temporary variables including matrices and vectors. Each memory region has two data ports that can be accessed independently. In addition, a multiplexer array is used to make other data ports accessible by the adder and multiplier. It is controlled by the DMA block for internal data delivery.

The basic structure of the ECQP solver is shown in Figure 10.16. It can be observed that except for the input variables and output variables, all other variables are temporary and can be stored locally instead of being transferred back and forth. Here the ECQP solver contains one matrix multiplier and one adder, which can be executed in parallel. Then the computing procedure of ECQP can be divided into 12 steps as listed in Table 10.2, where the meaning of the notations can be found in (10.13). Each step of the ECQP solver has at most one multiplication and one addition/subtraction, so they can be calculated at the same time.

Based on the ECQP solver design, the whole predictive controller can be completed by integrating some logical operations and I/O interface design on FPGA chips, as shown in Figure 10.17. For the hardware part, an ECQP solver calculates the optimal solution and the Lagrange multiplier at each sampling instant and constructs the QP solver together with cooperated embedded software. The FPGA microprocessor runs the embedded software, whereas the memory controllers connect external memory chips. All the components are connected via the FPGA internal bus. Furthermore, the QP solver should be able to handle different problem sizes so as to make the design flexible.

In [19], a prototype system is built to implement the predictive control on an Xilinx ML403 development board. This platform has a Virtex-4 series FPGA chip with a hard-core PowerPC 405 microprocessor. The PowerPC processor runs at a frequency of 200 MHz and other components at 100 MHz. The hardware ECQP solver is coded in Verilog HDL, and one ECQP solver module is instantiated in the implementation. Combined with other hardware modules provided by Xilinx, the whole hardware circuit design is downloaded into the FPGA.

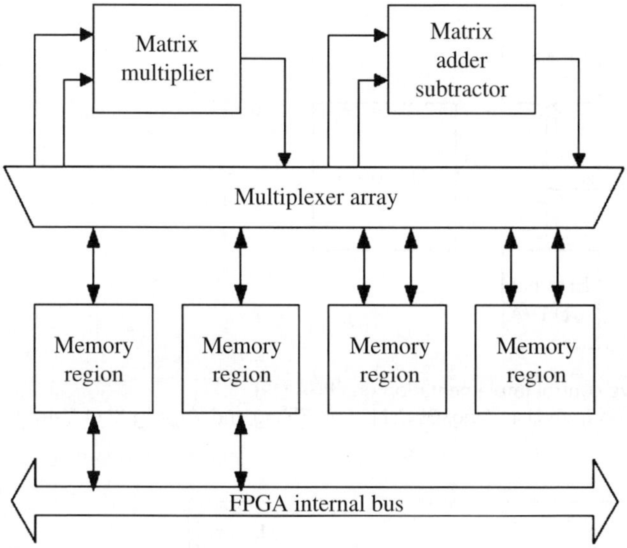

Figure 10.16 The structure of the ECQP solver [19]. *Source:* Reproduced with permission from Nan Yang, Dewei Li, Jun Zhang and Yugeng Xi of Elsevier.

Table 10.2 Execution sequence of the ECQP solver [19]. *Source:* Reproduced with permission from Nan Yang, Dewei Li, Jun Zhang and Yugeng Xi of Elsevier.

Number	Data transfer	Multi.	Add./Sub.	Complexity
1	In: H^{-1}, G			$O(n^2)$
2	In: c	$T = G \times H^{-1}$		$O(n^3)$
3	In: b	$P^{\mathrm{T}} = G \times T^{\mathrm{T}}$		$O(n^3)$
4		P^{-1} (by software)		$O(n^3)$
5		$R = P^{-1} \times T$		$O(n^3)$
6		$K = T^{\mathrm{T}} \times R$		$O(n^3)$
7		$m^{\mathrm{T}} = c^{\mathrm{T}} \times R$	$\mathrm{Q} = H^{-1} - K$	$O(n^2)$
8		$n = P^{-1} \times b$		$O(n^2)$
9		$o^{\mathrm{T}} = b^{\mathrm{T}} \times R$	$\lambda = m + n$	$O(n^2)$
10	Out: λ	$p^{\mathrm{T}} = -c^{\mathrm{T}} \times Q^{\mathrm{T}}$		$O(n^2)$
11			$x = o + p$	$O(n)$
12	Out: x			$O(n)$

This FPGA-based predictive controller is applied to controlling an angle servo system [19]. There are two DC motors, one as the master and the other as the slave, and two needles indicating their rotating angles. The control objective is to drive the slave motor to track the master motor's rotation, whereas the angular speed of the master motor is unknown and unmeasurable. A discrete-time state space model in increment form was first established with an angle difference and angular speed difference between two

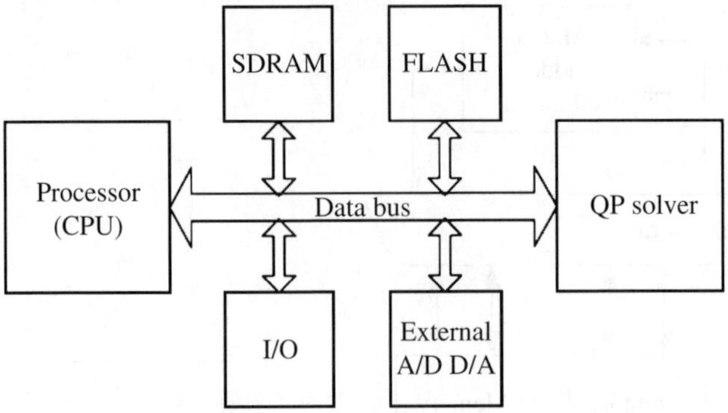

Figure 10.17 Scheme of predictive control implementation on FPGA [19].
Source: Reproduced with permission from Nan Yang, Dewei Li, Jun Zhang and Yugeng Xi of Elsevier.

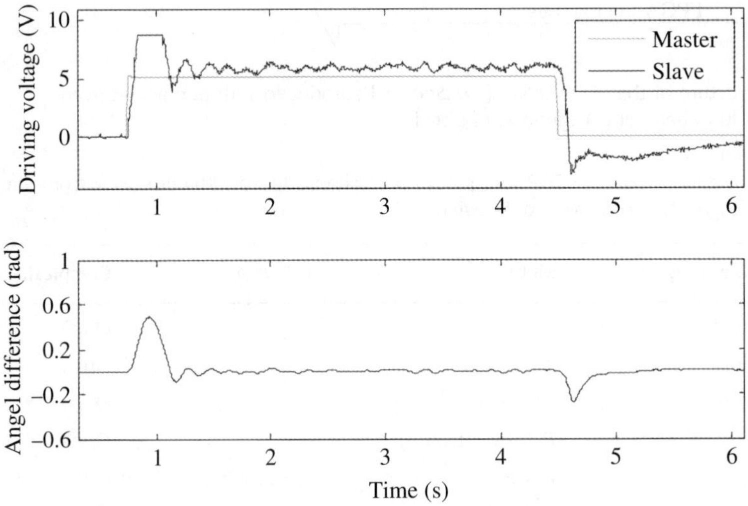

Figure 10.18 Experimental results when a master motor has a step input [19]. *Source:* Reproduced with permission from Nan Yang, Dewei Li, Jun Zhang and Yugeng Xi of Elsevier.

motors as the states and the voltage difference as the control input. The online optimization problem is formulated as a QP problem with constraints on the control input. With properly selected design parameters, the control result is shown in Figure 10.18.

The experiment result shows that when the driving voltage of the master motor has a step change from 0 to 5 V, the control signal of the predictive controller first hits the upper bound and then promptly gets out of the constraint boundary. The total regulation time is less than 0.5 s. Note that the driving voltages for two motors are different due to the different mechanical parameters. By a number of tests, the average online solving time of predictive control is 0.3 ms, which can satisfy the requirement of real-time control.

Although the above designed FPGA-based embedded predictive controller seems to be effective to field control, the matrix inversion was still solved by software in this design because it is resource-expensive. The software implementation results in large amounts of data transmission of intermediate computational results via the internal bus. An experimental investigation showed that the data transfer will waste a considerable amount of time, which sometimes even exceeds the computing time. To avoid extra data transmission as much as possible, Xu et al. [20] further improved the design by developing hardware matrix inversion.

In [20], a simple positive definite symmetric matrix inversion (SPMI) algorithm is adopted for matrix inversion, where the matrix to be inverted is first partitioned into a number of small matrices with lower dimensions. According to the inversion formula for partitioned matrices, the matrix inverse can be iteratively calculated through matrix multiplication and addition operations along with fewer division or selectable lower-dimensional matrix inversion operations. Compared with the matrix inversion based on Cholesky decomposition, the SPMI algorithm can get rid of the root-square operations, and the total number of iterations and division operations can be linearly decreased if a proper dimension is selected for the partitioned matrices, in which case a lower-dimensional matrix needs to be extra inverted during each iteration but the computation cost for it is very low. The SPMI algorithm can make full use of the parallelism and pipeline in hardware to further improve the computational efficiency. With similar calculating formulae at each iteration it is very suitable for block design in hardware.

By applying the SPMI algorithm to the ECQP solver, the overall inner architecture of the S-ECQP (the combination of the SPMI algorithm and ECQP) solver mainly consists of three parts: the process control module, the computation module, and the memory module, as shown in Figure 10.19. It transfers data via the internal bus only twice in solving an ECQP, one at the beginning to initialize the solver and the other at the end to read the feasible solution, bypassing extra frequent data delivery and saving time. Table 10.3 gives an experimental comparison of the time needed to solve the ECQP problem by the improved predictive controller with hardware matrix inversion and the QP solver with software matrix inversion in [19].

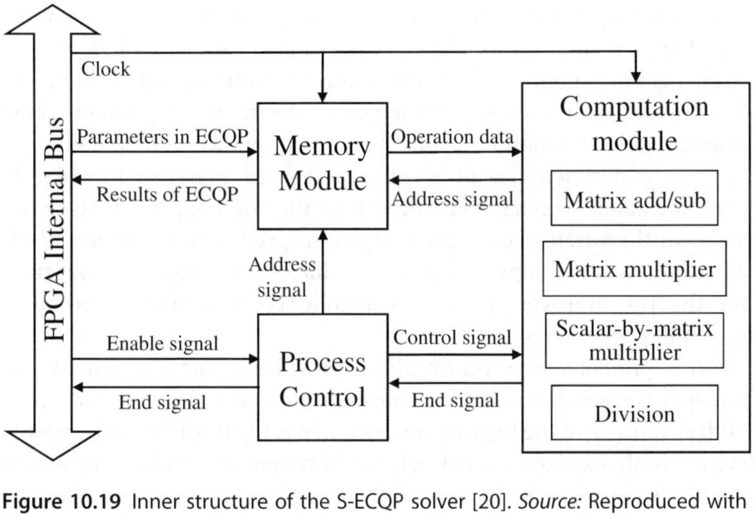

Figure 10.19 Inner structure of the S-ECQP solver [20]. *Source:* Reproduced with permission from Yunwen Xu, Dewei Li, Yugeng Xi, Jian Lan and Tengfei Jiang of IEEE.

Table 10.3 Performance comparison of two QP solvers (unit: ms) [20]. *Source:* Reproduced with permission from Yunwen Xu, Dewei Li, Yugeng Xi, Jian Lan and Tengfei Jiang of IEEE.

Matrix dimension/ No. of equality constraints	Improved predictive controller	QP solver in [19]
5/5	0.37	0.5
10/10	0.50	1.80
20/20	1.52	11.4
30/30	4.66	52.7

Connecting S-ECQP to the embedded microprocessor, an FPGA-based board is specially made for equipments with a low calculation capability and industrial field applications in [20]. Its functional blocks include the FPGA functional domain, regions for power, clock, configuration, AD/DA, communication, storage, and so on. The FPGA functional domain hosts the main chip XC4VFX12, which has an embedded hard core PowerPC 405 (PPC405) microprocessor and some logic resources. PPC405 performs software programs with a frequency of 200 MHz, while logic resources can be customized to fulfill specific function circuits at a frequency of 100 MHz. A prototype of the improved predictive controller is built on this board consisting of a hardware part of the S-ECQP solver, a software part running on PPC405, and several peripheral circuits.

This embedded predictive controller is applied to the holding process of an injection molding machine HTL68/JD, where the hydraulic valve must be maintained as a certain value to provide a holding pressure for the resistance of material shrinkage and hollows. The control purpose is to make the holding pressure kept on the set-point by controlling the hydraulic valve during the dwell time. By using the historical process data with a 10 ms sampling time, a discrete SISO ARMA model is established and then transformed into the state-space form. For predictive control, the online optimization aims at making the future holding pressure follow the set-point as closely as possible while minimizing the change of the valve opening. Furthermore, due to the physical characteristics of the injection molding, the range restriction for the valve opening is within 10–70% and for the valve successive changes is not beyond 10%. The formulated QP problem is solved by the improved predictive controller at each sampling time. During the experiment, field debugging is combined with simulation results to choose appropriate parameters. Figure 10.20 shows the experiment result when the desired pressure is equal to 60 bar. The control performance seems satisfactory where the holding pressure is maintained within ±1.5 bar. In the whole process, the improved predictive controller needs 1.4 ms in average to complete online optimization one time. Therefore, this application further demonstrates that the improved predictive controller is qualified for control problems with a high sampling frequency.

The implementation of predictive control on FPGA provides a portal to extend it to many industrial field applications. It can be customized to deal with different scale problems so as to take full advantage of the logic resources and has the flexibility to be ported to other FPGA devices. It also strikes a good balance between cost and performance, which makes it particularly appealing to field control where a large number of embedded yet capable controllers are required.

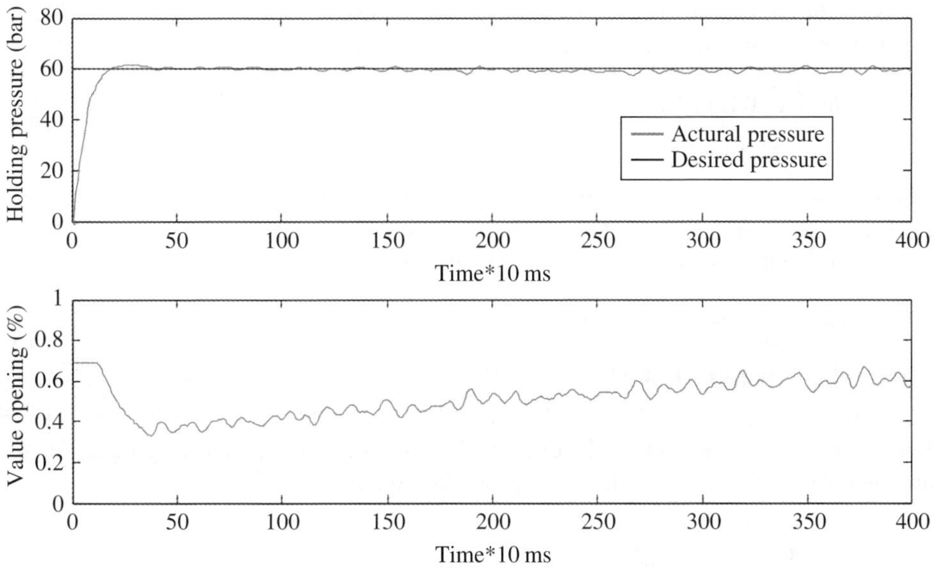

Figure 10.20 Experimental results when the desired pressure is equal to 60 bar [20].
Source: Reproduced with permission from Yunwen Xu, Dewei Li, Yugeng Xi, Jian Lan and Tengfei Jiang of IEEE.

10.3.2 Neural Network QP Implementation in DSP with Applications

FPGA chips have the advantage of a high computing speed, and its parallel processing ability by hardware is particularly attractive, but due to the restriction on hardware resources, an FPGA-based predictive controller is usually used for small-scale field control problems. Meanwhile, the cost of an FPGA is also higher than other digital devices. To implement a predictive controller with medium scale and general purpose on embedded systems, the hardware platform is required to have CPU processing ability to support on-site tuning and to have redundant capacity to deal with the variation of the constraint number when the control horizon changes. Taking a trade-off for computing speed, application requirement, chip capacity, and cost, a digital signal processor (DSP) seems to be a more suitable platform to implement predictive control algorithms for general problems.

As introduced in the last section, solving a QP problem by using the active set method has been successfully implemented on an FPGA platform. However, it is worth noting that the active set method needs to do matrix inversion online, which is rather time-consuming. Indeed, a number of promising alternative methods, such as recurrent neural networks, have been proposed to solve high-dimensional QP problems without online matrix inversion. Among those, the simplified dual neural network (SDNN) proposed in [21] has the least amount of neurons (equal to the number of inequality constraints in the QP problem) and is featuring in high parallelism, adaptivity, and circuit implement ability. Because the SDNN is superior to other neural networks in terms of hardware implementation, in the following, we introduce the predictive control implementation on DSP with SDNN as the QP solver presented in [22].

Firstly, the continuous-time SDNN presented and used to solve QP problem by [21] is briefly introduced. Consider the standard QP problem as

$$\begin{aligned} &\min \ \frac{1}{2}\boldsymbol{x}^T\boldsymbol{W}\boldsymbol{x}+\boldsymbol{c}^T\boldsymbol{x} \\ &\text{s.t.} \quad \boldsymbol{l}\le \boldsymbol{E}\boldsymbol{x}\le \boldsymbol{h} \end{aligned} \tag{10.14}$$

where $\boldsymbol{x}\in \mathrm{R}^{\gamma}$ is the decision variable, $\boldsymbol{W}$ is a $\gamma\times\gamma$ positive definite symmetric matrix, and $\boldsymbol{c}\in\mathrm{R}^{\gamma}$, $\boldsymbol{E}\in\mathrm{R}^{\eta\times\gamma}$, $\boldsymbol{h}, \boldsymbol{l}\in\mathrm{R}^{\eta}$, i.e. (10.14), has 2η inequality constraints.

The corresponding dual problem can be expressed by

$$\begin{aligned} &\max \ -\frac{1}{2}\boldsymbol{x}^{\mathrm{T}}\boldsymbol{W}\boldsymbol{x}+\boldsymbol{l}^{\mathrm{T}}\boldsymbol{r}-\boldsymbol{h}^{\mathrm{T}}\boldsymbol{s} \\ &\text{s.t.} \quad \boldsymbol{W}\boldsymbol{x}+\boldsymbol{c}-\boldsymbol{E}^{\mathrm{T}}\boldsymbol{v}=0 \end{aligned} \tag{10.15}$$

where $\boldsymbol{r}, \boldsymbol{s}\in\mathrm{R}^{\eta}$ are the decision variables of the dual problem and $\boldsymbol{v}=\boldsymbol{r}-\boldsymbol{s}$. From the Karush–Kuhn–Tucker (KKT) conditions, the solutions of the following equation set are identical with the solutions of the problem (10.14) and (10.15):

$$\begin{aligned} &\boldsymbol{W}\boldsymbol{x}+\boldsymbol{c}-\boldsymbol{E}^{\mathrm{T}}\boldsymbol{v}=\boldsymbol{0} \\ &\begin{cases} (\boldsymbol{E}\boldsymbol{x})_i=l_i, & \text{if } \beta v_i>0 \\ (\boldsymbol{E}\boldsymbol{x})_i=h_i, & \text{if } \beta v_i<0 \quad \beta>0,\quad i=1,\ldots,\eta \\ l_i\le(\boldsymbol{E}\boldsymbol{x})_i\le h_i, & \text{if } \beta v_i=0 \end{cases} \end{aligned} \tag{10.16}$$

where the resultant conditions from the KKT condition are slightly modified by [22] with an additional parameter β, which will not influence the original conditions because of $\beta>0$.

By the projection formulation, the above equation is equivalent to the following equation:

$$\begin{aligned} &\boldsymbol{W}\boldsymbol{x}+\boldsymbol{c}-\boldsymbol{E}^{\mathrm{T}}\boldsymbol{v}=\boldsymbol{0} \\ &\boldsymbol{E}\boldsymbol{x}=\boldsymbol{g}(\boldsymbol{E}\boldsymbol{x}-\beta\boldsymbol{v}) \end{aligned} \tag{10.17}$$

where

$$g_i(\boldsymbol{\theta})=\begin{cases} l_i, & \text{if } \theta_i<l_i \\ \theta_i, & \text{if } l_i\le\theta_i\le h_i \\ h_i, & \text{if } \theta_i>h_i \end{cases} \quad i=1,\ldots,\eta \tag{10.18}$$

From the first equation of (10.17), if $\boldsymbol{W}$ is invertible, it follows that

$$\boldsymbol{x}=\boldsymbol{W}^{-1}(\boldsymbol{E}^{\mathrm{T}}\boldsymbol{v}-\boldsymbol{c}) \tag{10.19}$$

Substituting it into the second equation of (10.17) yields

$$\boldsymbol{E}\boldsymbol{W}^{-1}(\boldsymbol{E}^{\mathrm{T}}\boldsymbol{v}-\boldsymbol{c})=\boldsymbol{g}\left(\boldsymbol{E}\boldsymbol{W}^{-1}(\boldsymbol{E}^{\mathrm{T}}\boldsymbol{v}-\boldsymbol{c})-\beta\boldsymbol{v}\right) \tag{10.20}$$

Based on (10.20), [21] designed the following differential equation of the continuous-time SDNN for solving the primal QP problem (10.14):

$$\varepsilon\frac{\mathrm{d}\boldsymbol{v}}{\mathrm{d}t}=-\boldsymbol{E}\boldsymbol{W}^{-1}(\boldsymbol{E}^{\mathrm{T}}\boldsymbol{v}-\boldsymbol{c})+\boldsymbol{g}\left(\boldsymbol{E}\boldsymbol{W}^{-1}(\boldsymbol{E}^{\mathrm{T}}\boldsymbol{v}-\boldsymbol{c})-\beta\boldsymbol{v}\right) \tag{10.21}$$

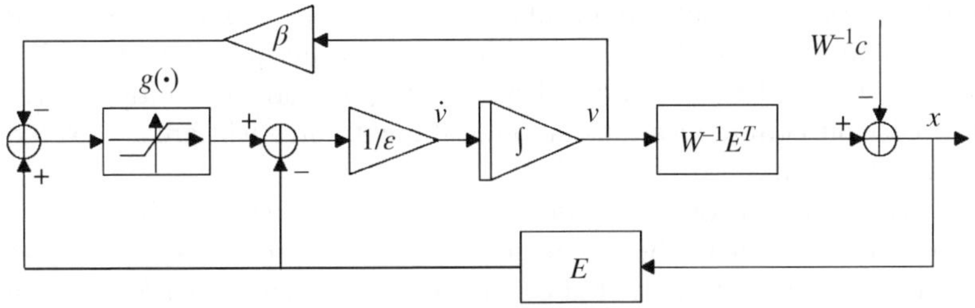

Figure 10.21 Block diagram of a continuous-time simplified dual neural network (SDNN) [21]. *Source:* Reproduced with permission from Shubao Liu and Jun Wang of IEEE.

where ε is a scaling parameter that controls the convergence rate of the neural network. The output of the network (10.19) gives the solution of the QP problem. The block diagram of the continuous-time SDNN is shown in Figure 10.21.

For implementation on an embedded hardware device, in [22] the continuous-time SDNN (10.21) is further transformed into the following discrete form through first-order discretization:

$$\boldsymbol{v}(n+1)-\boldsymbol{v}(n)=\alpha\left(-\boldsymbol{E}\boldsymbol{W}^{-1}\left(\boldsymbol{E}^{\mathrm{T}}\boldsymbol{v}(n)-\boldsymbol{c}\right)+\boldsymbol{g}\left(\boldsymbol{E}\boldsymbol{W}^{-1}\left(\boldsymbol{E}^{\mathrm{T}}\boldsymbol{v}(n)-\boldsymbol{c}\right)-\beta\boldsymbol{v}(n)\right)\right) \tag{10.22}$$

where $\alpha = \tau/\varepsilon$ with τ the sampling period. Meanwhile, from (10.19), the output equation is

$$\boldsymbol{x}(n)=\boldsymbol{W}^{-1}\left(\boldsymbol{E}^{\mathrm{T}}\boldsymbol{v}(n)-\boldsymbol{c}\right) \tag{10.23}$$

Reference [22] proved that the discretized SDNN (10.22) is convergent if

$$0<\alpha<\frac{2}{\beta+\lambda_{\max}}$$

where $\lambda_{\max}$ denotes the largest eigenvalue of $\boldsymbol{E}\boldsymbol{W}^{-1}\boldsymbol{E}^{\mathrm{T}}$.

The main calculations of the SDNN algorithm are matrix operations, which are carried out by the multiple accumulation resources on DSP. To make full use of the hardware computing resources, the two-way single-instruction multiple-data (SIMD) instructions*_dmpysp*, *_daddsp*, and *_dsubsp* are taken to execute matrix multiplication, addition, and subtraction operations, respectively. Using the two-way SIMD instructions, two calculations can be performed in a parallel manner at one time. Compared with one-way instructions, the data-type conversion operations are increased, but it is still beneficial because of saving computing time, particularly in the cases where the dimensions of matrices are large and calculations have to be done many times.

Based on the above design, a prototype system is built to implement the QP solver on a TMDSEVM6678L development board produced by TI. The key component of the platform is a TMS320C6678 DSP chip, which is integrated with eight C66x DSP core subsystems. Each core runs at a frequency of 1.00–1.25 GHz. On each core there is a local 32 kB L1 program (L1P) cache, a local 32 kB L1 data (L1D) cache, a local 512 kB L2 that can be configured as RAM or cache, and a 4 MB multicore shared memory SRAM (MSMSRAM) that can be shared by all eight cores. The basic functions of a predictive controller should have the following components: data acquisition, preprocessing, optimization, postprocessing, and actuation. All of these components including the QP solver are implemented in TMS320C6678. In particular the discrete-time SDNN is implemented on a single C66x DSP core subsystem rather than on multiple cores because the time cost on communications among different cores will greatly exceed the time cost on computation.

The DSP-based predictive controller is applied to an air separation unit (ASU) to show its high performance and efficiency. The cryogenic air separation process, also known as ASU, is used to produce large quantities of purified oxygen, nitrogen, and argon for the steel, chemical, food processing, semiconductor, and healthcare industries. It includes three unit operations: compression, purification, and air separation which feeds the air into the required gaseous and liquid oxygen, nitrogen, and argon product flows.

In many manufacturing processes, gaseous product demand is not fixed but periodical, stepwise, and intermittent, which leads to large changes in the product rate for the ASU (typically 75–105%). Therefore, the ASU must automatically and rapidly respond to the changing product demand from customers. Otherwise, excess oxygen has to be released, which leads to a high oxygen release ratio and operating cost. Hence, a predictive

Table 10.4 Controlled variables (CVs) and manipulated variables (MVs) of the ASU system [22]. *Source:* Reproduced with permission from Yang Lu, Dewei Li, Zuhua Xu, and Yugeng Xi of IEEE.

Symbol	Description
CV1	Oxygen content at the top of HPC (ppm)
CV2	Oxygen content of waste nitrogen from LPC (%)
CV3	Argon content at the feed of CAC-1 (%)
CV4	Oxygen content of product gaseous oxygen (%)
CV5	Oxygen content of product gaseous nitrogen (ppm)
CV6	Oxygen content at the top of CAC-1 (%)
CV7	Feed flow of CAC-1 (Nm^3/h)
CV8	Crude gaseous argon flow (Nm^3/h)
CV9	Product gaseous oxygen flow (Nm^3/h)
CV10	Product gaseous nitrogen flow (Nm^3/h)
CV11	Product liquid nitrogen flow (Nm^3/h)
MV1	Pure liquid nitrogen flow from HPC to LPC (Nm^3/h)
MV2	Total air feed flow (Nm^3/h)
MV3	Product gaseous oxygen flow (Nm^3/h)
MV4	Product gaseous nitrogen flow (Nm^3/h)
MV5	Product liquid nitrogen flow valve (%)
MV6	Pressure of waste gaseous nitrogen from LPC (kPa)
MV7	Bottom level of HPC (mm)
MV8	Bottom level of CAC-1 (mm)
MV9	Condenser liquid level of CAC-II (mm)
MV10	Crude gaseous argon flow valve (%)

controller for the cryogenic air separation process is designed to automatically and rapidly respond to the changing product demand from customers.

According to the real data, the ASU process can be described by the transfer function matrix. It has 11 outputs and 10 inputs and there are 32 correlative input–output pairs. The physical descriptions of the CVs and the MVs are given in Table 10.4. Part of the step responses of an ASU is shown in Figure 10.22. Details about the ASU model and constraints can be found in [22] and so are omitted here.

From the step response of the ASU system with a sampling period of 30 seconds, the dynamic matrix can be calculated and the online optimization problem can be formulated. Since the open-loop system is stable, a long prediction horizon $P = 100$ with $M = 5$ is chosen to enhance the closed-loop stability, whereas $\boldsymbol{Q}$ and $\boldsymbol{R}$ are chosen as unitary diagonal matrices with proper dimensions. However, a large P also brings a large amount of output constraints. In this case the QP problem has 50 decision variables and 2400 inequality constraints with 200 for inputs and 2200 for outputs. Such a constraint scale would cause the optimization problem to become overly complicated and difficult to solve. To effectively handle the large amount of output constraints, soft constraints are introduced by using the zone control technique. To do this, an 11-by-1 vector $\boldsymbol{\delta}(k)$ is introduced in [22], which is composed of the deviations of the outputs from their reference values and are handled as additional decision variables. Constraints on $\boldsymbol{\delta}(k)$ are put into the optimization problem instead of the large amount of output constraints. The objective function is also accordingly revised. In this way, the revised QP problem has 61 decision variables and 222 inequality constraints where the number for output constraints is greatly reduced.

In the experiment, the hardware QP solver takes 0.01 ms to calculate (10.22) once at the frequency of 1 GHz. In addition, the warm-start technique is utilized to set the initial values of the neural network state variables to speed up the convergence process. The ASU system is controlled for 2000 steps (1000 min). The product demand, i.e. the reference for CV9 (product gaseous oxygen flow) and CV10 (product gaseous nitrogen flow), has a step change during control. Other outputs are required to be kept within their feasible regions. The trajectories of control inputs and system outputs are shown in Figures 10.23 and 10.24, respectively.

This indicates that the proposed design is able to deal with high-dimensional problems at a low cost and achieve a good control performance. Furthermore, the longest time it takes for DSP to compute the control inputs is 0.0733 s, which is far less than the sampling period of 30 s and satisfies the real-time requirement.

10.4 Summary

This chapter introduces the application of predictive control technology in various fields, focusing on illustrating the fundamental ideas and implementation techniques when using predictive control to solve different kinds of practical problems, i.e. how to formulate the practical requirements on constrained optimization as a predictive control problem and which technical issues should be considered when implementing predictive control.

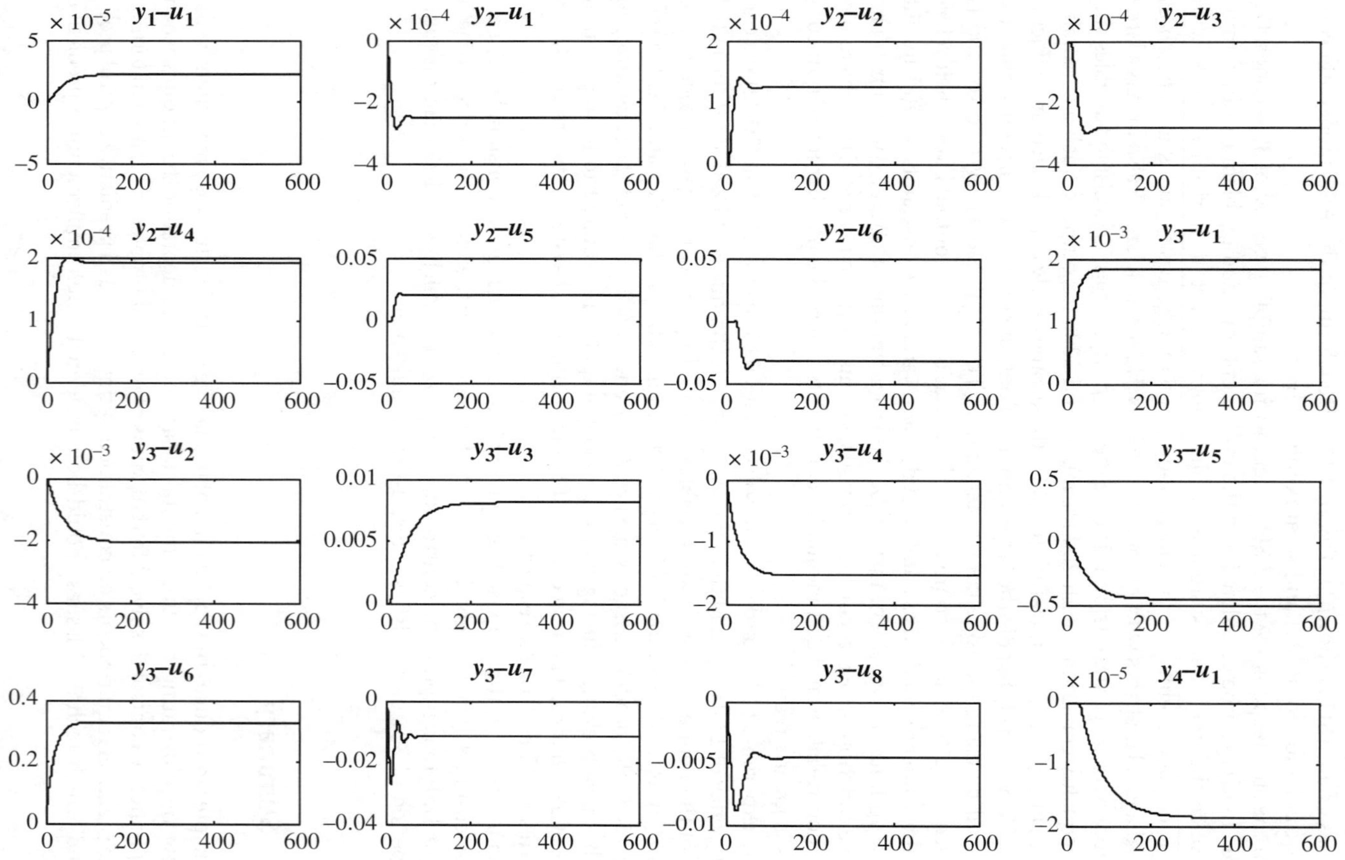

Figure 10.22 Part of the step responses of an air separation unit (ASU) system.

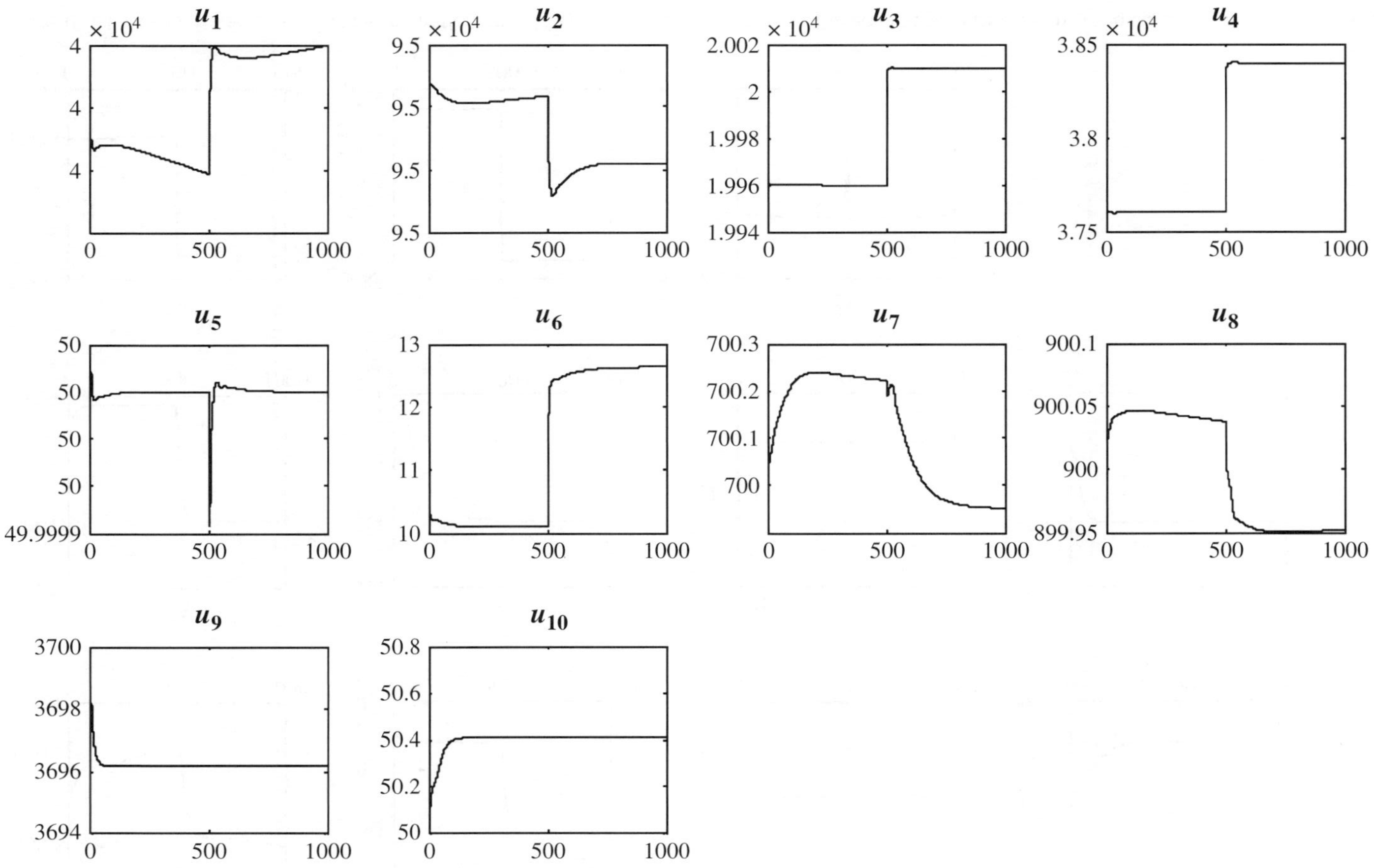

Figure 10.23 Input trajectories of an air separation unit (ASU) system [22]. *Source:* Reproduced with permission from Yang Lu, Dewei Li, Zuhua Xu, and Yugeng Xi of IEEE.

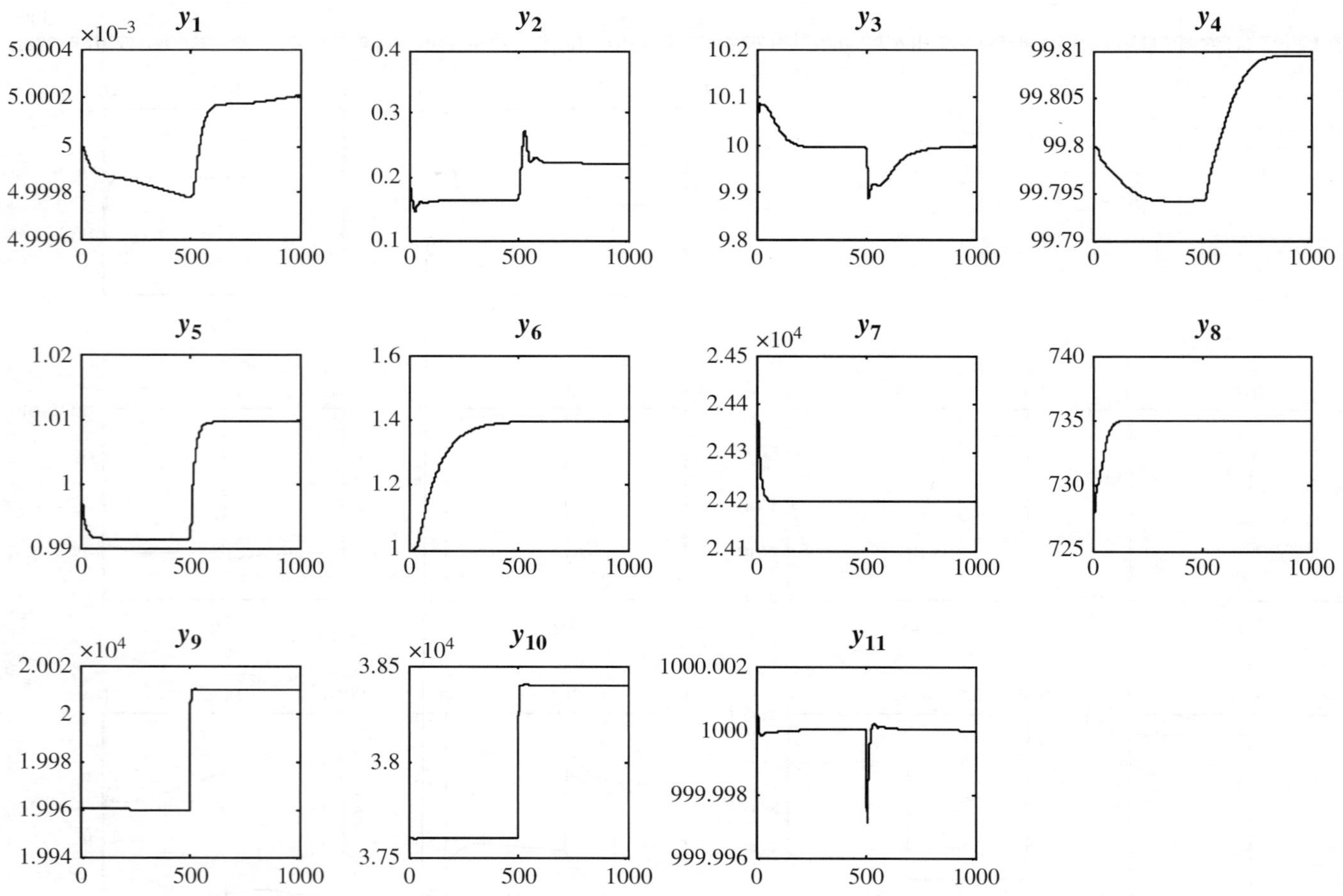

Figure 10.24 Output trajectories of an air separation unit (ASU) system [22]. *Source:* Reproduced with permission from Yang Lu, Dewei Li, Zuhua Xu, and Yugeng Xi of IEEE.

For industrial process control, the field where predictive control provides the most successful applications and the corresponding technology is developed most maturely, we introduce in detail the historical development of industrial predictive control technology and the characteristics of its commercial software, referring to [5]. The functional role of predictive control in industrial hierarchical control structure is analyzed and the key technologies involved in its implementation are introduced. The application example to a hydrocracking unit typically reflects the application details of industrial predictive control technology in large industrial processes.

With the popularization of predictive control technology, its application has been rapidly extended to many fields other than industrial processes, and the reports on predictive control applications in various fields are ever increasing. In this chapter, we analyze the reason of the growing demand on predictive control and give a brief description of extension of predictive control to several application fields. For illustration, two application examples for large-scale systems, i.e. the gas transportation network and the urban traffic network, as well as one example for electromechanical systems, i.e. modern automatic train control, are introduced. Different design ideas and strategies are given according to the specific application scenarios, showing the diversity of predictive control methods and strategies.

With the rapid development of microelectronic technology, high-performance chips appear in the market, which makes it possible to utilize the capacity of equipped hardware and software to develop embedded predictive controllers. Two application examples of such field controllers are introduced in this chapter, one with a QP solver on FPGA for small-scale problems and the other with a neural network solver on DSP for medium problems, both of which efficiently met the requirement on real-time computation.

The limited examples introduced in this chapter show the great potential of predictive control technology in a large variety of application fields. There is a big space for extending or developing the predictive control technique to different application cases where optimization is required but constraints exist and the environment is uncertain.

References

1 Richalet, J., Rault, A., Testud, J.L. et al. (1978). Model predictive heuristic control: applications to industrial processes. *Automatica* 14 (5): 413–428.
2 Cutler, C.R and Ramaker, B.L. Dynamic matrix control – a computer control algorithm, *Proceedings of the 1980 Joint Automatic Control Conference*. San Francisco, CA (13–15 August 1980), WP5-B.
3 Richalet, J. (1993). Industrial applications of model based predictive control. *Automatica* 29 (5): 1251–1274.
4 Rawlings, J.B. (2000). Tutorial overview of model predictive control. *IEEE Control Systems Magazine* 20 (3): 38–52.
5 Qin, S.J. and Badgwell, T.A. (2003). A survey of industrial model predictive control technology. *Control Engineering Practice* 11 (7): 733–764.
6 Froisy, J.B. (2006). Model predictive control – building a bridge between theory and practice. *Computers and Chemical Engineering* 30 (10–12): 1426–1435.

7 Kelly, S.J., Rogars, M.D., and Hoffman, D.W. Quadratic dynamic matrix control of hydro cracking reactors, *Proceedings of the 1988 American Control Conference*. Atlanta, GA (15–17 June 1988), 1: 295–300.

8 Vazquez, S., Leon, J.I., Franquelo, L.G. et al. (2014). Model predictive control: a review of its applications in power electronics. *IEEE Industrial Electronics Magazine* 8 (1): 16–31.

9 Kouro, S., Perez, M.A., Rodriguez, J. et al. (2015). Model predictive control: MPC's role in the evolution of power electronics. *IEEE Industrial Electronics Magazine* 9 (4): 8–21.

10 Rodriguez, J., Kazmierkowski, M.P., Espinoza, J.R. et al. (2013). State of the art of finite control set model predictive control in power electronics. *IEEE Transactions on Industrial Informatics* 9 (2): 1003–1016.

11 Killian, M. and Kozek, M. (2016). Ten questions concerning model predictive control for energy efficient buildings. *Building and Environment* 105: 403–412.

12 Brdys, M.A. (2014). Integrated monitoring, control and security of critical infrastructure systems. *Annual Reviews in Control* 38: 47–70.

13 Eren, U., Prach, A., Kocer, B.B. et al. (2017). Model predictive control in aerospace systems: current state and opportunities. *Journal of Guidance and Dynamics* 40 (7): 1541–1566.

14 Marques, D. and Morari, M. (1988). On-line optimization of gas pipeline networks. *Automatica* 24 (4): 455–469.

15 Yin, J., Tang, T., Yang, L. et al. (2017). Research and development of automatic train operation for railway transportation systems: a survey. *Transportation Research Part C* 85: 548–572.

16 Luo, Y., Li, D., and Xi, Y. Model predictive control with switching strategy for a train system, *Proceedings of the 13th International Conference on Control, Automation, Robotics and Vision* (ICARCV 2014). Singapore (10–12 December 2014), 913–918.

17 Zhou, Z., De Schutter, B., Lin, S., and Xi, Y. (2017). Two-level hierarchical model-based predictive control for large scale urban traffic networks. *IEEE Transactions on Control Systems Technology* 23 (2): 496–508.

18 Lin, S., De Schutter, B., Xi, Y., and Hellendoorn, H. (2011). Fast model predictive control for urban road networks via MILP. *IEEE Transactions on Intelligent Transportation Systems* 12 (3): 846–856.

19 Yang, N., Li, D., Zhang, J., and Xi, Y. (2012). Model predictive controller design and implementation on FPGA with application to motor servo system. *Control Engineering Practice* 20 (11): 1229–1235.

20 Xu, Y., Li, D., Xi, Y. et al. (2018). An improved predictive controller on the FPGA by hardware matrix inversion. *IEEE Transactions on Industrial Electronics* 65 (9): 7395–7405.

21 Liu, S. and Wang, J. (2006). A simplified dual neural network for quadratic programming with its KWTA application. *IEEE Transactions on Neural Networks* 17 (6): 1500–1510.

22 Lu, Y., Li, D., Xu, Z., and Xi, Y. (2006). Convergence analysis and digital implementation of a discrete-time neural network for model predictive control. *IEEE Transactions on Industrial Electronics* 61 (12): 7035–7045.

11

Generalization of Predictive Control Principles

11.1 Interpretation of Methodological Principles of Predictive Control

The application of predictive control was originally only in industrial processes, but today it is extended to many application fields. In addition to the technological reason that it might be the only control technique capable of explicitly incorporating constraints into the optimization problem and effectively solving it, more important is the universality of its methodological principles. As Richalet et al. early pointed out in [1], the basic ideas underlying this approach are related to the "scenario technique," which to some extent is similar to what the human operator is assumed to do with his internal model of the external world. The powerful methodological principles implied in predictive control should naturally have wider applicability.

In Section 1.2, the basic principles of predictive control, i.e. prediction model, rolling optimization, and feedback correction, have been explained in detail, which concretely embody the concepts of model, optimization, and feedback in cybernetics, respectively. For the prediction model, since the most emphasized issue is its function rather than structure, it can be flexibly established using known information in a most reasonable way according to the plant characteristics and the control requirements. Due to the use of error prediction of unmodeled information and the nonclassical rolling style optimization, various complex factors in the actual systems can be taken into account in the optimization process to form closed-loop optimization control, and various forms of constraints and objectives can be handled. Under the framework of predictive control principles, the flexibility for model types, optimization methods, and feedback strategies has strongly pushed forward its applications to different kinds of optimization-based control problems. The expansion of predictive control applications mentioned in the last chapter has demonstrated the diversity of predictive control algorithms in solving optimization control problems.

From the perspective of cybernetics and information theory, predictive control as a new type of control method has its distinct characteristics. It is a kind of model-based, rolling implemented, and feedback combined optimization control algorithm, embodying the reasonable combination of two fundamental mechanisms in cybernetics, i.e. optimization and feedback.

Predictive Control: Fundamentals and Developments, First Edition. Yugeng Xi and Dewei Li.

For control problems, the selection of the control scheme strongly depends on the available information. Two representative control categories, feedback control and optimal control, are initiated for two extreme cases of information environments, respectively.

For one extreme case where no prior knowledge on the plant and the environment is available, the control can only be based on the real-time feedback information, such as the widely used PID controller, which does not need the model and responds only after the appearance of an error constructed from feedback. Feedback control takes the real status of the controlled process into account and thus can overcome the influence of various unknown factors in the actual process in time. However, only relying on feedback shows a lack of foresight and without a model the performance optimization is out of the question.

For another extreme case where all the information on the plant and the environment is known in advance, the optimal control actions can be solved in advance with respect to some specified performance index according to the model. In this case the optimal control theory provides perfect theories and algorithms. When implementing the off-line solved control actions, it might be expected to get the optimal performance if the plant and environment information is accurate as prior known. However, this assumption is too idealistic. In practice, if a model mismatch exists or environment variation occurs, implementing the previously calculated optimal control will not achieve an optimum and even result in performance degeneration or unstable behavior.

In practical applications, prior information available for control is often between these two extreme cases. As a model-based control, predictive control needs certain prior causal knowledge of the plant that is embodied in the prediction model. However, because of the uncertainty of the plant and environment, the optimization in predictive control is based on the model, but not only on the model. Considering the existence of uncertainty both in the model and in the environment, solving the global open-loop optimization problem seems to make no sense and is replaced by repeatedly solving a local open-loop optimization problem combined with feedback correction. In this way, the feedback mechanism is naturally embedded into the control process, either by updating the prediction model or correcting the model prediction, making the optimization basis match the real status each time. Furthermore, with this rolling style, it seems unnecessary to solve the global optimization to get all future control actions. Instead the optimization will be solved for a finite future horizon, which makes the online optimization tractable.

By replacing one-time solving the global optimization by repeatedly solving the finite horizon optimization combined with feedback correction, predictive control combines optimization with feedback in a reasonable way. The whole control is based both on the model and optimization, as well as on feedback. Table 11.1 shows the differences of predictive control from optimal control and PID feedback control in the information requirement and control mechanism. Predictive control is a closed-loop optimal control method between the optimal control and the model-free PID control, both maintaining the advantage of optimization and introducing a feedback mechanism. Therefore, predictive control could be regarded as the compromise of ideal optimal control to real uncertainties, providing a reasonable way to pursue an optimum in an uncertain environment [2].

Table 11.1 Predictive control compared with optimal control and PID feedback control.

	Optimal control	Predictive control	PID feedback control
Information requirement	Accurate prior information on model and environment	Prior information on model and environment Real-time output information	Real-time output information
Control style	Off-line global optimization Online implementation	Online finite horizon optimization Online rolling implementation	Online immediate control
General performance	Ideal optimal, if uncertainty does not exist	Sustained local optimum Suit to uncertain environment	Suit to uncertain environment No optimization

Remark 11.1 Sometimes a global optimization problem with a long horizon can be solved by repeatedly solving a series of finite horizon optimization problems where the problem is definite without uncertainty and feedback. This is a technical strategy aiming at decreasing the scale of the optimization problem to be solved so as to reduce the computational complexity. It is efficient because no uncertainty exists, but will result in suboptimality. It is not the same as what happens in predictive control where the rolling optimization should be combined with feedback, not only for reducing the computational complexity but also, more importantly, as a control mechanism against the uncertainties.

11.2 Generalization of Predictive Control Principles to General Control Problems

11.2.1 Description of Predictive Control Principles in Generalized Form

The methodology implied in predictive control not only opens a broad development space for solving various optimization control problems, as mentioned above, but also provides a good inspiration for solving more general optimization and decision problems in a dynamic uncertain environment. In various application fields, there exist a large number of dynamic processes where optimization or decision is required, such as robot path planning, production planning and resource scheduling, discrete event dynamic system (DEDS) monitoring and so on. For quite a long time this kind of problem is often attributed to off-line solving an optimization problem, but it should not be ignored where the optimization result needs to be implemented in an actual dynamic environment where uncertainty inevitably exists. Therefore, different from optimization in a pure mathematical sense, we call this kind of optimization-based problem with implementation in the dynamic environment a general control problem.

Compared with traditional control problems, these general control problems have different structural characteristics and cannot be directly solved by existing control theory based on the dynamic mathematical model. Therefore, at first the focus is always on exploring effective optimization algorithms according to the structural characteristics of the problem. These studies often take a complete and accurate description of the problem under discussion as the premise and try to find the global optimal solution of the problem by developing an appropriate optimization strategy in terms of the prior information. Obviously, once the obtained optimal solution is put into practice, it would remain optimal if both the model and the environment are exactly as assumed. However, if the actual status deviates from the assumed one, such as the environmental map is inaccurate or an obstacle moves, the production conditions or demands change, the resource supply fails, the discrete event process varies with time, etc. then the optimum status could not be maintained, even if a serious deterioration of system performance resulted. Therefore this optimal solution off-line solved by prior information cannot guarantee an actual optimum in the dynamic uncertain environments. This is just the same problem encountered when applying the traditional optimal control to practical industrial processes.

In the real environment, the incomplete and inaccurate information, the uncertainty of the plant, the dynamic variation of the environment, and so on, are inevitable. For the general control problems, the offline global optimization takes into account the requirements of optimization, but no consideration is given to the dynamic uncertainty of the environment. In fact, in most cases complete information on the plant and the environment is difficult to obtain and accurately predict. The global optimization based on that is thus impossible or meaningless due to the unknown variation of the plant and the environment. In addition, global optimization often involves large-scale computation, which must also be considered from the viewpoint of implementation. Therefore, for various optimization-based general control problems, in order to obtain a practically available solution, the solving method should combine the optimization mechanism with a feedback mechanism, and remove the unnecessary global optimization style. It is obvious that the general control problems in a dynamic uncertain environment face a similar problem to that in complex industrial process control. Since predictive control can meet the requirements on control in a complex industrial environment with its specific methodological principles, its methodological thinking can be naturally applied to solving the general control problems in a dynamic uncertain environment. Therefore, it is necessary to generalize the basic principles of predictive control from being originally only applicable to control problems to a more inclusive methodological framework. The generalized principles of predictive control are shown in Figure 11.1 and can be described as follows [2].

1) **Scenario prediction**
   Although the structures of various general control problems are different, as a kind of problem to be solved with optimization, it is necessary to establish a prediction model to describe the dynamic scenario of the problem. This model should contain all the known information about the problem to be solved, especially the causality for the evolution of the scenario, the evolution law, the constraint conditions, and the objective of optimization. The input of the scenario prediction model is the required strategy and the output is the dynamic evolution of the system states. According to this model, if the current status is known and the future decision strategy is given, the dynamic evolution of the process scenario can be predicted, with which the satisfaction of the constraints can be judged and the corresponding performance index can be calculated. Therefore, the scenario prediction model is a generalization of the

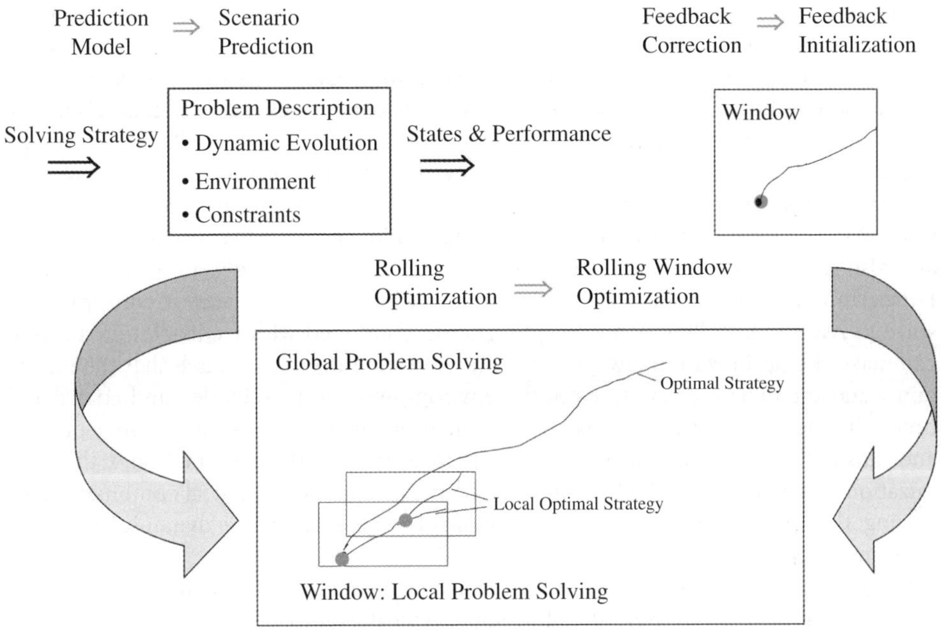

Figure 11.1 Generalized principles of predictive control.

prediction model in predictive control, with the strategy to be solved as the input and the performance to be optimized as the output.

2) **Rolling window optimization**
To solve the general control problem, the traditional method using off-line global optimization is replaced by a rolling window method using online local optimization repeatedly. At each step of the rolling procedure, a rolling window is first determined based on current status according to certain rules. Only the local optimization problem in the rolling window will be solved by using scenario prediction and optimization techniques. The optimal strategy in the rolling window can be obtained but only the current strategy is put into implementation. With the evolution of the dynamic process, the window is moved forward according to some driving mechanism; thus we call it rolling window optimization. The rolling window here can be defined in time domain, space domain, event domain, etc. according to the problem structure. Thus the driving mechanism of the rolling window can be defined as being periodic driven, event driven, etc. according to the characteristics of the problem.

3) **Feedback initialization**
At each step of the rolling procedure, before starting the local optimization, the status in the new rolling window should be firstly reinitialized (updated) according to real-time information. This means that the scenario description in the rolling window is not just a mapping of the evolution predicted by the global or the latest scenario model, but also includes the deviation information of the scenario prediction model as well as the scenario variations caused by uncertain and unpredictable events. The feedback initialization ensures that the local optimization at each step is based on the latest real-time information.

These three generalized principles promote the predictive control from only solving control problems to solving general control problems in a dynamic uncertain environment. Among these principles the scenario prediction is the generalization of the prediction model, which describes more general dynamic causality of general control problems instead of the input–output relation in a narrow sense. It exhibits the dynamic evolution process of the problem to be solved in a wider sense, and thus provides the basis for solving the optimal strategy according to the optimization performance index. Making local optimization in the rolling window is partly because it is impossible and also unnecessary to take into account the performance index in a global scope. Furthermore, reducing the scale of the optimization problem makes it convenient to solve in real time. Rolling window optimization combined with feedback initialization can make the optimization always based on real-time information such that the uncertainty and the dynamic variations of the environment can be included and considered timely in the optimization. Thus the overall problem solving is an ongoing process and runs in the closed-loop form. These principles embody the combination of the optimization mechanism and the feedback mechanism. It provides an ideal combination for solving the optimization-based general control problems in the dynamic uncertain environment.

In the following subsections, the above-mentioned generalized principles of predictive control are illustrated in more detail by some typical examples.

11.2.2 Rolling Job Shop Scheduling in Flexible Manufacturing Systems

Scheduling problems widely exist in many application fields where some specific requirements on cost, efficiency, etc., should be optimized with respect to constraints on general resources. Typical scheduling problems include production planning, job scheduling, pickup and delivery, etc. In particular, the job scheduling problem is widely investigated and used in modern manufacturing systems due to the requirement for production optimization.

In flexible manufacturing systems (FMSs), the job shop scheduling problem can be described as follows. A batch of jobs should be processed. Each of them has to go through a series of ordered operations to complete the processing. There are several machines. Each machine can process a set of operations, which may be different for different machines. The jobs belong to some job classes and different classes of jobs have different operations. The processing time of the same operation on different machines may also be different. Furthermore, each job is assigned a due date according to the class it belongs to. The scheduling problem is: when and on which machine to arrange each operation of each job such that the total makespan for all jobs, i.e. the minimum time to complete the operations of all jobs, is minimized, while the due date constraints for the jobs are satisfied. It is obviously a combinational optimization problem.

In the past, many researches on FMS job shop scheduling were focused on static scheduling, i.e. finding efficient scheduling algorithms with the assumption that all jobs are available and ready for processing. Once the schedule is prepared, the processing sequence of jobs is determined and not changed during processing. In the actual process, however, the obtained optimal schedule cannot guarantee the desired results because the assumption for static scheduling is often too idealistic. Indeed, the production is always in a dynamic uncertain environment where jobs arrive continuously and unpredictable events, such as machine breakdown, machine repair and due date changes of jobs, may occur. In this case, a dynamic scheduling strategy is needed and predictive control can provide an excellent methodology to construct a suitable dynamic scheduling strategy. Inspired by the generalized principles of predictive control, Fang and Xi [3] proposed a rolling horizon job shop scheduling strategy to the above problem in the dynamic uncertain environment, with specific implementations as follows.

(1) Scenario prediction

The prediction model of the job shop production can be established based on appropriate description of following two kinds of information:

- Prior information, including the operations required for each class of jobs and their order, the executable operations and corresponding processing time of each machine, due date of each class of job.
- Real-time information, including job status, machine status, due date change of jobs, etc.

This model is indeed an information set with the schedule to be solved as input and the production status as output. With this information set the job processing procedure can be simulated and the dynamic scenario of the manufacturing process can be exhibited. Once a schedule is given, the variation of production status could be calculated, with which it can be judged whether the schedule is reasonable, whether the due date constraints for the jobs are satisfied, and the corresponding total make-span can be calculated.

(2) Rolling window optimization

In the case when many jobs need to be processed but uncertain events or accident may occur, it is unnecessary to make a schedule for all the jobs. A job window is then constructed by selecting a number of jobs from jobs waiting for processing according to some specific rules. Only jobs in the job window are scheduled and partially processed according to the scheduling results. In the following some key points will be discussed in more detail.

1) **Job window**. According to the actual production environment, it is assumed that jobs arrive continuously and the future arrival time of some unavailable jobs is known. Therefore, the job set can be divided into three subsets: the available job set *Sa*, jobs in the current job window *Sw*, and jobs that have completed their operations *Sc*. For the selection rule, we first give a quantitative urgency index to each job in the available job set *Sa* with which a job with an earlier arrival time or with an estimated finish time much later than its due date may have a higher degree of urgency. Choose the job with the highest urgency in the available job set *Sa* and move it into the job window *Sw* until the number of jobs in *Sw* reaches the given number or *Sa* is empty. Only the jobs in the job window *Sw* are scheduled with the help of the scenario prediction, and the resultant schedule is used to arrange the current operations. This procedure will be repeated until all jobs have finished their operations.
2) **Scheduling problem formulation**. At each step, a scheduling problem is defined on the current job window *Sw*. The goal of optimization is to minimize the local production makespan, i.e. the completion time for processing all the jobs in the job window. Usually an additional term to penalize the tardy time of jobs is also needed in the performance index, which takes the due date constraints of jobs into consideration.

The constraints in the optimization problem include a series of equalities or inequalities due to technological conditions, such as the relationship between the makespan and the operation time of jobs on the machines, the condition that for each job the finish time of its any operation is not later than the start time of its following operation, etc. all of which are related to the information given by the prediction model. The optimal solution, i.e. the optimal schedule for the jobs in the job window, is given by the release time of each operation with a corresponding machine for all jobs.

3) **Scheduling problem solving**. For the above scheduling problem it is obvious that there are two key problems to be solved, i.e. to dispatch every operation of jobs to suitable machines and to determine the processing sequence and release time of jobs on each machine. With respect to the former problem, the size of search space is in general large because many jobs and multiple machines are often involved for job shop scheduling. For the latter problem, the constraints are often difficult to handle, such as the job operation sequence must be satisfied and at any time each job appears at most on one machine, etc. To avoid the difficulty from both a large search space and complicated constraints, [3] proposed a hybrid scheduling algorithm. Since the genetic algorithm (GA) is suitable for searching a large solution space, it is adopted to decide on which machine each operation should be processed. As the dispatching rule is simple where the operation sequence constraints are easily satisfied, it is used to arrange the processing sequence of jobs on each machine, and accordingly to determine the release time of jobs. The details on designing the algorithm can be found in [3]. It should be pointed out that this scheduling algorithm is a heuristic one and different dispatching rules may yield different optimal solutions.
4) **Rolling mechanism**. In general, the job window is rolling forward periodically with a fixed time or with completion of a fixed number of jobs. However, as mentioned above, the resultant actual performance may be poor due to the dynamic uncertain environment. To overcome the drawback, [3] proposed a periodic and event-driven mechanism. Some typical unpredictable events, such as machine breakdown, machine repair, and due date change of jobs, are defined as critical events. If a critical event occurs, rescheduling is performed immediately; otherwise periodic rescheduling is adopted. This rolling strategy combined with feedback initialization discussed below forms a specific and efficient mechanism against the dynamic uncertainties for the job shop scheduling problem.

(3) Feedback initialization

During rescheduling and before solving the optimal schedule, both the scenario prediction model and the real-time job status should be initialized according to the latest feedback information on the production process and possible critical events. New jobs will be moved into the available job set *Sa* when their arrival time becomes known. The jobs in *Sw* should be updated, i.e. jobs that have finished their operations are moved into *Sc* and new jobs are selected from *Sa* to *Sw* according to the selection rule. Furthermore, the scenario prediction model should be updated according to possible changes of the job due date and the machine status such that the scheduling for the jobs in the updated job window can be based on the latest processing capability and satisfy the updated requirements.

Fang and Xi [3] also gave an example of 30 jobs processed on four machines to illustrate the rolling job shop scheduling strategy. The 30 jobs belong to three classes J1, J2, and J3 with the job index 1–10 in J1, 11–20 in J2, and 21–30 in J3. The operations needed for each class of jobs with their sequences are given by: J1: A-C-B-D, J2: B-D-E, J3: B-A-F, and the operations that each machine can process are: m1: (A, B, C, F), m2: (B, C, D, E), m3: (B, D, F), m4: (E, F). The process times of operations on four machines and the set-up times from one operation to another on the four machines are known (not listed here). The due dates of three classes of jobs are: 300, 400, and 500, respectively. The jobs arrive dynamically over time and the arrival times of jobs are set randomly from the range of (0, 300). At any time only the arrival times of the incoming jobs that will arrive within a prediction period T_P can be exactly known and these jobs can then be added to the available job set. Some unpredictable events during processing are considered: at time $t = 100$ machine m2 breaks down and will be repaired at time $t = 200$, and at time $t = 150$ the due dates of J3 jobs will be changed from 500 to 350. In order to evaluate the dynamic scheduling results, the performance measure of the production makespan plus a weighted (taken as 0.65) penalty term on tardy time of jobs is taken as the criterion.

In the rolling window scheduling strategy, the maximum number of jobs in the job window is set as $L_W = 10$, the interval of the periodic rescheduling $\Delta T = 40$, and the predictive period $T_P = 60$. The scheduling result obtained is shown by grant charts in Figure 11.2.

In Figure 11.2, the square stands for the processing time of the related operations including the set-up time and the processing time of the operation, the character inside the square stands for the operation being processed, and the integer above the square is the job index the operation belongs to, where R stands for the rescheduling point. The related makespan $L = 579.23$ and the performance measure $C = 1813.03$. As a comparison, the static scheduling result using the same method solved once for the global scheduling problem is given in Figure 11.3, the related makespan $L = 643.77$, and the performance measure $C = 3040.86$.

The simulation results show that the rolling window scheduling strategy is more suitable for the dynamic uncertain environment than the static scheduling strategy. Firstly, after the due date change of J3 jobs from 500 to 350, with static scheduling, the J3 jobs are still processed later than J2 jobs although their due date is now earlier than that of J2 jobs. However, with rolling window scheduling, J3 jobs are processed earlier and J2 jobs are postponed to be processed later. Secondly, with static scheduling, jobs are processed according to the processing sequence decided by the schedule. After machine m2 breaks down, due to the postponement of scheduled job operations processing on m2, and the constraints on job operation sequences, the three other machines have a period of idle time and the production efficiency is considerably degraded. With rolling window scheduling, however, when machine m2 breaks down, rescheduling is performed immediately and the jobs waiting to be processed by machine m2 are dispatched to other machines, such as the operation D of job 9 being processed in machine m2 is moved to machine m3 to continue processing. After machine m2 is repaired, rescheduling is also performed at once and jobs are again dispatched to it for processing. Therefore the makespan derived by the rolling window scheduling strategy is shorter than that by the static one, and so is the whole cost.

The rolling window scheduling strategy has obvious advantages for FMS in a dynamic uncertain environment. The job window and its rolling mechanism is particularly suitable to real-time implementation because the scale of each scheduling is greatly reduced although the scheduling needs to be repeated online. The feedback initialization has a

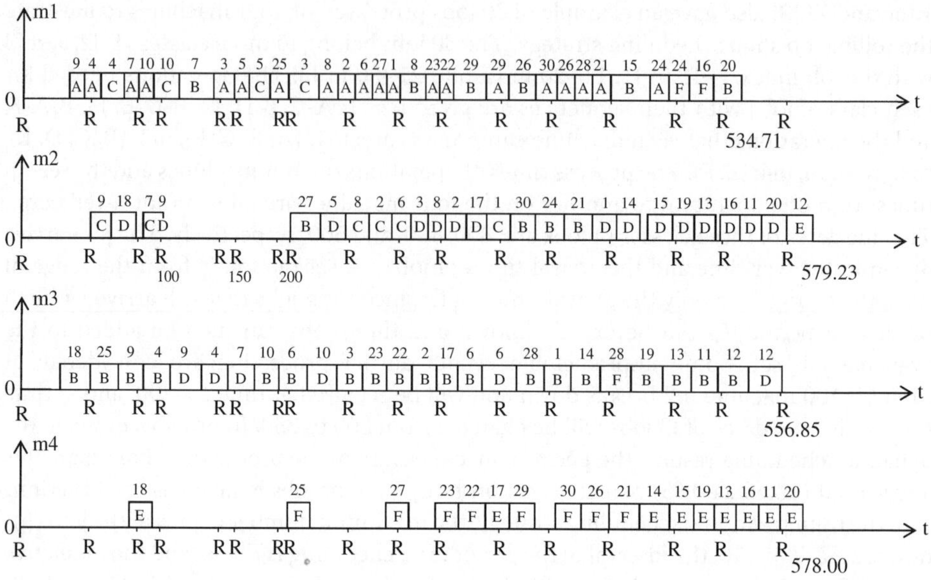

Figure 11.2 Grant charts for four machines with rolling window scheduling [3]. *Source:* Reproduced with permission from Jian Fang and Yugeng Xi of Springer Nature.

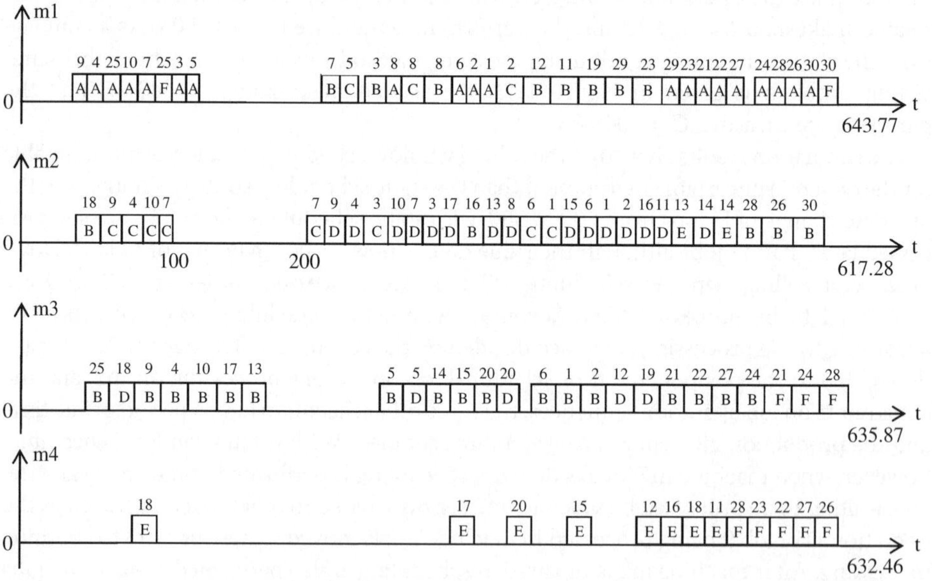

Figure 11.3 Grant charts for four machines with static scheduling [3]. *Source:* Reproduced with permission from Jian Fang and Yugeng Xi of Springer Nature.

quick response to unpredictable events, so that the scheduling at each time is always based on the real production status. In addition, this rolling mechanism is fully applicable to the cases where jobs have an unlimited number to arrive continuously, which cannot be handled by traditional scheduling methods. More details on the above algorithm and example can be found in [3]. A similar strategy also appeared in some later literature. In Wang et al. [4] a terminal penalty rolling scheduling strategy based on an initial schedule is proposed and investigated for a single-machine scheduling problem. Referring to the stable predictive controller design, a penalty function is added to the total completion time of the scheduling problem in the job window and this technique guarantees an improvement of the global performance over time.

11.2.3 Robot Rolling Path Planning in an Unknown Environment

The path planning problem of a mobile robot aims at finding a proper collision-free moving path for the mobile robot from a designated start point to a designated goal point in its workspace by using available environmental information. There are two extreme cases. One case is that the environmental information is fully known. In this case the feasible path or even the optimal path with respect to some performance index (such as the shortest path) can be solved by global planning methods. It has been extensively investigated and much literature with different solving methods have been published. Another case is that no prior knowledge of the global environment is available and the robot can only use the real-time environmental information detected by the sensors to move toward the goal by circumventing obstacles. In this case, optimizing the path is meaningless and the robot even possibly falls into a dead zone and fails to reach the goal. The above two extreme cases correspond, respectively, to the open-loop global planning based on complete environmental information and the closed-loop immediate navigation based on locally feedback information. However, the practical environmental information is often somewhere in between. On the one hand, there is always enough prior knowledge of the environment. For example, when an autonomous vehicle transports materials in a production workshop, the information on fixed obstacles such as machines, materials, stores, etc. can be obtained beforehand. On the other hand, the environmental information is often incomplete or inaccurate. For example, the obstacles from walking people or vehicles in the workshop are unpredictable and the piles might increase or decrease during the production.

Path planning of a mobile robot with incomplete information in an uncertain environment cannot be solved by traditional global planning methods. The navigation methods based on immediate feedback information are unable to provide a satisfactory solution due to lack of optimization, and even hardly guarantee the accessibility. However, note that the robot could detect some environmental information in a finite region under its sensor scope during motion, so this part of the information should be fully used. Therefore, a reasonable way to solve this problem is to fully use detected local environmental information and to adopt local planning repeatedly instead of the off-line global planning. Zhang and Xi [5] proposed a rolling window based method for robot path planning in globally unknown environments. It is along the lines of the generalized principles of predictive control mentioned in Section 11.2.1. In the following we introduce specific implementations of this method.

(1) Scenario prediction

To exhibit the scenario of the robot moving in the environment, a prediction model is established based on the information of the robot's motion and the environment, as well as the planning task. It includes some prior knowledge such as the robot kinematics, known static obstacles in the map of the environment, and the start point and the goal point of the planning task, as well as some real-time information, such as the detected current positions of the mobile robot itself and moving obstacles.

According to this model, the robot motion scenario in the environment can be demonstrated. For any given moving path from the start point to the goal point, whether the robot is collided with static or moving obstacles can be judged and, if not, the corresponding performance index can be calculated.

(2) Rolling window optimization

Since global environmental information is not available, a rolling window based path planning strategy is performed instead of the off-line global path planning. At each step of the rolling process, the rolling window is defined as a specific region around the robot and with the robot as its center, which should be included in the detectable region of the robot sensor system. Path planning is then performed within this rolling window based on the latest local information.

Information on the rolling window

The environmental information on the rolling window consists of two parts, coming from prediction model and feedback initialization, respectively. On the one hand, prior knowledge of the global environment included in the prediction model, such as robot kinematics, positions of known static obstacles (if any), should be mapped into the current window. On the other hand, the real-time detected new information in the rolling window, such as static obstacles that might be unknown before or moving obstacles with trajectory predictions, should be also included.

Subgoal determination and local path planning

For local path planning in the rolling window, the current position of the robot is naturally taken as the start point. However, the known goal point designated by the task may be out of the current window and cannot be directly taken as the goal point of the local path planning. Aiming at this specific problem, [5] proposed a heuristic method based on local detected information, which first determines a subgoal located at the border of the current window and then plans an optimal path from the current position of the robot to the subgoal in the rolling window.

Denote the rolling window and its border at time t as $Win(P_R(t))$ and $\partial Win(P_R(t))$, respectively, where $P_R(t)$ is the position of the robot, i.e. the center of the current window. Let P_{goal} be the known goal point designated by the task and $P_{sub}(t)$ be the subgoal to be determined. If the robot is very close to the goal point, i.e. $P_{goal} \subset Win(P_R(t))$, then P_{goal} can be directly taken as the goal point of local path planning. Otherwise, a heuristic function $f(P) = g(P) + h(P)$ is used to determine a point P on the border of the rolling window as the subgoal $P_{sub}(t)$ that minimizes $f(P)$, namely

$$\begin{aligned} &\min_P f(P) = g(P) + h(P) \\ &\text{s.t.} \;\; P \subset \partial Win(P_R(t)) \end{aligned}$$

where $g(P)$ denotes the cost (path length) from current position $P_R(t)$ to P and $h(P)$ denotes the cost from P to the known goal point P_{goal}. Since the environmental information out of the window is unknown, $h(P)$ is usually estimated by the distance from P to the goal P_{goal}. Solving the above optimization problem we can obtain both the subgoal $P_{sub}(t)$ and the optimal local path $\overline{P_R(t)P_{sub}(t)}$ in the window, with which and the straight path

$\overline{P_{sub}(t)P_{goal}(t)}$ out of the window $f(P)$ is minimized. This method reflects the compromise of the requirement on global optimization and the restriction of finite local information.

The above optimization problem can be solved using different strategies. To reduce the computational burden, in [5] the problems of determining the subgoal and planning the local path were separately solved. At first $g(P)$ is simply set as zero or infinite according to P in feasible region or not, then the subgoal $P_{sub}(t)$ can be easily taken as the point P at the feasible border of the rolling window which has the minimal distance to the goal point P_{goal}. After the subgoal has been determined, a strictly nearer feasible path planning algorithm is presented for local path planning, which guarantees that along the path the distance to the goal decreases with time.

Obstacle avoidance

Obstacle avoidance is the most critical issue in robot path planning. During local path planning, the environment map with static obstacles has been updated and the planned path can easily guarantee that the robot is collision free with these static obstacles. However, for any dynamic obstacle, although it may have been detected, collision might occur in a general sense due to the uncertainty of its future motion. To handle a dynamic obstacle, some conditions on its shape, motion, etc., should be given, such as with convex shape, constraints on changes of its direction or speed, and so on. Then the future path of the dynamic obstacle can be estimated and possible collision can be avoided by using a proper planning strategy. For details readers can refer to, for example, [6].

(3) Feedback initialization

Path planning based on the rolling window is performed periodically. After the local path has been determined, the robot will move one step forward along the planned path until the next period. At each new period, the robot first acquires real-time environmental information through its sensor system, and initializes the environment map with the latest information on obstacles in the window. It not only corrects the prior known information on static obstacles in the window but also identifies the unknown dynamic obstacles, and even predicts the future motion path of these dynamic obstacles. All this information is used to update the prediction model in the current window, with which the local path planning can be performed based on the latest and most practical information.

In Figure 11.4 a simulation example of rolling path planning for a mobile robot in a globally unknown dynamic environment is shown. The moving circle is the rolling window with the mobile robot at its center and the environmental information within it is fully detectable by the robot sensor system. S is the start point and G is the known goal point. The large gray convex bulks represent unknown static obstacles, while the small circles with trailing tails represent the unknown dynamic obstacles with irregular motions. The actual tracks of these four dynamic obstacles are marked by a series of short lines.

At the beginning of rolling path planning, the robot does not possess any prior information on static and dynamic obstacles. At each step, the robot firstly detects environmental information in the current window and updates the positions of the obstacles. Then an appropriate path in the rolling window is planned that guarantees collision avoidance and satisfies some optimization criterion. The robot moves along this local path one step forward and this process is repeated until the goal is reached. In Figure 11.4 this procedure for some specific instants are shown, where the black parts

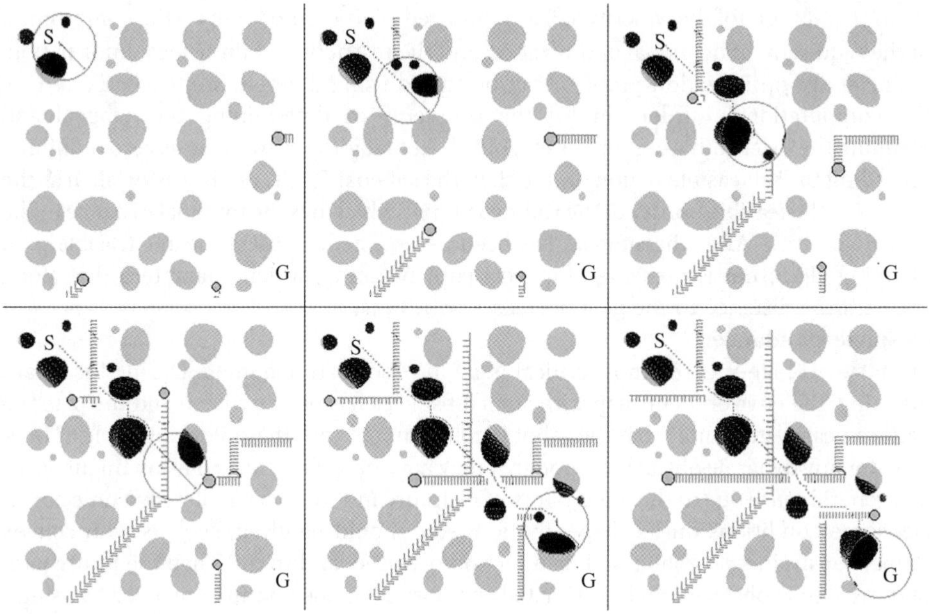

Figure 11.4 Example of robot rolling path planning in an unknown environment [6].

represent the obstacle information gradually detected by the robot during the procedure. It can be seen that with the rolling path planning method the robot can safely reach the goal point according to the repeatedly planned path.

The rolling path planning method for a mobile robot adopts repeated local optimization performed online instead of global optimization. It is particularly suitable to the path planning problem in a globally unknown environment, even with dynamic obstacles. Meanwhile, the path planning based on the local window reduces the scale of the optimization problem to be solved and is thus easy to implement in real time. Some theoretical aspects such as convergence of the algorithm, suboptimality of the resultant path, etc., were also discussed for the case with globally unknown static obstacles [5, 7]. The path planning strategy for dynamic obstacles in the local window and the safety analysis of the resultant path can be found in [6].

In addition to the above discussed robot path planning and FMS job shop scheduling problems in a dynamic environment, many other general control problems, such as network real-time optimization and management, DEDS real-time monitoring, transportation scheduling, investment decision, etc. can also be solved by similar principles. Although their implementation forms and solving algorithms may be quite different due to problem specifications and uncertainty differences, the idea is the same, i.e. to make full use of all causal information for prediction and optimization, and meanwhile to update the information continuously by rolling combined with feedback, so that the optimization and decision at each step can be based on the latest information. The generalization of the predictive control principles to the general control problems in the dynamic uncertain environment will also open up wider fields and provide richer modes for applications of predictive control.

11.3 Summary

In this chapter, the methodological principles of predictive control for solving control problems are analyzed from the viewpoint of cybernetics and information theory. In predictive control, due to the existence of uncertainty in the model and in the environment, not only model prediction is essential as the basis for optimization but also feedback correction is necessary to address uncertainties so as to put the optimization basis closer to the real system status. With rolling optimization, solving the global optimization problem is replaced by repeatedly solving local optimization problems, where the optimization and the feedback mechanism are naturally combined and form closed-loop control in global, providing a reasonable way for optimization-based control in an uncertain environment.

In addition to using predictive control to solve control problems, the universality of the methodological principles of predictive control also have a great potential for solving various optimization-based dynamic decision problems other than control. By defining the general control problems as a kind of decision problem with optimization demand and dynamic implementation, three principles, i.e. scenario prediction, rolling window optimization, and feedback initialization, are proposed as generalization of the methodological principles of predictive control. For many optimization-based dynamic decision problems in the areas of scheduling, planning, and decision making, if the solved optimal result should be implemented in a dynamic uncertain environment, these generalized principles can provide a new way to solve them more reasonably and efficiently. Examples of rolling path planning and rolling scheduling show how to understand and realize these principles according to the problem characteristics.

With respect to realization of these generalized principles, there may be quite diverse patterns for different kinds of problems. However, two issues are the same and should be particularly considered during design: firstly, how to define the optimization subproblem in the window, particularly how its goal or performance index is mapped from the global one. Secondly, how to solve the optimization problem in the window when the prediction model is now replaced by a scenario prediction model that may be entirely different from the usual mathematical description for input/output causality. It could be expected that with the increasing demand of various new fields on predictive control, more applications will appear, not only with existing predictive control algorithms but also with new algorithms developed for general control problems using the generalized principles of predictive control.

References

1 Richalet, J., Rault, A., Testud, J.L. et al. (1978). Model predictive heuristic control: applications to industrial processes. *Automatica* 14 (5): 413–428.

2 Xi, Y. (2000). Predictive control for general control problems under dynamic uncertain environment. *Control Theory and Applications* 17 (5): 665–670. (in Chinese).

3 Fang, J. and Xi, Y. (1997). A rolling horizon job shop rescheduling strategy in the dynamic environment. *International Journal of Advanced Manufacturing Technology* 13 (3): 227–232.

4 Wang, B., Xi, Y., and Gu, H. (2005). Terminal penalty rolling scheduling based on an initial schedule for single-machine scheduling problem. *Computers and Operations Research* 32 (11): 3059–3072.

5 Zhang, C. and Xi, Y. (2001). Robot path planning in globally unknown environments based on rolling window. *Science in China, Series E: Technological Sciences* 44 (2): 131–139.
6 Zhang, C. and Xi, Y. (2003). Rolling path planning and safety analysis of mobile robot in dynamic uncertain environment. *Control Theory and Applications* 20 (1): 37–44. (in Chinese).
7 Zhang, C. and Xi, Y. (2003). Sub-optimality analysis of mobile robot rolling path planning. *Science in China, Series F: Information Sciences* 46 (2): 116–125.

Index

Predictive Control: Fundamentals and Developments, First Edition. Yugeng Xi and Dewei Li.

d

q

r

S